Proceedings of an International Workshop

Stuttgart and Altensteig, May 1988,
organized by the Sonderforschungsbereich 228
(Special Collaborative Programme) of the
Deutsche Forschungsgemeinschaft

K. Linkwitz U. Hangleiter (Eds.)

High Precision Navigation

Integration of Navigational and Geodetic Methods

With 289 Figures

Springer-Verlag Berlin Heidelberg New York
London Paris Tokyo Hong Kong

Professor Dr.-Ing. Klaus Linkwitz
Dipl.-Ing. Ulrich Hangleiter

Institut für Anwendungen der
Geodäsie im Bauwesen
Universität Stuttgart
Keplerstraße 10
7000 Stuttgart 1, FRG

ISBN 3-540-50921-6 Springer-Verlag Berlin Heidelberg New York
ISBN 0-387-50921-6 Springer-Verlag New York Berlin Heidelberg

Printed in Germany

Printing: Druckhaus Beltz, 6944 Hemsbach/Bergstr.;
Binding: J. Schäffer GmbH & Co. KG., Grünstadt
2132/3145-543210 – Printed on acid-free paper

Preface

These proceedings are a complete collection of the contributions presented at the International Workshop High Precision Navigation in May 1988 in Stuttgart and at Altensteig. The workshop was organized by the Sonderforschungsbereich 228 (SFB 228) Hochgenaue Navigation - Integration navigatorischer und geodätischer Methoden (Special Collaborative Programme "High Precision Navigation - Integration of Navigational and Geodetic Methods"), and more than 100 scientists participated.

The Sonderforschungsbereich is an institution of the Deutsche Forschungsgemeinschaft (DFG) and the Universtity of Stuttgart. Its principal purpose is the integration of geodetic and navigational methods, based on the specific reasoning and methods of the two disciplines involved, navigation and geodesy. Through this integration real-time solutions for automatic positioning and high precision navigation should be obtained. Among the main topics of the SFB are the theory and application of nonlinear dynamic processes for analyzing, controlling, and steering of movements. By combination with geodetic methods techniques of classical navigation with a position accuracy of some decameter are intended to be improved to an accuracy of a few centimeters. In the field of geodesy, problems of geodynamic and local measurements are handled in global systems with the aid of satellite and inertial techniques; here the control of inertial systems with GPS techniques is of special interest. The field of image processing is especially concerned with sensors and imaging systems for optical systems as well as systems of the infrared and microwave area. The development of methods for the analysis of temporary or spatial image sequences is of increasing interest. In the field of engineering geodesy, navigational methods are applied to analyze processes on construction sites and to control the system behavior. Of common interest for all projects from a technical point of view are: the development of sensors, signal analysis, image analysis, model-supported measurements, and system dynamics.

At the workshop one part of the papers was presented in plenary sessions (overview papers), the other part was given in parallel sessions (special papers). The workshop was held under seclusive conditions leading to a highly concentrated working atmosphere. The different professions of the participants led to an interdisciplinary exchange of expert knowledge inspiring all involved. For the different fields many common tasks with respect to methods and algorithms became obvious as well as various unsolved technical problems concerning sensors, especially sensor integration. Special attention was given the increasing importance of image processing, especially in the field of car navigation. Here, the great interest of the industry became obvious to be satisfied not by final technical solutions but by dispositions of methodical solutions.

The proceedings presented here consist of two main parts: the first part contains papers with the character of overviews, the second part papers on special subjects. Such topics give the headlines for the other chapters of the proceedings: Satellite Techniques, Topographic Applications, Image Processing, Close Range Mensuration, and Inertial Techniques.

Klaus Linkwitz Ulrich Hangleiter

Contents

A. OVERVIEW ARTICLES

B. SATELLITE TECHNIQUES

C. TOPOGRAPHIC APPLICATIONS

D. IMAGE PROCESSING

E. CLOSE RANGE MENSURATION

F. INERTIAL TECHNIQUES

List of Contributors

Ackermann, Friedrich	Institut für Photogrammetrie Universität Stuttgart Keplerstr. 11 D 7000 Stuttgart 1, FRG
Appelt, Stefan	Physikalisches Institut Universität Stuttgart Pfaffenwaldring 57 D 7000 Stuttgart 80, FRG
Autenzeller, Peter	Messerschmitt-Bölkow-Blohm GmbH Abteilung KT 324 Postfach 80 11 69 D 8000 München 80, FRG
Bayer, Gerhard	Institut für Anwendungen der Geodäsie im Bauwesen Universität Stuttgart Keplerstr. 10 D 7000 Stuttgart 1, FRG
Bertiger, W.I.	Jet Propulsion Laboratory 4800 Oak Grove Drive Pasadena California 91109 USA
Blewitt, G.	Jet Propulsion Laboratory 4800 Oak Grove Drive Pasadena California 91109 USA
Bloch, Samuel	Standard Elektrik Lorenz AG Lorenzstr. 10 D 7000 Stuttgart 40, FRG
Braumandl, Walter	Messerschmitt-Bölkow-Blohm GmbH Abteilung KT 324 Postfach 80 11 69 D 8000 München 80, FRG
Burkhardt, Hans	TU Hamburg-Harburg Techn. Informatik I Harburger Schloßstr. 20 D 2100 Hamburg 90, FRG
Castanet, Yves	Société Anonyme de Télécommunications 41, Rue Cantagrel F-75631 Paris Cedex 13 FRANCE

Dickmanns, Ernst D.
Universität der Bundeswehr München
Fachbereich Luft- und
Raumfahrttechnik
Werner-Heisenberg-Weg 39
D 8014 Neubiberg, FRG

Diehl, Klaus
AEG Forschungsinstitut
Sedanstr. 10
D 7900 Ulm, FRG

Drew, Alice R.
NOAA, National Geodetic Survey
United States Dept. of Commerce
Rockville / Maryland 20852
USA

Dünnebeil, Dieter, Klaus
Physikalisches Institut
Universität Stuttgart
Pfaffenwaldring 57
D 7000 Stuttgart 80, FRG

Förstner, Wolfgang
Institut für Photogrammetrie
Universität Stuttgart
Keplerstr. 11
D 7000 Stuttgart 1, FRG

Frieß, Peter
Institut für Photogrammetrie
Universität Stuttgart
Keplerstr. 11
D 7000 Stuttgart 1, FRG

Grafarend, Erik
Geodätisches Institut
Universität Stuttgart
Keplerstr. 11
D 7000 Stuttgart 1, FRG

Gilles, Ernst-Dieter
Institut für Systemdynamik
und Regelungstechnik
Pfaffenwaldring 9
D 7000 Stuttgart 80, FRG

Griffiths, Hugh
University College London
Department of Electronic
and Electrical Engineering
Torrington Place
London WC1E 7JE
ENGLAND

Hartl, Philipp
Institut für Navigation
Universität Stuttgart
Keplerstr. 11
D 7000 Stuttgart 1, FRG

Hausch, Wolfgang
Institut für Physikalische Geodäsie
Technische Hochschule Darmstadt
Petersenstr. 13
D 6100 Darmstadt, FRG

Hebert, Martial	Department of Computer Science Carnegie Mellon University Pittsburgh Pennsylvania 15213-3890 USA
Huddle, James R.	Litton Guidance & Control Systems 5500 Canoga Avenue Woodland Hills California 91367-6698 USA
Jacob, Thomas	Institut für Flugführung Technische Universität Braunschweig Hans-Sommer-Str. 66 D 3300 Braunschweig, FRG
Kanade, Takeo	Department of Computer Science Carnegie Mellon University Pittsburgh Pennsylvania 15213-3890 USA
Keller, Dieter	Institut für Physikalische Geodäsie Technische Hochschule Darmstadt Petersenstr. 13 D 6100 Darmstadt, FRG
Kim, Yong-Hoon	Institut für Navigation Universität Stuttgart Keplerstr. 11 D 7000 Stuttgart 1, FRG
Kleusberg, Alfred	University of New Brunswick Dep. of Surveying Engineering Fredericton / N.B. PO Box 4400 New Brunswick CANADA
Kling, Thomas	Institut für Physikalische Geodäsie Technische Hochschule Darmstadt Petersenstr. 13 D 6100 Darmstadt, FRG
Knickmeyer, Elfriede	The University of Calgary Faculty of Engineering Division of Surveying Engineering 2500 University Drive N.W. Calgary, Alberta CANADA T2N 1N4
Komarek, Thomas	Institut für theoretische Nachrichtentechnik und Informationsverarbeitung Universität Hannover Appelstr. 9A D 3000 Hannover 1, FRG

Kories, Ralf	Fraunhofer Institut Sebastian-Kneipp-Str. 12-14 D 7500 Karlsruhe, FRG
Krysinski, Z.	Polish Academy of Sciences Space Research Centre Bartycka 18 01 - 237 Warszawa POLEN
Krzystek, Peter	Institut für Anwendungen der Geodäsie im Bauwesen Universität Stuttgart Keplerstr. 10, FRG D 7000 Stuttgart 1, FRG
Lanzl, Franz	Institut für Optoelektronik der DFVLR D 8031 Oberpfaffenhofen, FRG
Lei, Fang	Institut für Technische Optik Universität Stuttgart Pfaffenwaldring 9 D 7000 Stuttgart 80, FRG
Lichten, S.M.	Jet Propulsion Laboratory 4800 Oak Grove Drive Pasadena California 91109 USA
Lindenberger, Joachim	Institut für Photogrammetrie Universität Stuttgart Keplerstr. 11 D 7000 Stuttgart 1, FRG
Lindlohr, Wolfgang	Geodätisches Institut Universität Stuttgart Keplerstr. 11 D 7000 Stuttgart 1, FRG
Linkwitz, Klaus	Institut für Anwendungen der Geodäsie im Bauwesen Universität Stuttgart Keplerstr. 10 D 7000 Stuttgart 1, FRG
Manhart, Sigmund	Messerschmitt-Bölkow-Blohm GmbH Abteilung KT 324 Postfach 80 11 69 D 8000 München 80, FRG
Matthies, Larry	Department of Computer Science Carnegie Mellon University Pittsburgh Pennsylvania 15213-3890 USA

Mehring, Michael	Physikalisches Institut Universität Stuttgart Pfaffenwaldring 57 D 7000 Stuttgart 80, FRG
Menke, Bodo	Physikalisches Institut Universität Stuttgart Pfaffenwaldring 57 D 7000 Stuttgart 80, FRG
Mezger, Klaus	Institut für Systemdynamik und Regelungstechnik Universität Stuttgart Pfaffenwaldring 9 D 7000 Stuttgart 80, FRG
Möhlenbrink, Wolfgang	Institut für Anwendungen der Geodäsie im Bauwesen Universität Stuttgart Keplerstr. 10 D 7000 Stuttgart 1, FRG
Napier, Michael	University of Nottingham Dept. of Civil Engineering University Park Nottingham NG7 2RD ENGLAND
Neul, Reinhard	Institut für Systemdynamik und Regelungstechnik Universität Stuttgart Pfaffenwaldring 9 D 7000 Stuttgart 80, FRG
Pirsch, Peter	Institut für theoretische Nachrichtentechnik und Informationsverarbeitung Universität Hannover Appelstr. 9A D 3000 Hannover 1, FRG
Plocher, Thomas	Institut für Systemdynamik und Regelungstechnik Universität Stuttgart Pfaffenwaldring 9 D 7000 Stuttgart 80, FRG
Rehfeld, Nils	Fraunhofer Institut Sebastian-Kneipp-Str. 12-14 D 7500 Karlsruhe, FRG
Reid, Blake D.	Honeywell Limited 1345 Denison Street Markham/Ontario L3R 5V2 CANADA

Scherzinger, Bruno
Honeywell Limited
1345 Denbison Street
Markham/Ontario LR3 5V2
CANADA

Scheufler, Peter
Physikalisches Institut
Universität Stuttgart
Pfaffenwaldring 57
D 7000 Stuttgart 80, FRG

Schneck, Günther
Institut für Navigation
Universität Stuttgart
Keplerstr. 11
D 7000 Stuttgart 1, FRG

Schröder, Dietrich
Geodätisches Institut
Universität Stuttgart
Keplerstr. 11
D 7000 Stuttgart 1, FRG

Schüssler, Harald
Dornier System GmbH
Postfach 13 60
D 7990 Friedrichshafen, FRG

Schwarz, Klaus-Peter
The University of Calgary
Faculty of Engineering
Division of Surveying Engineering
2500 University Drive N.W.
Calgary, Alberta
CANADA T2N 1N4

Shafer, Steven
Department of Computer Science
Carnegie Mellon University
Pittsburgh Pennsylvania 15213-3890
USA

Snay, Richard
National Geodetic Survey
Horizontal Network Branch
United States Dept. of Commerce
Rockville / Maryland 20852
USA

Thorpe, Charles
Department of Computer Science
Carnegie Mellon University
Pittsburgh Pennsylvania 15213-3890
USA

Tiziani, Hans
Institut für Technische Optik
Universität Stuttgart
Pfaffenwaldring 9
D 7000 Stuttgart 80, FRG

Vetter, Thomas
Physikalisches Institut
Universität Stuttgart
Pfaffenwaldring 57
D 7000 Stuttgart 80, FRG

Vosselman, George	Institut für Photogrammetrie Universität Stuttgart Keplerstr. 11 D 7000 Stuttgart 1, FRG
Wehr, Aloysius	Institut für Navigation Universität Stuttgart Keplerstr.11 D 7000 Stuttgart 1, FRG
Wu, S.C.	Jet Propulsion Laboratory 4800 Oak Grove Drive Pasadena California 91109 USA
Yunck, Thomas P.	Jet Propulsion Laboratory 4800 Oak Grove Drive Pasadena California 91109 USA
Zielinski, Janusz B.	Polish Academy of Sciences Space Research Centre Bartycka 18 01 - 237 Warszawa POLEN
Zimmermann, Georg	Fraunhofer Institut Sebastian-Kneipp-Str. 12-14 D 7500 Karsruhe, FRG

A. OVERVIEW ARTICLES

High Precision Navigation with Satellites

Philipp Hartl
Institut für Navigation, Universität Stuttgart

0. Abstract

Navigation and geodesy were developed on seperate ways due to their applications. Now satellite navigation shows new possibilities in a global use and in an improvement in measurement accuracies. Thus navigation and geodesy moved towards each other. Several space systems are discussed in this paper and a brief description of the GPS-Navstar with its its possible applications in geodesy and navigation is given.

1. Introduction

Some scientists are convinced that the famous painting of the dwellers in the cave of Lascaux describes the direction and range to their hunting grounds by astronomical means. If this is true, then our ancestors applied astronavigation already 17 000 years ago, in the Stone Age. Historical findings teach us generally that astronomy and the skill to use this knowledge for identification of position, direction and time by means of stars was developed very early. Astronomical navigation was extremely important for hunting and orientation and for the individual as well as for mass migration, for merchandise etc. -

5 000 years ago surveying was well developed in Egypt. Surveyors measured the coordinates almost periodically in terms of latitude and longitude for real estate after the floodings of the Nile river. Also buildings like the Pyramides had to be planned and constructed with the help of terrestrial and astronomical measurement techniques.

Thus, the topic of navigation has a very long tradition and in the antiquity astronomy, geodesy and surveying had common roots and common tools. Astronomy had two areas, the mystic/religious one, and the practical one. The latter one had its application in navigation and surveying.

2. Navigation

The purpose of navigation is to guide objects, i.e. people or vehicles, from some location to some target point "safely" and "cost optimally". This means that navigation provides measurement tools to describe the coordinates of the own position and of the target as a function of time.

Navigational tasks are, therefore, positioning, heading, i.e. direction determination, velocity measurement, time determination, and in modern traffic active control of collision avoidance and proper route planning. Navigation is a continuous updating of such information. The updating rate of the individual data is a function of velocity with which the individual objects and involved processes occur. Rapid motion requires rapid reaction. The stationary process of geodesy is the one extreme, where, in principle, the measurement can last arbitrarily long, but should be very precise. The other extreme is the navigation of very rapid vehicles like aircrafts, where it is not the precision but the rapid measurement and control which is required for safety reasons. Ships and land vehicles also require instant measurements, but at more moderate updating rates.

The different requirements are the reason why geodesy and navigation went separate ways during the last 2 000 years. Only modern satellite navigation brings up new commonalities and combines these disciplines again. Up to now ships and planes used "conventional navigational tools" in practice, which have relative accuracies in the order of 10^{-3} to 10^{-4}. Modern navigational satellite techniques will offer in the near future at least 2 orders of magnitude improvement by delivering instantaneously and globally precise absolute coordinate values of several meters accuracy. This can lead to centimeter - and millimeter- relative accuracies for baselines of many kilometers by post processing.

3. Geodesy

Geodesy became a mathematical-physical science in the 17th century and describes in numbers and maps the figure of the Earth's surface and the laws and geometry of related forces, in particular the gravity field. One of its basic characteristics in the past was its stationary perspective. The relative accuracies have been 10^{-4} and better. The shorter distances could be measured with centimeter- and partly millimeter-accuracy. The distances from one continent to another were, however, worse.

In the last two decades the need for global reference systems and much higher accuracies came up for both geodesy and navigation. And even for geodesy the need for moderate measurement rates appeared. Much larger areas and denser networks have to be surveyed within shorter periods. Many changes due to the human activities have to be taken into account which require shorter updating periodes, such as years rather than decades. Many new tasks in the engineering field have arisen. Economy has become a main driver to reduce measurement time and cost. And, as regards accuracy, it turned out in the past two decades that the earth is far from being stationary. Many processes cause all kinds of parameters to describe " movements" within the earth's crust. From the scientific as well as from the practical point of view it is important to control these movements. Modern Geodesy is characterised by dynamic kinematic measurements.

4. The Space Age

Satellite navigation is marked by the following events:

- 1957 Sputnik was the first artificial satellite to orbit the earth. With Doppler, interferometric and optical camera measurements accuracies of some hundred meters could be achieved. The trajectories of the early satellites were mainly measured by optical determination of the local azimuth and elevation angle as a function of time, with the sky as background. It turned out that the coordinates of the observation stations were the main error sources. The relationship between the reference systems of the individual continents was inaccurate to several hundred meters. These uncertainties are, therefore, taken in the future as parameters to solve in the orbit equations. Considerable improvements are achieved. In the early VHF-radio orbit determination systems the ionospheric effects also caused large errors, which were later handled by using higher frequencies and two-frequency methods.

- In the early 60's the introduction of the TRANSIT system (Ref. 1) as the first navigational satellite system brought important improvements in terrestrial navigation for the submarine fleet of the US- Navy. In the early 70's this system also became available for the civil application, i.e. for commercial ships and later even for passenger ships. In addition, it became an important tool for the geodesist, who could improve the global network by this way to the 10 m level, and in the differential mode to the meter and even decimeter level. The benefit of the TRANSIT satellite navigation technique for satellite operators, navigators and geodesists became apparent. The coordinates of satellite ground stations and the knowledge about the gravitational field were considerably improved. The influences of atmosphere etc. could be modelled and thus the errors could be reduced. For many years and with TRANSIT the main method of satellite navigation became Doppler, i.e. the orbit determination was based on frequency shift due to the radial velocity vector groundstation - satellite. TRANSIT is still operational and will be available at least until the mid 90's. But the GPS technique is so superior to it that more and more people use this ranging method now.

- The optical systems were already used for observation of Sputnik and the following satellites. Originally the sunlight, the satellite and the star background were observed and photographed by means of special camera systems. This allowed to determine the angular position of the spacecraft as a function of time. Some geodetic satellites were later equipped with strong flashlights, clocked with a certain period. The satellite track was seen then on the film of the observing camera as a dotted line in the skymap.

- This type of optical observation of the angular position was no longer adequate in the 70's. Satellites became equipped with reflectors. Very strong pulse lasers were used to transmit optical pulses to these reflectors. The reflected signals were received with a telescope. The two-way signal delay for the propagation to and from the satellite was measured at the laser ground station on earth. The accuracy was originally in the order of a few nanoseconds and was improved for the second and the third generations to some fractions of nanoseconds. Accuracies in the order of some centimeters can now be achieved with the SLR (= Satellite Laser Tracking).

Dedicated geodetic satellites are compact and passive. They are equipped with many reflectors. They were so far launched from USA, France, USSR and Japan. Very important is LAGEOS, a satellite, which is supposed to be used for at least 100 years. It is an excellent reference. Another very important possibility for extraterrestrial laser ranging was offered by the Lunar programs of the NASA and the USSR. Several reflectors were posted on the lunar surface and are used by a few large laser stations to determine the distance between Earth and Moon to an accuracy of some centimeters as a function time. This is not only important for the geodesy of the Earth, but also for the determination of the dynamic forces between Earth, Moon, Sun and, to some extent, also other planets.

- Today the laser tracking network is the most reliable and accurate orbit determination system. The European Remote Sensing Satellite ERS- 1 will be equipped with some laser reflectors to allow very precise orbit determination. But it will also apply radio tracking by means of the Precise Range and Range Rate Equipment (PRARE) (Ref. 2). Laser ranging is costly and the operation is weather dependent. The number of stations is limited. Most of the stations are fixed. Only a few mobile laser stations exist. The orbit determination is based on observation and modelling, because there is no continuous observation possible with the limited number of stations.

- In the future microwave tracking systems, such as the Global Positioning System (GPS) with its 24 NAVSTAR satellites will be used extensively. A combination of GPS and PRARE will be as accurate as laser tracking.

5. NAVSTAR- GPS

The geodetic and scientific satellites cannot directly be used for terrestrial navigation. Even TRANSIT is only applicable to navigational tasks for vehicles which move very slowly, like ships. The Global Positioning Satellite System GPS-NAVSTAR is the first navigational system which is generally applicable for all kinds of vehicles, including rockets. It will consist, in its final deployment, of 24 satellites in orbit; Fig.1, (Ref. 3, 4).

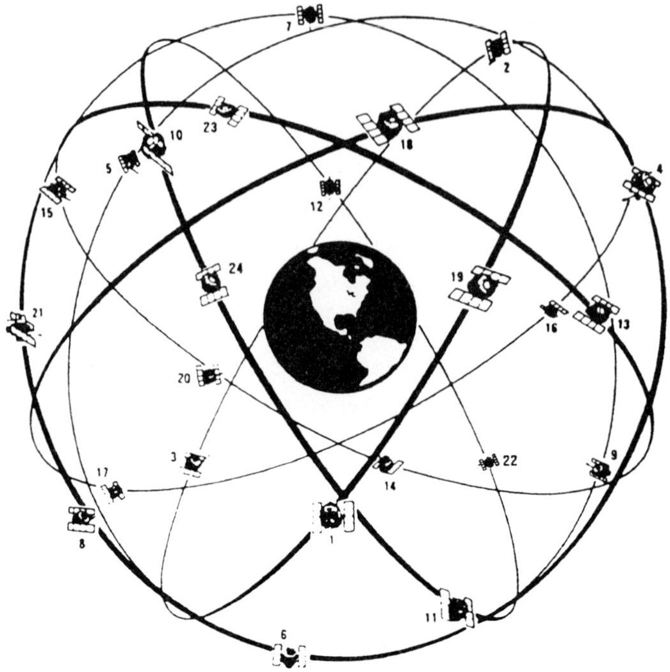

Fig. 1: The homogeneous distribution of the 24 GPS-Navstars

GPS has demonstrated with its first generation satellites, of which 6 units are still operating, that it can become the most effective system for navigation and geodesy. Numerous experimental and quasioperational tests have shown that GPS receivers can provide a very high accuracy fast, reliable, worldwide and with little demands to the receiver operator.

Due to the Shuttle disaster, the deployment of the operational GPS system was delayed. But it can be assumed that within the next three to five years the system will be fully operational with 24 satellites. At any place in the world and at any time it will then be possible to receive at least 4 to 6 satellite signals for position determination. We will have a sky of artificial satellites, which transmit relatively strong (20 W) and well defined, coherent signals, identify themselves and send their orbital elements and other data, which are needed to easily calculate the satellites's position as a function of time. The calculation is done in the processing unit of the GPS receiver.

The navigator can get from his GPS receiver the information at what time which satellites are received. The receiver calculates also the signal delay. This is easily possible: the satellites are equipped with atomic clocks and transmit signals very precisely, like a clock. For these signals the time of arrival at the receiver is measured, by means of the receiver's clock. Assume first that this clock is also extremely precise. Then the propagation time Ti of the satellite i is the difference between the time of arrival and the known time of transmission. Knowing the position of the satellite i, i.e. x_i, y_i, z_i, and the radial distance cT_i, one knows from the position of the receiver to be located at the sphere around the satellite i with radius cT_i. With the measurements from 3 such satellites, one can determine ones own position as the common point of the three spheres.

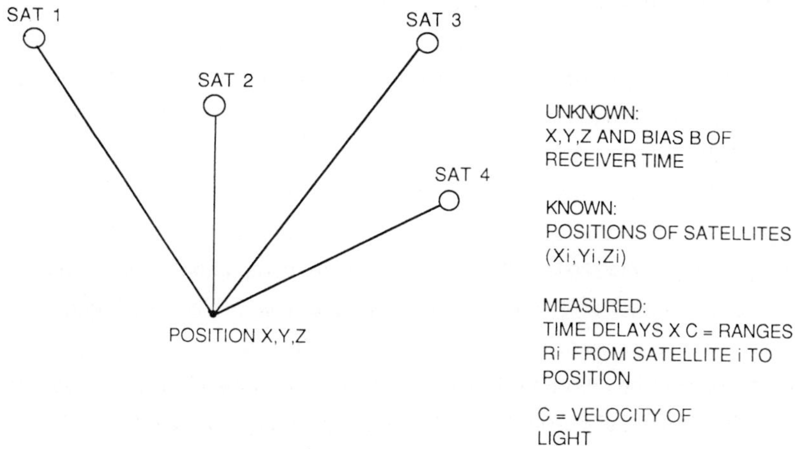

$$(Xi-X)^2 + (Y1-Y)^2 + (Z1-Z)^2 = (R1-B)^2$$

i = 1, 2, 3, 4

4 EQUATIONS, 4 UNKNOWNS

Fig. 2: Position determination with 4 GPS-satellites

Generally the receiver's clock is not as accurate and not synchronized to the GPS time. A one nanosecond deviation already means a 30 cm error, one microsecond a 300 m error. A bias B of the clock must, therefore, be taken into account. This can be tolerated, if ranging from at least 4 satellites is used. The set of 4 equations with the 4 unknowns x,y,z,B of the receiver location and time can then be solved. If more than 4 satellites are measured then this can be used for further improvements.

6. PRN- Ranging and Doppler

GPS applies pseudo-random-noise (PRN)-modulation. This means that a string of binary signals is phase modulated on a carrier frequency signal (L1, L2). The string of signals is periodic, but has an autocorrelation function , which is very similar to random noise. Details of this technique have been discussed in various papers, for example in Ref. 5.

- PRN-codes allow very efficient use of power.

- The ranging resolution is given by the so-called chip rate, i.e. the shortest interval between two changes. It is equivalent to the clock rate of the PRN generator. For the C/A-code, the ranging code of GPS, which is open for civilian users, is approximately 1 microsecond, i.e. 300 m. The P-code, which is the code for military applications of GPS, has a chip length of approximately 30 m. The ranging code can be resolved to about 1% of the chip length. Therefore, the resolution of the C/A-code is about 3 m and the P-code's resolution is about 30 cm. This is not the position accuracy, because the geometric factor of dilution and the propagation errors etc. contribute additionally to the error budget, Tab.1.

- Geodetic applications make use of the carrier phase measurement. This means that the measurement is not based on the PRN-code, but on the carrier frequency. Also here it is possible to make ranging measurements to about 1 % resolution. For the 19 cm wavelength of the L1 frequency this means that a resolution of about 2 mm can be achieved. Here we have, however, the problem of ambiguity. The phase measurement does not determine, how many wavelengths are to be added to this last fraction of the wavelength, where the phase is measured from. For the stationary and quasi-stationary applications in geodesy various ways of solving this ambiguity are applied. They can be used in particular for baseline measurements.

SEGMENT SOURCE	ERROR SOURCE	P-CODE PSEUDO RANGE ERROR 1 σ [m]	C/A-CODE PSEUDO RANGE ERROR 1 σ [m]
SPACE	CLOCK & NAV SUB-SYSTEM STABILITY	1.0	1.0
	PREDICTABILITY OF SATELLITE PERTUBATIONS	1.0	1.0
	OTHER	0.5	0.5
CONTROL	EPHEMERIS PREDICTION MODEL IMPLEMENTATION	4.2	4.2
	OTHER	0.5	0.5
USER	IONOSPHERIC DELAY COMPENSATION	2.3	5.0 - 10.0
	TROPOSPHERIC DELAY COMPENSATION	2.0	2.0
	RCVR NOISE & RESOLUTION	1.5	7.5
	MULTIPATH	1.2	1.2
	OTHER	0.5	0.5
1 σ SYSTEM UERE (m)	TOTAL	6.6	6.6

Tab. 1: GPS System Error Budget

- PRN-modulation allows recovery of the carrier frequency and from this the determination of the Doppler frequency shift of the received signal vs. the transmitted one is possible. Doppler measurement is based on the well-known fact that the radial velocity "v" changes the frequency proportionally to the absolute value of "v" and to the transmitted frequency. The received frequency becomes higher than the transmitted one, if the transmitter approaches the receiver and it becomes smaller when the transmitter increases its radial distance. Two effects arise in case of GPS: If the receiver is stationary then the Doppler effect is given by the motion of the satellite relative to the fixed position of the observer. Thus , it is a function of the GPS orbit which is known. If the observer moves, then an additional Doppler effect occurs. So, in measuring the resultant Doppler effect, with the GPS receiver and deducting the known orbital Doppler effect the remaining Doppler is due to the movement of the receiver. Accuracies in the order of dm/sec or even cm/sec can be achieved in 3 dimensions.

- GPS receivers for extremely precise measurements apply the two frequencies L_1 and L_2. This allows to compensate the ionospheric effect. The ionosphere delays the modulated signal, which means the "group velocity", and accelerates the carrier phase velocity. The product of group and phase velocities is equal to the c^2, where c is the velocity of light in vacuum. The effect on group and phase velocities is inversely proportional to the square of the carrier frequency, i.e. proportional to $1/L_1^2$ or $1/L_2^2$. It can easily be shown that by measuring the signal propagation delay for both frequencies, one can deduce the effect of the ionospheric electron density affecting both signals. Hence one can correct this effect. If this is done, then one can achieve by carrier phase ranging techniques an accuracy in the baseline measurement up to 10^{-8}. In the future, during the next 5 years, one intends to measure over several thousand kilometers the baseline distance to centimeter accuracies and in about 10 to 20 years it should be possible to measure any intercontinental baseline distance with some millimeter accuracy.

7. Directional measurements

We have already discussed that the carrier signals can be recovered in the receiver. In the interferometric setup (Fig.3) consisting of two receiver systems A and B, in baseline distance b, one can measure the path difference of the signals in fractions of a wavelength. For example, for the carrier frequency L_1 the wavelength is 19 cm and the resolution for the length measurement is in the order of 1 to 2 mm. The difference of the line of sight (LOS) determines the direction towards the satellite. The direction measurement to two satellites determines the platform orientation relative to the pair of satellites. With two baselines and two satellites one can so determine the orientation. Accuracies in the order of arc minutes and better are possible (Ref.6).

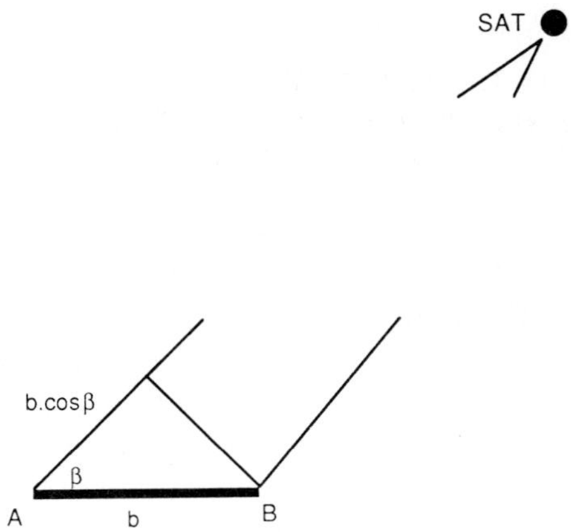

Fig.: 3 Interferometric mode

8. Practical applications

With GPS one can achieve in the future at any place in the world and at any time, accuracies in the order of some meters (P-code) or some ten meters (C/A-code) in real time and for very dynamic vehicles. Relative accuracies are much better. In the stationary case relative accuracies in the order of centimeter and decimeter can be reached now over distances several tens of kilometers.

The practical use of GPS-techniques will cover many fields:

- Surveying. In the future the 3-dimensional position determination of fixed points will be used. Of particular interest is the network determination, where direct LOS is not given. Today GPS-surveying is already practiced in Germany and elsewhere.

- Kinematic surveying is the next step. Many experiments have shown that the 3-dimensional measurements are possible by driving with a properly equipped car from one station to the next.

- Sea and air navigation are naturally very important applications, for en route as well as in harbour or in airport areas.

- The main driver for the development of new low-cost receivers will be the automobile market. It is expected that millions of receivers will be produced in the future for every-day applications.

We think that many new applications will come up in the future and that GPS receivers as small as a cigarette box will become almost as usual as radio sets and quartz watches.

References

(1) International Federation of Surveyors -FIG-Proceedings Inertial, Doppler and GPS Measurements for National and Engineering Surveys Schriftenreihe UniBw, Heft 20-1, Okt. 1985, München

(2) Wilmes H., Reigber Ch., Schäfer W., Hartl Ph. Precise Range and Range Rate Equipment, PRARE, on Board ERS-1 Proc. XiX I.U.G.G. General Assembly, International Association of Geodesy, Vancouver, B.C., Canada, 1987

(3) Global Positioning System, Vol.I Papers, published in NAVIGATION and reprinted by the INSTITUTE OF NAVIGATION, Washington, D.C. 1980

(4) Global Positioning Systems, Vol.II, dto. 1984

(5) Hartl, Ph., Schäfer, W. Nutzung der Spread-Spectrum-Technik bei Kombination der Aspekte Navigation und Telekommunikation in Satellitensystemen FREQUENZ 41 (1987) 11/12, S. 321 - 328 (Part 1) FREQUENZ 42 (1988) 1, S. 29 - 34 (Part 2)

(6) Hartl, Ph., Wehr, A. Coordinate Transformation Procedures for GPS-Attitude Control Proc. ISPRS Symp. 1986, pp. 216-225, ESA SP-252, Nov. 1986

Navigational Methods of Measurement in Geodetic Surveying

Klaus Linkwitz, Wolfgang Möhlenbrink
Inst. für Anwendungen der Geodäsie im Bauwesen, Universität Stuttgart

Summary

Geodesy reaches a new quality of performance with the development of computer-aided measurement systems. Compared to satellite techniques well suited for global tasks, close range area techniques are achievable with unseen accuracy and unseen sampling rates. Inertial measurement techniques based on purely dynamic principles will change the educational requirements of geodesists and surveyors. This paper summarizes principles of geodetic measurement techniques and shows with examples the developement to dynamic measurement techniques in close range area. The convergence of photogrammetric and surveying techniques with respect to realtime positioning is obvious.

1. Introduction

An outstanding characteristic in the development of geodetic surveying during the last ten years has been the increasing importance of position determination methods which were derived from navigational procedures for ships and aircraft. Contrary to classical geodetic methods of measurement which achieve high degrees of accuracy in static or quasi-static networks especially through repeated measurements, position determination in real time is the central issue in navigation. This means, that only one isolated information concerning position can be obtained for anyone point in time, and that each repetition of a measurement describes already another position (which, due to its correlation with the prior measurement may also be used for navigation).

For all types of geodetic measurements the utilization of the following techniques is of special importance

- inertial navigation systems (INS),
- global positioning system, e.g. NAVSTAR (navigation systems with time and ranging) and GPS,
- optical navigation aids: image processing.

Navigational and geodetic methods, developed separately in the past, will now be jointly treated within the special research area "High precision navigation - integration of navigational and geodetic methods" (SFB 228 - Sonderforschungsbereich 228 "Hochgenaue Navigation - Integration navigatorischer und geodätischer Methoden"). It is of particular significance in regard to geodetic surveying that methods developed for aircraft navigation in real time have cycle times of up to 10-20 ms. Consequently, moving objects will be very precisely measurable if the development of such navigational systems is persued with consequence. However, the dynamics of the measurement system and of the observed object assume to play a role far exceeding their importance in conventional geodesy. Thus the implementation of adequate methods originating from system theory and analytical mechanics, for the description of dynamic systems is an essential prerequisite for the further development of navigational position measurements intended to be used in geodesy. According to Hiller (1983), the term "system" is always then applicable when procedures are to be desribed which are complicated and difficult to comprehend. "A system is to be defined as a number of elements (parts, components), which influence each other mutually (principle of action-reaction), react upon outside influences (input) and release effects externally (outputs)." (Fig. 1)

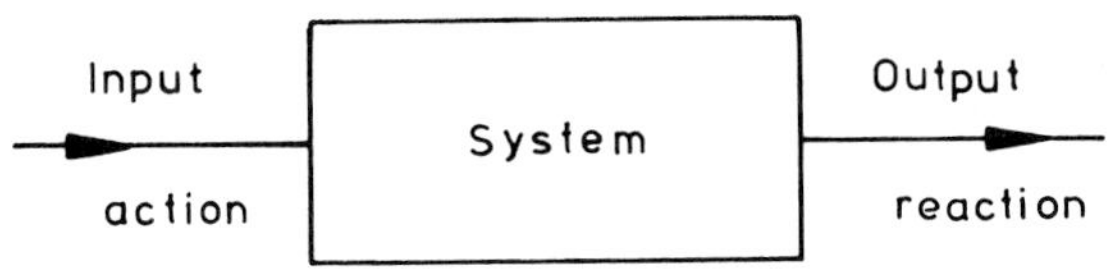

Fig. 1: System approach

Therefore, variability (movement) and influenceability (steerability) through an appropriate choice of inputs (forces) belong to the more important properties of a system. Following the reasoning of classical mechanics the interdependence between the influences upon the systems and the changes within the system are called "dynamics of systems" (Hiller,1983).

The geodesists using action-reaction analysis of dynamic systems in which their change of state is manifestated in the form of changes of position and changes of volume, may lead to a considerably expanded area of geodetic methods (Möhlenbrink/Krzystek 1984 and Möhlenbrink/Welsch 1984).

2. Navigational Measurement Techniques

2.1 Inertial navigation systems adopt a significant role as autonomous systems in vehicle navigation. The first geodetic position determination by this method became known since 1970. Up to now in inertial geodesy mostly gyro-orientated platforms have been used almost exclusively for high precision measurements. Here three accelerometers are mounted on a horizontal or space-stabilized platform. The orientation of the platform is achieved by mechanical gyros with one ore more axes. As a result of the measurements, coordinates which result from an integration of velocity increments and orientations are obtained.

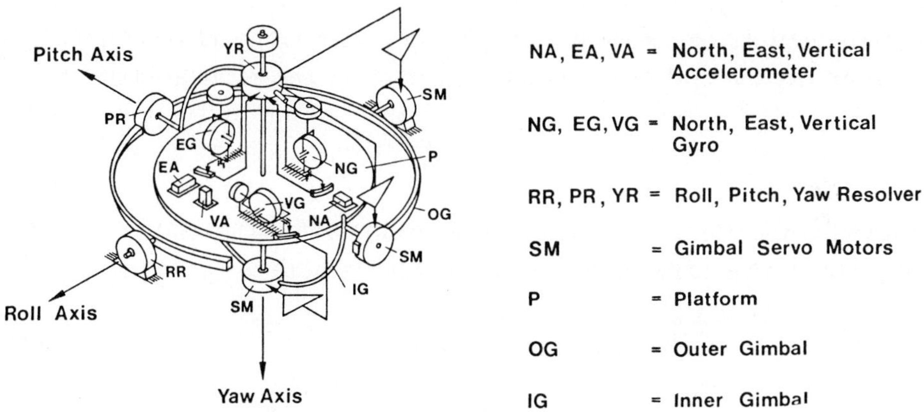

Fig. 2: Local level platform with three gyros and three accelerometers (Stieler, Winter 1982).

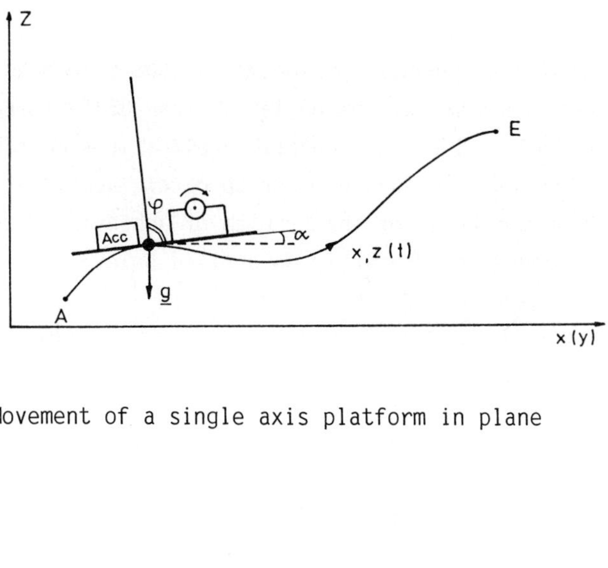

Fig. 3: Movement of a single axis platform in plane

The principle of inertial navigation systems can be demonstrated in a simplified manner using a system with a single axis moving in the x-z plane (Fig.3). Accelerometer and gyro to be mounted faultlessly on a platform "p". If the platform is moved from A to E, the following forces act upon the accelerometer:

force of inertia: $-m\ddot{x}$

gravity: $m \cdot g \sin\alpha \cdot$ (1)

The measured (specific) force

$$F_s = m\ddot{x} + m \cdot g \sin\alpha \tag{2}$$

therefore consists of components of the two forces exerting their influence. The aim of the inertial position determination is the ascertainment of

$$x(t) = \frac{1}{m}(F_s - g \cdot \sin\alpha(t))\, dt^2. \tag{3}$$

To determine the field of gravity the direction and the amount of the gravitational vector have to be determined in its horizontal and vertical components:

direction: $\approx \alpha \approx (+\ddot{x}).$ (4)

Obviously, the orientation - especially the inclination - of the platform relative to the plumb line has to been known if the output of accelerometers taken in the field of gravity is to be analysed further and used as primary data for the determination of position by double integration. In aircraft navigation inertial accelerations of up to several g occur. In the case of slowly moving systems for geodetic measurements, the precedence of gravity acceleration of the total signal may be larger than 95 %. Therefore the inclination of the platform in inertial-geodetic applications must be determined to within 1 to 2 " (angular seconds) in order to obtain acceleration accuracies of 10^{-5}g. Every geodesist knows of the problem of achieving such an accuracy when levelling a theodolite accurately on a fixed point.

Gyros are used as rotational sensors for the determination of the orientation of the vector of acceleration. Here the platform is either kept continuously in a horizontal position or stabilized in space, as explained above. With the development of strapdown inertial systems - the systems are hereby mounted on a carrier-fixed platform - the mechanical local level system of the gyro-stabilized platform can be dispensed with. Then the navigation calculation is referred to the so-called analytical platform. In our SFB 228, one focal point of research will be the development of geodetic measurement techniques using laser gyro strapdown systems. We hope and expect than from 1989 onwards this special field of research will disposed of a laser-gyro inertial navigation system and thus be able to verify different methods and models.

The basic equation (1) makes it very obvious that errors in the term of gravity $g \cdot \sin\alpha$ and errors in the measurement itself (e.g. scale, zeropoint, misalignment etc.) will grow more than proportionally due to the double time integration. Therefore, the error characteristic of a system with three axes is highly unfavourable with regard to calculated position (Britting 1971; Schwarz 1986; Stieler/Winter 1982). As a consequence, geodesists working with inertial systems use the technique of zero-velocity updating (ZUPT) and post-mission adjustment, and least square methods to incorporate inertially determined coordinates into a field of already existing fixed points. Those algorithms are used here which essentially consider geometrical parameters and drift effects.

However, the dynamical character of the measuring system - rotational and translational sensors are measurement instruments reacting to forces and momentums - so far are practically not considered in inertial geodesy. Also, a precise knowledge of the earths gravity field is of equal importance for the position determination as is the high accuracy of fixed geodetic points. For this reason different possibilities to increase the accuracy level will be investigated in the SFB in the following areas:

- further refinement of the models for the gravity field,
- improvement of the definition and determination of datums and coordinate systems,
- consideration of the dynamic characteristics of inertial navigation systems.

The global models of the gravity field have to be improved in detail and within close ranges so that the influence of an incorrectly known orientation can be kept down to a minimum.

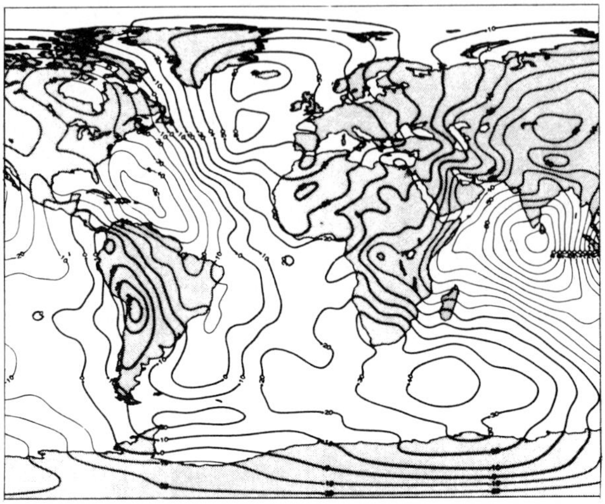

Fig. 4: Geoid (Reigber 1982)

The development of appropriate dynamic models for geodetic inertial navigation systems is handicapped by the inaccessibility of the primary outputs of gimbles and accelerometers and by the errors in a gyro-orientated platform. However, for the solution of many geodetic tasks and problems a realtime positioning is not necessary, and strap-down systems can be better modelled in a post-mission analysis than platform systems. Therefore the SFB intends to develop measurement techniques and models for inertial geodesy based on laser-gyro-strapdown systems supported by GPS measurements.

This type of inertial navigation system is interesting for two reasons:

- Since 1981 results obtained using local-level geodetic platform systems could pratically not be improved in their accuracy levels (Schwarz 1986). It seems that here the range of possible development has already reached an intermediate end.
- In virtue of their technical realisation laser-gyros are insensitive to vehicle vibrations and to shock effects. In addition, they allow larger angular velocities in the strapdown mode.

Prior to the extraordinarily precise GPS-results in differential modus, inertial geodesy was considered to be an efficient method for position determination, especially in the field of large-area geodetic surveying. However this former significance is now diminishing. Impeded by its time-quadratic error characteristic inertial navigation systems in the long run may range behind position determination with GPS.

2.2 The GPS-system NAVSTAR startled the professional world by its outstanding accuracy when used for geodetic fixed point determinations. It can be expected that especially in geodetic nets down to fourth order, GPS will in future play a dominant role. However, it has to be considered that NAVSTAR was developed for the navigation of ships and aircrafts, and that it needs an unimpeded view to at least four satellites as a pre-requisite for exact point determination.

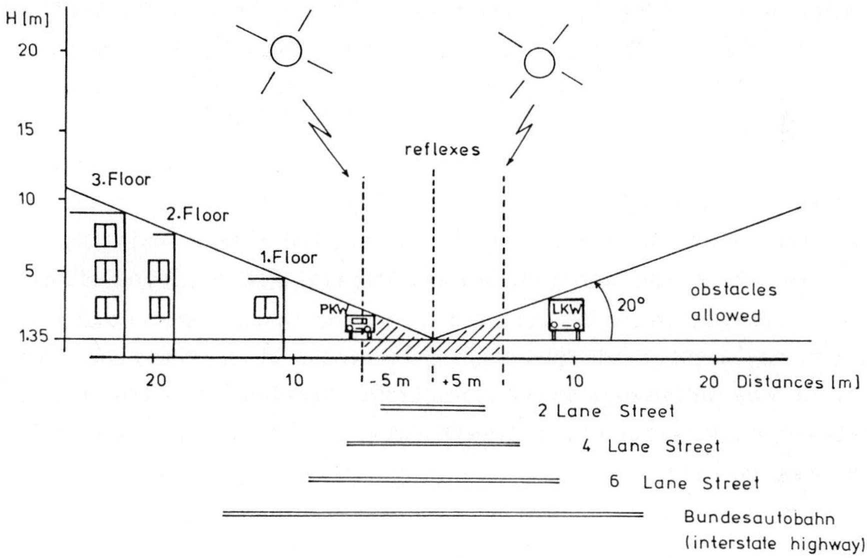

Fig. 5: Shadowing of GPS signals in urban areas

This means in practice that no impediments in site should appear atn elevation angle of upto 15°-20° (Lindlohr/Wells 1985; Kleusberg 1988), and that no metallic or reflecting objects should be situated within a radius of 5 km from the observed station. It is difficult to imagine such an environment in towns or on construction sites. Therefore, a combination of fixed points determined by GPS and of other methods of routine surveying could be an efficient combination within close ranges, executed by cadastrial offices and engineering or surveying consultants.

2.3 Optical position measurement systems

The development of CCD-sensors (Charged Coupled Device) and the rapid increase in computer efficiency have lead to the possibility to use computer-controlled object recognition and digital image processing increasingly as a navigational aid. (Computer vision). There are encouraging examples - steering of cars and of special experimental test vehicles (Rembold/Levi 1986) using laser and ultrasonic scanners, vehicle steering (Dickmanns 1988) using CCD-cameras in realtime and the first on the market available industrial measuring system "Space" by Kern - which lead to the expectation that computer-aided measurement systems with realtime capacities will eventually also be used in the classical geodetic field. As primary obervables these may use, as before, measurements of angles and distances. The provision of a modular optical position measurement system (MOPS) could fill a gap in close range measurement techniques. First steps in this direction are recognizable in the form of GEOROBOT (Kahmen) and SPACE (Kern).:

The combination of realtime measurement systems originating from navigation with geodetic methods will lead to the development of realtime surveying systems. Such systems will be applicable within the already existing net of fixed points - below the level of GPS and inertial methods - and allow the automatic determination of positions of moving and, of course, also non-moving points. One may question whether there is a demand at all for such surveying systems. To give an answer, several practical examples taken from papers and applications of the Institut für Anwendungen der Geodäsie im Bauwesen (Institute for Applied Geodesy - IAGB) will be presented.

3. Realtime Surveying Problems

The term "Real Time" originates from the area of process management, control technique and navigation. Realtime processes are characterized by the fact that immediate reactions to disturbances of a running procedure are wanted in such a way that the procedure remains always stable; for instance an autopilot has to react to a sudden momentary gust in such a way that aerodynamical safety is not endangered.

An analogous model for the control (steering) of construction machinery, e.g. a paving machine, can be imagined: the machine has to be steered along a pre-set line which up to now is realized by precisely staked out and then fixed wire.

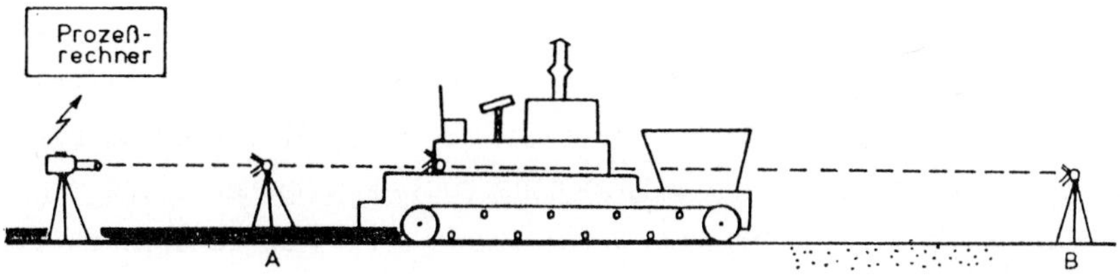

Fig. 6: A paving machine

Understanding the term navigation in its very sense it may be devided into the following sub-terms:

- position determination / course determination, present state
- comparison with given positions / given course
- calculation of the correcting parameters for the most favourable keeping on course
- correction of course.

In engineering geodesy control operations may be called "High Precision Navigation" if they are equipped with an accurate geodetic position determination system. Such an equipment could be realtime measurement systems. A position determination system then could be called a realtime measurement system when time consumption for date processing is so small that the process sequence itself is, if at all possible, not delayed.

The inclusion of a realtime measurement system into a control circuit for the control of the construction operation is shown in Fig.7. Such problems are sufficiently well-known to everybody from engineering geodesy. However, not all geodesists may be conscious of the fact that such models and system descriptions may lead to advantageous solutions.

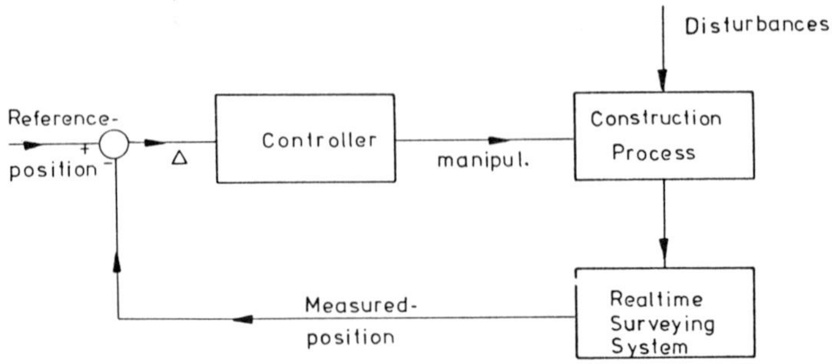

Fig. 7: Realtime surveying within construction process

4. High Precision Measurement of Moving Objects

4.1 Example deformation measurement

A continuous realtime control in an experiment of test loading of pairs of bridge pillars was carried out at several bridges along the new high-speed railway track Stuttgart-Mannheim of the German Bundesbahn.

In order to control the bending stiffness of the pillars and to control whether the maximum foundation pressures laid down in the design were adhered to, the bridge pillars were tight together horizontally at a time with steel cables and pulled towards one another with given forces.

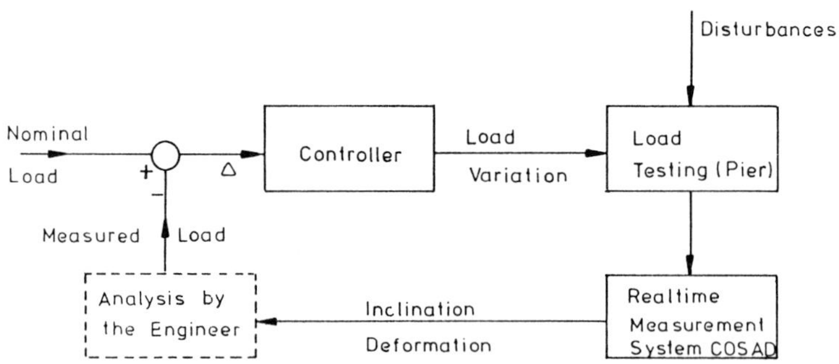

Fig. 8: Realtime system within test-procedure

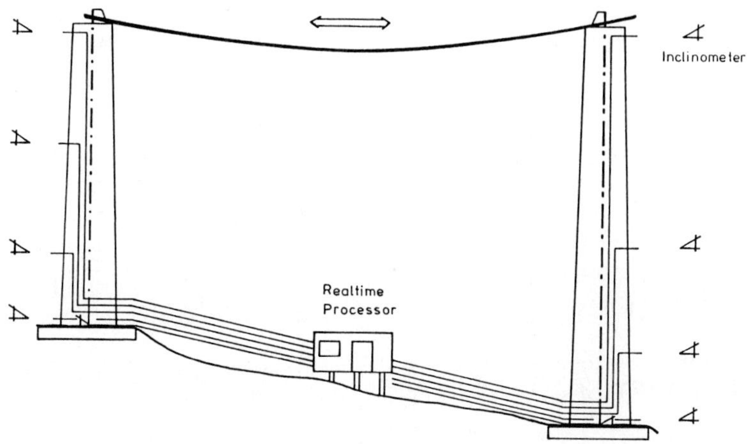

Fig. 9: Testload Bauerbach-Bridge

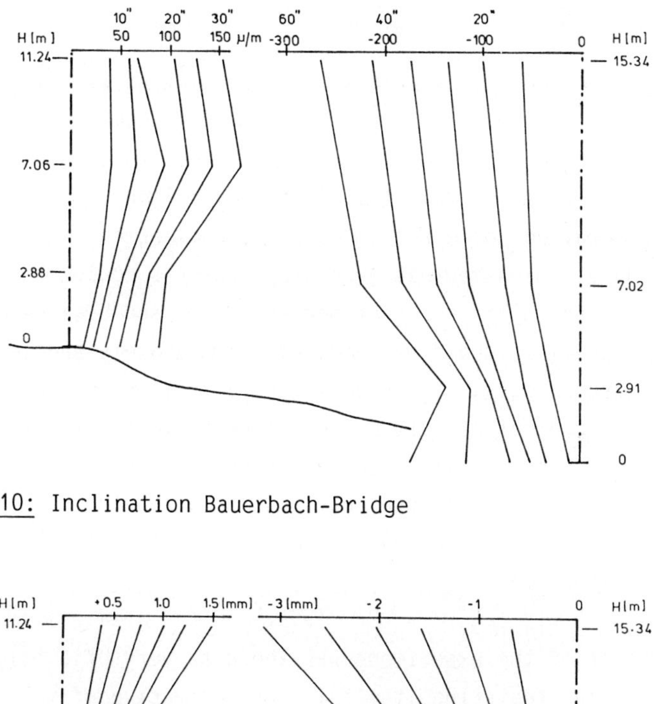

Fig. 10: Inclination Bauerbach-Bridge

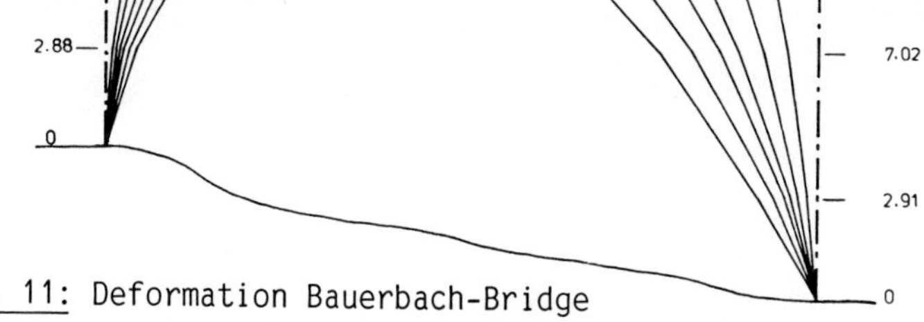

Fig. 11: Deformation Bauerbach-Bridge

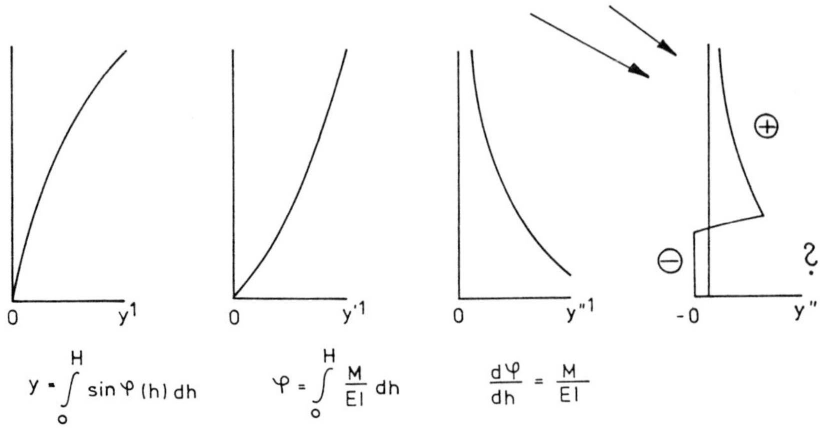

Fig. 12: Analysis of realtime measurement

In the experiment the continuous process of test loading with increasing increments of loads was controlled by the realtime measurement system COSAD. A connected computer processor showed on its screen deformations, inclinations changes of bending moments in realtime. In the experiment realtime measurement and analysis was required to detect immediately unexpected states of the pillars under continuous loads so that necessary measures could be taken in time to prevent damages. In the different test series bulges and breaches in the construction joints appeared, could be recognezed immediately and the experiments could be changed accordingly. It may be remarked that in this case we are concerned with an open control circuit where the deviations with respect to nominal values and the correcting variables are manually ascertained by expert engineers. The experiments demonstrate that also slow running processes may require realtime measurements in order to increase safety and efficiency in the execution of the experiment.

The essential result of the experiment was the discovery of a bulge at the foot of the pillar manifestating itself in the derivative of the inclination image through an edge in the inclination line. Only by using a realtime measurement system the recognition of such an occurance can influence the ongoing experiment in time.

4.2 Measurement of the trajectory of an industrial robot.

IAGB developed an optical position measurement system (OPM) to determine the position of moving objects in realtime. The essential components of this OPM are electronical cameras containing in their image planes lateral effect diodes sensitive to the position of a point image. Such Position Sensitive Detectors (PSDs) permit to determine precisely the coordinates of the images of actively lighted targets (e.g. infrared diodes) with an accuracy of a few micrometers (Bayer/Krzystek/Möhlenbrink 1988). For the dynamical measurement of moving objects the scanning speed and the capacity for data processing are of decisive importance in practical implementations. Compared to 512 x 512 pixels of a CCD camera which have an 8-bit gray value depth, the PSD has per image an amount of data 16000 times less. Therefore, this sensor was used for the OPM development for dynamical measurement applications. As disadvantage can be regarded the necessary signalization of control points. However, this is not of too much importance when steering construction machinery.

Prerequisite for the implementation of PSDs are the elimination of diffused light, the elimination of reflections of the measurement signal, and a sufficiently precise calibration of the highly non-linear sensor (Bayer/Krzystek/-Möhlenbrink 1988).

The OPM - coupled with the micro-processor - can be used as a realtime position measurement system as schematically outlined in Fig. 13.

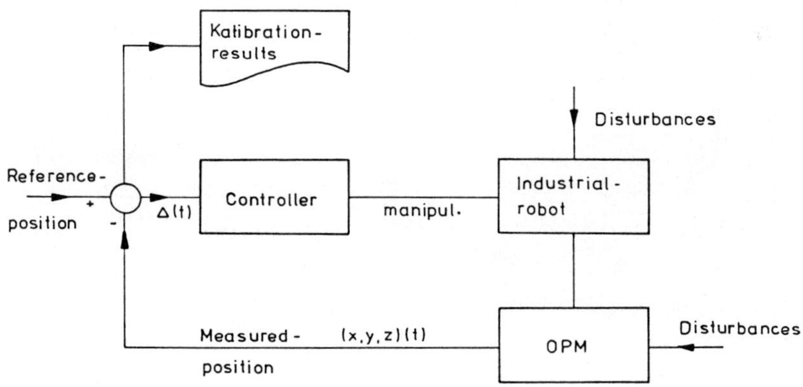

Fig. 13: Realtime robot guidance

An example is the online measurement of the movements of an industrial robot's arm (Krzystek 1988). For controlling and checking pre-programmed movements - circular movements in an oblique plane in space - static and dynamic determinations of position in realtime were carried out. As illustrated in Figures 14 and 15, the robot's arm was signalized with a luminous diode target. As a frame of reference and a field of control points in space was materialized using a set-up of carbon fibres. The results of the measurements are illustrated in fig. 16.

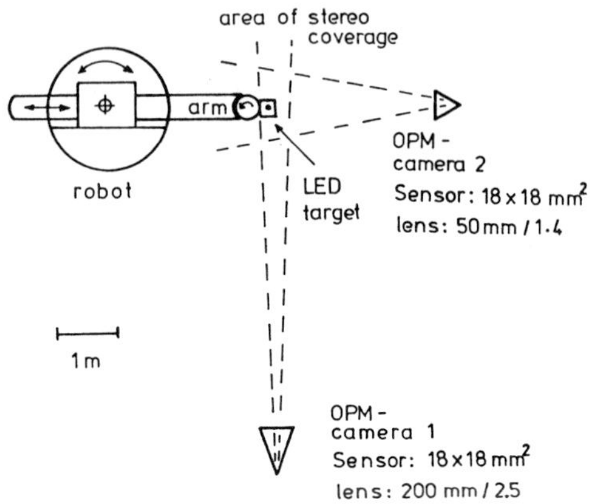

Fig. 14: Measurement

Fig. 15:

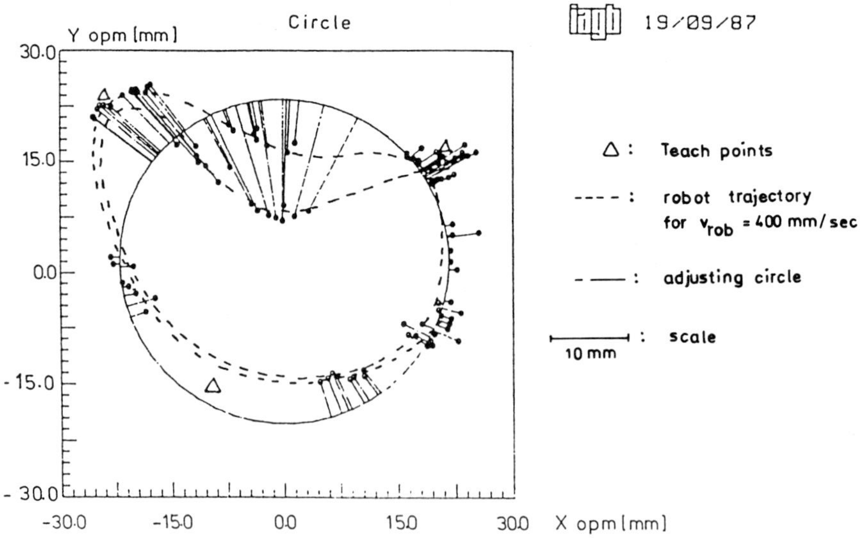

Fig. 16: Target tracking

It shows the deviations of the actual trajectory of the robot's arm against the prescibed movement of a circle. These deviations of the robot's arm could be directly fed into the control system as state variables. If then cartesian coordinates are used to steer the arm by an OPM multi-camera a trajectory accuracy of 10-15 micrometers in object space can be expected with a stationary installation.

4.3 Dynamic surveys with tracking theodolites

The above illustrated OPM camera system allows the realtime measurement of movements within predefined sections of space. The solution of certain problems, however, will only be possible if this restriction can be overcome. One solution would be to change the exterior orientation of the OPM camera by preset values. A practical solution adopted by IAGB was to mount an OPM camera on a computer guided WILD T2000 motor theodolite. This setup allows the observation of directions in sets.

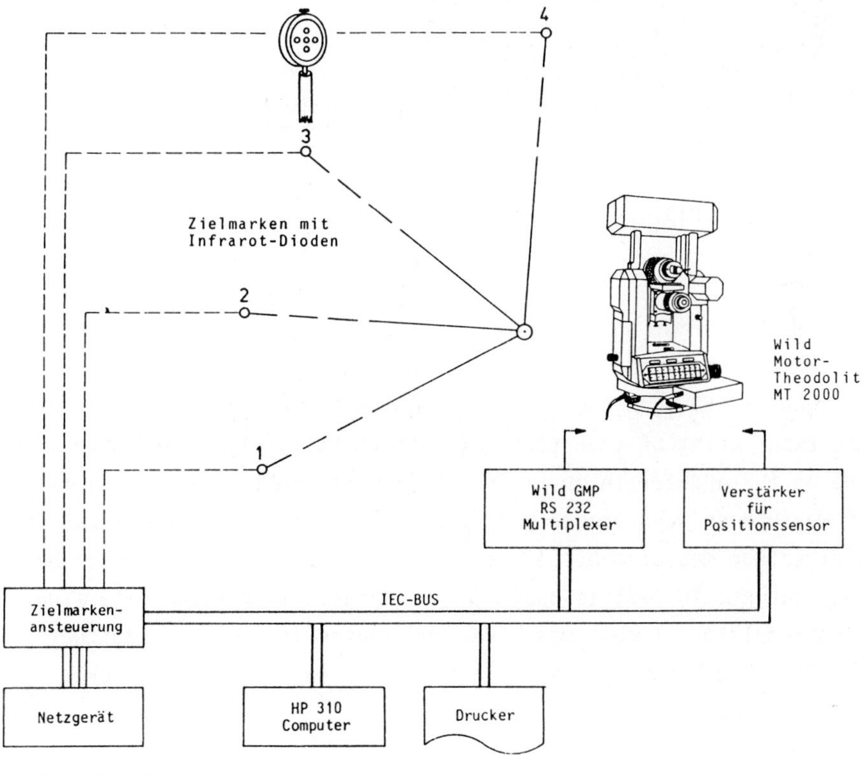

Fig. 17: Automatic resection with motorized theodolit

Fig. 17 illustrates the setup of equipment to determine the directions to 4 points installed in the laboratory. In a practical experiment it could be shown that accuracies in direction of ±0.03" can be obtained without difficulty.

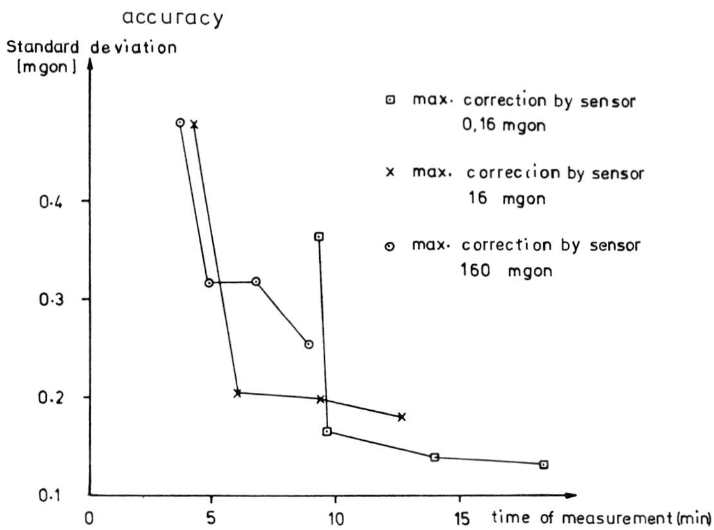

Fig. 18: Accuracy of angle-measurement

To intersect the target an ordinary photo lense with 50 mm focal distance was used. After presetting the theodolite to within an angular distance of 0.16 mgon, 16 mgon or 160 mgon the motor theodolite - in a regulating circuit - finds the exact point of intersection automatically; Figure 18 shown also how, depending on the angular interval in which the theodolite works automatically speed of measurement and accuracy of intersection vary. It is now possible to perform direction measurements in sets to 4 targets within a time of half a minute per target. To realize a system for dynamical position determinations, a tracking regulator circuit was developed (Bayer 1988). To investigate the accuracy of such a tracking system, a simulator was built which moves luminous diode target along a circular trajectory.

At present the accuracy achievable in the dynamical tracking mode of the system is about 80''. To increase this accuracy, especially the calibration of the sensor, the dynamical registration of the angle, and the realtime process software have to be investigated and to be improved. Since the simulator showed that angular velocities up to 60 /sec are possible, also rapid movement of objects can be tracked. This development is aimed at the steering of construction machinery.

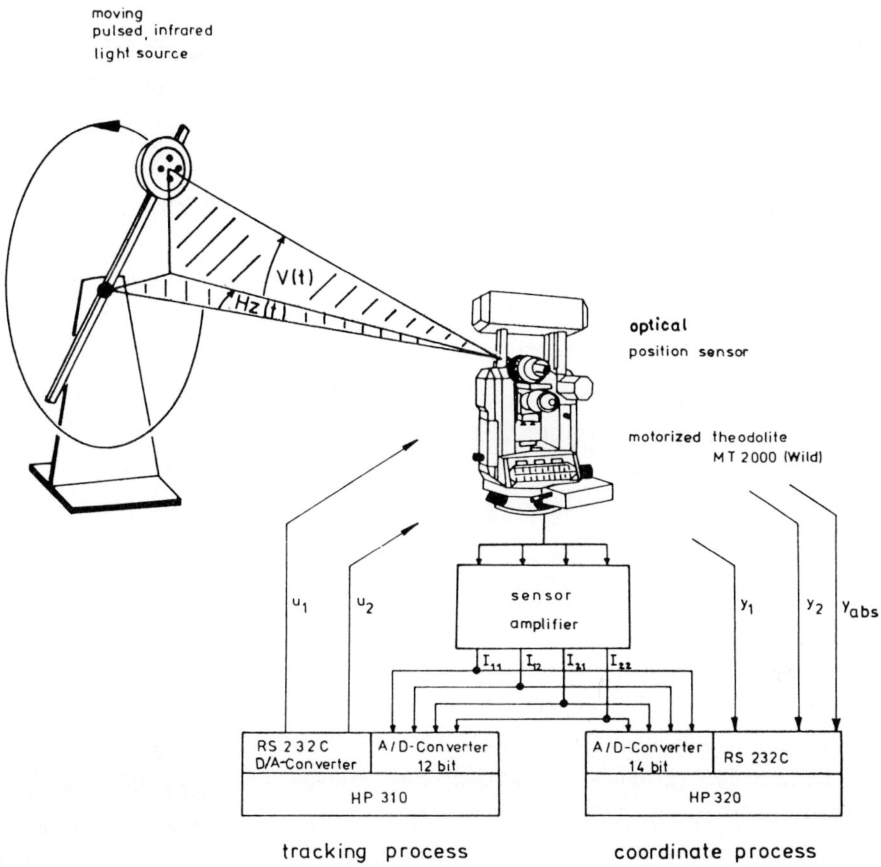

Fig. 19: Dynamic target tracking

Literature:

Bayer, G. (1988): Motorized Electronic Theodolites - High Precision Measurement Robots for Surveying, Int. Workshop High Precision Navigation, SFB 228, Stuttgart

Bayer, G., Krzystek, P. (1986): Arbeiten und Ergebnisse der Teilprojekte D1/D3. Arbeits- und Ergebnisbericht 1984 - 1986, Sonderforschungsbereich 228 "Hochgenaue Navigation", Stuttgart

Bayer, G., Krzystek, P., Möhlenbrink, W. (1988): Realtime Positioning of Moving Objects by Dynamic Target Tracking, ISPRS-Congress, Kyoto

Britting, K.R. (1971): Inertial Navigation System Analysis, Wiley-Interscience, New York

Conati, F. C. (1977): Real-Time Measurement of Threedimensional Multiple Rigid Body Motion, S. M. Thesis, Massachusetts Institute of Technology, Inne

Dickmanns, E.D. (1988): Vehicle Guidance by Computer Vision. Int. Workshop High Precision Navigation, SFB 228, Stuttgart

Hiller, M. (1983): Mechanische Systeme, Springer Verlag, Berlin

Hof, A.; Pfeiffer, T. (1985): Räumliches Wegmeßsystem, 3D-Interferometer GMR-Bericht Nr. 6, März, S. 151 - 165

Janocha, H., Marquardt, R. (1979): Universell einsetzbares Wegmeßsystem mit analog anzeigenden, positionsempfindlichen Photodioden. Teil 1: Eigenschaften der verwendeten zweiachsigen Photodioden. Technisches Messen tm 1979, Heft 10, S. 369 - 373

Kirschstein, M. (1980): Verfahren zur berührungslosen dreidimensionalen Form- und Verformungsmessung, Dissertation, Universität Stuttgart

Krauss, H. (1983): Das Bild-n-Tupel. Ein Verfahren für photogrammetrische Ingenieurvermessungen, Dissertation, Universität Stuttgart, DGK Reihe C, Heft 276, München

Krzystek, P. (1988): High Precision Surveying of Moving Objects by Electronic Cameras, Int. Workshop High Precision Navigation, SFB 228, Stuttgart

Lindlohr, W., Wells, D. (1985): GPS Design Using Undifferenced Carrier Beat Phase Observations, Manuscripta Geodaetica, Springer Verlag, Berlin - Heidelberg

Möhlenbrink, W. (1988): Realtime Surveying in Close Range Area with Inertial Navigation Systems and Optical Target Tracking Techniques, Int. Workshop High Precision Navigation, SFB 228, Stuttgart

Möhlenbrink, W., Krzystek, P. (1984): Messungen windinduzierter Bewegungen an Hochbauten - Erfahrungen und neue Erkenntnisse, IX. Int. Kurs für Ingenieurvermessung, Graz

Möhlenbrink, W., Welsch, D. (1984): Aktions-Reaktions-Analyse thermischer Verformungen an Ingenieurbauten, IX. Int. Kurs für Ingenieurvermessung, Graz

Reigber, Ch. (1982): Abschlußbericht zum Teilprojekt B3 "Figur- und Feldparameterbestimmung" des SFB 78 Satellitengeodäsie, DGK Reihe B, Heft 261, München

Rembold, U., Levi P. (1986): Sensors and Control for Autonomous Robots, Preprints Int. Conference Intelligent Autonomous Systems, Elsevier Science Publishers B.V.

Schwarz, K.P. (1986): The Error Model of Inertial Geodesy - A Study in Dynamic System Analysis, Lecture Notes, Fourth Int. Summer School, Admont, Austria

Stieler, B., Winter, H. (1982): Hyroscopie Instruments and their Application for Flight Testing, AGARD, Neuilly Sur Seine

Woltring, H. J. (1976): Calibration and Measurement in 3D-Monitoring of Human Motion by Opto-Electronic Means.II. Experimental Results and Discusion. Biotelemetry,3, pp.65-97

Schwidefsky, K., Ackermann, F. (1976): Photogrammetrie, Teubner Verlag, Stuttgart

Geodetic Positioning by Inertial and Satellite Systems: An Overview

Erik W. Grafarend
Geodätisches Institut, Universität Stuttgart

Summary

Once satellite and inertial systems are used for geodetic positioning they offer mainly the advantage of online coordination in geometry and gravity space as well as the unnecessary direct line-of-sight. Here their system analysis is based on the *threedimensional network* point of view for satellite systems, e.g. GPS or GLONASS, and the *traverse network philosophy* for inertial systems.

1. Satellite positioning systems

Example one : EDM

Assume the Euclidean distance in vacuo between a terrestrial point P_α (subindex) and a satellite point P^β (superscript) has been determined by comparing the geocentric clock readings T^β of the transmitted signal and T_α of the received signal whose travel geocentric time difference is $T^\beta - T_\alpha$.

$$S_\alpha^\beta = c(T^\beta - T_\alpha) = \sqrt{(X^\beta - X_\alpha)^2 + (Y^\beta - Y_\alpha)^2 + (Z^\beta - Z_\alpha)^2} \tag{1.1}$$

characterizes the *nonlinear* observational equation in Cartesian coordinates $(X_\alpha, Y_\alpha, Z_\alpha)$ of the ground point and $(X^\beta, Y^\beta, Z^\beta)$ of the target satellite point. Here we refer Cartesian coordinates to the Earth-fixed equatorial frame of reference being located in the Earth's mass centre.

A geodetic network is built up by g *ground* stations from which we observe h *high* targets, the satellites, simultaneously. Once we denote by n the number of observations, by m the number of estimable unknowns, e.g. coordinates, respectively, we are led to the identities

$$\boxed{n = gh\ , \quad m = 3g + 3h - 6.}$$

For *one* ground station there are h observed satellite points, for *two* ground stations 2h observed satellite points, in general n = gh observations. In contrast, per station and satellite point there are in general *three unknown coordinates*. The number of 3g + 3h unknown coordinates has to be reduced for *three translation parameters* and *three orientation parameters* as in general non-estimable quantities, in toto 6 parameters. These datum parameters remain free, since a threedimensional network may be translated or rotated without any change of the observed distances. If we just balance the *necessary observations* which determine the threedimensional *network configuration* (no configuration defects!) we arrive at

$$\boxed{n = m : gh = 3g + 3h - 6 \Longrightarrow h = \frac{3g-b}{g-3}} \tag{1.2}$$

which is a *diophantic equation* (only integers are admissible solutions) being solved for h. In case of $n > m$ we are led to an overdetermined system which may be solved by adjustment. If we postulate a minimum number of ground receivers, the solution of our diophantic equation is $g = 4$ ground stations which simultaneously observe $h = 6$ satellites. The 4 - 6 solution orginates from *E. Tsimis* (1973).

When we restrict ourselves to the case where all coordinates of satellite points P^{β} are perfectly known we derive the *diophantic equation*

$$\boxed{n = m : gh = 3g \Longrightarrow h = 3} \tag{1.3}$$

which teaches us that *three* satellite points have to be observed simultaneously.

There is a wellknown special case when we have to synchronize the terrestrial receiver with respect to "satellite time". Let us denote the corresponding time *offset* by a, being an unknown, we write

$$T_{\alpha} = a + t_{\alpha} \tag{1.4}$$

Now the balance of observations and unknowns results into

$$\boxed{n = gh \ , \quad m = 4g \ , \quad n = m : gh = 4g \Longrightarrow h = 4} \tag{1.5}$$

if in addition we again assume perfectly known satellite point coordinates. The classical solution $h = 4$ of the *diophantic equation* is known in kinematic positioning as the *navigation solution:* One terrestrial receiver tracks *four* satellites simultaneously.

Finally a comment with respect to critical configuration of the famous 4 - 6 general EDM solution: *E. Grafarend and V. Müller* (1985) have shown analytically that the 4 - 6 configuration approaches singularity as soon as the four ground stations become *planar*. ■

Example two : DOPPLER

With respect to the sympol ${}^{\beta-1}\nabla^{\beta}_{\alpha}$ *Doppler observations* from terrestrial points P_α to satellite points $P^{\beta-1}$ and P^{β} can be understood as distance differences

$$S^{\beta-1}_{\cdot\ \alpha\ \cdot}{}^{\beta} = S^{\beta}_{\alpha\cdot} - S^{\beta-1}_{\alpha\cdot} = \sqrt{(X^{\beta}-X_\alpha)^2 + (Y^{\beta}-Y_\alpha)^2 + (Z^{\beta}-Z_\alpha)^2} - \sqrt{(X^{\beta-1}-X_\alpha)^2 + (Y^{\beta-1}-Y_\alpha)^2 + (Z^{\beta-1}-Z_\alpha)^2} \,. \tag{1.6}$$

The corresponding *diophantic equation* which balances the observations n and the unknown coordinates m being estimable reads

$$\boxed{n = m : g(h-1) = 3g + 3h - 6 \implies h = \frac{4g-6}{g-3}\,.} \tag{1.7}$$

Thus g = 4 ground stations observe h = 10 satellite points simultaneously, the minimum g-solution in terms of integers of (1.7). Again as has been shown by *E. Grafarend and V. Müller* (1985) the satellite network approaches its critical configuration as soon as the four terrestrial points are located in a plane. ■

Example three : interferometric measurements

With respect to the symbol ${}_{\alpha-1}\Delta^{\beta}_{\alpha}$ *interferometric measurements* between terrestrial points $P_{\alpha-1}$ and P_α to satellite points P^{β} can be unterstood as *distance differences*

$$\begin{aligned} S_{\alpha-1\ \cdot\ \alpha}^{\ \ \ \ \ \beta} &= S_{\alpha-1\cdot}^{\ \ \ \ \beta} - S_{\alpha\cdot}^{\ \beta} = C(T_{\alpha-1}-T_\alpha) = \\ &= \sqrt{(X^{\beta}-X_{\alpha-1})^2 + (Y^{\beta}-Y_{\alpha-1})^2 + (Z^{\beta}-Z_{\alpha-1})^2} - \\ &- \sqrt{(X^{\beta}-X_\alpha)^2 + (Y^{\beta}-Y_\alpha)^2 + (Z^{\beta}-Z_\alpha)^2} \,. \end{aligned} \tag{1.8}$$

The corresponding *diophantic equation* which balances the observations n and the unknown estimable coordinates m reads

$$\boxed{n = m : (g-1)h = 3g + 3h - 6 \implies h = \frac{3g-6}{g-4}\,.} \tag{1.9}$$

For instance, the minimum g-solution is g = 5, h = 9: g = 5 ground stations observe h = 9 satellite points simultaneously. The 5 - 9 solution originates from *D. Delikaraoglou* (1984). *If* we assume perfectly known satellite points the *diophantic equation* for interferometric observations reduces to

$$\boxed{n = m : (g-1)h = 3g\ .} \tag{1.10}$$

Table 1: Solutions of the characteristic diophantic equation for phase-observations to GPS satellites; n = ghk, m = 3gk + 3hk + gh - 6k according to *W. Lindlohr and D. Wells* (1985)

g	h	k	m	hk	g+hk	n-m
4	7	28	784	196	200	0
4	8	16	512	128	132	0
4	9	12	432	108	112	0
4	10	10	400	100	104	0
5	5	25	625	125	130	0
5	6	10	300	60	65	0
5	7	7	245	49	54	0
5	8	6	238	48	53	2
5	9	5	225	45	50	0
6	5	10	300	50	56	0
6	6	6	216	36	42	0
6	7	5	207	35	41	3
6	8	4	192	32	38	0
7	4	28	784	112	119	0
7	5	7	245	35	42	0
7	6	5	207	30	37	3
7	7	4	193	28	35	3
7	9	3	189	27	34	0
8	4	16	512	64	72	0
8	5	6	238	30	38	2
8	6	4	192	24	32	0
8	8	3	190	24	32	2
9	4	12	432	48	57	0
9	5	5	225	25	34	0
9	7	3	189	21	30	0
10	4	10	400	40	50	0
11	4	9	395	36	47	1
12	4	8	384	32	44	0
12	5	4	240	20	32	0
12	6	3	216	18	30	0
12	10	2	240	20	32	0
14	4	7	392	28	42	0
14	9	2	252	18	32	0
18	4	6	432	24	42	0
18	8	2	288	16	34	0

Here we have assumed that at the epochs k = 1,2,... the terrestrial network may *deform*, the reason for 3gk terrestrial observations. In addition, for each observational epoch 3hk unknown satellite coordinates appear. gh is the total number of ambiguities, the unknown multiples of the electromagnetic wave length which fits into the distance from the terrestrial observation station to the satellite target point.

Thus g = 2 ground stations observe h = 5 satellite points simultaneously.

■

Example four : phase measurements

The most accurate satellite observations are those of type phase observations: Here the difference between the phase of the transmitted electromagnetic signal (*length of the Poynting vector*) and the received electromagnetic signal are compared. The most simple, therefore idealistic *observational phase equation*

$$\psi_{\alpha\cdot}^{\beta} = \omega c^{-1} S_{\alpha\cdot}^{\beta} + 2\pi\ell \qquad (1.11)'$$

contains the normalized oscillator frequency $\omega = 2\pi\nu$ and $\ell = 1,2,\ldots,$ the multiple of 2π. The term $2\pi\ell$ refers to the ℓ-multiple of the wave length λ of the transmitted signal which fits into the distance $S_{\alpha\cdot}^{\beta}$, a wellknown result from conventional terrestrial EDM. Let us again balance the observations n and the unknown estimable quantities m in order to derive the characteristic *diophantic equation*

$$n = m : ghk = 3gk + 3hk + gh - 6k \,. \qquad (1.12)$$

g satellite receivers on the ground coobserve h satellites at k instants. At least $k = 28$ observational epochs are necessary in order to measure from $g = 4$ *ground stations* to $h = 7$ *satellites*. Other solutions of the *diophantic equation* (1.12) are listed in *Table 1*.

The example of satellite pseudo-ranging, the so-called navigation solution (1.5), has introduced the terrestrial clock offset as a bias term. Only by the introduction of such a terrestrial clock offset with respect to satellite frame time a reasonable accuracy for kinematic positioning could be achieved. Similarly appears the modeling of satellite phase observations. A reasonable model, the phase observational equation, reads

$$\begin{aligned}\psi_{r\cdot}^{s}(t) = & -\frac{\nu}{c}\,[1-c^{-1}\dot{s}_{r}^{s}(t)]\; S_{r}^{s}(t) \;+\\ & + [1+c^{-1}\dot{s}_{r}^{\,s}(t)]\;\alpha_{r}(t) + \beta^{s}(t) + \gamma_{r\cdot}^{s} \\ & + (1+\delta_{r})d_{trop} - d_{ion} + e_{r\cdot}^{s}(t)\,.\end{aligned} \qquad (1.13)$$

$$s_{r\cdot}^{s}(t) := \|\underset{\sim}{x}^{s}(t) - \underset{\sim}{x}_{r}\|\,, \qquad (1.14)$$

$$\dot{s}_{r\cdot}^{s}(t) := \|\underset{\sim}{x}^{s}(t) - \underset{\sim}{x}_{r}\|^{-1}\,\langle \|\underset{\sim}{x}^{s}(t)-\underset{\sim}{x}_{r} | \dot{\underset{\sim}{x}}^{s}(t)\rangle \qquad (1.15)$$

$$\gamma_{r\cdot}^{s} := \psi^{s}(t_{o}) - \psi_{r}(t_{o}) + N_{r\cdot}^{s} \qquad (1.16)$$

ν denotes the GPS-L1 carrier frequency, c the speed of light in vacuo, d_{trop} the tropospheric model correction term, d_{ion} the single frequency ionospheric model correction term and e the stochastic measurement error. $\psi^s(t^s)$ denotes the *satellite oscillator phase* at the time t^s of transmission and $\psi_r(t_r)$ the ground receiver reference phase at the actual time instant t_r of reception. $N_{r.}^{\ s}$ refers to the *integer ambiguity* ("natural number"). Fundamental is the introduction of the *two-way classification model* for various biases: $\alpha_r(t)$ is a collection of the *receiver-specific bias*, $\beta^s(t)$ of the *satellite-specific bias* and $\gamma_{r.}^{\ s}$ of the *pair-specific bias*. Finally δ_r contains the *tropospheric bias term*. It is notable that the pair-specific bias parameters $\gamma_{r.}^{\ s}$ are *non-integer quantities*. Only after differencing with respect to receiver and satellite in reference or sequential mode, the *initial epoch ambiguities* are of *integer type*.

Once we apply the mechanism of *triple differences* between ground receivers ("Δ"), between satellites ("∇") and between epochs ("δ") - for more details we refer to *W. Lindlohr and D.E. Wells* (1985), *B.W. Remondi* (1984) and *B. Schaffrin and E. Grafarend* (1986) - we come up with the observational equation for *triple differences* of carrier beat phases of type (1.17) which explicitly has been computed by *K. Eren* (1987)

$$\boxed{\delta\nabla\Delta\psi_{r.}^{\ s}(t) = -c^{-1}\nu[1-c^{-1}\dot{s}_{r.}^{\ s}(t)]\delta\nabla\Delta S_{r.}^{\ s}(t) + \delta\nabla\Delta e_{r.}^{\ s}(t)} \tag{1.17}$$

This triple difference analytical technique has been used by *W. Pachelski et al.* (1988) in order to study the influence of stochastic prior information of orbital elements on ground station position measures.

In contrast *D. Wells et al.* (1987) have presented an extensive estimability analysis in terms of diophantic equations where the following unknowns appear:

(i)	Cartesian ground station coordinates
(ii)	Cartesian satellite position coordinates
(iii)	the family of bias terms α_r, β^s, $\gamma_{r.}^{\ s}$, δ_r
(iv)	ambiguity parameters $N_{r.}^{\ s}$
(v)	refraction parameters not included in the family of bias terms.

In case of a full estimation of all unknown parameters with the model equation (1.13) *W. Lindlohr* (1988) has shown that *dispersive prior information* about satellite and terrestrial positions is a *powerful analytical and numerical tool*. For the city network of Berlin he has computed all GPS Cartesian *coordinates* (ground as well as satellite position) and *bias terms* α_r, β^s, $\gamma_{r\cdot}^s$, δ_r.

Once GPS phase measurements can be based on *L1 and L2 frequencies* then another type of *differencing* can be applied in order to eliminate the refraction influence as much as possible. Here we refer to the contribution of *B. Schaffrin and Y. Bock* (1988). ■

2. Inertial positioning systems

Polygon type networks are built up by inertial positioning measuring devices: Three or more *accelerometers* measure in an operational frame of reference (e.g. one being defined by a gyroscopically fixed platform) the *relative acceleration* between the one of the moving vehicle and gravity. In case that the gravity field can be assumed to be sufficiently known a double time integration of the relative acceleration leads to position in *geometry space*. Or, vice versa, in case that the geometric positioning traverse is sufficiently known, the relative acceleration can be transformed into *gravity* or leads to positioning in *gravity space*.

Here we can only briefly outline an example for the system dynamics of an inertial positioning system, namely the gyroscopic platform slaved to the local horizontal frame of reference. For further details we refer to *E. Grafarend* (1981), *D. Schröder, S. Wiegner, E. Grafarend and B. Schaffrin* (1986), *A. Vassiliou and K.P. Schwarz* (1985), especially to *D. Schröder, N.C. Thong, S. Wiegner, E. Grafarend and B. Schaffrin* (1988) where a comparative study of geodetic inertial systems, also *strapdown systems*, are presented.

In the centre of mass P_d of the inertial measurement unit we fix the orthonormal triad of the sensor frame $d = \{\underset{\sim}{d}_1, \underset{\sim}{d}_2, \underset{\sim}{d}_3\}$, determined by the *ideal directions of the accelerometer input axes*. Besides this sensor frame there are three reference frames $e = \{\underset{\sim}{e}_1, \underset{\sim}{e}_2, \underset{\sim}{e}_3\}$, $f = \{\underset{\sim}{f}_1, \underset{\sim}{f}_2, \underset{\sim}{f}_3\}$ and $g = \{\underset{\sim}{g}_1, \underset{\sim}{g}_2, \underset{\sim}{g}_3\}$ as orthonormal base vectors which are located respectively in the *topocentre* P_e of a point, e.g. at the earth's surface, in the *centre of mass* P_f *of the earth* and in the *centre of mass* P_g of the entire *dynamical system*, the *inertial point*, for example in the frame of Newtonian mechanics approximated by the centre of mass of our solar system. The reference frames (d, P_d),

(e,P_e), (f,P_f), (g,P_g) form the *commutative diagram* of *Figure 1*

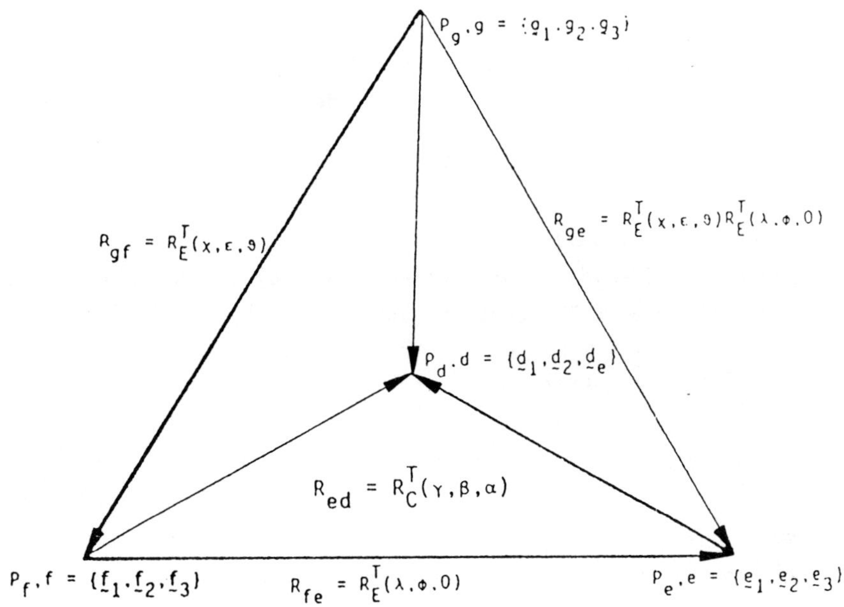

Fig. 1: Commutative diagram of the reference frames, the basic or inertial frame g *fixed* in space and time; the rotational or equatorial frame f *fixed* in space, *movable* in time; the south-east-vertical orientated horizontal frame e, *movable* in space and time; the sensor frame d.

The orthonormal triad $e = \{\underset{\sim}{e}_1, \underset{\sim}{e}_2, \underset{\sim}{e}_3\}$ is comprised of the *vertical* base vector $\underset{\sim}{e}_3$, the *eastern* base vector $\underset{\sim}{e}_2$ and the *southern* base vector $\underset{\sim}{e}_1$. To define more precisely the directions "south, east, vertical", we introduce *surface normal coordinates*, which determine a point at the earth's surface with respect to the centre of mass P_f. Let us denote by $\underset{\sim}{x}(\lambda,\phi)$ the position vector P_fP_d, which describes a point at the earth's surface as a function of the "longitude λ" and the "latitude ϕ". Usual geodetic surface normal coordinates are $\{\lambda,\phi,h(\lambda,\phi)\}$, the "height h" measured in the direction of the surface normal. For this purpose we establish a *sphere*, a *biaxial ellipsoid* (oblate rotational ellipsoid) or a *triaxial ellipsoid* with its major axes $a > b > c$. Concerning a point at the earth's surface we consider its orthogonal projection on the specified reference surface.

Thus there are three different representations for the coordinates of the position vector:

(a) *sphere* (radius a, spherical height h)

$$\underset{\sim}{x} = (a + h)\cos\lambda \cos\phi \, \underset{\sim}{f}_1 + (a + h)\sin\lambda \cos\phi \, \underset{\sim}{f}_2 + (a + h)\sin\phi \, \underset{\sim}{f}_3 \tag{1.18}$$

(b) *oblate rotational ellipsoid*

(semimajor axis a, semiminor axis c, eccentricity $e_{13} = 1-c^2a^{-2}$ in the (1,3) or (x,z) plane, ellipsoidal height h)

$$\underset{\sim}{x} = \left(\frac{a}{1-e_{13}^2\sin^2\phi} + h\right)\cos\lambda \cos\phi \, \underset{\sim}{f}_1 + \left(\frac{a}{1-e_{13}^2\sin^2\phi} + h\right)\sin\lambda \cos\phi \, \underset{\sim}{f}_2 + \left(\frac{a(1-e_{13}^2)}{1-e_{13}^2\sin^2\phi} + h\right)\sin\phi \, \underset{\sim}{f}_3 \tag{1.19}$$

(c) *triaxial ellipsoid*

(semimajor axis a, semimedian axis b, semiminor asix c, eccentricity e_{13} as in (b), eccentricity $e_{12} = 1-b^2a^{-2}$ in the (1,2) or (x,y) plane, ellipsoidal height h)

$$\underset{\sim}{x} = \left(\frac{a}{1-e_{13}^2\sin^2\phi-e_{12}^2\cos^2\phi\sin^2\lambda} + h\right)\cos\lambda \cos\phi \, \underset{\sim}{f}_1 + \left(\frac{a(1-e_{12}^2)}{1-e_{13}^2\sin^2\phi-e_{12}^2\cos^2\phi\sin^2\lambda} + h\right)\sin\lambda \cos\phi \, \underset{\sim}{f}_2 + \left(\frac{a(1-e_{13}^2)}{1-e_{13}^2\sin^2\phi-e_{12}^2\cos^2\phi\sin^2\lambda} + h\right)\sin\phi \, \underset{\sim}{f}_3 \tag{1.20}$$

Successively we introduce the *spherical* the bi- and the triaxial *ellipsoidal horizontal frame* e with the help of normalized tangential vectors.

(a) *spherical horizontal frame*

$$\underset{\sim}{e}_1 := \frac{-\partial\underset{\sim}{x}/\partial\phi}{\|\partial\underset{\sim}{x}/\partial\phi\|} = -\underset{\sim}{e}_\phi \qquad \text{"spherical south"} \tag{1.21i}$$

$$\underset{\sim}{e}_2 := \frac{-\partial\underset{\sim}{x}/\partial\lambda}{\|\partial\underset{\sim}{x}/\partial\lambda\|} = \underset{\sim}{e}_\lambda \qquad \text{"spherical east"} \tag{1.21ii}$$

$$\underset{\sim}{e}_3 := \underset{\sim}{e}_1 \wedge \underset{\sim}{e}_2 = \frac{\partial \underset{\sim}{x}/\partial h}{\|\partial \underset{\sim}{x}/\partial h\|} \quad \text{"spherical vertical"} \tag{1.21iii}$$

(b) *rotational ellipsoidal horizontal frame*

$$\underset{\sim}{e}_1 := \frac{-\partial \underset{\sim}{x}/\partial \phi}{\|\partial \underset{\sim}{x}/\partial \phi\|} = -\underset{\sim}{e}_\phi \quad \text{"ellipsoidal south"} \tag{1.22i}$$

$$\underset{\sim}{e}_2 := \frac{\partial \underset{\sim}{x}/\partial \lambda}{\|\partial \underset{\sim}{x}/\partial \lambda\|} = \underset{\sim}{e}_\lambda \quad \text{"ellipsoidal east"} \tag{1.22ii}$$

$$\underset{\sim}{e}_3 := \underset{\sim}{e}_1 \wedge \underset{\sim}{e}_2 = \frac{\partial \underset{\sim}{x}/\partial h}{\|\partial \underset{\sim}{x}/\partial h\|} \quad \text{"ellipsoidal vertical"} \tag{1.22iii}$$

(c) *ellipsoidal horizontal frame*

In the case of the triaxial ellipsoid we can obviously use the tangential vectors $\partial \underset{\sim}{x}/\partial \phi$ and $\partial \underset{\sim}{x}/\partial \lambda$ as the definition of an ellipsoidal horizontal plane as above. But in this case the tangential vectors do not form an orthogonal triad. Therefore, the following definition of an *orthonormal horizontal frame* is introduced.

$$\underset{\sim}{e}_1 := \frac{-\partial \underset{\sim}{x}/\partial \phi}{\|\partial \underset{\sim}{x}/\partial \phi\|} = -\underset{\sim}{e}_\phi \quad \text{"ellipsoidal south"} \tag{1.23i}$$

$$\underset{\sim}{e}_2 := \underset{\sim}{e}_3 \wedge \underset{\sim}{e}_1 \quad \text{"ellipsoidal east"} \tag{1.23ii}$$

$$\underset{\sim}{e}_3 := \frac{\partial \underset{\sim}{x}/\partial h}{\|\partial \underset{\sim}{x}/\partial h\|} \quad \text{"ellipsoidal vertical"} \tag{1.23iii}$$

Here we wish to continue with the special cases of the spherical and rotational ellipsoidal horizontal frame. The advantage of parametrizing the earth's surface with the help of surface normal coordinates will appear later. We want to add that from the point ov view of the inertial frame (g,P_g), fixed in space and time, the horizontal frame (e,P_e) moves in space and time.

The orthonormal triad $f = \{\underset{\sim}{f}_1, \underset{\sim}{f}_2, \underset{\sim}{f}_3\}$ is defined by the base vector $\underset{\sim}{f}_3$, that is the normalized *vector of the angular velocity* $\underset{\sim}{\omega}$ of the earth, the base vector $\underset{\sim}{f}_1$ *"of Greenwich"* in the case of the spherical surface normal coordinates and the *direction of the major axis* of the ellipsoid of rotation in the case of ellipsoidal surface normal coordinates, finally the orthonormal supplement $\underset{\sim}{f}_2$ of the base vectors $\underset{\sim}{f}_1$ and $\underset{\sim}{f}_3$. Summarizing, the triad f is called the *equatorial frame*, more precisely the earth-fixed equatorial frame.

Finally the *inertial frame* is defined by the orthogonal triad $g = \{g_1, g_2, g_3\}$, with its origin in the mass center P_g of our solar system, here located in the eliptic plane. Disregarding linear motion we connect the orthonormal reference triads by

$$g \to f = gR_{gf} , \tag{1.24i}$$

$$R_{gf} = R_E^T(\chi,\varepsilon,\vartheta) = R_3^T(\chi)\, R_2^T(\tfrac{\pi}{2} - \varepsilon)\, R_3^T(\vartheta) = \tag{1.24ii}$$

$$= \begin{bmatrix} -\sin\vartheta\sin\chi+\cos\vartheta\cos\chi\sin\varepsilon & -\cos\vartheta\sin\chi-\sin\vartheta\cos\chi\sin\varepsilon & +\cos\chi\cos\varepsilon \\ +\sin\vartheta\cos\chi+\cos\vartheta\sin\chi\sin\varepsilon & +\cos\vartheta\cos\chi-\sin\vartheta\sin\chi\sin\varepsilon & +\sin\chi\cos\varepsilon \\ -\cos\vartheta\cos\varepsilon & +\sin\vartheta\cos\varepsilon & \sin\varepsilon \end{bmatrix}$$

$$f \to e = fR_{fe} , \tag{1.25i}$$

$$R_{fe} = R_E^T(\lambda,\phi,0) = R_3^T(\lambda)\, R_2^T(\tfrac{\pi}{2} - \phi) = \tag{1.25ii}$$

$$= \begin{bmatrix} +\sin\phi\cos\lambda & -\sin\lambda & +\cos\phi\cos\lambda \\ +\sin\phi\sin\lambda & +\cos\lambda & +\cos\phi\sin\lambda \\ -\cos\phi & 0 & +\sin\phi \end{bmatrix}$$

with the Eulerian matrices of rotation R_E, each depending on the Eulerian rotational angles $(\chi,\varepsilon,\vartheta)$- for instance in the case of the regular precession $\chi = \chi_0 + pt$ and $\vartheta = \vartheta_0 + qt$ as functions of time in contrast to $\varepsilon = 23°27'08{,}26''$ = const. as the obliquity of ecliptic - and (λ,ϕ) as functions of position and time as above.

After introducing geodetic coordinates, mainly to present a spherical or ellipsoidal representation of the position vector, we want now to derive the proper *observational equation of linear motion* of an inertial measurement unit. With the help of the second Newtonial theorem we obtain

$$\underset{\sim}{y} = \ddot{\underset{\sim}{x}} + d^2(\overrightarrow{P_g P_f})/dt^2 - \underset{\sim}{\gamma} , \tag{1.26}$$

where $\underset{\sim}{y}$ is the observation vector of the measured *acceleration difference*, $\ddot{\underset{\sim}{x}}$ is the *vehicle's acceleretion* with respect to the earth's mass center and γ is the *vector of gravitation* (to be compensated). We now have to integrate the observation equation with the help of a *given gravitation field*, to obtain from $\ddot{\underset{\sim}{x}} = \underset{\sim}{y} - (\underset{\sim}{\gamma} - d^2(P_g P_f)/dt^2)$ the position vector $\underset{\sim}{x}$. The term

$d^2(P_gP_f)/dt^2$ describes the acceleration of the earth's mass centre with respect to the origin P_g of the inertial frame. In the following we want to point out the particularities which will occur, in particular a non-linear system of differential equations.

In a first step we construct the position vector $\underset{\sim}{x} = P_fP_d$ and its first and second time derivatives $\dot{\underset{\sim}{x}}$, $\ddot{\underset{\sim}{x}}$ (velocity and acceleration) with respect to the *earth-fixed equatorial frame* (P_f, f). Then in a second step we transform the velocity vector $\dot{\underset{\sim}{x}}$ and the acceleration vector $\ddot{\underset{\sim}{x}}$ into the *horizontal frame* (P_e, e).

1st step

$$x = fx_f,\ \dot{x} = f\dot{x}_f + \dot{f}x_f\ ,\ \ddot{x} = f\ddot{x}_f + 2\dot{f}\dot{x}_f + \ddot{f}x_f \tag{1.27}$$

$x^T = [x^1, x^2, x^3] = [x, y, z]$ represents the Cartesian coordinates of the relative position vector $\underset{\sim}{x}$. "T" is the transposition sign. From the relation (1.24) we have

$$f = gR_{gf}\ ,\quad g = fR^T_{gf}\ , \tag{1.28i}$$

$$\dot{f} = g\dot{R}_{gf} = f(R^T_{gf}\dot{R}_{gf}) = f(\Omega^g_f)^T\ ,\quad \dot{g} = 0\ , \tag{1.28ii}$$

$$\Omega^g_f = \dot{R}^T_{gf} R_{gf} =$$

$$= \begin{bmatrix} 0 & +\dot{\vartheta}+\dot{\chi}\sin\varepsilon & +\dot{\varepsilon}\cos\vartheta-\dot{\chi}\sin\vartheta\cos\varepsilon \\ -\dot{\vartheta}-\dot{\chi}\sin\varepsilon & 0 & -\dot{\varepsilon}\sin\vartheta-\dot{\chi}\cos\vartheta\cos\varepsilon \\ -\dot{\varepsilon}\cos\vartheta+\dot{\chi}\sin\vartheta\cos\varepsilon & +\dot{\varepsilon}\sin\vartheta+\dot{\chi}\cos\vartheta\cos\varepsilon & 0 \end{bmatrix}_{3\times 3} \tag{1.29i}$$

Ω^g_f describes as the *anti-symmetric* or *Cartan-matrix* the transformation between the base f and the base $\dot{f}$ according to (1.28), with respect to the base g. The coordinates $\omega_1 = \Omega_{23}$, $\omega_2 = \Omega_{31}$, $\omega_3 = \Omega_{12}$ are dually assigned to a Cartan-Matrix Ω, in particular

$$\omega^g_f = \begin{bmatrix} -\dot{\varepsilon}\sin\vartheta-\dot{\chi}\cos\vartheta\cos\varepsilon \\ -\dot{\varepsilon}\cos\vartheta+\dot{\chi}\sin\vartheta\cos\varepsilon \\ +\dot{\vartheta}+\dot{\chi}\sin\varepsilon \end{bmatrix} = \left(\begin{bmatrix} 0 & 1 & 0 \\ 0 & 0 & 1 \\ 1 & 0 & 0 \end{bmatrix} \odot \begin{bmatrix} 0 & 0 & 1 \\ 1 & 0 & 0 \\ 0 & 1 & 0 \end{bmatrix} \right)^T \mathrm{vec}\,\Omega^g_f = \tag{1.29ii}$$

$$= \mathrm{vecdiag} \left(\begin{bmatrix} 0 & 1 & 0 \\ 0 & 0 & 1 \\ 1 & 0 & 0 \end{bmatrix} \Omega^g_f \begin{bmatrix} 0 & 1 & 0 \\ 0 & 0 & 1 \\ 1 & 0 & 0 \end{bmatrix} \right)\ ,$$

where "⊙" denotes the "Khatri-Rao-Product" of matrices and "vec" transforms a matrix into a vector by stacking one column under the other. Substituting (1.28) and (1.29) into (1.27), we obtain

$$\boxed{\begin{aligned} \underset{\sim}{x} &= f x_f \\ \dot{\underset{\sim}{x}} &= f[\dot{x}_f + (\Omega_f^g)^T x_f] \\ \ddot{\underset{\sim}{x}} &= f[\ddot{x}_f + 2(\Omega_f^g)^T \dot{x}_f + (\dot{\Omega}_f^g)^T x_f + (\Omega_f^g)^2 x_f] \end{aligned}} \tag{1.30}$$

Herein the relative terms occur as accelerations, namely

- the relative acceleration $\ddot{x}_f$
- the Coriolis acceleration $2(\Omega_f^g)^T \dot{x}_f$
- the Eulerian or rotational acceleration $(\dot{\Omega}_f^g)^T x_f$
- the centrifugal acceleration $(\Omega_f^g)^2 x_f$.

<u>2nd step</u>

Now we transform the relative terms, given in the equatorial frame f (variable in time), into the horizontal frame e (variable in space), namely $\underset{\sim}{e}_1 = -\underset{\sim}{e}_\phi = \underset{\sim}{e}_S$ ("south"), $\underset{\sim}{e}_2 = \underset{\sim}{e}_\lambda = \underset{\sim}{e}_E$ ("east"), $\underset{\sim}{e}_3 = \underset{\sim}{e}_h = \underset{\sim}{e}_V$ ("vertical pointing outward") based on (1.30).

$$e = fR_{fe} \; , \quad f = eR_{fe}^T \; , \tag{1.31i}$$

$$e' = fR'_{fe} = e(R_{fe}^T \, R'_{fe}) = e(\Pi_e^f)^T \; , \quad f' = 0 \; , \tag{1.31ii}$$

$$\Pi_e^f = (R'_{fe})^T \, R_{fe} = \tag{1.32i}$$

$$= \begin{bmatrix} 0 & +\lambda'\sin\phi & +\phi' \\ -\lambda'\sin\phi & 0 & -\lambda'\cos\phi \\ -\phi' & +\lambda'\cos\phi & 0 \end{bmatrix}_{9\times 3}$$

$$\lambda' = \left[\frac{\partial\lambda}{\partial x^i}\right] , \quad \phi' = \left[\frac{\partial\phi}{\partial x^i}\right] , \qquad (i = 1,2,3) \tag{1.32ii}$$

$$\dot{\underset{\sim}{x}} = f\dot{x}_f + f(\Omega_f^g)^T x_f = e[R_{fe}^T \dot{x}_f + R_{fe}^T(\Omega_f^g)^T x_f] = \tag{1.34}$$

$$= e[(\dot{x}_e + R_{fe}^T \, \dot{R}_{fe} x_e) + R_{fe}^T(\Omega_f^g)^T R_{fe} x_e] \; , \tag{1.35i}$$

$$e(R_{fe}^T \dot{x}_f) = \underset{\sim}{e}_\phi \cdot (a+h)\dot{\phi} + \underset{\sim}{e}_\lambda \cdot (a+h)\dot{\lambda}\cos\phi + \underset{\sim}{e}_h \cdot \dot{h}$$

(spherical horizontal frame)

$$e(R^T_{fe}\dot{x}_f) = \underset{\sim}{e}_\phi \cdot \left(\frac{a(1-e^2)}{(1-e^2\sin^2\phi)^{3/2}} + h\right)\dot{\phi} + \qquad (1.35ii)$$

$$+ \underset{\sim}{e}_\lambda \cdot \left(\frac{a}{(1-e^2\sin^2\phi)^{1/2}} + h\right)\dot{\lambda}\cos\phi + \underset{\sim}{e}_h \cdot \dot{h}$$

ellipsoidal horizontal frame, $e^2 = e^2_{13}$).

Thus in the *spherical horizontal frame*

$$v_S = -(a+h)\dot{\phi}\ , \quad v_E = (a+h)\dot{\lambda}\cos\ , \quad v_V = \dot{h} \qquad (1.36i)$$

and

$$\left[\begin{array}{l} v_S = -\left(\dfrac{a(1-e^2)}{(1-e^2\sin^2\phi)^{3/2}} + h\right)\dot{\phi}\ , \\ v_E = +\left(\dfrac{a}{(1-e^2\sin^2\phi)^{1/2}} + h\right)\dot{\lambda}\cos\phi\ , \quad v_V = \dot{h} \end{array}\right. \qquad (1.36ii)$$

in the *ellipsoidal horizontal frame* ($e^2 = e^2_{13}$) specify respectively the *southern velocity* v_S, the *eastern velocity* e_E and the *vertical velocity* v_V (pointing outward). Thus the relative acceleration vector based on (1.30) is rewritten as

$$\underset{\sim}{\ddot{x}} = e[R^T_{fe}\ddot{x}_f + 2R^T_{fe}(\Omega^g_f)^T\dot{x}_f + R^T_{fe}(\dot{\Omega}^g_f)^T x_f + R^T_{fe}(\Omega^g_f)^2 x_f] \qquad (1.37i)$$

$$\boxed{\begin{aligned} \underset{\sim}{\ddot{x}} = {} & e[\ddot{x}_e + 2(\Omega^f_e)^T\dot{x}_e + (\dot{\Omega}^f_e)^T x_e + (\Omega^f_e)^2 x_e] + \\ & + e[2R^T_{fe}(\Omega^g_f)^T R_{fe}(\dot{x}_e + (\Omega^f_e)^T x_e)] + \\ & + e[R^T_{fe}(\dot{\Omega}^g_f)^T R_{fe}x_e + R^T_{fe}(\Omega^g_f)^2 R_{fe}x_e] \end{aligned}} \qquad (1.37ii)$$

Implicitly we used the time derivative $\dot{e}$, which is obtained with the help of the representation $e = fR_{fe}$:

$$\dot{e} = f\dot{R}_{fe} + \dot{f}R_{fe} = e[(\Omega^f_e)^T + R^T_{fe}(\Omega^g_f)^T R_{fe}] \qquad (1.38i)$$

$$\Omega^f_e = \dot{R}^T_{fe}\ R_{fe} = \qquad (1.38ii)$$

$$= \begin{bmatrix} 0 & +\dot{\lambda}\sin\phi & +\dot{\phi} \\ -\dot{\lambda}\sin\phi & 0 & -\dot{\lambda}\cos\phi \\ -\dot{\phi} & +\dot{\lambda}\cos\phi & 0 \end{bmatrix}_{3\times3} =$$

$$= \left(\begin{bmatrix} 1 & 1 & 1 \\ 1 & 1 & 1 \\ 1 & 1 & 1 \end{bmatrix}_{3\times3} \otimes \begin{bmatrix} \frac{dx^i}{dt} \end{bmatrix}_{3\times1} \right)^T \Pi_e^f \quad ,$$

where "$\otimes$" denotes the "Kronecker-Zehfuss-product" of matrices. Finally we have the following representation of the relative acceleration vector

$$\ddot{\underset{\sim}{x}} = e[\partial(R_{fe}^T \dot{x}_f)/\partial t + (\Omega_e^f)^T (R_{fe}^T \dot{x}_f) + 2(R_{fe}^T (\Omega_f^g)^T R_{fe})(R_{fe}^T \dot{x}_f) + \\ + R_{fe}^T (\dot{\Omega}_f^g)^T x_f + R_{fe}^T (\Omega_f^g)^2 x_f] \tag{1.39}$$

in the *spherical horizontal frame*

$$\ddot{\underset{\sim}{x}} = -\underset{\sim}{e}_\phi\{-(a+h)\ddot{\phi} - 2\dot{h}\dot{\phi} - (a+h)(\dot{\lambda}+\Omega_{12})^2 \sin\phi\cos\phi\} + \\ +\underset{\sim}{e}_\lambda\{(a+h)\ddot{\lambda}\cos\phi + 2\dot{h}(\dot{\lambda}+\Omega_{12})\cos\phi - 2(a+h)\dot{\phi}(\dot{\lambda}+\Omega_{12})\sin\phi\} + \\ +\underset{\sim}{e}_h\{\ddot{h} - (a+h)\dot{\phi}^2 - (a+h)(\dot{\lambda}+\Omega_{12})^2 \cos^2\phi\} \quad , \tag{1.40i}$$

in the *ellipsoidal horizontal frame* ($e^2 = e_{13}^2$)

$$\ddot{\underset{\sim}{x}} = -\underset{\sim}{e}_\phi\{-(\frac{a(1-e^2)}{(1-e^2\sin^2\phi)^{3/2}} + h)\ddot{\phi} - 2\dot{h}\dot{\phi} - \frac{3ae^2(1-e^2)}{(1-e^2\sin^2\phi)^{5/2}} \dot{\phi}^2 \sin\phi\cos\phi - \\ -(\frac{a}{(1-e^2\sin^2\phi)^{1/2}} + h)(\dot{\lambda}+\Omega_{12})^2 \sin\phi\cos\phi\} + \\ +\underset{\sim}{e}_\lambda \{(\frac{a}{(1-e^2\sin^2\phi)^{1/2}} + h)\ddot{\lambda}\cos\phi + 2\dot{h}(\dot{\lambda}+\Omega_{12})\cos\phi - \\ -2(\frac{a(1-e^2)}{(1-e^2\sin^2\phi)^{3/2}} + h)\dot{\phi}(\dot{\lambda}+\Omega_{12})\sin\phi\} + \\ +\underset{\sim}{e}_h\{\ddot{h} - (\frac{a(1-e^2)}{(1-e^2\sin^2\phi)^{3/2}} + h)\dot{\phi}^2 - \\ -(\frac{a}{(1-e^2\sin^2\phi)^{1/2}} + h)(\dot{\lambda}+\Omega_{12})^2 \cos^2\phi\} \; . \tag{1.40ii}$$

Based on the coordinates of the velocity vector introduced in (1.35) we finally have

$$\begin{aligned} \ddot{\underset{\sim}{x}} = \underset{\sim}{e}_S\{\dot{v}_S - \dot{\phi}v_V - \dot{\lambda}^{-1}(\dot{\lambda} + \Omega_{12})^2 v_E \sin\phi\} + \\ +\underset{\sim}{e}_E\{\dot{v}_E + [\dot{\lambda}^{-1}(\dot{\lambda}+\Omega_{12})^2 - \dot{\lambda}^{-1}\Omega_{12}^2](v_S \sin\phi + v_V \cos\phi)\} + \\ +\underset{\sim}{e}_V\{\dot{v}_V + \dot{\phi}v_S - \dot{\lambda}^{-1}(\dot{\lambda} + \Omega_{12})^2 v_E \cos\phi\} \end{aligned} \tag{1.41}$$

in the *spherical* as well as in the *ellipsoidal horizontal frame*. (1.39) up to (1.40) are only valid with the assumption

$$\Omega_{12} = \dot{\vartheta}\ , \quad \Omega_{13} = \Omega_{23} = 0\ (i.e.\ \dot{\chi} = 0\ ,\ \dot{\varepsilon} = 0),\ \dot{\Omega}^g_f = 0\ . \tag{1.42}$$

Remember once again, the local spherical and ellipsoidal horizontal frame each with the directions "*south*", "*east*" and "*vertical*" is constructed by the base vectors $\{\underset{\sim}{e}_S, \underset{\sim}{e}_E, \underset{\sim}{e}_V\}$.

To describe completely the translational observation equation (1.26), the *vector of gravitation* $\underset{\sim}{\gamma}$ enclosing the *second time derivative of the position vector* $\overrightarrow{P_g P_f}$ and the measured *specific force vector* $\underset{\sim}{y}$ have to be described with respect to the local horizontal frame. For this purpose we decompose the gravitation vector into a part of the *earth's gravitation* $\underset{\sim}{\gamma}_1$ and a second part $\underset{\sim}{\gamma}_2$, representing the gravitation effects of the extraterrestrial masses (*tidal effect*). In the corresponding decomposition

$$\ddot{\underset{\sim}{x}} - \underset{\sim}{y} = \underset{\sim}{\gamma}_1 + [\underset{\sim}{\gamma}_2 - d^2(\overrightarrow{P_g P_f})/dt^2] \tag{1.43}$$

we obtain the vectors $\underset{\sim}{\gamma}_1$ and $(\underset{\sim}{\gamma}_2 - d^2(\overrightarrow{P_g P_f})/dt^2)$ respectively from the (mass normalized) gradient of the *terrestrial gravitation potential* and of the *tidal potential*, the latter already related to the origin P_f of the equatorial frame. Thus the representation

$$\ddot{\underset{\sim}{x}} - \underset{\sim}{y} = \underset{\sim}{e}_S\gamma_S + \underset{\sim}{e}_E\gamma_E + \underset{\sim}{e}_V\gamma_V \tag{1.44}$$

is obtained in the local spherical and ellipsoidal horizontal frame respectively, where $[\gamma_S, \gamma_E, \gamma_V]$ are the coordinates of the modified gravitation vector according to (1.42); γ_S and γ_E are closely related to the vector of the *deflection of the vertical*, γ_V is approximated by the *absolute value of the gravitation vector* itself.

The coordinates of the specific force vector are measured with respect to the sensor frame d:

$$y = dy_d = gR_{gd}y_d = eR_{eg}R_{gd}y_d \tag{1.45}$$

the determination of the sensor frame's orientation with respect to the inertial frame g is carried out with the help of the *rotational sensors*. Of these only mechanical gyros are considered here.
The *rotational observation equation* is then based on the *theorem of angular momentum*. Refering to the mass centre of the rotor of a gyro, the theorem is written as

$$\dot{\tilde{D}} = \tilde{M} \; , \tag{1.46}$$

where $\tilde{D}$ is the *vector of angular momentum* with respect to the inertial frame and $\tilde{M}$ the *vector of torques*. The latter implies the proper rotational observations. We now introduce a reference frame fixed to the rotor $k = \{\tilde{k}_1, \tilde{k}_2, \tilde{k}_3\}$, with its origin located at the mass centre P_k of the rotor. With respect to this base the *inertia matrix* T_k is constant. The inertia matrix is defined by

$$T_k = \int_{rotor} (X_k X_k^T) dm \; . \tag{1.47}$$

The elements of the skew-symmetric matrix

$$X_k = \begin{bmatrix} 0 & x_3 & -x_2 \\ -x_3 & 0 & x_1 \\ x_2 & -x_1 & 0 \end{bmatrix} \tag{1.48}$$

are the coordinates of the mass points dm of the rotor. The *angular momentum* with respect to the base k is written as

$$\tilde{D} = K\{T_K \, \omega_k^g\} \; , \tag{1.49}$$

where $\tilde{\omega}_k^g$ is the *angular velocity vector* of the rotor with respect to the inertial frame g. Therefore the theorem of angular momentum is rewritten as

$$K\{(\Omega_k^g)^T \, T_k \, \omega_k^g + T_k \, \dot{\omega}_k^g\} = kM_k \; . \tag{1.50}$$

The *inertia matrix* and the *angular velocity* of the rotor with respect to a frame fixed to the housing of the gyro, for instance the sensor frame d, are *constants* and are in the following considered as given quantities. The angular velocity of the sensor frame d with respect to the local horizontal frame e is obtained by (1.49) after transforming the observation equation into the sensor frame d.

The evaluation of the *rotational observation equation* (1.49) in the case of the mechanical slaved platform system is done by feeding additional *slaving torques* to the platform. They are chosen so that the angular velocity of the sensor frame d is equal to the angular velocity of the horizontal frame e, i.e.

$$\underset{\sim}{\omega}_d^e = 0 \ .$$

The axes of the sensor frame d are aligned at the beginning of a measurement operation with the aces of the horizontal frame e, i.e. given an ideal alignment and an ideal slaving of the platform it holds that $R_{ed} = I$. Assuming a misalignment in the initial point and an erroneous slaving, the sensor frame d and the horizontal frame e differ by an infinitesimal rotation:

$$e \to d = eR_{ed} \ , \tag{1.51i}$$

$$R_{ed} = I + E^T \ , \tag{1.51ii}$$

$$E = \begin{bmatrix} 0 & -\varepsilon_V & -\varepsilon_E \\ -\varepsilon_V & 0 & +\varepsilon_S \\ +\varepsilon_E & -\varepsilon_S & 0 \end{bmatrix} = R_1(\varepsilon_S)R_2(\varepsilon_E)R_3(\varepsilon_V) - I \tag{1.51iii}$$

where $(\varepsilon_S, \varepsilon_E, \varepsilon_V)$ denote the three infinitesimal rotation angles, which can be associated with the directions "*south, east, vertical*". The order of the three single rotations is arbitrary.

The coordinates of the specific force vector are measured in the sensor frame d:

$$\underset{\sim}{y} = dy_d = eR_{eyd} \tag{1.52}$$

By the *mechanical slaving* with respect to the horizontal frame it is achieved that the axes of the sensor frame d remain almost parallel to the instantaneous horizontal frame e.

On account of sensor errors, including the torque generators, and of gravitation field modelling errors, the slaving of the platform can only be approximately performed. As already mentioned in (1.45) the rotation matrix R^T_{ed} comprises the transposed rotation matrix R_{gd} (rotation of the sensor frame d with respect to the inertial frame g) and the rotation matrix R_{ge} (computed rotation of the horizontal frame e with respect to the inertial frame g):

$$e = d\, R^T_{ed} = d\, R^T_{gd}\, R_{ge} = d\, R^T_{gd}(R_{gf}\, R_{fe}) \tag{1.53}$$

The temporal change of the rotation matrix R_{ed} is a function of the *earth rate* Ω^g_f and the *transport rate* Ω^f_e as well as a function of the *angular velocity* Ω^g_e:

$$\dot{R}_{ed}\, R_{dg} + R_{ed}\, \dot{R}_{dg} = \dot{R}_{ef}\, R_{fg} + R_{ef}\, \dot{R}_{fg} = \dot{R}_{eg} \tag{1.54i}$$

$$\dot{R}_{ed} = (\Omega^f_e + R_{ef}\, \Omega^g_f\, R^T_{ef})R_{ed} - R_{ed}\, \Omega^g_d \tag{1.54ii}$$

Based on (1.51), the matrix R_{ed} is a matrix of infinitesimal rotation, (1.54) yields

$$-\dot{E} = \dot{E}^T = \dot{R}_{ed} = (\Omega^f_e + R_{ef}\Omega^g_f R^T_{ef})(I+E)^T - (I+E)^T\, \Omega^g_d \tag{1.55i}$$

with the approximation

$$\begin{aligned}(I+E)^T\, \Omega^g_d &\approx \Omega^g_d + E^T\, \dot{R}_{eg}\, R^T_{eg} = \\ &= \Omega^g_d + E^T(\Omega^f_e + R_{ef}\, \Omega^g_f\, R^T_{ef}) \;.\end{aligned} \tag{1.55ii}$$

Thus we finally obtain in the *vectorial representation* the *differential relation*

$$e\dot{\varepsilon} = e\{-(\omega^f_e + R_{ef}\omega^g_f) + (\Omega^f_e + R_{ef}\Omega^g_f R^T_{ef})\varepsilon + \omega^g_d\} \tag{1.56}$$

The *complete mechanization equations* of an inertial measurement unit *mechanically slaved* relative to the spherical or ellipsoidal horizontal frame are thus given by (1.56) and (1.44) together wit (1.41) and (1.52), that is

$$\underset{\sim}{e}_S\dot{\varepsilon}_S + \underset{\sim}{e}_E\dot{\varepsilon}_E + \underset{\sim}{e}_V\dot{\varepsilon}_V = \underset{\sim}{e}_S\{(\dot{\lambda}+\Omega_{12})(\cos\phi + \varepsilon_E\sin\phi) + \dot{\phi}\varepsilon_V + (\omega^g_d)^1\} + \tag{1.57i}$$

$$+ \underset{\sim}{e}_E\{\dot{\phi} - (\dot{\lambda}+\Omega_{12})(\varepsilon_S \sin\phi + \varepsilon_V \cos\phi) + (\omega_d^g)^2\} + $$
$$+ \underset{\sim}{e}_V\{(\dot{\lambda}+\Omega_{12})(-\sin\phi + \varepsilon_E \cos\phi) - \dot{\phi}\varepsilon_S + (\omega_d^g)^3\}$$

in connection with (1.36) and

$$\begin{aligned} &\underset{\sim}{e}_S\{y^1 - y^2\varepsilon_V + y^3\varepsilon_E\} + \underset{\sim}{e}_E\{y^2 + y^1\varepsilon_V - y^3\varepsilon_S\} + \underset{\sim}{e}_V\{y^3 - y^1\varepsilon_E + y^2\varepsilon_S\} = \\ &= \underset{\sim}{e}_S\{\dot{v}_S - \dot{\phi}v_V - (\dot{\lambda}+\Omega_{12})^2\dot{\lambda}^{-1}\ v_E \sin\phi - \gamma_S\} + \\ &+ \underset{\sim}{e}_E\{\dot{v}_E + [(\dot{\lambda}+\Omega_{12})^2 - \Omega_{12}^2]\dot{\lambda}^{-1}(v_S \sin\phi + v_V \cos\phi) - \gamma_E\} + \\ &+ \underset{\sim}{e}_V\{\dot{v}_V + \dot{\phi}v_S - (\dot{\lambda}+\Omega_{12})^2\dot{\lambda}^{-1}\ v_E \cos\phi - \gamma_V\} \quad . \end{aligned} \qquad (1.57ii)$$

Let us now discuss the differences in the mechanization equations of a *strapdown mechanization* and of a *slaved platform mechanization*.
In the case of a strapdown mechanization the specific forces, measured with respect to the sensor frame d, have to be *transformed* with the help of the *computed* transformation matrix into the approximated frame e_o, whereas in the case of a platform mechanization the sensor frame d already almost *coincides* with the horizontal frame e.

The mechanization equations (1.57) form a system of *coupled, nonlinear, inhomogeneous differential equations of first and second order*, where *prior information* has to be made available for $[(\omega_{e_o}^g)^1, (\omega_{e_o}^g)^2, (\omega_{e_o}^g)^3, (y_{e_o})^1, (y_{e_o})^2, (y_{e_o})^3]$ with $y_{e_o} = R_{e_o d} y_d$ in the case of the strapdown mechanization, for $[y_d^1, y_d^2, y_d^3]$ in the case of the platform mechanization and for $[\gamma_S, \gamma_E, \gamma_V]$ in the case of both mechanizations.

The solution of the mechanization equations is performed in two steps:
In the first step the equations are solved by the help of a simple gravitation field model by numerical integration. We obtain an *approximated trajectory*, which will differ from the *real trajectory* of the inertial measurement unit. The small difference between both trajectories is the base of the computation of corresponding *increments* in order to improve the approximated trajectory. If the difference between real and approximated

trajectory is small, we are able to derive an *incremental system of differential equations* by the help of (1.57). This incremental system is linear with respect to the position increments and their time derivatives. In a second step, this system of linear differential equations is solved.

The system of incremental differential equations is derived from the non-linear system by applying linear calculus of variation. Assuming that the difference between the real and approximated trajectory is small, we only have to consider first variations. Products of first variations will also be neglected. Thus we will substitute into the mechanization equations

$$\left.\begin{array}{llll} \dot{v}_S = \dot{v}_{S0} + \delta\dot{v}_S , & \dot{v}_E = \dot{v}_{E0} + \delta\dot{v}_E , & \dot{v}_V = \dot{v}_{V0} + \delta\dot{v}_V , \\ v_S = v_{S0} + \delta v_S , & v_E = v_{E0} + \delta v_E , & v_V = v_{V0} + \delta v_V , \\ \ddot{\phi} = \ddot{\phi}_0 + \delta\ddot{\phi} , & \ddot{\lambda} = \ddot{\lambda}_0 + \delta\ddot{\lambda} , & \ddot{h} = \ddot{h}_0 + \delta\ddot{h} , \\ \dot{\phi} = \dot{\phi}_0 + \delta\dot{\phi} , & \dot{\lambda} = \dot{\lambda}_0 + \delta\dot{\lambda} , & \dot{h} = \dot{h}_0 + \delta\dot{h} , \\ \phi = \phi_0 + \delta\phi , & \lambda = \lambda_0 + \delta\lambda , & h = h_0 + \delta h , \end{array}\right] \quad (1.58)$$

In the case of the *platform mechanization* the *linearized* equations are

$$\left.\begin{array}{l} \underset{\sim}{y} = \underset{\sim}{y}_0 + \delta\underset{\sim}{y} = d(y_{d0} + \delta y_d) = \\ \qquad = d(R_{de}\, y_{e0} + R_{de}\, \delta y_e) = e(y_{e0} + \delta y_e) , \\ \underset{\sim}{\omega}_d^g = (\underset{\sim}{\omega}_d^g)_0 + \delta\underset{\sim}{\omega}_d^g = d[(\omega_d^g)_0 + \delta\omega_d^g] \approx e[(\omega_d^g)_0 + \delta\omega_d^g] \end{array}\right] \quad (1.59)$$

with the assumption

$$(\omega_d^g)_0 \approx (\omega_{e_0}^g)_0 .$$

With (1.42) we obtain

$$(\omega_e^g)_0 = [-(\dot{\lambda}+\Omega_{12})\cos\phi_0 ,\ -\phi_0 , (\dot{\lambda}_0+\Omega_{12})\sin\phi_0]^T .$$

The vectors $\delta\underset{\sim}{y}$ and $\delta\underset{\sim}{\omega}_d^g$ are the variations describing on the one hand an erroneous modelling of the *sensor output* (i.e. of the "observations") and on the other hand an erroneous *analytical* or *mechanical slaving* of the sensor frame with respect to the horizontal frame. Thus we have

$$\delta \underset{\sim}{y} = d\delta y_d = e\delta y_e \approx eR_{e_0 d}\, \delta y_d$$

and

$$\delta \underset{\sim}{\omega}^g_d = d\delta\omega^g_d = e(\delta\omega^g_d)_e \approx eR_{e_0 d}\delta\omega^g_d \, .$$

In the first step the mechanization equations are solved assuming a simple model $\underset{\sim}{\gamma}_M = \underset{\sim}{\gamma}_M(\underset{\sim}{x})$ of the gravitation field. The representation of the gravitation field of the earth is completed by the perturbation $\delta\underset{\sim}{\gamma}$

$$\underset{\sim}{\gamma} = \underset{\sim}{\gamma}_M + \delta\underset{\sim}{\gamma} = \underset{\sim}{\gamma}_M(\underset{\sim}{x}_0 + \delta\underset{\sim}{x}) + \delta\underset{\sim}{\gamma}(\underset{\sim}{x}_0 + \delta\underset{\sim}{x}) \tag{1.60)i}$$

Evaluating the variations we thus obtain

$$\underset{\sim}{\gamma} = \underset{\sim}{\gamma}_{M0} + \left(\frac{\partial\underset{\sim}{\gamma}_M}{\partial\underset{\sim}{x}}\right)_0 \delta\underset{\sim}{x} + \delta\underset{\sim}{\gamma}_0 + \left(\frac{\partial\delta\underset{\sim}{\gamma}}{\partial\underset{\sim}{x}}\right)_0 \delta\underset{\sim}{x} \; , \; \underset{\sim}{\gamma}_{M0} := \underset{\sim}{\gamma}_M(\underset{\sim}{x}_0) \; , \tag{1.60ii}$$

As an approximation only $\underset{\sim}{\gamma}_{M0}$ is considered in the first step, thus we define in the following

$$\delta\underset{\sim}{\gamma} := \left(\frac{\partial\underset{\sim}{\gamma}_M}{\partial\underset{\sim}{x}}\right)_0 \delta\underset{\sim}{x} + \delta\underset{\sim}{\gamma}_0 + \left(\frac{\partial\delta\underset{\sim}{\gamma}}{\partial\underset{\sim}{x}}\right)_0 \delta\underset{\sim}{x} \; . \tag{1.60iii}$$

If the gravitation perturbation $\delta\underset{\sim}{\gamma}$ is given for example in the form of a digital model, the perturbations have to be evaluated at the *real* trajectory and not at the approximated one.
Considering the equations (1.36) we obtain after some algebra

$$\left.\begin{aligned}
\delta\dot\phi &= \dot\phi_0\, v_{S0}^{-1}[\delta v_S + \dot\phi_0\delta h - \dot\phi_0^{-1}(\dot v_{S0} - \dot\phi_0^{-1} v_{S0}\ddot\phi_0 + \dot\phi_0 v_{V0})\delta\phi] = \\
&= \dot\phi_0\, v_{S0}^{-1}[\delta v_S + \dot\phi_0\delta h - 3(v_{S0} + \dot\phi_0 h_0)\tan\phi_0 (v_{E0} - h_0\dot\lambda_0\cos\phi_0)^{-1} \cdot \\
&\qquad\qquad \cdot (v_{e0} + \dot\phi_0^{-1} v_{S0}\dot\lambda_0\cos\phi_0)\delta\phi] \; , \\
\delta\dot\lambda &= \dot\lambda_0\, v_{E0}^{-1}[\delta v_E - \dot\lambda_0\cos\phi_0\delta h - \dot\phi_0^{-1}(v_{E0} - \dot\lambda_0^{-1} v_{E0}\ddot\lambda_0 - \dot\lambda_0 v_{V0}\cos\phi_0)\delta\phi] = \\
&= \dot\lambda_0 v_{E0}^{-1}(\delta v_E - \dot\lambda_0\cos\phi_0\delta h - \dot\phi_0^{-1} v_{S0}\tan\phi_0\dot\lambda_0\cos\phi_0\delta\phi) \; , \\
\delta\dot h &= \delta v_V
\end{aligned}\right] \tag{1.61}$$

and thus

$$\left.\begin{aligned}
\delta\{\dot\lambda\cos\phi\} &= \cos\phi_0\delta\dot\lambda - \dot\lambda_0\sin\phi_0\delta\phi = \\
&= v_{E0}^{-1}\dot\lambda_0\cos\phi_0(\delta v_E - \dot\lambda_0\cos\phi_0\delta h) - \dot\lambda_0\sin\phi_0(1 + \dot\phi_0^{-1} v_{S0} v_{E0}^{-1}\dot\lambda_0\cos\phi_0)\delta\phi \; , \\
\delta\{\dot\lambda\sin\phi\} &= \sin\phi_0\delta\dot\lambda + \dot\lambda_0\cos\phi_0\delta\phi = \\
&= v_{E0}^{-1}\dot\lambda_0\sin\phi_0(\delta v_E - \dot\lambda_0\cos\phi_0\delta h) + \dot\lambda_0\cos\phi_0(1 - \dot\phi_0^{-1} v_{S0}\tan\phi_0 v_{E0}^{-1}\dot\lambda_0\sin\phi_0)\delta\phi
\end{aligned}\right] \tag{1.62i}$$

with

$$\begin{aligned} &\dot\lambda_0\cos\phi_0(1-\dot\phi_0^{-1}v_{S0}\tan\phi_0 v_{E0}^{-1}\dot\lambda_0\sin\phi_0) = \\ &= \dot\lambda_0\cos\phi_0(1+\dot\phi_0^{-1}v_{S0}v_{E0}^{-1}\dot\lambda_0\cos\phi_0) - \dot\phi_0^{-1}v_{S0}v_{E0}^{-1}\dot\lambda_0^2 \end{aligned} \tag{1.62ii}$$

Thus the remaining *incremental differential equations* are rewritten as

(1.63i)

$$\begin{aligned} &\underset{\sim}{e}_S\varepsilon_S + \underset{\sim}{e}_E\varepsilon_E + \underset{\sim}{e}_V\varepsilon_V = \\ &= \underset{\sim}{e}_S\{(\dot\lambda_0+\Omega_{12})\sin\phi_0\varepsilon_E+\dot\phi_0\varepsilon_V - \\ &\quad -[\dot\lambda_0(1+\dot\phi_0^{-1}v_{S0}v_{E0}^{-1}\dot\lambda_0\cos\phi_0)+\Omega_{12}]\sin\phi_0\delta\phi + \\ &\quad + v_{E0}^{-1}\dot\lambda_0\cos\phi_0(\delta v_E-\dot\lambda_0\cos\phi_0\delta h) + (\delta\omega_d^g)^1\} + \\ &+ \underset{\sim}{e}_E\{-(\dot\lambda_0+\Omega_{12})(\sin\phi_0\varepsilon_S+\cos\phi_0\varepsilon_V) - \\ &\quad -3\dot\phi_0\tan\phi_0(1+\dot\phi_0^{-1}v_{S0}v_{E0}^{-1}\dot\lambda_0\cos\phi_0)(1-h_0v_{E0}^{-1}\dot\lambda_0\cos\phi_0)^{-1}(1+h_0v_{S0}^{-1}\dot\phi_0)\delta\phi + \\ &\quad +v_{S0}^{-1}\dot\phi_0(\delta v_S+\dot\phi_0\delta h) + (\delta\omega_d^g)^2\} + \\ &+ \underset{\sim}{e}_V\{-\dot\phi_0\varepsilon_S+(\dot\lambda_0+\Omega_{12})\cos\phi_0\varepsilon_E - \\ &\quad -([\dot\lambda_0(1+\dot\phi_0^{-1}v_{S0}v_{E0}^{-1}\dot\lambda_0\cos\phi_0)+\Omega_{12}]\cos\phi_0-\dot\phi_0^{-1}v_{S0}v_{E0}^{-1}\dot\lambda_0^2)\delta\phi - \\ &\quad -v_{E0}^{-1}\dot\lambda_0\sin\phi_0(\delta v_E-\dot\lambda_0\cos\phi_0\delta h) + (\delta\omega_d^g)^3\} \end{aligned}$$

in connection with

(1.63ii)

$$\begin{aligned} &\underset{\sim}{e}_S\{\delta y^1-y_0^2\varepsilon_V+y_0^3\varepsilon_E\}+\underset{\sim}{e}_E\{\delta y^2+y_0^1\varepsilon_V-y_0^3\varepsilon_S\}+\underset{\sim}{e}_V\{\delta y^3-y_0^1\varepsilon_E+y_0^2\varepsilon_S\} = \\ &= \underset{\sim}{e}_S\{\delta\dot v_S-\dot\phi_0v_{S0}^{-1}v_{V0}\delta v_S-2(\dot\lambda_0+\Omega_{12})\sin\phi_0\delta v_E-\dot\phi_0\delta v_V - \\ &\quad -[(\dot\lambda_0+\Omega_{12})^2\dot\lambda_0^{-1}v_{E0}\cos\phi_0-(\dot\lambda_0^2-\Omega_{12}^2)\dot\phi_0^{-1}v_{S0}\sin^2\phi_0 - \\ &\qquad -3v_{V0}\dot\phi_0\tan\phi_0(1+\dot\phi_0^{-1}v_{S0}v_{E0}^{-1}\dot\lambda_0\cos\phi_0)(1-h_0v_{E0}^{-1}\dot\lambda_0\cos\phi_0)^{-1}(1+h_0v_{S0}^{-1}\dot\phi_0)]\delta\phi + \\ &\quad +[(\dot\lambda_0^2-\Omega_{12}^2)\sin\phi_0\cos\phi_0-\dot\phi_0^2v_{S0}^{-1}v_{V0}]\delta h - (\delta\gamma)^1\} + \\ &+ \underset{\sim}{e}_E\{\delta\dot v_E+(\dot\lambda_0+2\Omega_{12})(\sin\phi_0\delta v_S+\cos\phi_0\delta v_V)+(v_{S0}\sin\phi_0+v_{V0}\cos\phi_0)\dot\lambda_0v_{E0}^{-1}\delta v_E + \\ &\quad +([\dot\lambda_0(1+\dot\phi_0^{-1}v_{S0}v_{E0}^{-1}\dot\lambda_0\cos\phi_0)+2\Omega_{12}](v_{S0}\cos\phi_0-v_{V0}\sin\phi_0)-\dot\phi_0^{-1}v_{S0}^2v_{E0}^{-1}\dot\lambda_0^2)\delta\phi - \\ &\quad -v_{E0}^{-1}\dot\lambda_0^2\cos\phi_0(v_{S0}\sin\phi_0+v_{V0}\cos\phi_0)\delta h - (\delta\gamma)^2\} + \\ &+ \underset{\sim}{e}_V\{\delta\dot v_V+2\dot\phi_0\delta v_S-2(\dot\lambda_0+\Omega_{12})\cos\phi_0\delta v_E + \\ &\quad +[(\dot\lambda_0+\Omega_{12})^2\dot\lambda_0^{-1}v_{E0}\sin\phi_0+(\dot\lambda_0^2-\Omega_{12}^2)\dot\phi_0^{-1}v_{S0}\sin\phi_0\cos\phi_0 - \\ &\qquad -3v_{S0}\dot\phi_0\tan\phi_0(1+\dot\phi_0^{-1}v_{S0}v_{E0}^{-1}\dot\lambda_0\cos\phi_0)(1-h_0v_{E0}^{-1}\dot\lambda_0\cos\phi_0)^{-1}(1+h_0v_{S0}^{-1}\dot\phi_0)]\delta\phi + \\ &\quad +[(\dot\lambda_0^2-\Omega_{12}^2)\cos^2\phi_0+\dot\phi_0^2]\delta h - (\delta\gamma)^3\} \end{aligned}$$

Substituting (1.36) into the generally linearized mechanization equations (1.61) and (1.63), in the *ellipsoidal horizontal frame* ($e^2 = e_{13}^2$)

$$\left.\begin{aligned}
\dot\phi_0^{-1} v_{S0} &= -[a(1-e^2)(1-e^2\sin^2\phi_0)^{-3/2} + h_0] ,\\
\dot\lambda_0^{-1} v_{E0} &= +[a(1-e^2\sin^2\phi_0)^{-1/2} + h_0] \cos\phi_0 ,\\
\dot h_0^{-1} v_{V0} &= 1
\end{aligned}\right] \quad (1.64\text{i})$$

and

$$\dot\phi_0^{-1} v_{S0} = -(a+h_0) , \quad \dot\lambda_0^{-1} v_{E0} = (a+h_0)\cos\phi_0 , \quad \dot h_0^{-1} v_{V0} = 1 \quad (1.64\text{ii})$$

in the *spherical horizontal frame*, respectively, we easily obtain the following relations with respect to the *ellipsoidal horizontal frame* ($e^2 = e_{13}^2$):

$$\left[\begin{aligned}
&-\dot\phi_0^{-1}(\dot v_{S0} - \dot\phi_0^{-1} v_{S0}\ddot\phi_0 + \dot\phi_0 v_{V0}) = \quad (1.65\text{i})\\
&\quad = -3v_{S0}\tan\phi_0(1+\dot\phi_0^{-1} v_{S0} v_{E0}^{-1}\dot\lambda_0\cos\phi_0)(1-h_0 v_{E0}^{-1}\dot\lambda_0\cos\phi_0)^{-1}(1+h_0 v_{S0}^{-1}\dot\phi_0) = \\
&\quad = +3\dot\phi_0\tan\phi_0[(1-e^2\sin^2\phi_0)^{-3/2}\, ae^2\cos^2\phi_0] \cdot \\
&\qquad \cdot [(1-e^2\sin^2\phi_0)^{1/2}\, a^{-1}]\cdot[(1-e^2\sin^2\phi_0)^{-3/2}\, a(1-e^2)] = \\
&\quad = 3[a(1-e^2)(1-e^2\sin^2\phi_0)^{-5/2}\, e^2\cos^2\phi_0]\,\dot\phi_0\,\tan\phi_0 ,\\
&-\dot\phi_0^{-1}(\dot v_{E0} - \dot\lambda_0^{-1} v_{E0}\ddot\lambda_0 - \dot\lambda_0 v_{V0}\cos\phi_0) = \\
&\quad = -\dot\phi_0^{-1} v_{S0}\tan\phi_0\dot\lambda_0\cos\phi_0 = [a(1-e^2)(1-e^2\sin^2\phi_0)^{-3/2} + h_0]\dot\lambda_0\sin\phi_0 ,\\
&-\dot\phi_0^{-1} v_{S0} v_{E0}^{-1}\dot\lambda_0^2 = [a(1-e^2\sin^2\phi_0)^{-1/2} + h_0]^{-1}[a(1-e^2)(1-e^2\sin^2\phi_0)^{-3/2} + h_0] \cdot \\
&\qquad \cdot \dot\lambda_0\cos^{-1}\phi_0 ,\\
&1+\dot\phi_0^{-1} v_{S0} v_{E0}^{-1}\dot\lambda_0\cos\phi_0 = [a(1-e^2\sin^2\phi_0)^{-1/2} + h_0]^{-1}\, a(1-e^2\sin^2\phi_0)^{-3/2}\, e^2\cos^2\phi_0
\end{aligned}\right.$$

and by specializing with respect to the *spherical horizontal frame*

$$\left[\begin{aligned}
&-\dot\phi_0^{-1}(v_{S0} - \dot\phi_0^{-1} v_{S0}\ddot\phi_0 + \dot\phi_0 v_{V0}) = 0 ,\\
&-\dot\phi_0^{-1}(\dot v_{E0} - \dot\lambda_0^{-1} v_{E0}\ddot\lambda_0 - \dot\lambda_0 v_{V0}\cos\phi_0) = (a+h_0)\dot\lambda_0\sin\phi_0 ,\\
&-\dot\phi_0^{-1} v_{S0} v_{E0}^{-1}\dot\lambda_0^2 = \dot\lambda_0\cos^{-1}\phi_0 ,\\
&1 + \dot\phi_0^{-1} v_{S0} v_{E0}^{-1}\dot\lambda_0\cos\phi_0 = 0
\end{aligned}\right. \quad (1.65\text{ii})$$

The final incremental differential equations with respect to the *ellipsoidal base* are specified in the appendix, whereas the basic behaviour of the solution with respect to the *spherical base* is discussed in the following.
The system of linear differential equations is rewritten

$$
\begin{aligned}
&\underset{\sim}{e}_S\dot{\varepsilon}_S + \underset{\sim}{e}_E\dot{\varepsilon}_E + \underset{\sim}{e}_V\dot{\varepsilon}_V = \\
&= \underset{\sim}{e}_S\{(\dot{\lambda}_0+\Omega_{12})\sin\phi_0\varepsilon_E+\dot{\phi}_0\varepsilon_V-\Omega_{12}\sin\phi_0\delta\phi - \\
&\quad -\dot{\lambda}_0\cos\phi_0(a+h_0)^{-1}\,\delta h + (a+h_0)^{-1}\,\delta v_E + (\delta\omega_d^g)^1\} + \\
&+ \underset{\sim}{e}_E\{-(\dot{\lambda}_0+\Omega_{12})(\sin\phi_0\varepsilon_S+\cos\phi_0\varepsilon_V) - \\
&\quad -\dot{\phi}_0(a+h_0)^{-1}\,\delta h - (a+h_0)^{-1}\,\delta v_S + (\delta\omega_d^g)^2\} + \\
&+ \underset{\sim}{e}_V\{-\dot{\phi}_0\varepsilon_S+(\dot{\lambda}_0+\Omega_{12})\cos\phi_0\varepsilon_E-(\Omega_{12}\cos\phi_0+\dot{\lambda}_0\cos^{-1}\phi_0)\delta\phi + \\
&\quad +\dot{\lambda}_0\sin\phi_0(a+h_0)^{-1}\,\delta h - \tan\phi_0(a+h_0)^{-1}\,\delta v_E + (\delta\omega_d^g)^3\}
\end{aligned}
\tag{1.66i}
$$

in connection with

$$
\begin{aligned}
\delta\dot{\phi} &= -\dot{\phi}_0(a+h_0)^{-1}\,\delta h -(a+h_0)^{-1}\,\delta v_S\ , \\
\delta\dot{\lambda} &= \dot{\lambda}_0\tan\phi_0\delta\phi-\dot{\lambda}_0(a+h_0)^{-1}\delta h+(a+h_0)^{-1}\cos^{-1}\phi_0\delta v_E\ , \\
\delta\dot{h} &= \delta v_V
\end{aligned}
\tag{1.66ii}
$$

and

$$
\begin{aligned}
&\underset{\sim}{e}_S\delta\dot{v}_S + \underset{\sim}{e}_E\delta\dot{v}_E + \underset{\sim}{e}_V\delta\dot{v}_V = \\
&= \underset{\sim}{e}_S\{(\delta\gamma)^1+\delta y^1+y_0^3\varepsilon_E-y_0^2\varepsilon_V + \\
&\quad +[(\dot{\lambda}_0^2-\Omega_{12}^2)+2\Omega_{12}(\dot{\lambda}_0+\Omega_{12})\cos^2\phi_0](a+h_0)\delta\phi - \\
&\quad -[(\dot{\lambda}_0^2-\Omega_{12}^2)\sin\phi_0\cos\phi_0+\dot{\phi}_0\dot{h}_0(a+h_0)^{-1}]\delta h - \\
&\quad -\dot{h}_0(a+h_0)^{-1}\delta v_S+2(\dot{\lambda}_0+\Omega_{12})\sin\phi_0\delta v_E+\dot{\phi}_0\delta v_V\} + \\
&+ \underset{\sim}{e}_E\{(\delta\gamma)^2+\delta y^2-y_0^3\varepsilon_S+y_0^1\varepsilon_V + \\
&\quad +[2\Omega_{12}(\dot{\phi}_0\cos\phi_0+\dot{h}_0(a+h_0)^{-1}\sin\phi_0)+\dot{\phi}_0\dot{\lambda}_0\cos^{-1}\phi_0](a+h_0)\delta\phi - \\
&\quad -\dot{\lambda}_0[\dot{\phi}_0\sin\phi_0-\dot{h}_0(a+h_0)^{-1}\cos\phi_0]\delta h-(\dot{\lambda}_0+2\Omega_{12})\sin\phi_0\delta v_S + \\
&\quad +[\dot{\phi}_0\tan\phi_0-\dot{h}_0(a+h_0)^{-1}]\delta v_E-(\dot{\lambda}_0+2\Omega_{12})\cos\phi_0\delta v_V\} + \\
&+ \underset{\sim}{e}_V\{(\delta\gamma)^3+\delta y^3+y_0^2\varepsilon_S-y_0^1\varepsilon_E - \\
&\quad -2\Omega_{12}(\dot{\lambda}_0+\Omega_{12})\sin\phi_0\cos\phi_0(a+h_0)\delta\phi - \\
&\quad -[(\dot{\lambda}_0^2-\Omega_{12}^2)\cos^2\phi_0+\dot{\phi}_0^2]\delta h-2\dot{\phi}_0\delta v_S+2(\dot{\lambda}_0+\Omega_{12})\cos\phi_0\delta v_E\}
\end{aligned}
\tag{1.66iii}
$$

Considering that the approximated frame e_0 corresponds to the sensor frame d in the case of the platform mechanization, the equations (1.66) hold also for the platform system, if the following terms are substituted

$$
\begin{aligned}
&(y_{e_0})_0 && \rightarrow (y_d)_0 \\
&\delta y_e \approx R_{e_0 d}\delta y_d && \rightarrow \delta y_d \\
&(\delta\omega_d^g)_e \approx R_{e_0 d}\delta\omega_d^g && \rightarrow \delta\omega_d^g \ .
\end{aligned}
\tag{1.67}
$$

We at once realize, that (1.66) is a *system of linear first order differential equations*, which is given in its general form

$$
\boxed{\dot{z} = F(t)z + G(t)\,u(t)}
\tag{1.68i}
$$

wherein we have to provide proper prior information for the initial *state vector* $z(t_o)$ with

$$
z(t) := [\varepsilon_S,\varepsilon_E,\varepsilon_V,\delta v_S,\delta v_E,\delta v_V,\delta\phi,\delta\lambda,\delta h]^T \ ,
\tag{1.68ii}
$$

and the *disturbing or forcing vector*

$$
\begin{aligned}
u(t) := [&(\delta\omega_d^g)^1,(\delta\omega_d^g)^2,(\delta\omega_d^g)^3,\ (\delta y_d)^1,(\delta y_d)^2, \\
&(\delta y_d)^3,(\delta\gamma)^1,(\delta\gamma)^2,(\delta\gamma)^3]^T
\end{aligned}
\tag{1.68iii}
$$

We now want to discuss the solution of the linear equations by an example. To simplify the linear error-equations (1.66) of a strapdown mechanization we assume that

- $\dot{\lambda}_0 = 0$, $\dot{\phi}_0 = 0$, $\dot{h}_0 = 0$, i.e. the inertial measurement unit is not moved with respect to the earth-fixed frame,
- the Coriolis acceleration and the rotational acceleration are neglected, i.e. $\Omega_{12} = 0$
- only a central field is introduced as a model of the gravitation field, i.e.

$$
\underset{\sim}{\gamma}_M = e[0,0,-\frac{Gm}{(a+h_0)^2}]^T \ , \ \underset{\sim}{\gamma} = \underset{\sim}{\gamma}_M+\delta\underset{\sim}{\gamma}_M+\delta\underset{\sim}{\gamma}_R \ ,
$$

$$
\delta\underset{\sim}{\gamma}_M = e[0,0,\frac{2Gm}{(a+h_0)^3}\,\delta h]^T, \ \delta\underset{\sim}{\gamma}_R = \delta\underset{\sim}{\gamma}-\delta\underset{\sim}{\gamma}_M \ .
$$

The rotation matrix $R_{e_0 d}$ is determined by an analytical alignment. With the help of the rotation matrix the vector of the specific forces $d(y_d)$ is transformed into the approximated horizontal base e_0. With the assumptions above we obtain

$$(y_{e_0})_0^1 = (y_{e_0})_0^2 = 0 \; ; \; (y_{e_0})_0^3 = \frac{Gm}{(a+h_0)^2} \, .$$

Thus the system of differential equations (1.66) is rewritten as

$$\begin{aligned} &\underset{\sim}{e}_S\dot{\varepsilon}_S + \underset{\sim}{e}_E\dot{\varepsilon}_E + \underset{\sim}{e}_V\dot{\varepsilon}_V = \\ &= \underset{\sim}{e}_S\{\cos\phi_0\delta\dot{\lambda}+(\delta\omega_d^g)^1\}+\underset{\sim}{e}_E\{\delta\dot{\phi}+(\delta\omega_d^g)^2\}+\underset{\sim}{e}_V\{-\sin\phi_0\delta\dot{\lambda}+(\delta\omega_d^g)^3\} \, . \end{aligned} \tag{1.69i}$$

$$\delta\dot{\phi} = -(a+h_0)^{-1}\delta v_S \, , \quad \delta\dot{\lambda} = (a+h_0)^{-1}\cos^{-1}\phi_0\delta v_E \, , \quad \delta\dot{h} = \delta v_V \, ,$$

$$\underset{\sim}{e}_S\{\delta y^1+y_0^3\varepsilon_E\} + \underset{\sim}{e}_E\{\delta y^2-y_0^3\varepsilon_S\} + \underset{\sim}{e}_V\{\delta y^3\} = \tag{1.69ii}$$

$$\begin{aligned} &= \underset{\sim}{e}_S\{-(a+h_0)\delta\ddot{\phi}-(\delta\gamma)_R^1\}+\underset{\sim}{e}_E\{(a+h_0)\cos\phi_0\delta\ddot{\lambda}-(\delta\gamma)_R^2\} + \\ &+ \underset{\sim}{e}_V\{\delta\ddot{h} - \frac{2Gm}{(a+h_0)^3}\,\delta h - (\delta\gamma)_R^3\} \, . \end{aligned} \tag{1.69iii}$$

(1.69i) we can integrate at once:

$$\begin{aligned} &\underset{\sim}{e}_S\varepsilon_S + \underset{\sim}{e}_E\varepsilon_E + \underset{\sim}{e}_V\varepsilon_V = \\ &= \underset{\sim}{e}_S\{\cos\phi_0\delta\lambda+ \int_0^t (\delta\omega_d^g)^1 dt\}+\underset{\sim}{e}_E\{\delta\phi+ \int_0^t (\delta\omega_d^g)^2 dt\}+\underset{\sim}{e}_V\{-\sin\phi_0\delta\lambda+ \int_0^t(\delta\omega_d^g)^3 dt\} \end{aligned}$$

The solution is substituted into (1.69iii)

$$\begin{aligned} &\underset{\sim}{e}_S\{\delta y^1 + \frac{Gm}{(a+h_0)^2}\,\delta\phi + \frac{Gm}{(a+h_0)^2}\int_0^t (\delta\omega_d^g)^1\, dt\} + \\ &+ \underset{\sim}{e}_E\{\delta y^2 - \frac{Gm}{(a+h_0)^2}\cos\phi_0\delta\lambda - \frac{Gm}{(a+h_0)^2}\int_0^t (\delta\omega_d^g)^2 dt\} + \underset{\sim}{e}_V\{\delta y^3\} = \\ &= \underset{\sim}{e}_S\{-(a+h_0)\delta\ddot{\phi}-(\delta\gamma)_R^1\}+\underset{\sim}{e}_E\{(a+h_0)\cos\phi_0\delta\ddot{\lambda}-(\delta\gamma)_R^2\}+ \underset{\sim}{e}_V\{\delta\ddot{h}- \frac{2Gm}{(a+h_0)^3}\,\delta h-(\delta\gamma)_R^3\} \end{aligned}$$

Rewriting this equation, we obtain

$$\underset{\sim}{e}_S\{\frac{\delta y^1}{(a+h_0)} + \frac{Gm}{(a+h_0)^3}\int_0^t (\delta\omega_d^g)^1 dt + \frac{(\delta\gamma)_R^1}{(a+h_0)}\} +$$
$$+ \underset{\sim}{e}_E\{\frac{\delta y^2}{(a+h_0)\cos\phi_0} - \frac{Gm}{(a+h_o)^3\cos\phi_0}\int_0^t (\delta\omega_d^g)^2 dt + \frac{(\delta\gamma)_R^2}{(a+h_0)\cos\phi_0}\} + \underset{\sim}{e}_V\{\delta y^3 + (\delta\gamma)_R^3\} =$$
$$= \underset{\sim}{e}_S\{-\delta\ddot{\phi} - \frac{Gm}{(a+h_0)^3}\delta\phi\} + \underset{\sim}{e}_E\{\delta\ddot{\lambda} + \frac{Gm}{(a+h_0)^3}\delta\lambda\} + \underset{\sim}{e}_V\{\delta\ddot{h} - \frac{2Gm}{(a+h_0)^3}\delta h\} \tag{1.70}$$

The general solution of the system of differential equations consists of the general solution of the *homogeneous system* and a particular solution of the *inhomogeneous system*. We have to distinguish between the position error of the horizontal directions $\underset{\sim}{e}_S$ and $\underset{\sim}{e}_E$ and of the vertical direction $\underset{\sim}{e}_V$. The characteristic equation yields for the horizontal position two conjugate complex roots; the two *fundamental solutions* are $\cos\omega_s t$ and $\sin\omega_s t$ with a suitable constant ω_s. The two integration constants are determined by the given initial values $\delta\phi(t_o)$, $\delta\lambda(t_o)$ and $\delta\dot{\phi}(t_o)$, $\delta\dot{\lambda}(t_o)$, thus the *general solution of the homogeneous system of differential equations* is specified by

$$\delta\phi_{hom}(t) = \delta\phi(t_0)\cos\omega_s t + \frac{\delta\dot{\phi}(t_0)}{\omega_s}\sin\omega_s t\ , \tag{1.71i}$$
$$\delta\lambda_{hom}(t) = \delta\lambda(t_0)\cos\omega_s t + \frac{\delta\dot{\lambda}(t_0)}{\omega_s}\sin\omega_s t\ ,$$

where $\omega_s := \frac{Gm}{(a+h_0)^3}$ is the Schuler rate with a period of $T \approx 84{,}4$ min.

The characteristic equation yields for the vertical direction $\underset{\sim}{e}_V$ two real different roots. The two fundamental solutions are $\exp(2\ \omega_s t)$ and $\exp(-2\ \omega_s t)$. With the initial values $\delta h(t_0)$ and $\delta\dot{h}(t_0)$, we obtain the *general solution of the homogeneous system of differential equations*

$$\delta h_{hom}(t) = \frac{1}{2}\delta h(t_0)\sinh\sqrt{2}\,\omega_s t + \frac{1}{2}\frac{\delta\dot{h}(t_0)}{\sqrt{2}\,\omega_s}\cosh\sqrt{2}\,\omega_s t \tag{1.71ii}$$

A particular solution of the *inhomogeneous system* is obtained if the *forcing functions* $f_S(t)$, $f_E(t)$ and $f_V(t)$ are known,

$$\delta\phi_{inhom}(t) = -\frac{1}{\omega_s}\int_{t_0}^{t} \sin \omega_s(t-\tau)\, f_S(\tau)\, d\tau \;,$$

$$\delta\lambda_{inhom}(t) = \frac{1}{\omega_s}\int_{t_0}^{t} \sin \omega_s(t-\tau)\, f_E(\tau)\, d\tau \;, \qquad (1.72)$$

$$\delta h_{inhom}(t) = -\frac{2}{\omega_s}\int_{t_0}^{t} \sinh \omega_s(t-\tau)\, f_V(\tau)\, d\tau \;.$$

As a result of the example we can state:

- The position error of an inertial measurement unit is *stable* with a *periodic* behaviour in the two *horizontal directions*
- The position error of the *vertical direction* is *not stable.*

These statements hold for a strapdown system as well as for a platform system in the case of the *decoupled* system of differential equations. If, in contrast, all terms in equation (1.) are considered, the unstable vertical direction influences the horizontal directions after a short period of time.

The forcing functions $f_S(t)$, $f_E(t)$ and $f_V(t)$ appearing in the mathematical description of an inertial measurement unit and modelled by the components of the disturbing vector u(t) as introduced in chapter two, are characterized by two different effects. On the one hand these are the *sensor errors* $d(\delta y_d)$ and $d(\delta\omega_d^g)$ and on the other hand that is the effect of the *perturbation of the gravitation field* $e(\delta\gamma)$. In the following the error model used in the simulation is described.

a) <u>error model of the accelerometers</u>

The precise modelling depends on the type of accelerometer used in the inertial measurement unit. To obtain comparable simulation results for different mechanizations a generalized error model is preferred. Thus we assume that the error vector $d(\delta y_d)$ is modelled by a fixed *bias*, a fixed *scale factor* and an additional *stochastic error* for each accelerometer. Further we consider the *non-orthogonality* of the accelerometer input axes. The stochastic properties of the error vector $d(\delta y_d)$ are described by its first and second stochastical moments. $E\{\cdot\}$ denotes the *expectation operator*, while $D\{\cdot\}$ denotes the *dispersion*.

The matrices

$$S_y := \begin{bmatrix} S_y^1 & 0 & 0 \\ 0 & S_y^2 & 0 \\ 0 & 0 & S_y^3 \end{bmatrix} \quad , \quad T_y = \begin{bmatrix} 0 & -\theta_y^{13} & \theta_y^{12} \\ \theta_y^{23} & 0 & -\theta_y^{21} \\ -\theta_y^{32} & \theta_y^{31} & 0 \end{bmatrix} \tag{1.73}$$

comprise successively the errors of the *fixed* scale factors and the *fixed* parameters of non-orthogonality of the three accelerometer input axes $\{\underset{\sim}{a}_1, \underset{\sim}{a}_2, \underset{\sim}{a}_3\}$. The parameters θ_y^{ij} describe the transformation

$$a \rightarrow d = a\, T_y$$

that is the inversion of an affine transformation of the sensor base vectors d into the *non-orthogonal* base $a = \{\underset{\sim}{a}_1, \underset{\sim}{a}_2, \underset{\sim}{a}_3\}$ defined by the real accelerometer input axes. The vectors

$$b_y := (b_y^1,\ b_y^2,\ b_y^3)^T \quad , \quad v_y := (v_y^1,\ v_y^2,\ v_y^3)^T \tag{1.74}$$

denote the *fixed* biases and the *stochastic* noise of the three accelerometers. Thus we obtain

$$\begin{aligned} E\{d(\delta y_d)\} &= E\{d[(S_y+T_y)y_a+b_y+v_y]\} = \\ &= d[(S_y+T_y)E\{y_a\}+b_y] \;, \end{aligned} \tag{1.75i}$$

$$\begin{aligned} D\{d(\delta y_d)\} &= (d \otimes d^T)E\{\delta y_d(t)\delta y_d^T(\tau)\} = \\ &= (d \otimes d^T)[(S_y+T_y)E\{y_a(t)y_a^T(\tau)\} \cdot \\ &\quad \cdot (S_y+T_y)^T+Q_y\delta(t-\tau)] \end{aligned} \tag{1.75ii}$$

The matrix Q_y is the spectral density matrix of the stochastic process $v_y(t)$, $\delta(t-\tau)$ denotes the Dirac delta function.

b) <u>error model of the gyros</u>

The vector $d(\delta\omega_d^g)$ denotes in the case of the strapdown mechanization the error (by the gyros) in the determination of the angular velocity while in the case of the platform mechanization it denotes the error of the unit gyro/torquer. The evetor is modelled by a similar error model as used for the translational motion. The matrices

$$S_\omega := \begin{bmatrix} S_\omega^1 & 0 & 0 \\ 0 & S_\omega^2 & 0 \\ 0 & 0 & S_\omega^3 \end{bmatrix} \quad , \quad T_\omega := \begin{bmatrix} 0 & -\theta_\omega^{13} & \theta_\omega^{12} \\ \theta_\omega^{23} & 0 & -\theta_\omega^{21} \\ -\theta_\omega^{32} & \theta_\omega^{31} & 0 \end{bmatrix} \tag{1.76}$$

denote successively the *fixed* scale factor errors of the three gyros or of the torque generators and the *fixed* parameters of non-orthogonality of the three input axes $r = \{\underset{\sim}{r}_1, \underset{\sim}{r}_2, \underset{\sim}{r}_3\}$ of the gyros. The affine transformation $d \to r$, or rather its inversion

$$r \to d = r\, T_\omega$$

is necessary because of the deviation of the real gyro input axes from the sensor frame d.

The vectors

$$b_\omega := (b_\omega^1,\ b_\omega^2,\ b_\omega^3)^T \quad , \quad v_\omega := (v_\omega^1,\ v_\omega^2,\ v_\omega^3)^T \tag{1.77}$$

denote the *fixed* gyro drifts and the *stochastical* drift. Thus we obtain

$$\begin{aligned} E\{d(\delta\omega_d^g)\} &= E\{d[(S_\omega + T_\omega)(\omega_d^g)_r + b_\omega + v_\omega]\} = \\ &= d[(S_\omega + T_\omega)E\{(\omega_d^g)_r + b_\omega] \end{aligned} \tag{1.78i}$$

$$\begin{aligned} D\{d(\delta\omega_d^g)\} &= (d \otimes d^T)E\{(\delta\omega_d^g)(t)(\delta\omega_d^g)^T(\tau)\} = \\ &= (d \otimes d^T)[(S_\omega + T_\omega)E\{(\omega_d^g)_r(t)(\omega_d^g)_r^T(\tau)\} \cdot \\ &\quad \cdot (S_\omega + T_\omega)^T + Q_\omega\, \delta(t-\tau)] \end{aligned} \tag{1.78ii}$$

The matrix Q_ω denotes the spectral density matrix of the stochastic process $v_\omega(t)$.

c) Disturbance of the gravitation field

The vector $e(\delta\gamma)$ is a function of time, because the measurement unit is moved with respect to the equatorial frame. Thus, given the components of the deflection of the vertical and the gravitation disturbance in the form of a digital model (i.e. as *a function of position*) they have to be transformed into a *function of time* by the instantaneous relative velocity. Thus the effect of the disturbing of the gravitation field depends on the one hand on the *specific trajectory* of the measurement unit and on the other hand on the *relative velocity* with which the unit is moved along the trajectory.

Introducing the stochastical parts of the sensor error models the incremental differential equations remain no longer *deterministic* but become *stochastical differential equations* or *stochastical difference equations* in the digital simulations. Thus especially the state variables have to be treated as *stochastic variables* for $t > t_0$. On the other hand, the disturbing of the gravitation field has to be treated as a *deterministic input* into the dynamical system because of its deterministic character.

It is well-known, that any system of differential equations of *higher order* can be rewritten as a system of differential equations of *first order*. Our system of first order has the general explicit form

$$\dot{z} = F(t)z + G(t)u(t) \tag{1.79}$$

Such a representation of a dynamical system is called a *state-space* representation. In system theory z is commonly called the *state vector*, F(t) the *dynamic matrix* and G(t) the *coefficient matrix* of the *forcing vector* u(t).

The state-space representation of a dynamical system is in a mathematical sense a question of finding a set of values such that for all future moments the dynamical behaviour of the autonomous system (or if the forcing functions are known of the nonautonomous system) can be determined. To do this the system of differential equations mentioned above has to be solved. The matrices F(t) and G(t) describe completely any time-dependent system, if suitable initial or boundary conditions are given. In contrast to this the matrices are not definitely assigned to a system. It is possible by a suitable choice of the state variables to give the matrix F(t) a special form, for instance a diagonal form. In general, it is difficult to interpret the corresponding state variables. But the choice of the state variables concerns only the *mathematical* description of a system, the *physical* properties are invariant against a modification in the representation. Another choice of the state variables implies a change of base in the representation of the physical properties.

One possible choice of the state variables of our dynamical system is already given by the definition (1. ii):

$$z = \{\varepsilon_S, \varepsilon_E, \varepsilon_V, \delta v_S, \delta v_E, \delta v_V, \delta\phi, \delta\lambda, \delta h\}^T$$

The dynamic matrix F(t) is given by the linear incremental system of differential equations (1.63) with (1.65). The state-space representation is done with respect to the base $\underset{\sim}{e}_S, \underset{\sim}{e}_E, \underset{\sim}{e}_V$. In the *spherical case* the dynamic matrix is given implicitely by (1.66), in the *ellipsoidal case* it is specified explicitely in Table 2.
The elements of the coefficient matrix G(t) consist in the case of the strapdown mechanization of the elements of the transformation matrix $R_{e_0 d}$. The 9×9 matrix G(t) has the elements

$$G(t) = \begin{bmatrix} \underset{(3\times3)}{R_{e_0 d}(t)} & \underset{(3\times3)}{0} & \underset{(3\times3)}{0} \\ \underset{(3\times3)}{0} & \underset{(3\times3)}{R_{e_0 d}(t)} & \underset{(3\times3)}{I} \\ \underset{(3\times3)}{0} & \underset{(3\times3)}{0} & \underset{(3\times3)}{0} \end{bmatrix} \tag{1.80}$$

if the forcing vector is chosen as defined in (2. iii). In the case of the slaved platform mechanization, the matrix G(t) reduces to the 9×9 identity matrix.
To simplify the computation of the state vector z(t), the forcing vector can be included into the state-vector, if the forcing functions are *deterministic*. For this purpose the forcing term G(t)u(t) is modified by

$$G(t)u(t) = G_1(t)A(t)u_S + G_2\,u_\gamma(t) \tag{1.81}$$

where the 24 × 1 vector u_S includes the 24 *constant* error parameters of the accelerometer and gyro error model. The 3 × 1 vector $u_\gamma(t)$ includes the remaining in general time dependent gravitation perturbation defined by (1.60iii). The functional matrix A(t) is defined by

$$A(t) = \begin{bmatrix} \partial(\delta\omega_d^g)/\partial u_S^T \\ \partial(\delta y_d)/\partial u_S^T \\ 6 \times 24 \end{bmatrix} \tag{1.82}$$

and the matrices $G_1(t)$ and G_2 are given by

$$G_1(t) = \begin{bmatrix} \underset{3\times3}{R_{e_0}d} & \underset{3\times3}{0} \\ \underset{3\times3}{0} & \underset{3\times3}{R_{e_0}d} \\ \underset{3\times3}{0} & \underset{3\times3}{0} \end{bmatrix}_{9\times6} \tag{1.83i}$$

$$G_2 = \begin{bmatrix} \underset{3\times3}{0} \\ \underset{3\times3}{I} \\ \underset{3\times3}{0} \end{bmatrix}_{9\times3} \tag{1.83ii}$$

The state vector is now augmented by the constant forcing vector u_S

$$z^*(t) = (z^T, u_S^T)^T \ , \tag{1.84i}$$

at which the second part of z^* is no longer time dependent, i.e.

$$\dot{u}_S = 0 \ .$$

The matrix F(t) is thus extended by rows with zero elements and by columns with the corresponding elements of the matrix product of $G_1(t)$ and the functional matrix A(t)

$$F^*(t) = \begin{bmatrix} \underset{9\times9}{F(t)} & \underset{9\times24}{G_1(t)A(t)} \\ \underset{24\times9}{0} & \underset{24\times24}{0} \end{bmatrix}_{33\times33} \tag{1.84ii}$$

The modified system of differential equations is rewritten as

$$\dot{z}^* = F^*(t)z^* + G_2^* \, u_\gamma(t) \tag{1.85}$$

with G_2^* extended by the required zero elements.

Given suitable *initial conditions* $z(t_0)$ the system of differential equations (1.) has the solution

$$z(t) = \Phi(t,t_0)z(t_0) + \int_{t_0}^{t} \Phi(t,\tau)G(\tau)u(\tau)d\tau \, . \tag{1.86}$$

The *fundamental matrix* $\Phi(t,\tau)$ has the following properties

$$\begin{aligned} &(i) && \Phi(t_3,t_2)\Phi(t_2,t_1) && = \Phi(t_3,t_1) \\ &(ii) && \Phi(t,\tau)^{-1} && = \Phi(\tau,t) \\ &(iii) && \frac{d\Phi(t,\tau)}{dt} && = F(t)\Phi(t,\tau) \end{aligned}$$

Extending the general theory of boundary value problems of differential equations of higher order to our initial condition problem of a system of differential equations of first order, the *fundamental matrix* $\Phi(t,\tau)$ corresponds to the *Green function* or *Green matrix*.

In general the fundamental matrix of a time dependent system, i.e. $F = F(t)$, cannot be determined analytically. Remember that even for a scalar linear differential equation with variable coefficients there exists no closed-form solution. But in the case of *constant* coefficients of the dynamical matrix, the fundamental matrix can be determined by a *matrix exponential function*. In this special case the fundamental matrix is no longer a function of the two independent variables t and τ, but a function of the difference $t - \tau$. The solution of the system of differential equations (1.) in this special case is obtained by

$$\boxed{z(t) = e^{F(t-t_0)} \, z(t_0) + \int_{t_0}^{t} e^{F(t-\tau)} \, Gu(\tau)d\tau \quad .} \tag{1.87}$$

The matrix function $e^{F(t-\tau)}$ is defined analogously to the series of polynoms of the scalar function $e^{f(t-\tau)}$, in particular

$$\boxed{e^{F(t-\tau)} = \sum_{n=0}^{\infty} \frac{F^n(t-\tau)^n}{n!} \quad .} \tag{1.88}$$

The computation of the fundamental matrix is only approximately performed by (1.), nevertheless there exists the *theorem of Cayley-Hamilton* of matrix algebra, which states that any power F^n of the matrix $F_{m\times m}$ with $n \geqq m$ can be represented by a linear combination of the powers $F^0, F^1, \ldots, F^{m-1}$. Thus the fundamental matrix in the special case of constant coefficients has a closed form. As possibilities of the *analytical* computation of the fundamental matrix one can mention here the method of *spectral representation* and the method of *inversed Laplace-transformation*.

In the digital simulation, the method of *truncated series* has been applied: Based on the dynamic matrix (Table 2), i.e. no terms will be neglected, the time interval $t_j - t_i$ is decomposed into the smaller intervals Δt:

$$n \cdot \Delta t = t_j - t_i \; .$$

For each interval Δt the coefficients of the dynamic matrix are computed again. The fundamental matrix is computed with (1.), wherein the series development is truncated at a point previously defined. Assuming for each interval Δt that the coefficients of the dynamic matrix are constants, it holds

$$\boxed{\Phi(t_j,t_i) = \Phi(t_j,t_j-\Delta t)\Phi(t_j-\Delta t,t_j-2\Delta t) \;\ldots\; \Phi(t_i+\Delta t,t_i)}$$

Further we have

$$\Phi(t_j-k\Delta t,t_j-(k+1)\Delta t) \neq \Phi(\Delta t)$$

i.e. the property of the solution with constant coefficients $\Phi(t,\tau) = \Phi(t-\tau)$ is not valid for the whole interval $t_j t_i$. It holds only for the smaller interval Δt.

Finally we give the basic dynamic matrix F in the rotational *ellipsoidal reference frame* in *Table 2*.

The eigenvalues of the dynamic matrix F have been computed for the spherical approximation $e^2 = e_{13}^2 = 0$ by *symbolic formula* manipulation. The leading terms are presented in *Table 3*.

$$
\begin{bmatrix}
0 & (\dot\vartheta+\dot\lambda)\cdot\sin\phi & \dot\phi & 0 & \frac{1}{N+h} & 0 & -(\dot\vartheta+\frac{(N-M)\dot\lambda}{N+h})\sin\phi & 0 & -\frac{\dot\lambda\cos\phi}{N+h} \\
-(\dot\vartheta+\dot\lambda)\cdot\sin\phi & 0 & -(\dot\vartheta+\dot\lambda)\cdot\cos\phi & -\frac{1}{M+h} & 0 & 0 & -\frac{3\dot\phi\tan\phi M(N-M)}{N(M+h)} & 0 & -\frac{\dot\phi}{M+h} \\
-\dot\phi & (\dot\vartheta+\dot\lambda)\cdot\cos\phi & 0 & 0 & -\frac{\tan\phi}{N+h} & 0 & -\dot\vartheta\cos\phi-\frac{(M+h)\dot\lambda-(N-M)\dot\lambda\cos^2\phi}{(N+h)\cos\phi} & 0 & \frac{\dot\lambda\sin\phi}{N+h} \\
0 & s_3 & -s_2 & -\frac{\dot h}{M+h} & 2(\dot\vartheta+\dot\lambda)\cdot\sin\phi & \dot\phi & (\dot\vartheta+\dot\lambda)^2(N+h)\cos^2\phi-(\dot\vartheta^2-\dot\lambda^2)(M+h)\sin^2\phi-\frac{3\dot h\dot\phi\tan\phi M(N-M)}{N(M+h)} & 0 & (\dot\vartheta^2-\dot\lambda^2)\sin\phi\cos\phi-\frac{\dot h\dot\phi}{M+h} \\
-s_3 & 0 & s_1 & -(2\dot\vartheta+\dot\lambda)\cdot\sin\phi & \frac{\dot\phi\tan\phi(M+h)-\dot h}{N+h} & -(2\dot\vartheta+\dot\lambda)\cdot\cos\phi & (\frac{\dot\lambda(N-M)}{N+h}+2\dot\vartheta)(\dot\phi\cos\phi(M+h)+\dot h\sin\phi)+\frac{(M+h)^2\dot\phi\dot\lambda}{(N+h)\cos\phi} & 0 & -\frac{\dot\lambda\dot\phi\sin\phi(M+h)+\dot\lambda\dot h\cos\phi}{N+h} \\
s_2 & -s_1 & 0 & -2\dot\phi & 2(\dot\vartheta+\dot\lambda)\cdot\cos\phi & 0 & -(\dot\vartheta+\dot\lambda)^2(N+h)\sin\phi\cos\phi-(\dot\vartheta^2-\dot\lambda^2)(M+h)\sin\phi\cos\phi-\frac{3\dot\phi^2\tan\phi M(N-M)}{N} & 0 & (\dot\vartheta^2-\dot\lambda^2)\cos^2\phi-\dot\phi^2 \\
0 & 0 & 0 & -\frac{1}{M+h} & 0 & 0 & \frac{-3\dot\phi\tan\phi(N-M)M}{N(M+h)} & 0 & -\frac{\dot\phi}{M+h} \\
0 & 0 & 0 & 0 & \frac{1}{(N+h)\cos\phi} & 0 & \frac{\dot\lambda\tan\phi(M+h)}{N+h} & 0 & -\frac{\dot\lambda}{N+h} \\
0 & 0 & 0 & 0 & 0 & 1 & 0 & 0 & 0
\end{bmatrix}
$$

<u>Table 2:</u> The dynamic matrix F of order 9×9 in the ellipsoidal horizontal frame of reference, $M = a(1-e^2)(1-e^2\sin^2\phi)^{-\frac{3}{2}}$, $N = a(1-e^2\sin^2\phi)^{-\frac{1}{2}}$

Table 3: First order approximation of the eigenvalues of the dynamic matrix F according to D. Schröder (1988)

$$\lambda_1 = -i(\omega_S^2 + 2\omega_E^2\sin^2\phi - (2\omega_E\sin\phi(\omega_E^2\sin^2\phi + \omega_S^2)^{1/2}))^{1/2}$$

$$\lambda_2 = \ \ i(\omega_S^2 + 2\omega_E^2\sin^2\phi - (2\omega_E\sin\phi(\omega_E^2\sin^2\phi + \omega_S^2)^{1/2}))^{1/2}$$

$$\lambda_3 = -i(\omega_S^2 + 2\omega_E^2\sin^2\phi + (2\omega_E\sin\phi(\omega_E^2\sin^2\phi + \omega_S^2)^{1/2}))^{1/2}$$

$$\lambda_4 = \ \ i(\omega_S^2 + 2\omega_E^2\sin^2\phi + (2\omega_E\sin\phi(\omega_E^2\sin^2\phi + \omega_S^2)^{1/2}))^{1/2}$$

$$\lambda_5 = -i\omega_E$$

$$\lambda_6 = \ \ i\omega_E$$

$$\lambda_7 = \ \ 0$$

$$\lambda_8 = \ \ \ 2\omega_S$$

$$\lambda_9 = - \ 2\omega_S$$

The eigenvalues are fundamental for the dynamic system behavior, namely stability. The eigenvalue λ_9 contains the famous *Schuler frequency* ω_S (84.4 min) with a negative sign, thus generating a damped system reply. In contrast, the eigenvalue λ_8 appears positive: This eigenvalue produces instability. The zero eigenvalue λ_7 expresses the longitude ambiguity of the dynamic system. The eigenvalues λ_1 up to λ_6 are imaginary: They produce a periodic system response leading to instability, too. These eigenvalues contain the *earth rotation frequency* ω_E(24 h = 1,440 min) and the *Foucault frequency* $\omega_E\sin\phi$ (e.g. $\phi = 60°$, $\omega_E\sin$ = 20.785 h = 1,247 min).

Literature

Budde, K. (1988): Adjusting large threedimensional networks: strategies and computations. Submitted Paper, Third SIAM Conference on Applied Linear Algebra, Madison, May 23-26.

Delikaraoglou, D. (1985): Estimability analysis of free networks of differential range observations to GPS satellites, in: Optimization and Design of Geodetic Networks, ed. E.W. Grafarend and F. Sanso, Springer-Verlag, p. 196-220, Berlin - Heidelberg - New York - Tokyo 1985.

Eren, K. (1987): Geodetic network adjustment using GPS triple difference observations and a priori stochastic information. Technical Report No. 1, Institute of Geodesy, University of Stuttgart.

Grafarend, E. (1981): From kinematical geodesy to inertial positioning. Bulletin géodésique 55 (1981) 286-299.

Grafarend, E. and W. Lindlohr (1988): World: A multipurpose GPS-network computer package. Proceedings, GPS-Workshop, ed. E. Groten, Darmstadt 1988.

Grafarend, E. and V. Müller (1985): The critical configuration of satellite networks, especially of Laser and Doppler type, for planar configurations of terrestrial points, manuscripta geodaetica 10 (1985) 131-152.

Grafarend, E.W., W. Lindlohr and D.E. Wells (1985): GPS redundancy design using the undifferenced phase observation approach. Proceedings, Second Meeting of the European Working Group on Satellite Radio Positioning, Saint-Mandé, pp. 100-107.

Grafarend, E.W. and B. Schaffrin (1988): Von der statischen zur dynamischen Auffassung geodätischer Netze. Zeitschrift für Vermessungswesen, Vol. 112, No. 2, pp. 79-103.

Heitz, S. (1984): Mechanik fester Körper, Bd. 2: Dynamik elastischer Körper, mechanische Grundlagen der Geodäsie, Dümmler-Verlag, Bonn 1984.

Lindlohr, W. (1988a): Alternative modeling of GPS carrier phases for geodetic network analysis. Submitted Paper, International Workshop High Precision Navigation, Stuttgart-Altenstein/Wart, May 17-20.

Lindlohr, W. (1988b): PUMA: processing of undifferenced GPS carrier beat phase measurements and adjustment computations. Technical Report No. 5, Institute of Geodesy, University of Stuttgart, in preparation.

Lindlohr, W. (1988c): Dynamische Analyse geodätischer Netze auf der Basis von GPS-Phasenbeobachtungen, Dissertation, Universität Stuttgart 1988.

Lindlohr, W. and D. Wells (1985): GPS design using undifferenced carrier beat phase observations, manuscripta geodaetica 10 (1985) 255-295.

Magnus, K. (1986): Die Beschleunigungsabhängigkeit der Vertikalen-Anzeige von Schwerependel und Lotkreisel. Ingenieur-Archiv, 35. Band, Heft 3, 143-149, 1986.

Pachelski, W., D. Lapucha and K. Budde (1988): GPS network analysis: the influence of stochastic prior information of orbital elements on ground station position measures. Technical Report No. 4, Institute of Geodesy, University of Stuttgart, Stuttgart 1988.

Remondi, B.W. (1984): Using the Global Positioning System (GPS) phase observable for relative geodesy: modeling, processing, and results. CSR-84-2, Center for Space Research, The University of Texas at Austin.

Schaffrin, B. (1985): Das geodätische Datum mit stochastischer Vorinformation. Deutsche Geodätische Kommission, C-13, München.

Schaffrin, B. (1987): Statistical tests for an improved deformation analysis based on GPS and due to stochastic prior information, EOS 68 (1987), No. 44, p. 1237 (Abstract).

Schaffrin, B. and Y. Bock (1988): A unified scheme for processing GPS dual-band phase observations, Bull. Géod. 62 (1988), to appear.

Schaffrin, B. and E.W. Grafarend (1986): Generating classes of equivalent linear models by nuisance parameter elimination: applications to GPS observations. Manuscripta Geodaetica, Vol. 11, No. 4, pp. 262-271.

Schröder, D., S. Wiegner, E. Grafarend and B. Schaffrin (1986): Simulation eines Geodätischen Inertialsystems. Schriftenreihe Studiengang Vermessungswesen, Universität der Bundeswehr München, Heft 22, 197-235, 1986.

Schröder, D., N. Chi Thong, S. Wiegner, E.W. Grafarend and B. Schaffrin (1988): A comparative study of local level and strapdown inertial systems. Manuscripta Geodaetica, Vol. 13, No. 4, in print.

Schuler, M. (1923): Die Störung von Pendel- und Kreiselapparaten durch die Beschleunigung des Fahrzeugs. Physikalische Zeitschrift, Band 24, 344-350, 1923.

Schwarz, K.P. (1983): Inertial surveying and geodesy. Reviews of Geophysics and Space Physics, Vol. 21, No. 4, 878-890, 1983.

Schwarz, K.P. (1986): Inertial technology for surveying and geodesy. Proceedings Third International Symposium, Banff/Canada, Sept. 16-20, 1985, ed. K.P. Schwarz, 850 pages, 2 volumes, Calgary 1986.

Tsimis, E. (1973): Critical configurations (determinantal loci) for range and range-difference satellite networks. Department of Geodetic Science, Ohio State University, Report 191, Columbus 1973.

Vassiliou, A. and K.P. Schwarz (1985): Eigenvalues of the dynamics matrix used in inertial geodesy, manuscripta geodaetica 10 (1985) 213-221.

GPS Geodesy: A Status Report

G. Blewitt, T.P. Yunck, S.M. Lichten, W.I. Bertiger, S.C. Wu
Jet Propulsion Laboratory, Pasadena, California, USA

Summary

Accuracies of a centimeter or better in earth baseline measurements are crucial for many investigations in crustal dynamics. This implies measurement accuracies of the order of a few parts in 10^8 of baseline length for regional geodesy. In recent years, techniques exploiting the Global Positioning System (GPS) have advanced rapidly toward this accuracy goal. The latest GPS techniques developed at JPL have resulted in centimeter-level agreement with independent measurements made by very long baseline interferometry (VLBI) for baselines up to 2000 km in length. Techniques that have shown promise for GPS-based geodesy include: (1) carrier phase ambiguity resolution, (2) multi-day GPS orbit determination, (3) stochastic estimation of the zenith tropospheric delay, and (4) simultaneous use of carrier phase and pseudorange. The order of importance depends upon the scale of the network, and the approaches are often synergistic. For example, ambiguity resolution can depend upon the ability of the other techniques to improve precision. These results have been acheived despite a partial GPS constellation, regional tracking only, and pseudorange data plagued by multipath. A full GPS constellation and a permanent global tracking network will be in place in the early 1990's, and a new receiver/antenna prototype at JPL is now producing smoothed pseudorange observables accurate to better than 5 cm.

1. Introduction

Our team at NASA's Jet Propulsion Laboratory (JPL) has recently produced GPS-based measurements of geodetic baselines which agree with measurements by very long baseline interferometry (VLBI) at the level of about 1 part in 10^8 (1 cm per 1000 km of baseline length) in the horizontal and length components. Agreement in the vertical is typically 2-3 times worse. These results reflect the effectiveness of recently developed estimation techniques, which include new approaches to carrier phase ambiguity resolution, multi-day GPS orbit determination, stochastic troposphere estimation, and the use of simultaneous group and phase delay. The goal of achieving centimeter-level geodesy has thus been realized for regional GPS networks having good fiducial control. Centimeter-level geodetic accuracy should soon be achievable worldwide as a result of expected improvements in hardware, software, experiment design, and tracking networks.

2. Data Analysis Techniques

2.1 Cycle Ambiguity Resolution

GPS carrier phase delay measurements are biased by an integer number of wavelengths. Determining these integer biases is the well-known problem of cycle ambiguity resolution. Covariance studies show that if the biases are not resolved and instead are estimated as real-valued parameters, there is a factor of 3-5 degradation in baseline measurement precision [Melbourne, 1985]. Successful ambiguity resolution techniques have generally been limited to relatively short baselines by the difficulty in modeling ionospheric delay. Our method to some extent relaxes this limitation by using information from simultaneous measurements of the pseudorange and carrier phase observables [Melbourne, 1985; Wubbena, 1985]. We start by using a time-average of a linear combination of the pseudorange data (P1 and P2) and carrier phase data (L1 and L2) to first determine the integer offsets, N, between the two carrier channels. We define the "widelane" bias

$$N \equiv n_1 - n_2,$$

where n_1 and n_2 are the individual biases on L1 and L2. With sufficient smoothing of pseudorange measurements, N can be determined from the real-valued estimate N', where

$$N' = \frac{L1(cm)}{19.029} - \frac{L2(cm)}{24.421} - \frac{P1(cm)}{153.35} - \frac{P2(cm)}{196.80} \; .$$

With the currently popular TI-4100 receivers, roughly one-half hour is needed to smooth pseudorange noise to a level at which the widelane biases are resolvable. With future high precision receivers, widelane resolution should be possible in less than a minute. To complete the ambiguity resolution, we next proceed to form the ionosphere-free linear combination of the phase data:

$$L_c = L1 + \alpha(L1 - L2) \, ,$$

$$L_c(cm) = \rho + 48.444N + 10.695n_c + \nu \, ,$$

where $\alpha \cong 1.5457$, ρ is the non-dispersive delay, and ν is the data noise. Each phase-connected data arc is now biased by n_c, an integer number of 10.695 cm wavelengths.

In order to resolve these "narrowlane" biases, we estimate them simultaneously with ρ, which is modeled in terms of station coordinates, satellite ephemerides, clocks, and tropospheric delays. An algorithm based on the square-root information filter (SRIF) formalism is then applied which sequentially adjusts the biases to the nearest integer value in such a way that the global solution is updated at each step. In this way, ambiguity resolution over long baselines can be achieved by using information from resolved biases on shorter baselines.

At present [Blewitt et al., 1987; Counselman, 1987; Dong and Bock, 1988], and perhaps for some time to come, a well-designed GPS network must contain several shorter length baselines (100-200 km) if ambiguity resolution is to be reliably acheived over longer baselines. With such a network, we have demonstrated ambiguity resolution over distances up to 2000 km [Blewitt, 1988]. In the future, with sufficiently clean pseudorange data, it may be possible to achieve model-independent resolution of the narrowlane bias just as we now do for the widelane bias. Then, direct resolution of long baseline cycle ambiguities will be possible without the need for preliminary resolution steps on shorter baselines. That, however, will require extremely tight control of multipath around the receiving antenna, which will be difficult to achieve in most settings. Until such measurement precision is routinely available, network design to ensure ambiguity resolution, including a mixture of longer and shorter baselines, should be an important consideration when planning GPS experiments.

2.2 Multi-Day Arc GPS Orbit Determination

The GIPSY software used at JPL for GPS data processing uses a pseudo-epoch state process noise filter to estimate all geodetic and orbital parameters simultaneously [Bierman, 1977; Thornton and Bierman, 1980]. Spacecraft motion is modeled in geocentric J2000 coordinates with a numerical integration of the variational equations and equations of motion. The GEM-L2 geopotential harmonic expansion (complete to degree and order 8 for GPS models) and the effects from solar, lunar, and planet point masses are modeled. The ROCK4 solar radiation pressure model, developed by Rockwell for the GPS satellites, is also used. General relativistic corrections for clock models and gravitational bending are included. These models are sufficient to enable accurate estimation and propagation of satellite motion over arcs of more than two weeks.

The filtering strategy for multi-day arc GPS orbit determination has evolved at JPL over the past two years through the analysis of data taken during the 1985 and 1986 GPS experiments in North America [Lichten and Border, 1987]. Three solar pressure coefficients (including the so-called Y-bias parameter) are estimated as constant parameters along with the orbital states. Preliminary solutions indicate that with arcs of longer than two weeks results improve when the solar pressure coefficients are estimated stochastically. This may be evidence for the time-variation of solar pressure or of other time-varying unmodeled forces on the GPS satellites. These possibilities are now being actively investigated.

2.3 Stochastic Troposphere Estimation

Water vapor radiometers (WVR's) can accurately calibrate the wet tropospheric delay, provided it isn't raining during observation [Elgered et al., 1987; Tralli et al., 1988]. Occasionally it does rain, however, and in any case WVR's are bulky, expensive, and scarce, and tend to complicate field operations. Consequently, many GPS observing sites are not equipped with WVR's and those that are cannot always obtain accurate calibrations. Standard tropospheric models employing surface meteorological data alone have consistently proven to be of little or no value, and can be in error by more than 10 cm.

With the GIPSY software, the tropospheric zenith delay can be estimated at every station as part of the grand solution for orbits and baselines. For stations with accurate WVR calibrations, a tightly constrained constant calibration bias is estimated from the GPS data; typical corrections are 1-3 cm at zenith. When WVR's are not available, we have found that estimating the time-varying tropospheric delays using a random walk stochastic model can be very effective. The stochastic model is tuned so that the tropospheric delay can wander ~4 cm over an 8-hour daily tracking session. From one day to the next, the troposphere zenith delay is reset to be independent of the previous value. With this strategy, daily repeatability is consistently improved for baseline and mapped orbit solutions by up to a factor of 2 compared with estimating a constant residual zenith delay [Lichten and Border, 1987; Tralli et al., 1988]. Most of these results were obtained with data taken in comparatively dry North America. We are currently investigating the performance of stochastic troposphere estimation in considerably moister regions, principally Central America and the Caribbean.

2.4 Simultaneous Group and Phase Delay

Although the pseudorange (group delay) data type is typically two orders of magnitude less precise than phase delay, it provides an unambiguous differential range measurement. If the pseudorange and carrier phase measurements are made simultaneously (to within 1 msec), the pseudorange data can strengthen the carrier phase solution in two ways: (1) by providing a precise estimate of receiver synchronization offsets and (2) by constraining the carrier phase bias estimates. With the meter-level pseudorange obtained from current TI-4100 receivers, baseline accuracy is improved by roughly a factor of 2 over carrier-only solutions [Lichten and Border, 1987; Tralli and Dixon, 1988]. This advantage will increase with the advent of high precision pseudorange receivers. With the inclusion of pseudorange, moreover, carrier phase biases are estimated more accurately, enabling more reliable ambiguity resolution. Covariance studies show that if ambiguous carrier phase is combined with high precision pseudorange (10 cm precision for 6 minute points), the solution will be nearly as accurate as with unambiguous carrier phase [Lichten and Border, 1987]. This may be important for longer baselines where ambiguity resolution is unreliable.

3. Recent Experimental Results

3.1 Daily Repeatability

A crucial measure of system performance is the repeatability of solutions. Repeatability gives a direct measure of system precision and a lower bound on system accuracy. It is useful to compare estimates of system precision based on baseline measurement repeatability with the formal error estimates. If the data contain systematic errors that change slowly between experiments, observed repeatability may be substantially worse than the estimated formal error. On the other hand, if rapidly oscillating multipath error is present in the data—error which will average down faster than random data noise—the observed repeatability may be better than a formal error estimate based on an assumption of random data noise. For baselines which have not been previously measured with VLBI or satellite laser ranging, repeatability tests may be the best way to assess performance.

First we summarize the analysis of data from 18 sites in the western U.S. which were occupied for up to 4 days in June 1986 [Blewitt et al., 1987; Skrumeda et al., 1987]. A solution for both station locations and satellite orbits was performed independently for each of the four days using combined carrier phase and pseudorange data. The residual zenith tropospheric delay was estimated stochastically each day. We were able to resolve 94% of the carrier phase ambiguities in the network, resulting in subcentimeter baseline repeatability in the horizontal plane for all baselines up to 600 km in length. With the ambiguities resolved, repeatability is improved by a factor of 3 over solutions without ambiguity resolution. In the vertical direction, repeatability is about 3 cm.

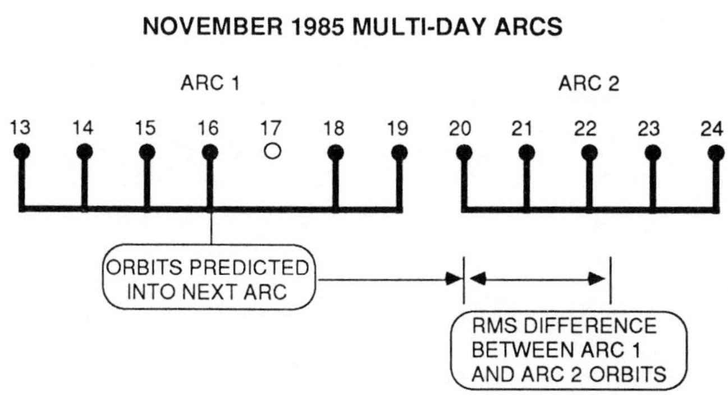

Fig 1. Orbit repeatability for November 1985 was measured by mapping orbit solutions from arc 1 forward and computing the RMS difference over a 6-72 hour intervals in arc 2.

Multi-day orbit determination has proved valuable in improving GPS orbit accuracy and hence the precision of longer baseline measurements [Bertiger and Lichten, 1987]. Well-tracked GPS satellites show sub-meter orbit agreement in comparisons of multi-day solutions using fully independent data sets, even when one solution is mapped forward several days to compare with the other [Lichten and Border, 1987]. The strategy for these orbit comparisons is depicted in Fig 1. Figure 2 shows the effect of different solution strategies on orbit agreement. To examine daily baseline repeatability we employ a strategy in which a single multi-day orbit solution is formed together with separate daily baseline solutions. Again, combined carrier phase and pseudorange data are used and the zenith troposphere is estimated stochastically. The daily repeatability with this strategy is better than 3 cm in all components for baselines 2000 km in length, or better than 1.5 parts in 10^8. Figure 3 shows the daily repeatability of each component for both the single- and multi-day solution strategies.

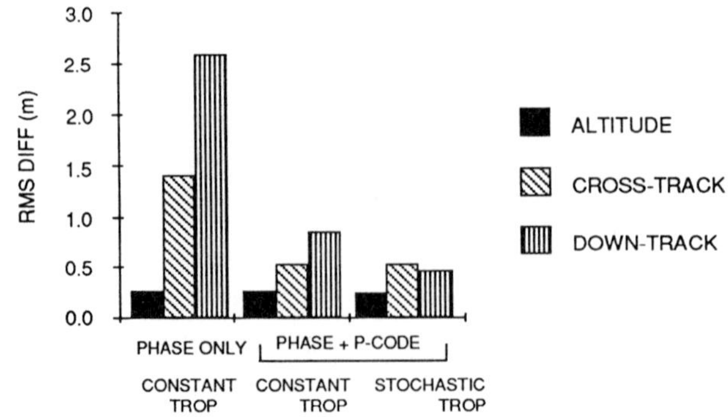

Fig 2. Orbit repeatability of GPS 8 for three estimation strategies. The solution from the first arc was mapped ahead 24 hours; RMS difference with 2nd arc computed over 6-hour interval.

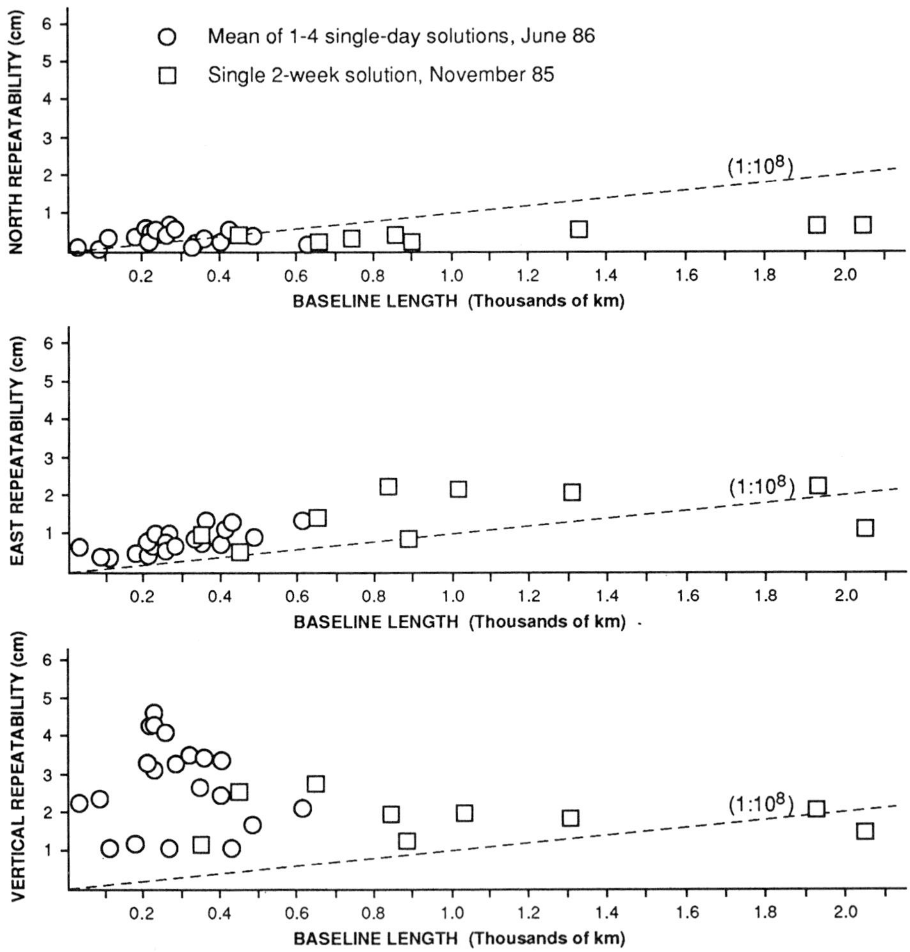

Fig 3. Daily repeatability of baseline measurements ranging from 20 to 2000 km with data from North American experiments in November 1985 and June 1986.

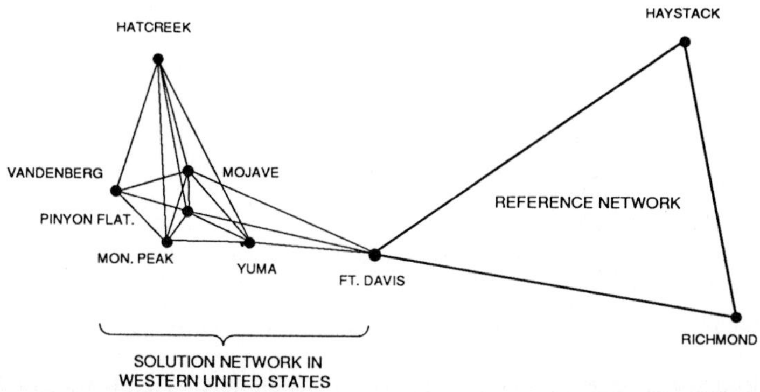

Fig 4. Network configuration of the June 1986 GPS experiment for sites collocated with VLBI. The fiducial network was held fixed at the VLBI-inferred coordinates, and the other station locations were estimated; 94% of the ambiguities were resolved.

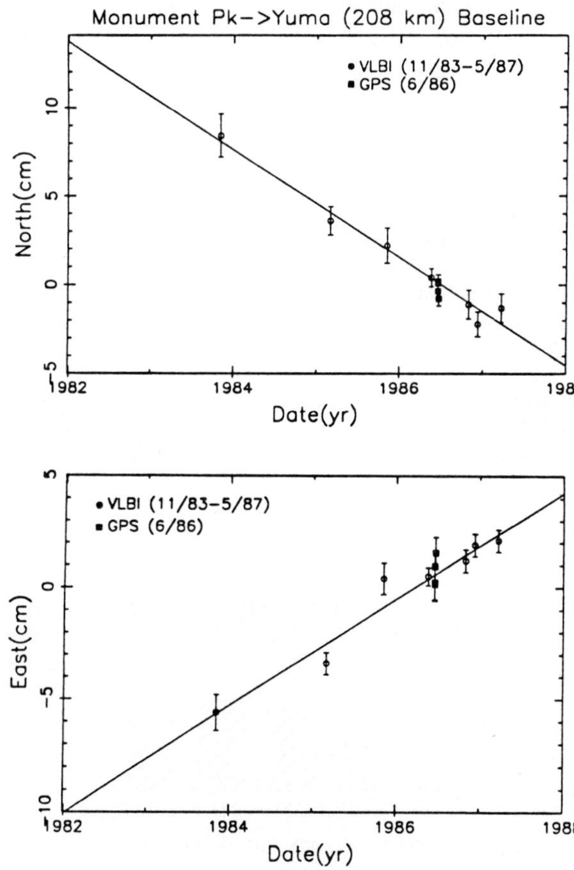

Fig 5. Example of daily GPS baseline solutions from the June 1986 test superimposed on a time history of VLBI solutions for Monument Peak, California, to Yuma, Arizona (208 km).

3.2 Comparison with VLBI

Direct comparisons between GPS and VLBI measurements of the same baselines provide a valid test of absolute system performance. A comparison of GPS solutions with both mobile and fixed VLBI solutions for 13 baselines in the western U.S. [Skrumeda et al., 1987] shows a root-mean square (RMS) difference of less than 2 cm in the horizontal components. The longest of these was the 1086 km baseline between Hatcreek, California and Yuma, Arizona. The network configuration is shown in Figure 4. Taking the mean over all baselines, the RMS difference between GPS and VLBI coordinates is 1.4 cm (east), 1.6 cm (north), 3.6 cm (vertical), and 1.2 cm (length). Without ambiguity resolution, this agreement degrades by about a factor of 2. Figure 5, shows representative examples of GPS baseline solutions along with the time history of VLBI solutions.

For centimeter accuracy on long baselines, multi-day GPS orbit determination is currently required. The comparison of GPS with VLBI is believed to be limited by uncertainties in the fiducial baselines and in the local ties between the VLBI and GPS observing sites. Nevertheless, multi-day solutions for baselines of 2000 km agree with the VLBI solutions to better than 2.5 cm (about 1 part in 10^8) in *all* vector components. Figure 6 shows the GPS-VLBI agreement for two long baselines: Hatcreek, California to Ft. Davis, Texas (1933 km), and Haystack, Massachussets to Richmond, Florida (2046 km). The GPS solutions were obtained with a 2-week arc of data acquired in November of 1985. Note that the Hatcreek-Ft Davis baseline was computed with two different fiducial sets. In the first case the fiducial set is more compact (Mass/Florida/Texas) and the floating end of the baseline, in California, extends far outside the network. In the second case the fiducial set spans the continent (Calif/Mass/Florida) and the floating end of the baseline is essentially interior to the network—a far more favorable geometry. As a result, the GPS-VLBI agreement is about 50% better in the second case and the difference in all components is less than 2 cm.

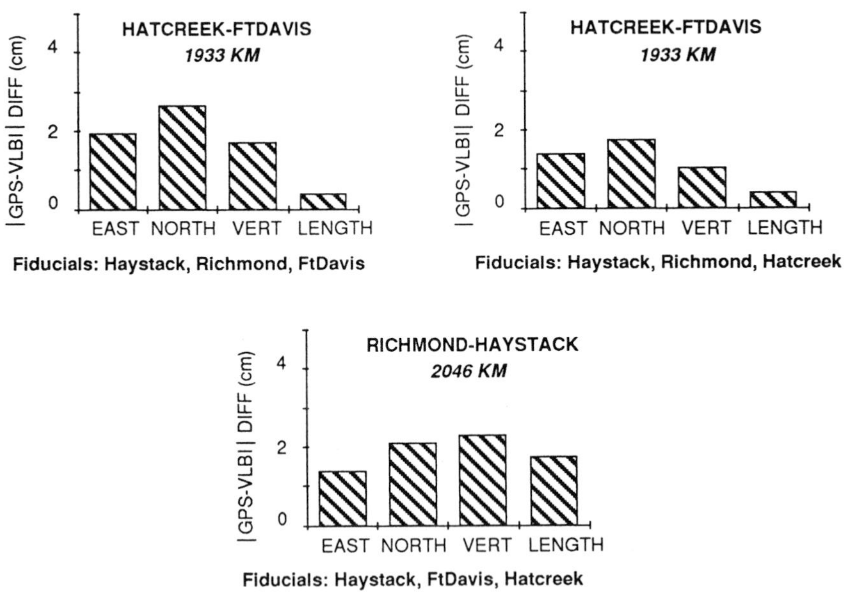

Fig 6. *Results of GPS-VLBI comparisons for two long baselines in Nov 85. The second Hatcreek-Ft Davis solution (upper right) has much stronger fiducial geometry. The east coast Richmond-Haystack baseline has somewhat weaker fiducial and GPS observing geometry.*

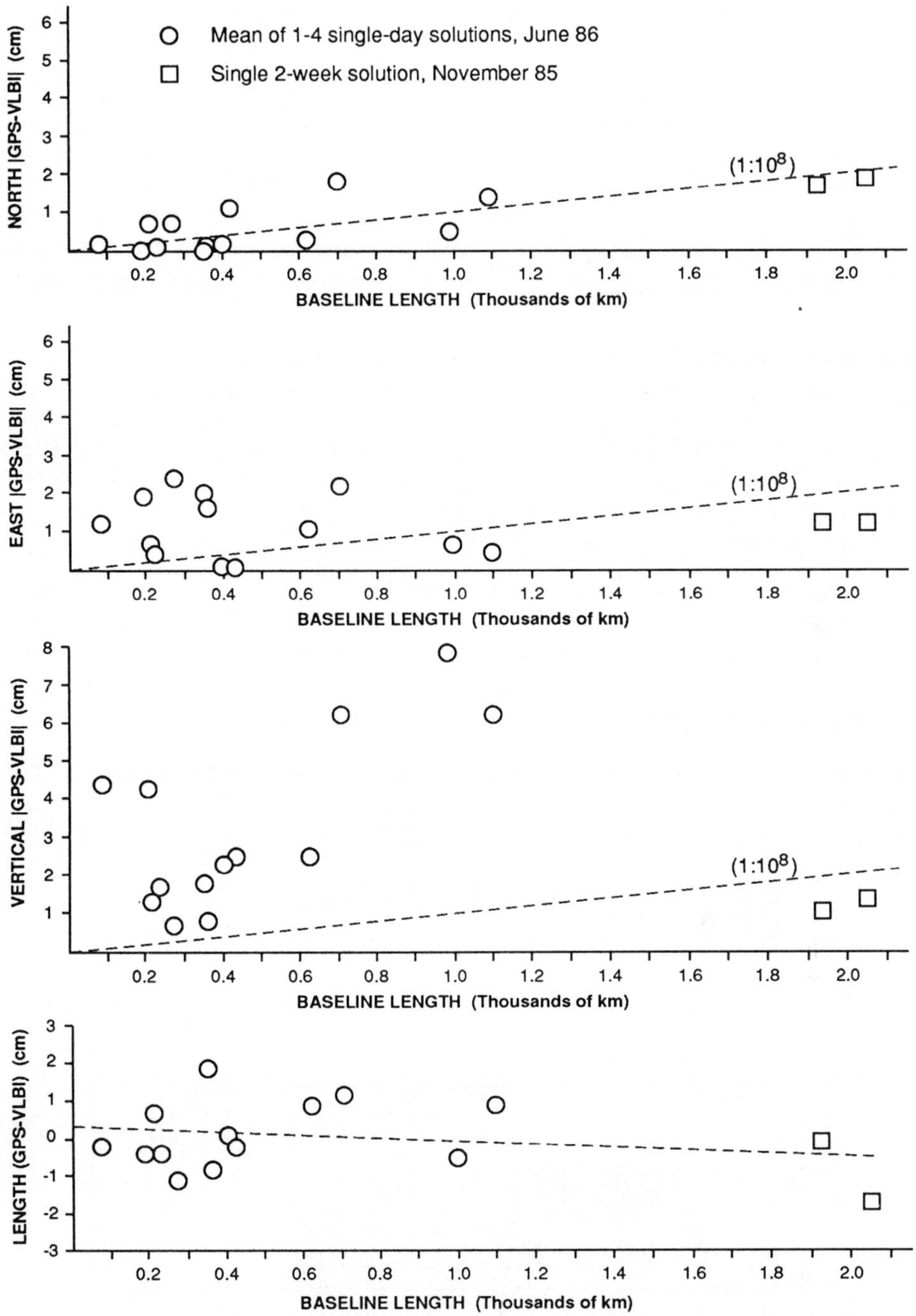

Fig 7. Results of GPS-VLBI comparisons from Nov 85 and Jun 86. The first three plots show the magnitude of the East, North, and Vertical component differences. The last plot shows the observed length difference (GPS-VLBI), indicating no scale offset between GPS and VLBI.

Richmond-Haystack is a virtually north-south baseline at the far eastern end of the continent. It is not particularly well positioned with respect to either the fiducial set (Calif/Texas/Mass) or the GPS constellation, which is optimized for viewing from the southwestern U. S. Consequently, the formal errors are greater for this baseline, which may explain the slightly greater disagreement compared with the Hatcreek-Ft Davis solution. Hatcreek-Ft Davis gives a fairer idea of what will be achieved globally with a full GPS constellation and worldwide tracking. Figure 7 collects the GPS-VLBI comparisons for all of the baselines examined. Note that the final plot (length) gives the actual difference (GPS minus VLBI) rather than the magnitude of the difference and indicates that there is little or no scale difference between the GPS and VLBI reference frames. Note also that both the GPS-VLBI agreement and the GPS daily repeatability (Fig 3) are to a large extent independent of baseline length.

4. Prospects for the Future

4.1 JPL Receiver & Antenna Development

A GPS receiver known as the Rogue has been developed for NASA at JPL expressly for high precision geodesy and precise orbit determination of low earth satellites [Meehan et al., 1987a]. The random measurement error from receiver thermal noise is about 0.5 mm for 1-second carrier phase measurements, and 10-14 cm for 1-second P-code pseudorange. The precision of 2-minute smoothed pseudorange is about 1.5 cm. Often, however, pseudorange data are degraded by multipath signals at the antenna [Meehan et al., 1987b], substantially increasing the effective measurement error. The pseudorange data used in the 1985-86 field experiments, for example, contained multipath with typical amplitude of about 1 meter (in some cases much worse). In conjunction with the Rogue receiver development, considerable effort has therefore gone into the design of antenna and backplane configurations to reduce multipath.

One such configuration, developed at JPL by Don Spitzmesser, reduces long term multipath peaks on pseudorange to less than 5 cm. This configuration uses a drooped cross-dipole antenna to receive the L1 and L2 satellite signals and a conducting backplane consisting of annular choke rings. Two prototype antenna/

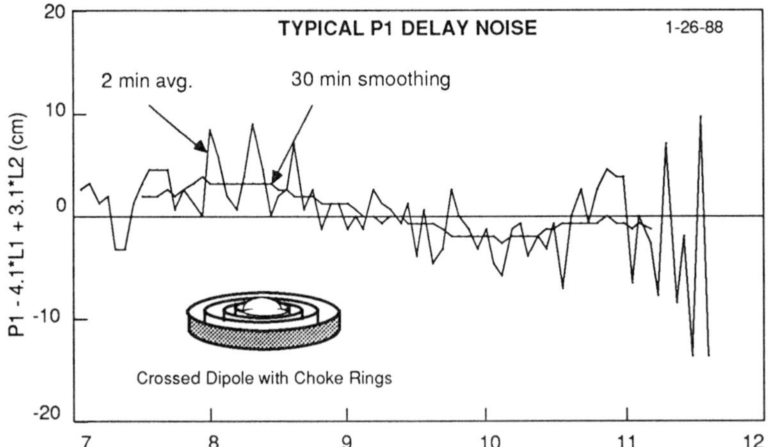

Fig 8. A sample of single-band pseudorange data taken at Owens Valley by the Rogue receiver with a crossed dipole antenna and choke-ring backplane. Geometry and ionosphere variations are removed by a linear combination of L1 and L2 carrier phase.

backplane combinations were used with Rogue receivers at the Mojave and Owens Valley Radio Observatory sites in California during the recent CASA Uno experiment (see below). Figure 8 shows a sample of the pseudorange noise for data taken at Owens Valley. A linear combination of the data, which removes the signatures from changing geometry and ionosphere, is plotted with time for 2-minute and 30-minute smoothed samples. Note that after 30 minutes of smoothing the higher frequency oscillations due to multipath are virtually eliminated, leaving sub-centimeter data noise on top of a low frequency sinusoid having an amplitude of 3 or 4 cm and a period of several hours. The origin of this low frequency signature, which is not believed to be multipath, is now under investigation.

4.2 Global Tracking: The CASA Uno Experiment

The CASA Uno (Central And South America #1) experiment provided three weeks of GPS data for an initial epoch measurement of a geodetic network in Central and South America. Managed by JPL, CASA Uno involved participants from about 30 institutions worldwide. In all, 24 sites in Central and South America were occupied by 16 receivers. An additional 13 receivers occupied 20 sites in the U.S. To provide effective fiducial control for this extended receiver array, a nearly global network of 12 reference receivers was deployed, most collocated with VLBI sites (Figure 9). The non-collocated reference site positions will be estimated simultaneously with the GPS orbit parameters. Covariance studies show that the consequences of not placing all tracking sites at well-known reference points are insignificant because of the abundance and distribution of the GPS data [Freymueller and Golombek, 1987]. In all, about 600 station-days of data are now being reduced.

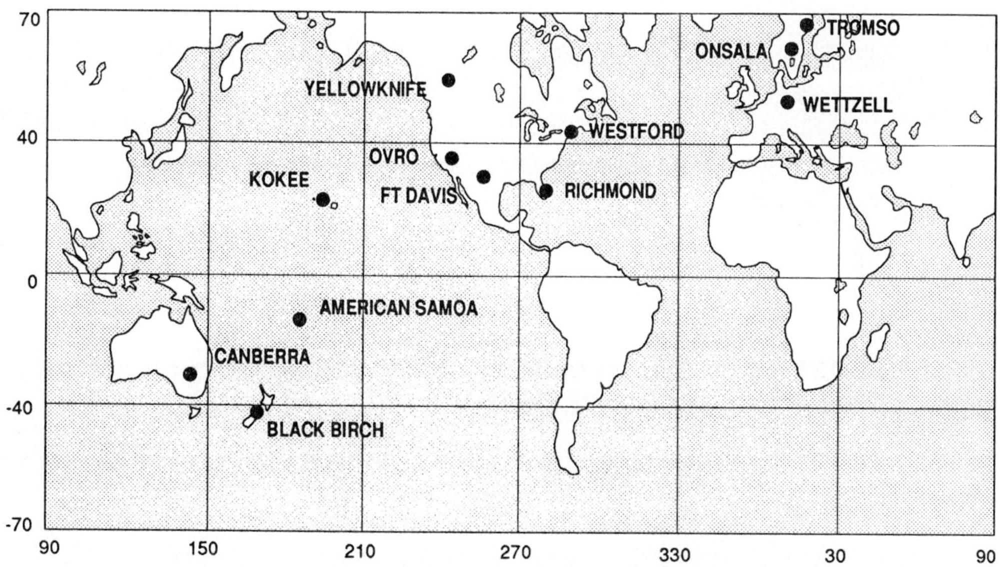

Fig 9. Near-global network of 12 reference receivers used during the 3-week CASA Uno experiment in January, 1988.

The CASA Uno experiment provides a unique opportunity to investigate the benefits of global tracking using real data. In previous experiments, fiducial networks were limited to about 3000 km in extent. Results from CASA Uno will provide a valuable benchmark to calibrate covariance studies, so that we can more accurately predict the performance of future global tracking systems. Intensive investigations are now being undertaken to determine the best estimation strategies for global networks. In addition to carrying out the usual orbit and baseline solutions, we will use the CASA Uno data in preliminary experiments to determine how accurately earth orientation and the location of the geocenter can be recovered with GPS data.

5. Conclusions

In areas of good fiducial control, GPS-based geodetic techniques can now routinely achieve centimeter-level accuracy in the horizontal components for baselines less than 1000 km in length. Accuracies of about 1 part in 10^8 have been demonstrated in all components for baselines of 2000 km. This performance requires careful modeling of the observables, stochastic estimation techniques, and either resolution of carrier phase ambiguities or multi-day arc orbit determination. (Early results indicate that there is little additional improvement when multi-day arcs and ambiguity resolution are combined.) Accuracies of a part in 10^8 or better should become routine worldwide in the 1990's as GPS data systems and solution techniques are improved. Important developments soon to appear include receivers capable of very precise pseudorange measurement, antenna designs that effectively eliminate local multipath, continuously operating global tracking networks, a 21-24 satellite GPS constellation, experiment design to enable reliable ambiguity resolution, orbiting GPS receivers to supplement ground tracking sites, and refined estimation techniques.

Acknowledgements

The work described in this paper was carried out at the Jet Propulsion Laboratory, California Institute of Technology, under contract with the National Aeronautics and Space Administration. We thank the many people who have made key contributions, including William Melbourne, Catherine Thornton, Larry Young, Peter Kroger, John Davidson, Tom Meehan, Don Spitzmesser, Ojars Sovers, Lisa Skrumeda, Ruth Neilan, Jim Border, and the late Gerald J. Bierman.

References

Bertiger, W.I. and S.M. Lichten (1987): Demonstration of 5 to 20 parts per billion repeatability for a continental baseline estimated with multi-day GPS orbits, EOS Trans. Am. Geo. U., Vol. 68, No. 44, p. 1238

Bierman, G.J. (1977): Factorization Methods for Discrete Sequential Estimation, Academic Press, Orlando, Florida

Blewitt, G. et al. (1987): Improved GPS network solutions using bias-optimizing techniques, EOS Trans. Am. Geo. U., Vol. 68, No. 44, p. 1236

Blewitt, G. (1988): Successful GPS carrier phase ambiguity resolution for baselines up to 2000 km in length, EOS Trans. Am. Geo. U., Vol. 69, No. 16, p. 325

Dong, D. and Y. Bock (1988): GPS network analysis: Ambiguity resolution, EOS Trans. Am. Geo. U., Vol. 69, No. 16, p. 325

Counselman III, C.C. (1987): Resolving carrier phase ambiguity in GPS orbit determination EOS Trans. Am. Geo. U., Vol. 68, No. 44, p. 1238

Tralli D.M. and T.H. Dixon (1988): A few parts in 10^8 geodetic baseline repeatability in the Gulf of California using the Global Positioning System, Geophysical Research Letters, Vol. 15, pp. 353-356

Elgered G. et al. (1987): On the Weather Dependence of Baseline-Lengths Estimated by Very-Long-Baseline Interferometry, EOS Trans. Am. Geo. U., Vol. 68, No. 44, p. 1239

Freymueller J.T. and M.P. Golombek (1987): Effect of Fiducial Network Geometry on GPS Baseline Accuracy in South America, EOS Trans. Am. Geo. U., Vol. 68, No. 44, p. 1237

Lichten, S.M. and J.S. Border (1987): Strategies for high-precision global positioning system orbit determination, Journal of Geophysical Research, Vol. 92, No. B12, pp. 12751-12762

Meehan T.K. et al. (1987a): ROGUE: A new high accuracy, digital GPS receiver, Proceedings of the IUGG Conference, Vancouver, August 1987

Meehan T.K. et al. (1987b): GPS multipath reduction using absorbing backplanes, EOS Trans. Am. Geo. U., Vol. 68, No. 44, p. 1238

Melbourne, W.M. (1985): The case for ranging in GPS based systems, Proceedings of the First Symposium on Precise Positioning with the Global Positioning System, Positioning with GPS—1985, Ed. C.C. Goad, Rockville, Maryland, pub. U.S. Department of Commerce, NOAA

Skrumeda, L.S. et al. (1987): Baseline results from the June 1986 Southern California and Caribbean GPS Experiments, EOS Trans. Am. Geo. U., Vol. 68, No. 44, p. 1236

Thornton, C.L., and G.J.Bierman (1980): UDU^T covariance factorization for Kalman filtering, Control and Dynamic Systems, Advances in Theory and Application, Ed. C.T. Leondes, Academic Press, Vol. 16, pp. 177-248

Tralli, D.M. et al. (1988), The effect of wet tropospheric path delays on estimation of geodetic baselines in the Gulf of California using the Global Positioning System, accepted for publication in Journal of Geophysical Research

Wubbena, G. (1985): Software developments for geodetic positioning with GPS using TI-4100 code and carrier measurements, Proceedings of the First Symposium on Precise Positioning with the Global Positioning System, Positioning with GPS—1985, Ed. C.C. Goad, Rockville, Maryland, pub. U.S. Department of Commerce, NOAA

Vehicle Guidance by Computer Vision

Ernst-Dieter Dickmanns
Fachb. Luft- und Raumfahrttechnik, Universität der Bundeswehr, München

Summary

The 4D-approach for machine vision using integral spatio-temporal models is applied to road vehicle guidance. Both for road parameter estimation in a look ahead range and for vehicle state estimation relative to the road, dynamical models are given. A system architecture for active machine vision in road vehicle guidance is discussed. Based on the recursive, feature-based state estimation, numerically efficient situation recognition and process control has been achieved. With 0.1 second cycle time, speeds up to 60 mph have been driven autonomously over distances up to 24 km on a standard highway with the 5-ton test vehicle VaMoRs.

Introduction

Automatic control of vehicles by computer vision is attracting considerable attention since the "Autonomous Land Vehicle (ALV)" has been selected as one of the demonstrator projects in the US-DARPA Program on "Strategic Computing" in 1982 <DARPA 82> and since in 1986 the European car manufacturing companies in agreement with their governments started launching the "Prometheus"-project, a program for future European traffic with highest efficiency and unprecedented safety.

Early research in the field originated from ideas on a Mars rover around 1970 <Gennery 77>. Pioneering work has been performed by H. Moravec in connection with the Stanford cart and later at CMU with small outdoor vehicles <Moravec 83> (pure indoor-carts are not considered here, see e.g. <Giralt 84>). From Japan a project is reported in <Tsugawa e.a. 79>. At the authors institution the problem has been picked up, on a simulation level first, in 1977.

Three truck-sized wheeled vehicles with the capability of fully autonomous locomotion have been in operation around the world in 1987: The Martin Marietta "ALV", an 8-ton, 4-axle specially built vehicle <Klass 85>, and two modified vans: the 8-ton "NavLab" at CMU <Thorpe 86> (both funded by DARPA in the program mentioned above) and the 5-ton "VaMoRs" at our laboratory.

In the US-projects emphasizing "artificial-intelligence (AI)"-approaches and looking for cross-country capabilities, several seconds cycle time for image evaluation and vehicle control have been achieved <Turk 86, Thorpe 86>. In our program, confined to guidance along well built roads as the initial goal, a control cycle time of at most 0.1 second has been deliberately adopted as a side constraint in order to focus attention on the real-time aspects when looking for image sequence processing hardware and methods. For this reason, recursive estimation techniques well known from modern control theory have been selected as the basis for image sequence processing around 1980.

The method has been steadily improved during the period of application to four completely different problem areas in simulation and field tests since then: 1. The balance of an (unstable) inverted pendulum on an electrocart <Meissner 82>, 2. road vehicle guidance <Meissner 82; Dickmanns, Zapp 85; 86; 87>, 3. planar vehicle docking using reaction jet control <Wünsche 86; 87> and 4. simulated landing of an aircraft in six degrees of freedom <Eberl 87>, all with real-time image sequence processing hardware in the loop.

In 1987 a general scheme for feature based dynamic scene analysis emerged <Dickmanns 87; Wünsche 87>. This 4D-approach using integrated spatio-temporal world models allows efficient image sequence processing algorithms for the on-line interpretation of dynamic scenes. Only moderate computer power is necessary to handle well structured scenes with dynamics in a range manageable by a human operator. This general approach to vision is described in the next section.

4D-approach based on integral spatio-temporal world models

The dynamical models for motion of and around the center of gravity (c.g.) are combined with shape representations of the 3D-object emphasizing the position of (contour element-) features relative to the c.g.. Feature groupings (aggregations) in the perspective map are used to recognize objects in a scene. However, this is done in the usual nonunique inverse way only in the initialisation phase, if at all; if the type of scene is known a priori, forward perspective mapping of generic models and adjustment of model- and relative position-parameters is done until the measured image is matched or until the model pool is exhausted. Once the real-time phase has been initiated, only the model based approach is used. Up to now, features used have been limited to linear contour elements (edge elements) and corners. A theory for efficient curvilinear shape representation has been developed and is presently under investigation.

Figure 1 shows a juxtaposition of the conventional method in image sequence processing (top), where a considerable amount of computation is done in image coordinates and in inter-image comparisons, and the integrated 4D-approach, where a 4D-representation of a model world is being maintained in the interpretation process and servo controlled by measurements in the last image of the sequence only.

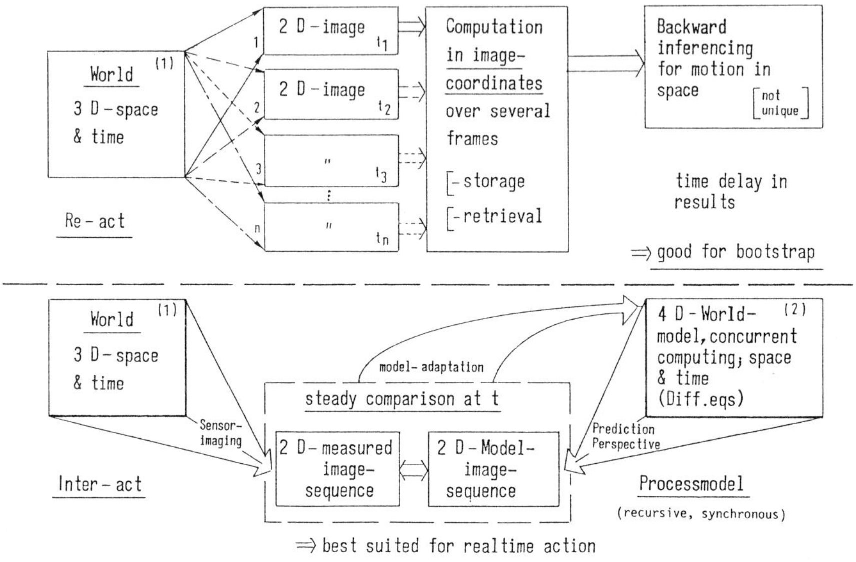

Figure 1: Two basically different ways in image sequence processing

This model-based prediction-error-feedback of feature positions has several advantages: no differencing between noise corrupted images of a sequence is required for obtaining velocity components (like in optical flow computation); motion interpretation does not have to be done in image coordinates first and to be reinterpreted in space afterwards, instead, all interpretation is immediately performed in the 4D-space-time-continuum.

The recursive estimation based on the smoothing numerical process of integration proceeds as follows: An estimate of the complete state vector $\hat{\mathbf{x}}$ describing the process to be interpreted is assumed to be given; this hypothesis generation in the initialization phase is a hard task in partially unknown environments. Knowing the n-vector $\hat{\mathbf{x}}$ and a dynamical model of the process in the discrete form for sampled measurement and control, a state prediction $\mathbf{x}^*$ for the next measurement can be made

$$\mathbf{x}^*[(k+1)T] = \mathbf{A}(T)\hat{\mathbf{x}}[kT] + \mathbf{B}\mathbf{u}[kT] \ . \tag{1}$$

$\mathbf{A}$ is a n*n state transition matrix over one sampling period T and $\mathbf{B}$ is the n*q control effectiveness matrix for the q components of the control vector $\mathbf{u}$.

If the shape of the objects observed and the relative geometry is known, may be even described in terms of state components x_i^*, the predicted position of features $\mathbf{y}^*$ in the next image can be derived by forward application of the laws of perspective projection exploiting a model of the actual camera used for measurement. In general, this will be a nonlinear relationship containing measurement parameters $\mathbf{p}$

$$\mathbf{y}^* = \mathbf{f}(\mathbf{x}^*,\mathbf{p}) \ . \tag{2}$$

Both the process modeled in (1) and the measurements will be corrupted by noise, designated $\mathbf{v}(kT)$ for (1) and $\mathbf{w}(kT)$ for the measurements. The problem is to determine best estimates for the state

$$\mathbf{x} = \mathbf{x}^* + \delta\mathbf{x} \tag{3}$$

given the measured quantities

$$\mathbf{y} = \mathbf{f}(\mathbf{x},\mathbf{p}) + \mathbf{w} \ . \tag{4}$$

Assuming that the influence of the noise is small and its average is zero, a linearized relationship between $\mathbf{y}$, $\mathbf{y}^*$ and $\mathbf{x}$, $\mathbf{x}^*$ may be a good approximation to reality

$$\delta\mathbf{y} = \mathbf{y} - \mathbf{y}^* = \mathbf{f}(\mathbf{x}^* + \delta\mathbf{x},\mathbf{p}) + \mathbf{w} - \mathbf{f}(\mathbf{x}^*,\mathbf{p})$$

$$= \frac{\partial \mathbf{f}(\mathbf{x}^*,\mathbf{p})}{\partial \mathbf{x}^*}\,\delta\mathbf{x} + \mathbf{w} = \mathbf{C}\delta\mathbf{x} + \mathbf{w} \ . \tag{5}$$

Note that the Jacobian matrix $\mathbf{C}$ contains the m*n partial derivatives of the m measurement quantities $\mathbf{y}^*$ as predicted, relative to the state variables $\mathbf{x}^*$ in

3D-space including the spatial velocity components. This rich information is used for doing a recursive least squares state estimation using well known techniques. If the covariance matrices of the noise processes **v** and **w** are known, Kalman filter techniques or derivatives thereof may be used <Maybeck 79>. The new best estimate $\hat{\mathbf{x}}$ then becomes

$$\hat{\mathbf{x}} = \mathbf{x}^* + \mathbf{K}(\mathbf{y} - \mathbf{y}^*) \ , \qquad (6)$$

where the gain matrix **K** for innovation is determined depending on the method used; sequential formulations well suited for time varying measurement vector lengths, e.g. due to occlusion, are available <Wünsche 87>.

<u>The road vehicle guidance task</u>

A road vehicle with autonomous locomotion capability up to higher velocities has to perform the following recognition and control tasks: 1. Determine the road parameters in a look ahead range so that a safe control to a stop is always possible and speed may be controlled for efficient and safe mission performance, 2. determine own complete motion state vector relative to the road, and 3. recognize obstacles of any kind, static or moving, relative to the egomotion and to the road, react properly.

For achieving both wide angle coverage in the near range and good resolution farther away for early obstacle recognition, a combination of two cameras fixed relative to each other seems to be favourable and has been selected. Given very limited image processing power, two black and white CCD-TV-cameras with a resolution of 244 lines have been selected, with different focal lengths. In order to further increase visual flexibility, the two cameras are mounted closely together in combination with rate-gyros on a two-axis, fast pan and tilt platform ZP which is controlled by a microprocessor ZPP integrated into the image sequence processing system BVV (see fig 2). This allows both fast inertial stabilization of the viewing direction and optical gaze anchoring on features or feature combinations (fixation) via image sequence processing. Active gaze control is considered to be an essential aspect of efficient real-time vision systems since it allows flexible control and allocation of high resolution sensing capability combined with reduction of motion blur during fixation periods.

Vision system architecture for feature based dynamic scene analysis

In the upper right of figure 2 a typical road scene is depicted; only the wide angle image will be treated in the sequel. Those areas of utmost importance to the task are covered by rectangular "windows" marked 1 through 6. The video signal is digitized into 256 pel per line and given onto a video bus in the image sequence processing system BVV developed by our partner-institute at UniBwM <Graefe 84>. A number of parallel microprocessors (PP_i) do have access to the video bus through a special interface allowing them to fetch up to 4K pel data belonging to a rectangular window of any shape, size, position and sparseness; these parameters may be adjusted by software control during run-time. PP's 1 to 6, each independently programmable, are ordered to acquire image data in windows 1 to 6. Each PP searches for linear contour elements under a preferred slope in its window. The slope is prescribed by the higher levels of the system depending on the interpretation context. The positions of the contour elements found by ternary correlation are passed to a general purpose processor (GPP) or the central processing system which combine them in order to recognize objects or the egomotion state.

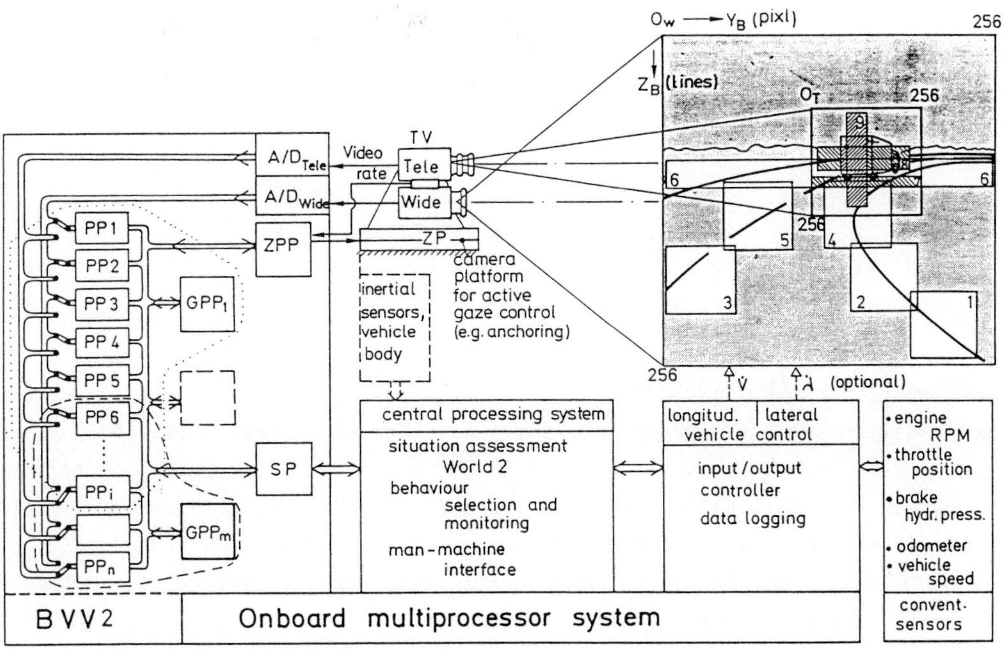

Figure 2: System architecture for active machine vision in road vehicle guidance

PP's may be ordered to track "their" features, i.e. to keep their position centered in the window. Windows of several PP may freely overlap; they may be repositioned or newly initialized on command of the interpretation process when this one prefers to receive data from different image regions or from the same region, but processed by a different algorithm.

The interpretation process may be spread out over the multiprocessor system: Single GPP's are usually allocated to the recognition and parameter- or state-estimation of one object, e.g. GPP1 to road recognition. Some of the data feeding GPP1 may, however, also be sent to the central processing system in the lower center of figure 2; for example, the data from window 1 may there be taken for driving the dynamical model used for estimating the complete vehicle state relative to the road. Estimated state variables or parameters like road curvature (in GPP1) may be exchanged between the units of the interpretation process by message passing through the communication processor SP. The central processing system is provided all the results from the GPP's in order to arrive at a situation assessment. This then is the basis for parameter adjustment in the behavioural mode running, or for mode switching.

The actual control actuation depends on the estimated and directly measured (like speed V) state variables and is implemented by an input/output controller.

In the interpretation process, exploiting the predicted state and feature positions $\mathbf{y}^*$, the incoming measurement data are screened, outliers are removed and the image sequence analysis process by the PP's is supervised and adjusted to the present interpretation status.

<u>Dynamical model for road recognition</u>

Well built roads obey (piecewise) the linear curvature law ("clotoide")

$$C = C_0 + C_1 l, \quad \text{with } C_1 = dc/dl = \text{const}, \tag{7}$$

where $c = 1/R$ is the local curvature at runlength l. Driving on the road at speed V yields the temporal curvature change

$$\dot{C} = dc/dl \cdot dl/dt = C_1 V$$

$$\dot{C}_1 = \begin{cases} 0 \text{ on each clothoide arc,} \\ \delta(l-\lambda_i)\text{, a Dirac impulse at each clothoide boundary } \lambda_i; \\ \text{for practical purposes } C_1 \text{ is assumed to be a finite driving} \\ \text{term to be determined from visual measurements.} \end{cases} \tag{8}$$

Integrating eq. (7) twice with respect to l yields the lateral offset y_c due to curvature relative to the road tangent at distance L_i backwards, for small relative heading changes ($\sin \chi \approx \chi$)

$$y_{C_i} = C_o\, L_i^2/2 + C_1\, L_i^3/6 \ . \tag{9}$$

From figure 3 there follows

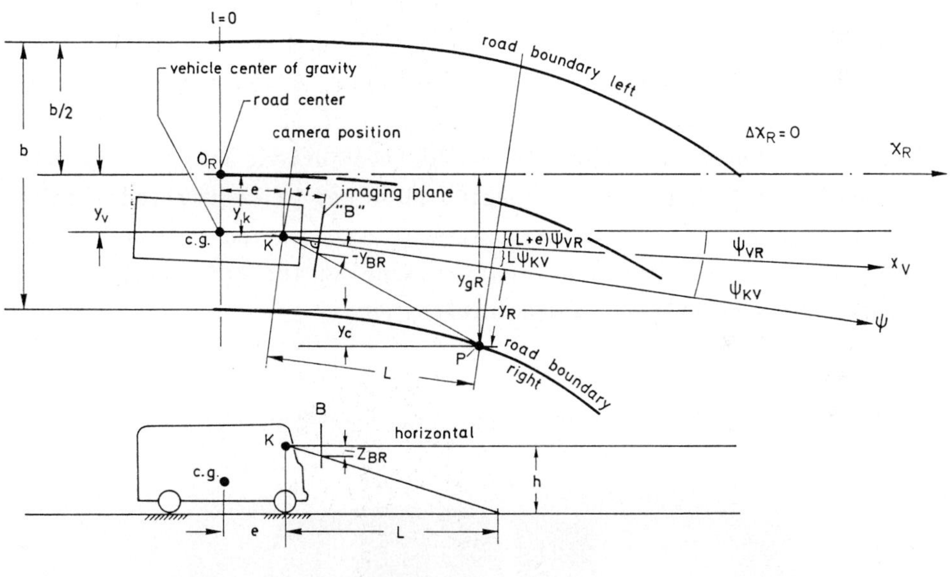

Figure 3: Vehicle/camera/road-geometry

$$y_{gR} \approx y_V + (L+e)\psi_{VR} + L\psi_{KV} + y_R \approx b/2 + y_c \ , \tag{10}$$

where y_R is the indirectly measured quantity in the image (through y_{BR}). With the focal length f there follows from perspective projection

$$y_{BR}/f = y_R/L \ ; \quad z_{BR}/f = h/L \ . \tag{11}$$

Together with eqs. (9,10) one obtains from figure 3

$$y_{BR} = f[b/(2L) - y_V/L - (1+e/L)\psi_{VR} - \psi_{KV} + C_o a_o + C_1 a_1] \tag{12}$$

where $a_o = e^2/(2L) + e + L/2$ and $a_1 = [e^3/L + 3e(e+L) + L^2]/6$

which are the perspective mapping eqs. (2) governing the vision process. The vehicle position on the road y_V and the relative heading ψ_{VR} are constrained by the dynamics of the vehicle, modeled by another set of differential equations as given in <Dickmanns, Zapp 85>.

VaMoRs experimental results

Our autonomous test vehicle VaMoRs, a 5-ton van (see fig. 4), is providing results with different measurement and control schemes from the class described above since the last quarter of 1986. In 1987 many hours of testing on a construction site for a new stretch of Autobahn near Dingolfing (north/east of Munich) have been performed without and with lane markings and under different weather conditions including light rain. Speed has been increased steadily up to the maximum speed of the vehicle of 60 mph. In an endurance test, the total length available (24 km) has been driven autonomously without human intervention from the start at rest. Speed may be controlled according to road curvature in order to limit lateral acceleration to a preset limit. Lane changes, also from the acceleration lane at the entrance, have been demonstrated. More details are given in <Zapp 88>.

Figure 4: VaMoRs test vehicle of UniBwM

Conclusions

Now classical methods for state estimation combined with feature carrying shape representations and the laws of perspective projection do provide a good basis for dynamic scene analysis. It seems favourable to restrict AI-methods to higher levels where they can make use of the powerful state variable representations provided by the 4D-estimation process and where they can control both the measurement process and the use of models and methods.

Literature

<DARPA 82> Strategic Computing. New Generation Computing Technology: A Strategic Plan for its Development and Application to Critical Problems in Defense. DARPA AD-A141982.

<Dickmanns 85> E.D. Dickmanns: 2D-Object recognition and representation using normalized curvature functions. In M.H. Hamza (ed): Proc. IASTED Int. Symp. on Robotics Automation '85, Acta Press, 1985, pp 9-13.

<Dickmanns 87a> E.D. Dickmanns: 4D-dynamic scene analysis with integral temporal models. 4-th International Symposium on Robotics Research, Santa Cruz, Aug. 1987 (to appear in MIT Press, R. Bolles (ed)).

<Dickmanns 87b> E.D. Dickmanns: Object Recognition and Real-Time Relative State Estimation Under Egomotion. In: A.K. Jain (ed.) "Real-Time Object Measurement and Classification" Springer-Verlag, Berlin, 1988, pp 41-56.

<Dickmanns 87c> E.D. Dickmanns: 4D-Szenenanalyse mit integralen raum-/zeitlichen Modellen. In E. Paulus (ed): Mustererkennung 1987, 9. DAGM-Symposium, Proc., Informatik Fachberichte 149, Springer-Verlag 1987.

<Dickmanns, Zapp 85> E.D. Dickmanns, A. Zapp: Guiding Land Vehicle along Roadways by Computer Vision. In: Proc. AFCET, Congres Automatique 1985, The Tools for Tomorrow, Toulouse, Oct. 1985, pp 233-244.

<Dickmanns, Zapp 86> E.D. Dickmanns, A. Zapp: A Curvature-based Scheme for Improving Road Vehicle Guidance by Computer Vision. SPIE Conf. 727 on "Mobile Robots", Cambridge, Mass., USA, Oct. 1986.

<Dickmanns, Zapp 87> E.D. Dickmanns, A. Zapp: Autonomous High Speed Road Vehicle Guidance by Computer Vision. Preprint, Vol. 4, IFAC-Congress 1987, Munich, pp 232-237.

<Dickmanns, Wünsche 86> E.D. Dickmanns, H.J. Wünsche: Satellite Rendezvous Maneuvers by Means of Computer Vision. Jahrestagung der Deutschen Gesellschaft für Luft- und Raumfahrt e.V., München, Okt. 1986. In: Jahrbuch 1986 Bd. 1 der DGLR, Bonn, pp 251-259.

<Eberl 87> G. Eberl: Automatischer Landeanflug durch Rechnersehen. Diss. UniBw München, LRT, Juli 1987.

<Gennery 77> D.B. Gennery: A stereo vision system for autonomous vehicle. Proc. 5th IJCAI, Cambridge, Mass., USA.

<Giralt 84> G. Giralt: Mobile Robots. In Brady et.al. (eds): Robotics and Artificial Intelligence. Springer-Verlag, 1984, pp 365-393.

<Graefe 84> V. Graefe: Two Multi-Processor systems for Low-Level Real-Time Vision. In J.M. Brady, L.A. Gerhardt and H.F. Davidson (eds): Robotics and Artificial Intelligence, Springer, 1984, pp 301-308.

<Klass 85> P.J. Klass: DARPA Envisions New Generation of Machine Intelligence Technology. Aviation Week & Space Techn. April 1988, pp 46.

<Maybeck 79> P.S. Maybeck: Stochastic Models, Estimation and Control. Vol 1, Academic Press, 1979.

<Meissner 82> H.G. Meissner: Steuerung dynamischer Systeme aufgrund bildhafter Information. Diss. HSBw München, LRT, Juli 1982.

<Meissner 83> H.G. Meissner, E.D. Dickmanns: Control of an Unstable Plant by Computer Vision. In T.S. Huang (ed): Image Sequence Processing and Dynamic Scene Analysis. Springer-Verlag, 1983.

<Moravec 83> H.P. Moravec: The Stanford Cart and the CMU Rover. Proc. IEEE, Vol 71, No 7, July 1983, pp 872-884.

<Mysliwetz, Dickmanns 88> B. Mysliwetz, E.D. Dickmanns: Ein verteiltes System zur Echtzeitinterpretation von Straßenszenen für die autonome Fahrzeugführung. Fachtagung Prozeßrechensysteme '88, Stuttgart, März 1988.

<Thorpe 86> C.E. Thorpe, T. Kanade: Vision and Navigation for the CMU NavLab. SPIE Conf. 727 on Mobile Robots, Cambridge, Mass., Oct. 1986.

<Tsugawa 79> S. Tsugawa et.al.: An Automobile with Artificial Intelligence. Proc. 6th IJCAI, Tokyo, 1979, pp 893-895.

<Tsugawa 84> S. Tsugawa, T. Hirose, T. Yatabe: An Intelligent vehicle with obstacle detection and navigation functions. Proc. IECON'84, Vol 1, Tokyo 1984, pp 303-308.

<Turk 86> M. Turk: Colour road segmentation and video obstacle detection. SPIE Conf. 727 on Mobile Robots, Cambridge, Mass., Oct. 1986.

<Wünsche 86> H.J. Wünsche: Detection and Control of Mobile Robot Motion by Real-Time Computer Vision. SPIE Conf. 727 on Mobile Robots, Cambridge, Mass., Oct. 1986, pp 100-109.

<Wünsche 87> H.J. Wünsche: Erfassung und Steuerung von Bewegungen durch Rechnersehen. Diss. UniBw München, LRT, Aug. 1987.

<Zapp 85> A. Zapp: Automatische Fahrzeugführung mit Sichtrückkopplung; Erste Fahrversuche mit einer Fernsehkamera als Echtbauteil im Simulationskreis. UniBw München, LRT/WE 13/ FB/85-1.

<Zapp 88> A. Zapp: Automatische Straßenfahrzeugführung durch Rechnersehen. Diss. UniBw München, LRT (to appear 1988).

Vision and Navigation for the Carnegie Mellon Navlab*

Charles Thorpe, Martial Hebert, Takeo Kanade, Steven Shafer
Dept. of Computer Science, Carnegie Mellon University, Pittsburgh, USA

1. Introduction

Robotics is where Artificial Intelligence meets the real world. AI deals with symbols, rules, and abstractions, reasoning about concepts and relationships. The real world, in contrast, is tangible, full of exceptions to the rules, and often stubbornly difficult to reduce to logical expressions. Robots must span that gap. They live in the real world, and must sense, move, and manipulate real objects. Yet to be intelligent, they must also reason symbolically. The gap is especially pronounced in the case of outdoor mobile robots. The outdoors is constantly changing, due to wind in trees, changing sun positions, even due to a robot's own tracks from previous runs. And mobility means that a robot is always encountering new and unexpected events. So static models or preloaded maps are inadequate to represent the robot's world.

The tools a robot uses to bridge the chasm between the external world and its internal representation include sensors, image understanding to interpret sensed data, geometrical reasoning, and a concept of time and of the vehicle's motion over time. We are studying those issues by building a mobile robot, the Carnegie Mellon Navlab, and giving it methods of understanding the world. The Navlab has perception routines for understanding color video images and for interpreting range data. CODGER, our *whiteboard*, proposes a new paradigm for building intelligent robot systems. The CODGER tools, developed for the Navlab and its smaller cousin the Terregator, handle much of the modeling of time and geometry, and provide for synchronization of multiple processes. Our architecture coordinates control and information flow between the high-level symbolic processes running on general purpose computers, and the lower-level control running on dedicated real-time hardware. The system built from these tools is now capable of driving the Navlab along narrow asphalt paths near campus while avoiding trees and pausing for joggers that get in its way.

This report describes the Navlab [5] and the software we have built over the past year: color vision, for finding and following roads [6]; 3-D perception, for obstacle avoidance [3]; and the CODGER whiteboard [4].

2. Navlab: Navigation Laboratory

The Navigation Laboratory, Navlab, is a self-contained laboratory for navigational vision system research (see figures 1 and 2). The motivation for building the Navlab came from our earlier experience with the Terregator, a six-wheeled vehicle teleoperated from a host computer through a radio link. The Terregator had been a reliable workhorse for small-scale experiments, such as the Campus Sidewalk navigation system [2]. However, we have outgrown its capabilities. As we began to experiment with sensor fusion, the Terregator ran out of space and power for multiple sensors. When we wanted to expand our test areas, communications to a remote computer in the lab became more difficult. And as the experiments became more sophisticated, we found it more productive for the experimenters to test or

Figure 1: The Navlab

debug new programs near or in the vehicle, instead in a remotely located laboratory. All these factors culminated in the design and construction of the Navlab [5].

Navlab is based on a commercial van chassis, with hydraulic drive and electric steering. Computers can steer and drive the van by electric and hydraulic servos, or a human driver can take control to drive to a test site or to override the computer. The Navlab has room for researchers and computers on board, and has enough power and space for all our existing and planned sensors. This gets the researchers close to the experiments, and eliminates the need for video and digital communications with remote computers.

Features of the Navlab include:

- **Onboard computers:** We have five computer racks, one for low-level controllers and power smoothing, one for video distribution, VCRs, communications and miscellaneous equipment, two racks for general-purpose processors (currently Sun workstations), and one for a Warp processor.
- **Onboard researchers:** There is always a safety driver in the driver's seat. There is room for four researchers in the back, with a terminal or workstation for each. An overhead shelf holds video monitors and additional terminals. The researchers can monitor both their programs and the vehicle's motion.
- **Onboard power:** The Navlab carries two 5500-W generators, plus power conditioning and battery backup for critical components.

- **Onboard sensors:** Above the cab is a pan mount carrying our laser scanner and a mounting rail for a color TV camera. There will eventually be a separate pan/tilt mount for stereo cameras.
- **Evolving controller:** The first computer controller for the Navlab is adequate for our current needs. It steers the van along circular arcs, and has commands to set speed and acceleration, and to ask for the current dead reckoned position estimate. The controller will evolve to do smoother motion control, and to interface with an inertial guidance system possibly even with GPS satellite navigation. It will also eventually watch vital signs such as computer temperature and vehicle hydraulic pressure.

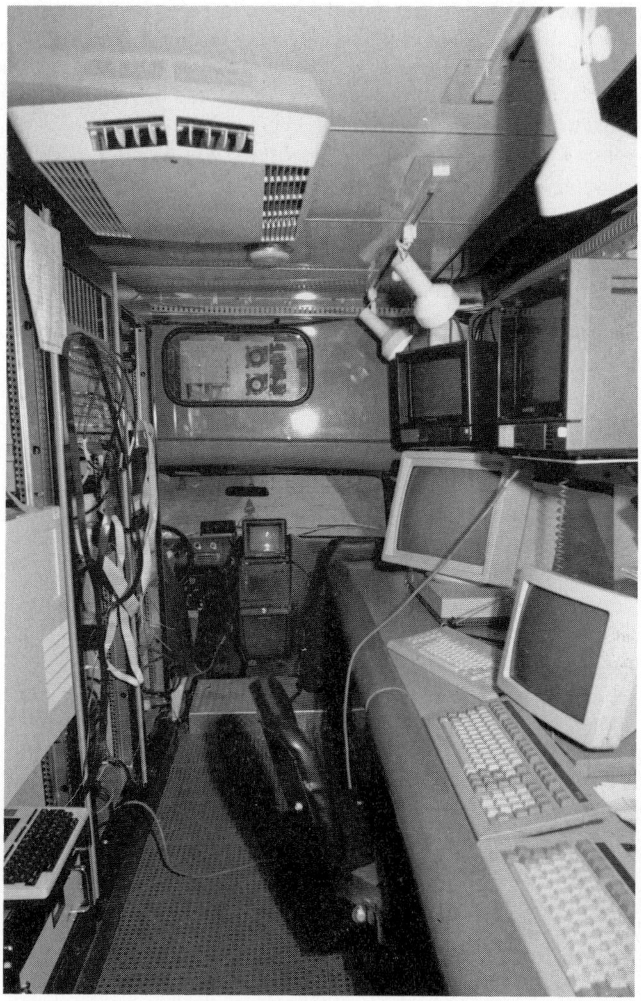

Figure 2: Navlab interior

3. Color Vision

The Navlab uses color vision, specifically multi-class adaptive color classification, to find and follow roads. Image points are classified into "road" or "non-road" principally on the basis of their color. Since the road is not a uniform color, color classification must have more than one road model, or class, and more than one non-road class. Because conditions change from time to time and from place to place over the test course, the color models must be adaptative. Once the image is classified, the road is identified by means of an area-based voting technique that finds the most likely location for the road in the image.

3.1. Vision Principles for the Real World

We based the development of our vision system on the following principles:

Assume variation and change. On sunny days there are shadowed areas, sunlit areas, and patches with dappled sunlight. On rainy days, there are dry patches and wet patches. Some days there are wet, dry, sunny and shadowed areas all in the same image. The road has clean spots and other places covered with leaves or with drips of our own hydraulic fluid. And as the sun goes behind a cloud or as the vehicle turns, lighting conditions change. We therefore need more than one road and non-road color model at any one time, those color models must adapt to changing conditions, and that we need to process images frequently so that the change from one image to the next will be moderate.

Use few geometric parameters. A complete description of the road's shape in an image can be complex. The road can bend gently or turn abruptly, can vary in width, and can go up- or downhill. However, the more parameters there are, the greater the chance of error in finding those parameters. Small misclassifications in an image could give rise to fairly large errors in perceived road geometry. Furthermore, if all the road parameters can vary, there are ambiguous interpretations: Does the road actually rise, or does it instead get wider as it goes? We describe the road with only two free parameters: its orientation and its distance from the vehicle. Road width is fixed, we assume a flat world, and we decree that the road is straight. While none of these assumptions is true over a long stretch of the road, they are nearly true within any one image; and the errors in road position that originate in our oversimplifications are balanced by the smaller chance of bad interpretations. If our system classifies a few pixels incorrectly as road, the worst it will do is to find a slightly incorrect road. A method that tries to fit more parameters, on the other hand, may interpret parts of the road perfectly, but could find an abrupt turn or sudden slope near any bad pixels.

Work in the image. The road can be found either by projecting the road shape into the image and searching in image coordinates, or by back projecting the image onto the ground and searching in world coordinates. The problem with the latter approach comes in projecting the image onto an evenly spaced grid in the world. The points on the world grid close to the vehicle correspond to a big area in the lower part of the image; points farther away may correspond to one or a few pixels near the top. Unless one uses a complex weighting scheme, some image pixels (those at the top that project to distant world points) will have more weight than other (lower) points. A few noisy pixels can then have a big or a small effect, depending on where in the image they lie. On the other hand, working directly in the image makes it much easier to weight all pixels evenly. We can directly search for the road shape that has the most road pixels and the fewest non-road pixels. Moreover, projecting a road shape is much more efficient than back projecting all the image pixels.

Calibrate directly. A complete description of a camera must include its position and orientation in space, its focal length and aspect ratio, lens effects such as fisheye distortion, and nonlinearities in the optics or sensor. The general calibration problem of trying to measure each of these variables is difficult. It is much easier, and more accurate, to calibrate the whole system than to tease apart the individual parameters. The easiest method is to take a picture of a known object and build a lookup table that relates each world point to an image pixel and vice versa. Projecting road predictions into the image and back projecting detected road shapes onto the world are done by means of table lookup (or table lookup for close-by values with simple interpolations). Such a table is straightforward to build and provides good accuracy, and there are no instabilities in the calculations.

Use outside constraints. Even without a map of our test course or an expensive inertial navigation system, we know, based on the previous image and on vehicle motion, approximately where the road should be. Our *whiteboard*, described in section 5, can predict where the road should appear if the road were straight and vehicle navigation were perfect. Adding a suitable margin for curved roads and sloppy navigation still gives useful limits on where in the image to look for the road.

Test with real data. We ran our VCR nearly every time we took the vehicle out, to collect images under as many conditions as possible. We recorded sunny days, cloudy days, rainy days, leaves on trees, leaves turning color, leaves falling, early morning, noon, after dusk, even a partial solar eclipse. Strategies that worked well on one set of images did not always work on the others. We selected the toughest images, ran our best algorithms and printed the classification results, changed parameters or algorithms, reran the data set, and compared results. This gave us the best chance of being methodical and of not introducing new bugs as we progressed. When the image processing worked to our satisfaction, we ran simulations in the lab that included the whiteboard, range processing, path planning, and a vehicle simulator, with the vision component processing stored images and interacting with the rest of the system. When the simulations worked in the lab, we moved them to the vehicle. Only after the simulations worked on the vehicle's computers, and we were sure that all necessary software was on the van, did we go into the field for real tests. Even then not everything worked, but there were many fewer bugs than there would have been without the simulations and tests.

3.2. Road Following Algorithm

We followed the same principles in building and tuning adaptive color classification for following roads. Figure 3 shows a relatively simple scene to help explain our algorithm. As shown in figure 4, the algorithm involves three stages:

1. Classify each pixel.
2. Use the results of classification to vote for the best-fit road position.
3. Collect new color statistics based on the detected road and non-road regions.

Pixel classification is done by standard pattern classification. Each class is represented by the means, variances, and covariances of red, green, and blue values, and by its a priori likelihood based on expected fraction of pixels in that class. For each pixel, calculating the class to which it most likely belongs involves finding how far the pixel's values lie from the mean of each class, where distance is measured in standard deviations of that class. Figures 5 and 6 show how each pixel is classified and how well it matches.

Figure 3: Original Image

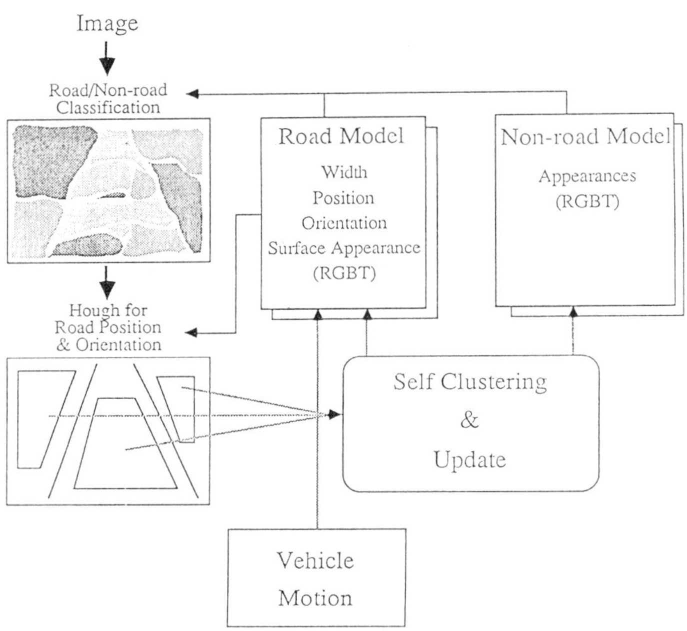

Figure 4: Color vision for road following, including color classification, Hough transform for road region detection, and updating multiple road and non-road models.

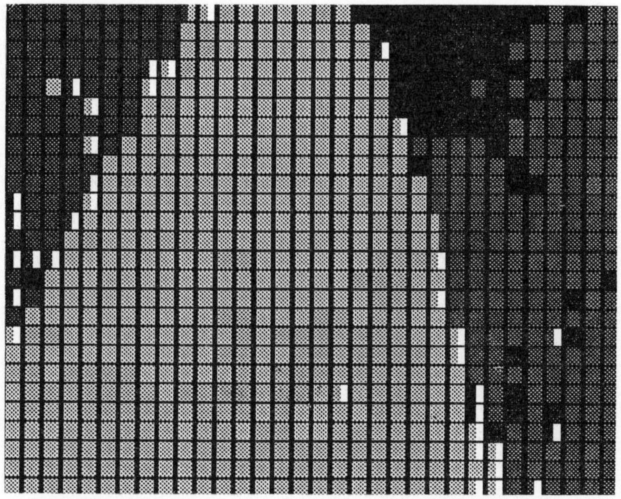

Figure 5: Segmented image. Color and texture cues are used to label points below the horizon into two road and two offroad classes

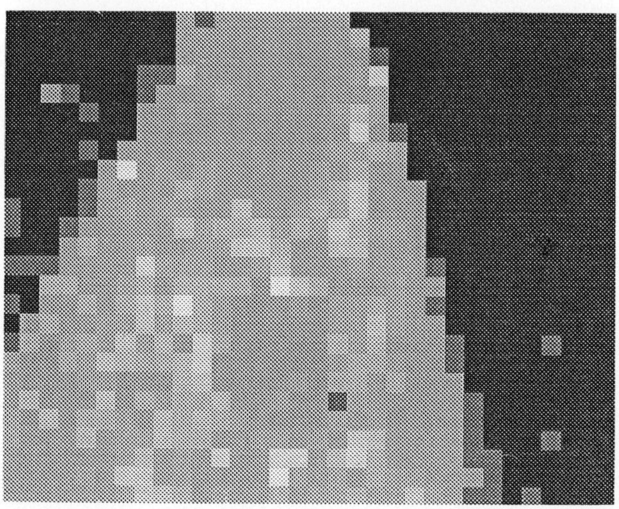

Figure 6: Road probability image. The pixels that best match typical road colors are displayed brightest.

Once each point has been classified, we must find the most likely location of the road. We assume the road is locally flat, straight, and has parallel sides. The road geometry can then be described by two parameters as shown in figure 7:

1. The intercept, which is the image column of the road's *vanishing point*. This is where the road centerline intercepts the horizon (or more precisely the vanishing line of the locally flat plane of the road; since the camera is fixed to the vehicle this vanishing line is constant independent of the vehicle's pitch, roll, and yaw). The intercept gives the road's direction relative to the vehicle.
2. The orientation of the road in the image, which tells how far the vehicle is to the right or left of the centerline.

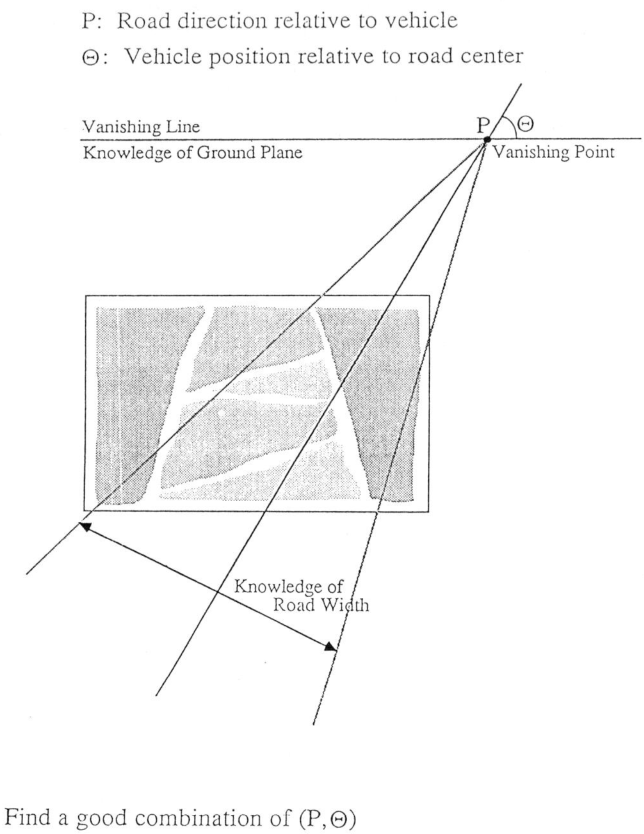

Figure 7: Hough Transform that considers the geometry of road position and orientation. Geometry of locally flat, straight, and parallel road regions can be described by only P and θ. Point A classified as road could be a part of the road with the shown combination of (P, θ), and thus casts a positive vote for it. Point B classified as off-road, however, will cast a negative vote for that (P, θ) combination.

We set up a two-dimensional parameter space, with intercept as one dimension and orientation as the other. Each point classified as road votes for all road orientation/intercept combinations to which it could belong, while nonroad points cast negative votes, as shown in figure 9. The orientation/intercept pair that receives the most votes is the one that contains the most road points, and it is reported as the road. For the case of figure 3, the votes in orientation/intercept space look like figure 10. Figure 11 shows the detected position and orientation of the road. It is worth noting that since this method does not rely on the exact local geometry of the road, it is very robust. The road may actually curve or not have parallel edges, or the segmentation may not be completely correct. But since this method does not rely on exact geometry, the answer it produces is adequate to generate an appropriate steering command.

Once the road has been found in an image, the color statistics of the road and offroad models are modified for each class by resampling the detected regions (figure 12) and updating the color models. The updated color statistics will gradually change as the vehicle moves into a different road color, as lighting conditions change, or as the colors of the surrounding grass, dirt, and trees vary. As long as the processing time per image is low enough to provide a large overlap between images, the statistics adapt as the vehicle moves. The road is picked out by hand in the first image. Thereafter, the process is automatic, using the segmentation from each image to calculate color statistics for the next.

There are several variations on this basic theme. One variation is to smooth the images first. This throws out outliers and tightens the road and non-road clusters. Another is to have more than one class for road and for non-road, for instance one for wet road and one for dry, or one for shadows and one for sun. Other variations change the voting for best road. Besides adding votes for road pixels, we subtract votes for non-road points. Votes are weighted according to how well each point matches road or non-road classes. Finally, image contains clues other than color: for instance, visual texture. Roads tend to be smooth, with less high-frequency variation than grass or leaves, as shown in figure 13. We calculate a normalized texture measure, and use that in addition to color in the road classification.

3.3. Implementation, Details, and Results

The implementation of road following runs in a loop of six steps: image reduction, color classification, texture classification, combining color and texture results, voting for road position, and color update. These steps are shown in figure 14, and are explained in detail below.

Image Reduction. We create a pyramid of reduced resolution R, G, and B images. Each smaller image is produced by simple 2 x 2 averaging of the next larger image. Other reduction methods, such as median filtering, are more expensive and produce no noticeable improvement in the system. We start with 480 x 512 pixel images, and typically use the images reduced to 30 x 32 for color classification. We use less reduced versions of the images for texture classification. Image reduction is used mainly to improve speed, but as a side effect the resulting smoothing reduces the effect of scene anomalies such as cracks in the pavement.

Color Classification. Each pixel (in the 30 x 32 reduced image) is labeled as belonging to one of the road or non-road classes by standard maximum likelihood classification. We usually have two road and two non-road classes. Each class is represented by the mean R, G, and B values of its pixels, by a 3 x 3 covariance matrix, and by the fraction of pixels expected a priori to be in that class. The classification procedure calculates the probability that a pixel belongs to each of the classes, assigns the label of the most probable class, and records the maximum road and non-road probabilities for each pixel.

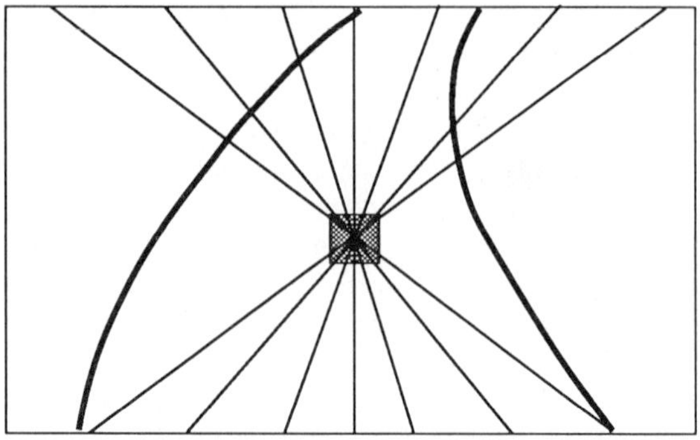

Figure 8: A point classified as road could be a part of roads with different orientations and vanishing points.

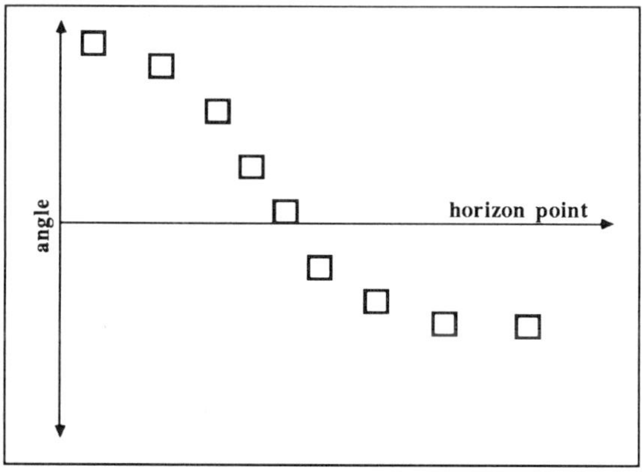

Figure 9: The point from figure 8 would vote for these orientation / intercept values in the parameter space.

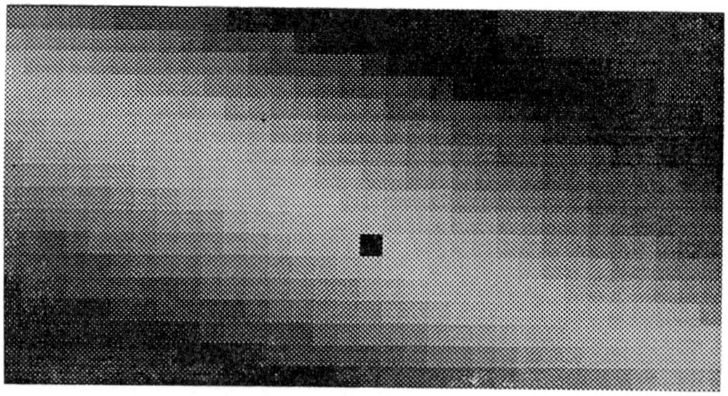

Figure 10: Votes for best road orientation and intercept, and point with most votes (dark square), for road in figure 3.

Figure 11: Detected road, from the point with the most votes shown in figure 10.

Figure 12: Updating road and nonroad model colors, leaving a safety zone around the detected road region.

Figure 13: Zoomed picture of road-nonroad boundary. The road (at left) is much less textured than the grass (at right).

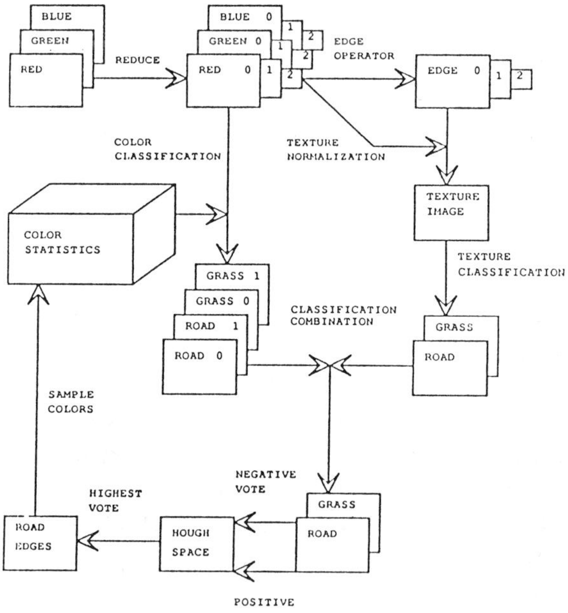

Figure 14: Processing cycle for color vision.

Texture Calculation. This is composed of six substeps:

- Calculate texture at high resolution by running a Robert's operator over the 240 x 256 image.
- Calculate a low resolution texture by applying a Robert's operator to the 60 x 64 image.
- Normalize the texture by dividing the high resolution texture by a combination of the average pixel value for that area (to handle shadow interiors) and the low resolution texture (to remove the effect of shadow boundaries). The average pixel value is the value from the corresponding pixel in the 120 x 128 reduced image.

$$\text{normalized gradient} = \frac{\text{high-freq gradient}}{\alpha \times \text{low-freq gradient} + \beta \times \text{mean pixel value}}$$

  Typical values for the coefficients are $\alpha = 0.2$ and $\beta = 0.8$.

- Threshold. Produce a binary image of "microedges" by thresholding the normalized gradient. A fairly low threshold, such as 1, is usually adequate.
- Count Edges. Count the number of edges in each pixel block. This gives a 30 x 32 pixel texture magnitude image. Figure 15 shows the texture image derived from figure 3. Each texture pixel has a value between 0 and 256, which is the number of pixels in the corresponding area of the full-resolution image that are microedges.
- Texture Classification. Classify each pixel in the 30 x 32 image as road or non-road on the basis of texture, and calculate a confidence for each label. We found experimentally that a fixed mean and standard deviation for road and non-road textures were better than adaptive texture parameters. Our best results were with road mean and standard deviation of 0 and 25, and non-road values of 175 and 100. Effectively, any pixel block of the image with more than 35 microedges above threshold is considered textured, and is therefore classified as nonroad.

Figure 15: Low resolution texture image, showing textures from figure 3. The brighter blocks are image areas with more visual texture.

Combination of Color and Texture Results. Color is somewhat more reliable than texture, so the color probabilities are weighted somewhat more than the probabilities calculated by texture. The result of this step is a final classification into road or non-road, and a "confidence" calculated by

Max(road confidence, non-road confidence) — Min(road confidence, non-road confidence)

Vote for Best Road Position. This step uses a 2-D parameter space similar to a Hough transform. Parameter 1 is the column of the road's vanishing point, quantized into 32 buckets because the image on which the classification and voting are based has 32 columns. Parameter 2 is the road's angle from vertical in the image, ranging from -1 to 1 radian in 0.1 radian steps. A given road point votes for all possible roads that would contain that point. The locus of possible roads whose centerlines go through that point is an arctangent curve in the parameter space. Because the road has a finite width, the arctan curve has to be widened by the width of the road at that pixel's image row. Road width for a given row is not a constant over all possible road angles but is nearly constant enough that it doesn't justify the expense of the exact calculation. Each pixel's vote is weighted by its calculated confidence. Pixels classified as non-road cast negative votes (with their weights reduced by a factor of 0.2) while road pixels add votes. In pseudo C code, the voting for a pixel at (row, col) is

```
for (theta = -1; theta <= 1; theta+= 0.1) {
    center = col + arctan (theta);
    for (c = center - width/2; c <= center + width/2; c++) {
        parameter_space[theta][c] += confidence;
```

At the end of voting, one road intercept/angle pair will have the most votes. That intercept and angle describe the best road shape in the scene.

Color Update. The parameters of the road and non-road classes need to be recalculated to reflect changing colors. We divide the image into four regions plus a "safety zone": left offroad, right offroad, upper road, and lower road. We leave a 64-pixel wide "safety zone" along the road boundary, which allows for small errors in locating the road, or for limited road curvature. For each of the four regions, we calculate the means of red, green, and blue. We use the calculated parameters to form four classes, and reclassify the image using a limited classification scheme. The limited reclassification allows road pixels to be classified as either of the two road classes, but not as non-road, and allows non-road pixels to be reclassified only as one of the non-road classes. The reclassified pixels are used as masks to recalculate class statistics. The loop of classify pixels/recalculate statistics is repeated, typically 3 times, or until no pixels switch classes. The final reclassified pixels are used to calculate the means, variances, and covariances of R, G, and B for each of the classes, to be used to classify the next image. Limited reclassification is based on distance from a pixel's values to the mean values of a class, rather than the full maximum likelihood scheme used in classifying a new image. This tends to give classes based on tight clusters of pixel values, rather than lumping all pixels into classes with such wide variance that any pixel value is considered likely.

Calibration. There is no need for complete geometric calibration. The vision algorithms calculate the road's shape (road width and location of the horizon) from the first training image. We also take two calibration graphics, with a meter stick placed perpendicular to the vehicle, 8 and 12 ms in front. Then during the run, given the centerline of a detected road in image coordinates, it is easy to get the x position of the road at 8 and 12 m, and then to calculate the vehicle's position on the road.

Performance. This algorithm is reliable. Running on the Navlab, with predictions of where the road should appear, our failure rate is close to 0. The occasional remaining problems come from one of three causes:

- The road is covered with leaves or snow, so one road color class and one non-road color class are indistinguishable.
- Drastic changes in illumination occur between pictures (*e.g.* the sun suddenly emerges from behind a cloud) so all the colors change dramatically from one image to the next.
- The sunlight is so bright and shadows are so dark in the same scene that we hit the hardware limits of the camera. It is possible to have pixels so bright that all color is washed out, and other pixels in the same image so dark that all color is lost in the noise.

Not every image is classified perfectly, but almost all are good enough for navigation. We sometimes find the road rotated in the image from its correct location, so we report an intercept off to one side and an angle off to the other side. But since the path planner looks ahead about the same distance as the center of the image, the steering target is still in approximately the correct location, and the vehicle stays on the road. This algorithm runs in about 10 s per image on a dedicated Sun 3/160, using 480 x 512 pixel images reduced to 30 rows by 32 columns. We currently process a new image every 4 m, which gives about three fourths of an image overlap between images. Ten seconds is fast enough to balance the rest of the system but is slow enough that clouds can come and go and lighting conditions change between images. We are porting this algorithm to the Warp, Carnegie Mellon's experimental high-speed processor. On that machine, we hope to process an image per second and to use higher resolution.

4. Perception in 3-D

Our obstacle detection starts with direct range perception using an ERIM scanning laser rangefinder. Our ERIM produces, every half second, an image containing 64 rows by 256 columns of range values; an example is shown in figure 16. The scanner measures the phase difference between an amplitude-modulated laser and its reflection from a target object, which in turn provides the distance between the target object and the scanner. The scanner produces a dense range image by using two deflecting mirrors, one for the horizontal scan lines and one for vertical motion between scans. The volume scanned is 80 degrees wide and 30 high. The range at each pixel is discretized over 256 levels from zero to 64 feet.

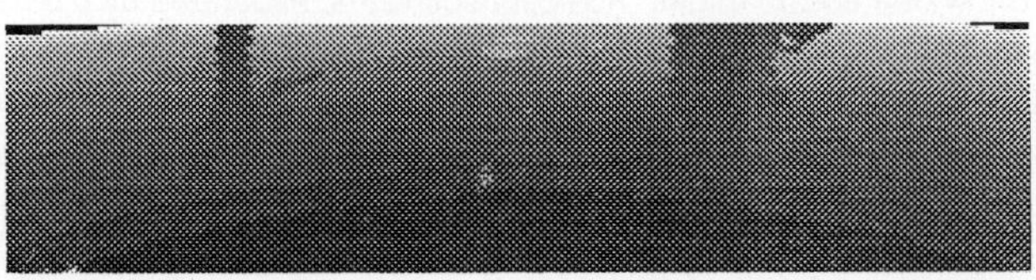

Figure 16: Range image of two trees on flat terrain. Gray levels encode distance; nearer points are painted darker.

Our range processing begins by smoothing the data and undoing the peculiarities of the ranging geometry. The *ambiguity intervals*, where range values wrap around from 255 to 0, are detected and unfolded. Two other undesirable effects are removed by the same algorithm. The first is the presence of mixed points at the edge of an object. The second is the meaninglessness of a measurement from a surface such as water, glass, or glossy pigments. In both cases, the resulting points are in regions limited by considerable jumps in range. We then transform the values from angle-angle-range, in scanner coordinates, to x-y-z locations. These 3-D points are the basis for all further processing.

We have two main processing modes: obstacle detection and terrain analysis. Obstacle detection starts by calculating surface normals from the x-y-z points. Flat, traversable surfaces will have vertical surface normals. Obstacles will have surface patches with normals pointed in other directions. This analysis is relatively fast, running in about 5 s on a Sun 3/75, and is adequate for smooth terrain with discrete obstacles.

Simple obstacle maps are not sufficient for detailed analysis. For greater accuracy we do more careful terrain analysis and combine sequences of images corresponding to overlapping parts of the environment into an *extended obstacle map*. The terrain analysis algorithm first attempts to find groups of points that belong to the same surface and then uses these groups as seeds for the region growing phase. Each group is expanded into a smooth connected surface patch. The smoothness of a patch is evaluated by fitting a surface (plane or quadric). In addition, surface discontinuities are used to limit the region growing phase. The complete algorithm is:

1. Edges: Extract surface discontinuities, pixels with high jumps in x-y-z.
2. Clustering: Find clusters in the space of surface normals and identify the corresponding regions in the original image.
3. Region growing: Expand each region until the fitting error is larger than a given threshold. The expansion proceeds by iteratively adding the point of the region boundary that adds the minimum fitting error.

The clustering step is designed so that other attributes such as color or curvature can also be used to find potential regions on the object. The primitive surface used to compute the fitting error can be either a plane or a quadric surface. The decision is based on the size of the region. Figure 17 shows the resultant description of 3-D terrain and obstacles for the image of figure 16. The flat, smooth, navigable region is the meshed area, and the detected 3-D objects (the two trees) are shown as polyhedra.

Obstacle detection works at longer range than terrain analysis. When the scanner is looking at distant objects, it has a very shallow depression angle. Adjacent scanlines, separated by 0.5 degree in the range image, can strike the ground at widely different points. Because the grazing angle is shallow, little of the emitted laser energy returns to the sensor, producing noisy pixels. Noisy range values, widely spaced, make it difficult to do detailed analysis of flat terrain. A vertical obstacle, such as a tree, shows up much better in the range data. Pixels from neighboring scanlines fall more closely together, and with a more nearly perpendicular surface the returned signal is stronger and the data cleaner. It is thus much easier for obstacle detection to find obstacles than for terrain analysis to certify a patch of ground as smooth and level.

When neither video nor range information alone suffices, we must fuse data to determine mobility or recognize an object. One such case occurs in navigating the smaller Terregator vehicle around campus sidewalks. At one spot, a sidewalk goes up a flight of stairs and a bicycle path curves around. Video alone has a tough time distinguishing between the cement stairs and the cement bicycle path. Range data cannot tell the difference between the smooth rise of the grassy hill and the smooth bicycle ramp. The only way to identify the safe vehicle path is to use both kinds of data.

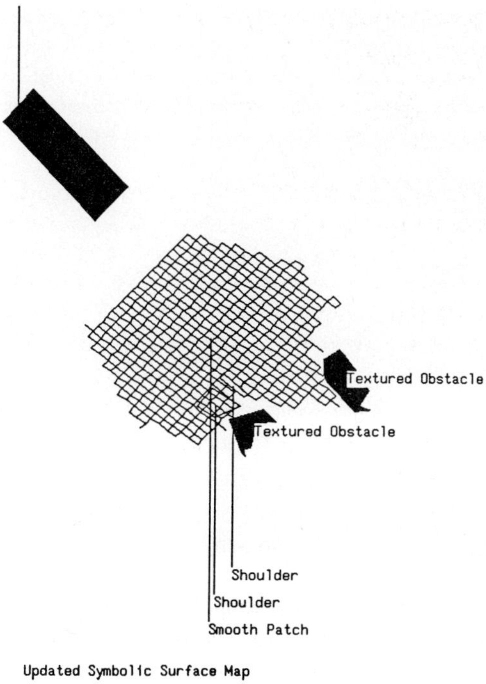

Figure 17: The resultant description of 3D terrain and obstacles from the image in figure 16. The navigable area is shown as a mesh, and the two trees are detected as "textured obstacles" and shown as black polygons

We start by fusing the data at the pixel level. For each range point, we find the corresponding pixel in the video image. We produce a painted range image in which each pixel is a {red, green, blue, x, y, z} 6-vector. Figure 18 shows the painted range image, rotated and projected from a different angle. We can then run our standard range segmentation and color segmentation programs, producing regions of smooth range or constant color. For the stairs in particular, we have a special-purpose step detection program that knows about vertical and horizontal planes and how they are related in typical stairs. It is easy to combine the regions from these separate processes, since they are all in the same coordinates of the painted range image. The final result is a smooth concrete region in which it is safe to drive, and a positive identification and 3-D location of the stairs, for updating the vehicle position.

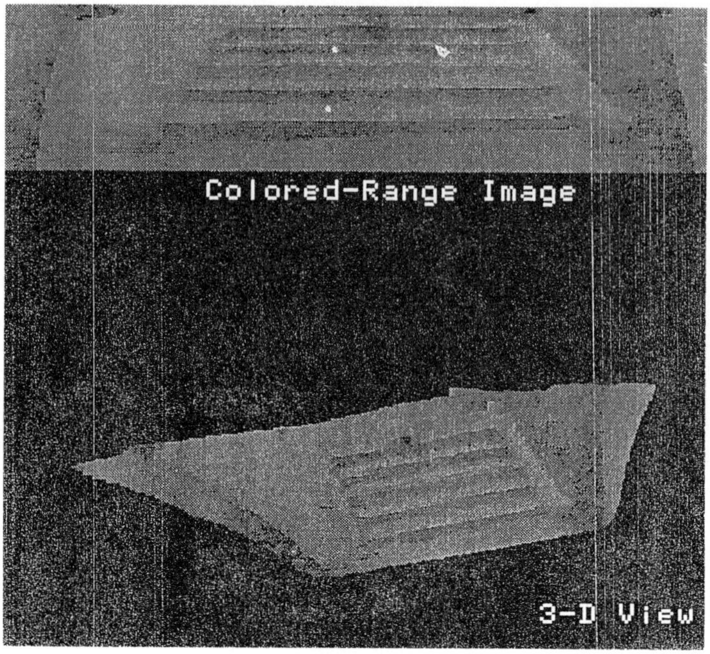

Figure 18: Painted range image of campus stairs. Each point is a {red, green, blue, x, y, z} 6-vector. This image has been rotated and projected from a different viewpoint. The color and range images are registered, so the color edges and regions line up with range edges and regions.

5. System Building

5.1. Artificial Intelligence for Real World Robots

We have developed a new paradigm for intelligent robot system building. Artificial Intelligence systems, including intelligent mobile robots, are symbol manipulators. Indeed, the very definition of intelligence, artificial or otherwise, includes symbol manipulation. But the manipulation used by most AI systems is based on inference, either by the logic of predicate calculus or by probabilities. The bulk of the work of a mobile robot, in contrast, is based on geometry and on modeling time. Inference may be a part of a mobile robot system, but geometry and time are pervasive. Consequently, intelligent mobile robots need a new kind of expert system *shell*, one that provides tools for handling 3-D locations and motion.

This fits into the context of changes in the field of AI as a whole. Early systems, such as the Logic Theorist or GPS [1], were search engines that had no domain knowledge. They could solve problems such as the Towers of Hanoi or Missionaries and Cannibals that are essentially logic puzzles. "Expert systems" brought lots of knowledge to bear on a problem. A system such as R1 or MYCIN [1] has thousands of rules of the form "if P then try Q" or "if X is true then Y is true with confidence 0.7". This type of knowledge allows these programs to deal with many real world problems. However, it is "shallow" knowledge in the sense that it deals with externally visible input-output behavior, with no knowledge of internal structure or mechanisms. MYCIN is like a doctor who has never taken Anatomy or Physiology, but has seen a lot of cases. Its knowledge is adequate for handling things it has already seen, but, because it does not understand the underlying mechanisms and structures of its domain, there is a limit to its competence in reasoning about new or unexpected behavior. The newest generation of expert systems is beginning to embed more "deep knowledge." For instance, the Aladdin aluminum alloy design system includes both shallow knowledge rules ("If the alloy is too heavy, try adding lithium") and deep knowledge of crystal structure and chemical interactions.

The evolution of mobile robot systems is following an analogous course. Early systems such as SRI's Shakey were based on deduction. Shakey could decide which light switch to flip and in what order to traverse a sequence of rooms; it was a success with respect to logical action, but it lacked the deep knowledge needed to move and live in a complicated environment. Its home was a series of empty rooms with flat floors and uniform walls that allowed Shakey to function with very simple perception and motion capabilities. In contrast, a robot that must move through the real outdoor world, needs a vast reservoir of deep knowledge of perception, object models, motion, path planning, terrain models, navigation, vehicle dynamics, and so forth.

The deep knowledge needed by a mobile robot must be supported by the system architecture and by the system building tools. We have developed and followed the following tenets of mobile robot system design in building our system:

Use separate modules. Much of the deep knowledge can be limited to particular specialist modules. The effects of lighting conditions and viewing angle on the appearance of an object, for instance, are important data for color vision but are not needed by path planning. So one principle of mobile robot system design is to break the system into modules and minimize the overlap of knowledge between modules.

Provide tools for geometry and time. Much of the knowledge that needs to be shared between modules has to do with geometry, time, and motion. An object may be predicted by one module (the lookout), seen separately by two others (color vision and 3-D perception), and used by two more (path planner and position update). During the predictions, sensing, and reasoning, the vehicle will be moving, new position updates may come in, and the geometrical relationship between the vehicle and the object will be constantly changing. Moreover, there may be many different frames of reference: one for each sensor, one for the vehicle, one for the world map, and others for individual objects. Each module should be able to work in the coordinate frame that is most natural; for instance, a vision module should work in camera coordinates and should not have to worry about conversion to the vehicle reference frame. The system should provide tools that handle as many as possible of the details of keeping track of coordinate frames, motion, and changing geometry.

Provide tools for synchronization. A system that has separate modules communicating at a fairly coarse grain will be loosely coupled. Lock-step interactions are neither necessary nor appropriate. However, there are times when one module needs to wait for another to finish, or when a demon module needs to fire whenever certain data appear. The system should provide tools for several different kinds of interaction and for modules to synchronize themselves as needed.

Handle real-time vs symbolic interface. At one level, a mobile robot reasons symbolically about perceived objects and planned paths, probably on a slow time scale. At the same time, the vehicle is constantly moving, and low-level servo processes are controlling steering and motion. The top level processes need to be free to take varying amounts of time to process scenes of varying difficulty. They are often event driven, running when a particular object is seen or a particular event occurs. The servo processes, though, must run continuously and in real time (not "simulated real time" or "real time not counting garbage collection"). The system should provide for both real-time and asynchronous symbolic processes, and for communications between them.

Provide a ***virtual vehicle.*** As many as possible of the details of the vehicle should be hidden. At Carnegie Mellon, we have one robot (the Terregator) that has six wheels, steers by driving the wheels on one side faster than those on the other side, and carries a camera mount approximately 6 ft high. A second robot (the Navlab) is based on a commercial van, steers and drives conventionally, and mounts its camera 2 ft higher. We need to be able to use one system to drive either of the vehicles, with only minor modifications. This requires hiding the details of sensing and motion in a "virtual vehicle" interface, so a single "move" command, for instance, will use the different mechanisms of the two vehicles but will produce identical behavior.

Plan for big systems. It takes good software engineering to build a mobile robot. The system may be written in a mixture of programming languages, will probably run on multiple processors, and may use different types of processors including specialized perception machines. System tools must bridge the gaps between languages, data formats, and communications protocols.

In addition to these tenets of good design, we have identified certain approaches that are inappropriate. Many good ideas in other areas of AI present difficulties for mobile robots. Specifically, we avoid the following.

Do not throw away geometric precision. Mobile robots need all the information they can get. It is often important to know as precisely as possible where an object is located, either for planning efficient paths or for updating vehicle location. There is no need to turn a measured distance of 3.1 m into *fairly close.* Given the relative costs and speeds of computers and vehicles, it is more efficient to spend extra computing effort (if any) to handle precise data than to plan fuzzy paths that take the vehicle unnecessarily far out of its way.

Do not concentrate on explanations. It is important to have hooks inside the vehicle's reasoning, for debugging and for learning about the system behavior. However, the prime output of the vehicle is its externally observable behavior. Producing explanations is nice, but is not the primary product as it is in expert systems for diagnosis or in intelligent assistants.

Do not build an omniscient master process. In some systems (notably early blackboards) a single master process "knows" everything. The master process may not know the internal working of each module, but it knows what each module is capable of doing. The master controls who gets to run when. The master itself becomes a major AI module and can be a system bottleneck. In contrast, the individual modules in a mobile robot system should be autonomous, and the system tools should be slaves to the modules. The module writers should decide when and how to communicate and when to execute. The system support should be as unobtrusive as possible.

We have followed these tenets in building the Navlab system. At the bottom level, we have built the CODGER "whiteboard" to provide system tools and services. On top of CODGER we have built an architecture that sets conventions for control and data flow. CODGER and our architecture are explained below.

5.2. Blackboards and Whiteboards

The program organization of the NAVLAB software is shown in figure 19. Each of the major boxes represents a separately running program. The central database, called the *Local Map*, is managed by a program known as the Local Map Builder *(LMB)*. Each module stores and retrieves information in the database through a set of subroutines called the *LMB Interface* which handle all communication and synchronization with the LMB. If a module resides on a different processor than the LMB, the LMB and LMB Interface will transparently handle the network communication. The Local Map, LMB, and LMB Interface together comprise the CODGER (COmmunications Database with GEometric Reasoning) system.

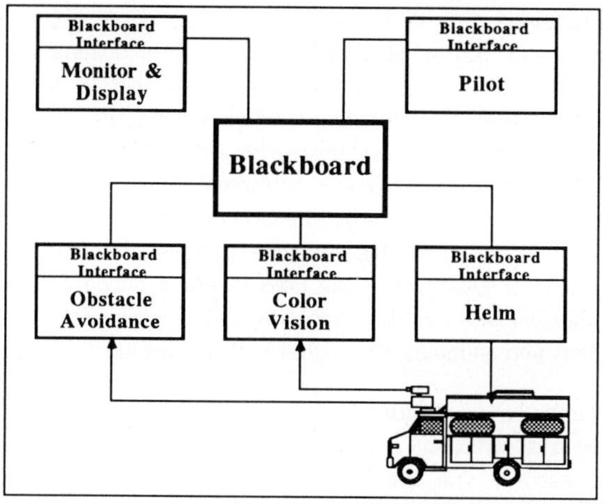

Figure 19: Navlab software architecture

The overall system structure—a central database, a pool of knowledge-intensive modules, and a database manager that synchronizes the modules—is characteristic of a traditional blackboard system. Such a system is called "heterarchical" because the knowledge is scattered among a set of modules that have access to data at all levels of the database (i.e. low-level perceptual processing ranging up to high-level mission plans) and may post their findings on any level of the database; in general, heterarchical systems impose de facto structuring of the information flow among the modules of the system. In a traditional blackboard, there is a single flow of control managed by the database (or blackboard) manager. The modules are subroutines, each with a predetermined precondition (pattern of data) that must be satisfied before that module can be executed. The manager keeps a list of which modules are ready to execute. In its central loop it selects one module, executes it, and adds to its ready-list any new modules whose preconditions become satisfied by the currently executing module. The system is thus synchronous and the manager's function is to focus the attention of the system by selecting the "best" module from the ready-list on each cycle.

We call CODGER a *whiteboard* because although it implements a heterarchical system structure, it differs from a blackboard in several key respects. In CODGER, each module is a separate, continuously running program; the modules communicate by storing and retrieving data in the central database. Synchronization is achieved by primitives in the data retrieval facilities that allow, for example, for a module to request data and suspend execution until the specified data appears. When some other module stores the desired data, the first module will be reactivated and the data will be sent to it. With CODGER a module programmer thus has control over the flow of execution within his module and may implement real-time loops, demons, data flows among cooperating modules, etc. CODGER also has no precompiled list of data retrieval specifications; each time a module requests data, it provides a pattern for the data desired at that time. Such a system a whiteboard is heterarchical like a blackboard, but each module runs in parallel, with the module programmer controlling the synchronization and data retrieval requests as best suited for each module. Like other recent distributed AI architectures, whiteboards are suited to execution on multiple processors.

5.3. Data Storage and Retrieval

Data in the CODGER database (Local Map) is represented in *tokens* consisting of classical *attribute-value pairs*. The types of tokens are described in a *template file* that tells the name and type of each attribute in tokens of each type. The attributes themselves may be the usual scalars (integers, floating-point values, strings, enumerated types), arrays (or sets) of these types (including arrays of arrays), or geometric locations (as described below). CODGER automatically maintains certain attributes for each token: the token type and id number, the *generation number* as the token is modified, the time at which the token was created and inserted into the database, and the time at which the sensor data was acquired that led to the creation of this token. The LMB Interface provides facilities for building and dissecting tokens and attributes within a module. Rapid execution is supported by mapping the module programmer's names for tokens and attributes onto globally used index values at system startup time.

A module can store a token by calling a subroutine to send it to the LMB. Tokens can be retrieved by constructing a pattern called a *specification* and calling a routine to request that the LMB send back tokens matching that specification. The specification is simply a Boolean expression in which the attributes of each token may be substituted; if a token's attributes satisfy the Boolean expression, then the token is sent to the module that made the request. For example, a module may specify:

tokens with **type** *equal to "intersection" and* **traffic-control** *equal to "stop-sign"*

This would retrieve all tokens whose **type** and **traffic-control** attributes satisfy the above conditions. The specification may include computations such as mathematical expressions, finding the minimum value within an array attribute, comparisons among attributes, etc. CODGER thus implements a general database. The module programmer constructs a specification with a set of subroutines in the CODGER system.

One of the key features of CODGER is the ability to manipulate geometric information. One of the attribute types provided by CODGER is the *location*, which is a 2-D or 3-D polygon and a reference to a *coordinate frame* in which that polygon is described. Every token has a specific attribute that tells the location of that object in the Local Map, if applicable, and a specification can include geometric calculations and expressions. For example, a specification might be:

tokens with **location** *within 5 units of (45,32) [in* world *coordinates]*

or

tokens with **location** *overlapping X*

where *X* is a description of a rectangle on the ground in front of the vehicle. The geometric primitives currently provided by CODGER include calculation of centroid, area, diameter, convex hull, orientation, and minimum bounding rectangle of a location, and distance and intersection calculations between a pair of locations. We believe that this kind of geometric data retrieval capability is essential for supporting spatial reasoning in mobile robots with multiple sensors. We expect geometric specifications to be the most common type of data retrieval request used in the NAVLAB.

CODGER also provides for automatic coordinate system maintenance and transformation for these geometric operations. In the Local Map, all coordinates of location attributes are defined relative to **WORLD** or **VEHICLE** coordinates; **VEHICLE** coordinates are parameterized by time, and the LMB maintains a time-varying transformation between **WORLD** and **VEHICLE** coordinates. Whenever new information (i.e. a new **VEHICLE**-to-**WORLD** transform) becomes available, it is added to the "history" maintained in the LMB; the LMB will interpolate to provide intermediate transformations as needed. In addition to these basic coordinate systems, the LMB Interface allows a module programmer to define *local coordinates* relative to the basic coordinates or relative to some other local coordinates. Location attributes defined in a local coordinate system are automatically converted to the appropriate basic coordinate system when a token is stored in the database. CODGER provides the module programmer with a conversion routine to convert any location to any specified coordinate system.

All of the above facilities need to work together to support asynchronous sensor fusion. For example, suppose we have a vision module A and a rangefinder module B whose results are to be merged by some module C. The following sequence of actions might occur:

1. A receives an image at time 10 and posts results on the database at time 15. Although the calculations were carried out in the camera coordinate system for time 10, the results are automatically converted to the **VEHICLE** system at time 10 when the token is stored in the database.
2. Meanwhile, B receives data at time 12 and posts results at time 17 in a similar way.
3. At time 18, C receives A's and B's results. As described above, each such token will be tagged with the time at which the sensor data was gathered. C decides to use the vehicle coordinate system at time 12 (B's time) for merging the data.
4. C requests that A's result, which was stored in **VEHICLE** time 10 coordinates, be transformed into **VEHICLE** time 12 coordinates. If necessary, the LMB will automatically interpolate coordinate transformation data to accomplish this. C can now merge A's and B's results since they are in the same coordinate system. At time 23, C stores results in the database, with an indication that they are stored in the coordinate system of time 12.

5.4. Synchronization Primitives

CODGER provides module synchronization through options specified for each data retrieval request. Every time a module sends a specification to the LMB to retrieve tokens, it also specifies options that tell how the LMB should respond with the matching tokens:

- *Immediate Request.* The module requests all tokens currently in the database that match this specification. The module will block (i.e. the "request" subroutine in the LMB Interface will not return control) until the LMB has responded. If there are no tokens that match the specification, the action taken is determined by an option in the module's request:

- *Non-Blocking.* The LMB will answer that there are no matching tokens, and the module can then proceed. This would be used for time-critical modules such as vehicle control. Example: "Is there a stop sign?"
- *Blocking.* The LMB will record this specification and compare it against all incoming tokens. When a new token matches the specification, it will be sent to the module and the request will be satisfied. Meanwhile, the module will remain blocked until the LMB has responded with a token. This is the type of request used for setting up synchronized sets of communicating modules: each one waits for the results from the previous module to be posted to the database. Example: "Wake me up when you see a stop sign."

- *Standing Request.* This provides a mechanism for the LMB to generate an interrupt for a running module. The module gives a specification along with the name of a subroutine. The module then continues running; the LMB will record the specification and compare it with all incoming tokens. Whenever a token matches, it will be sent to the module. The LMB Interface will intercept the token and execute the specified subroutine, passing the token as an argument. This has the effect of invoking the given subroutine whenever a token appears in the database that matches the given specification. It can be used at system startup time for a module programmer to set up "demon" routines within the module. Example: "Execute that routine whenever you see a stop sign."

5.5. Architecture

Several modules use the CODGER tools and fit into a higher level architecture. The modules are:

- Pilot: Looks at the map and at current vehicle position to predict road location for Vision. Plans paths.
- Map Navigator: Maintains a world map, does global path planning, provides long-term direction to the Pilot. The world map may start out empty, or may include any level of detail up to exact locations and shapes of objects.
- Color Vision: Waits for a prediction from the Pilot, waits until the vehicle is in the best position to take an image of that section of the road, returns road location.
- Obstacle Detection: Gets a request from the Pilot to check a part of the road for obstacles. Returns a list of obstacles on or near that chunk of the road.
- Helm: Gets planned path from Pilot, converts polyline path into smooth arcs, steers vehicle.
- Graphics and Monitor: Draws or prints position of vehicle, obstacles, predicted and perceived road.

There are two other modules in our architecture. These have not yet been implemented:

- Captain: Talks to the user and provides high-level route and mission constraints such as *avoid area* or *go by road B.*
- Lookout: Looks for landmarks and objects of importance to the mission.

These modules use CODGER to pass information about *driving units.* A driving unit is a short chunk of the road or terrain (in our case 4 m long) treated as a unit for perception and path planning. The Pilot gives driving unit predictions to Color Vision, which returns an updated driving unit location. Obstacle Detection then sweeps a driving unit for obstacles. The Pilot takes the driving unit and obstacles, plans a path, and hands the path off to the Helm. The whole process is set up as a pipeline, in which Color Vision is looking ahead 3 driving units, Obstacle Detection is looking 2 driving units ahead, and path planning at the next unit. If for any reason some stage slows down, all following stages of the pipeline must wait. So, for instance, if Color Vision is waiting for the vehicle to come around a bend so it can see down the road, Obstacle Detection will finish its current unit and will then have to wait for Color Vision to proceed. In an extreme case, the vehicle may have to come to a halt until everything clears up. All planned paths include a deceleration to a stop at the end, so if no new path comes along to overwrite the current path the vehicle will stop before driving into an area that has not been seen or cleared of obstacles.

In our current system and test area, 3 driving units is too far ahead for Color Vision to look, so both Color Vision and Obstacle Detection are looking at the same driving unit. Obstacle Detection looks at an area sufficiently larger than the Pilot's predicted driving unit location to guarantee that the actual road is covered. Another practical modification is to have Obstacle Detection look at the closest driving unit also, so a person walking onto the road immediately in front of the vehicle will be noticed. Our system will try to plan a path around obstacles while remaining on the road. If that is not possible, it will come to a halt and wait for the obstacle to move before continuing.

6. Conclusions and Future Work

The system described here works. It has successfully driven the Navlab many tens of times, processing thousands of color and range images without running off the road or hitting any obstacles. CODGER has proved to be a useful tool, handling many of the details of communications and geometry. Module developers have been able to build and test their routines in isolation, with relatively little integration overhead. Yet there are several areas that need much more work.

Speed. We drive the Navlab at 10 cm/sec, a slow shuffle. Because our test road is narrow and winding, and we deliberately concentrate on competence rather than speed. But faster motion is always more interesting, so we are pursuing several ways of increasing speed. One bottleneck is the computing hardware. We are mounting a Warp, Carnegie-Mellon's experimental high-speed processor, on the Navlab. The Warp will give us a factor of 100 more processing power than a Sun for color and range image processing. At the same time, we are looking at improvements in the software architecture. We need a more sophisticated path planner, and we need to process images that are more closely spaced than the length of a driving unit. Also, as the vehicle moves more quickly, our simplifying assumption that steering is instantaneous and that the vehicle moves along circular arcs becomes more seriously flawed. We are looking at other kinds of smooth arcs, such as clothoids. More important, the controller is evolving to handle more of the low-level path smoothing and following.

Map. One reason for the slow speed is that the Pilot assumes straight roads. We need to have a description that allows for curved roads, with some constraints on maximum curvature. The next steps will include building maps as we go, so that subsequent runs over the same course can be faster and easier.

Cross-country travel. Travel on roads is only half the challenge. The Navlab should be able to leave roads and venture cross-country. Our plans call for a fully integrated on-road/off-road capability.

Intersections. Current vision routines have a built-in assumption that there is one road in the scene. When the Navlab comes to a fork in the road, vision will report one or the other of the forks as the true road depending on which looks bigger. It will be important to extend the vision geometry to handle intersections as well as straight roads. We already have this ability on our sidewalk system and will bring that over to the Navlab. Vision must also be able to find the road from offroad.

Landmarks. Especially as we venture off roads, it will become increasingly important to be able to update our position based on sighting landmarks. This involves map and perception enhancements, plus understanding how to share limited resources, such as the camera, between path finding and landmark searches.

Software Development. Our current blackboard system can manipulate primitive data elements but has no concept of data structures made up of tokens on the blackboard. We need aggregate data types for representing complex 3-D geometric descriptions of objects for recognition. We will also be implementing a Lisp interface to our blackboard. All current modules are written in C, but we will soon want to write higher-level modules in Lisp.

Integration with Work from Other Sites. Other universities and research groups cooperating with Carnegie-Mellon through DARPA Strategic Computing Vision program. We plan to incorporate some of their programs into the Navlab system in the coming years as it evolves into the "new generation vision system" that is the goal of that program.

Acknowledgments

This research is supported in part by the Advanced Research Projects Administration of the Department of Defense (DARPA), under contract DACA 76 85-C-003, monitored by the US Army Engineer Topographic Laboratories. We are part of the Strategic Computing Vision program.

The Terregator and Navlab were built by William Whittaker's group in the Construction Robotics Laboratory, and the Warp group is led by H. T. Kung and Jon Webb. The real work gets done by an army of eight staff, nine graduate students, five visitors, and three part time programmers.

References

[1] Cohen, P., Barr, A., Feigenbaum, E., eds.
The Handbook of Artificial Intelligence.
William Kaufman, 1982.

[2] Goto, Y., Matsuzaki, K., Kweon, I., Obatake, T.
CMU sidewalk navigation system.
In *Fall Joint Computer Conference.* ACM/IEEE, 1986.

[3] Hebert, M., Kanade, T.
Outdoor scene analysis using range data.
In *IEEE International Conference on Robotics and Automation.* 1986.

[4] Shafer, S., Stentz, A., Thorpe, C.
An architecture for sensor fusion in a mobile robot.
In *IEEE International Conference on Robotics and Automation.* 1986.

[5] Singh, J., et al.
NavLab: an autonomous vehicle.
Technical Report, Carnegie Mellon Robotics Institute, 1986.

[6] Thorpe, C.
Vision and navigation for the CMU Navlab.
In *SPIE.* Society of Photo-Optical Instrumentation Engineers, October, 1986.

Error Characteristics of Inertial Systems from the Viewpoint of Oscillation Theory

Elfriede Knickmeyer, Klaus-Peter Schwarz
The University of Calgary, Faculty of Engineering, Devision of Surveying Engineering, Calgary, Canada

Summary

The error behaviour of inertial systems is usually described using first and second order differential equations. These can be interpreted in the framework of oscillation theory. For unaided inertial systems these oscillations are determined by the response to initial conditions and to sensor errors. For aided systems they are strongly influenced by update measurements which are used as a state feedback. The effect of this feedback is discussed using a numerical example.

1. Introduction

In a way, the topic of this contribution goes back to Schuler's (1923) seminal paper in which he showed that a physical pendulum or, equivalently, a gyrocompass or a gyroscopic pendulum with an oscillation period of 84 minutes is insensitive to horizontal accelerations when operated on the surface of a spherical Earth. The frequency of oscillation is due to the gravity feedback of the system. It is determined by the ratio g/r, where g is the magnitude of gravity at the measurement point and r is its distance from the centre of mass of the Earth. The so-called Schuler oscillation is therefore a dynamic characteristic of any inertial navigation system with gravity feedback. It dominates the error behaviour of an unforced inertial system for mission times of less than one and a half hours. For longer mission times, the other two natural frequencies of a terrestrial inertial navigation system, the Earth and the Foucault frequencies, come into play, see for instance Vassiliou and Schwarz (1985) for a discussion.

While the analysis of the unforced dynamic system is important for an understanding of the general system behaviour, it is the analysis of the forced system which gives insight into the error characteristics of a specific mechanization. A simplified discussion is given in Schwarz (1986). It is interesting to note that most sensor errors generate velocity and position errors which are dominated by the natural oscillations of the system and that higher frequencies as e.g. those generated by local variations of the anomalous gravity field are low-pass filtered by the Schuler frequency.

However, there are sensor errors, like platform drift and height accelerometer bias, which will lead to instabilities of the system when not controlled by external information. This information is supplied to the system in the form of update measurements. It is the objective of this paper to study the effects of update measurements and their accuracy on the error behaviour and to interpret them in the framework of oscillation theory. The approach is tutorial. Emphasis will therefore be on the principles involved rather than on a complete discussion of all relevant details. The analogies provided by oscillation theory are useful for a deeper understanding of the error characteristics of inertial systems.

2. Inertial Navigation Systems as Oscillators

The errors of an inertial navigation system are usually described using the state space approach. It reduces higher order differential equations to a system of first-order differential equations. They are linearized using a reference trajectory. The linearized system equations are then written as

$$\dot{x}(t) = Fx(t) + Gu(t) \tag{1a}$$

where x(t) is the state vector, F is the dynamics matrix, Gu(t) is the forcing function and for this investigation

$$x^T = (\delta\varphi, \delta\lambda, \delta h; \delta\dot{\varphi}; \delta\dot{\lambda}, \delta\dot{h}; \varepsilon_E, \varepsilon_N, \varepsilon_u) \tag{1b}$$

where

$\delta\varphi, \delta\lambda, \delta h$ are position errors, $\delta\dot{\varphi}, \delta\dot{\lambda}, \delta\dot{h}$ are velocity errors, $\varepsilon_E, \varepsilon_N, \varepsilon_u$ are attitude errors, all of them expressed in the local-level frame. A formulation which is completely equivalent to (1) expresses the six first equations of the previous system by three second-order differential equations, with the vector

$$q^T = (\delta\varphi, \delta\lambda, \delta h; \varepsilon_E, \varepsilon_N, \varepsilon_u) \ .$$

This system is of the general form

$$A\ddot{q}(t) + B\dot{q}(t) + Cq(t) = d \tag{2}$$

describing a damped oscillator. For mechanical oscillators A is called the matrix of inertia, B is the damping matrix and C the stiffness matrix; d is the forcing function. The matrices A, B, C and the vector d on the right hand side are readily obtained from (1):

$$A = \begin{pmatrix} I & 0 \\ 0 & 0 \end{pmatrix}, \quad B = \begin{pmatrix} -F_{22} & 0 \\ -F_{32} & I \end{pmatrix}, \quad C = \begin{pmatrix} -F_{21} & -F_{23} \\ -F_{31} & -F_{33} \end{pmatrix}$$

where the subscripts 1 to 3 refer to submatrices of F obtained by partitioning the state vector as done in (1b). The left-hand side of equation (2) is the same for all system mechanizations. The right-hand side is different for each mechanization and can also be obtained from (1). It contains the non-zero elements of the forcing function Gu(t) which are specific for each system mechanization.

Following Britting (1971), we can therefore write

$$\Lambda x = d_j$$

where the left-hand side is the same for all system mechanizations while d_j is different for each mechanization. Thus, for a

Space stable system (i):

$$d_i = \begin{pmatrix} D\{R_i^l u_a - \Delta g^l + R_i^l (\Delta C_a^p)^T R_l^i f^l\} \\ R_i^l u_g \end{pmatrix}$$

Local level system (l):

$$d_l = \begin{pmatrix} D\{u_a - \Delta g^l + (\Delta C_a^p)^T f^l\} \\ u_g + T^l \omega_{il}^l + \Delta C_g^p \omega_{il}^l \end{pmatrix}$$

Strapdown system (p):

$$d_p = \begin{pmatrix} D\{R_p^l u_a - \Delta g^l + R_p^l (\Delta C_a^p)^T R_l^p f^l\} \\ R_p^l u_g + R_p^l T^p \omega_{ip}^p + R_p^l \Delta C_g^p \omega_{ip}^p \end{pmatrix}$$

where

$$D = \begin{pmatrix} \frac{1}{R} & 0 & 0 \\ 0 & \frac{1}{R\cos\phi} & 0 \\ 0 & 0 & 1 \end{pmatrix}$$

R^{β}_{α} rotation matrix from the α frame to the β frame where α and β can denote the inertial, local level, or platform frame. In the case of strapdown systems p denotes the body frame

$\omega^{\gamma}_{\alpha\beta}$ angular velocity of the β frame relative to the α frame, expressed in the γ frame

u_a accelerometer uncertainty which includes bias, scale factor and random noise

u_g gyro or torquer uncertainty

Δg anomalous gravity vector

ΔC^p_a matrix describing accelerometer misalignment

ΔC^p_g matrix describing gyro axis misalignment

T gyro torquer scale factor uncertainty matrix

f specific force.

As the left-hand side is independent of the system mechanization, the unforced oscillations of all systems are completely described by the left-hand side and nine initial conditions. They are dominated by the Earth, Schuler, and Foucault frequencies.

The actual oscillations of an inertial system are described by the response of the system to the nine initial conditions and to the forcing function d. Looking at the forcing functions of the different mechanizations one can see that the effect of the uncertainty in the gravitational model and the misalignment of the accelerometers is the same for all mechanizations. The first will generate mainly high frequency oscillations which will be low-pass filtered through the Schuler frequency, while the second one will change the equilibrium. However, the effects of accelerometer and gyro uncertainty and gyro misalignment are different for each mechanization because the errors are frequency modulated at a rate dependent on the angular motion between the platform frame and the local-level frame.

A forced oscillation shows frequencies which are equal to the ratio of the forcing frequency and the natural frequency. If this ratio is close to 1, as in the case of the space stable platform, this leads to undesirable resonance effects. In the strapdown case high frequency disturbances are typical; they will, however, be low-pass filtered through the Schuler frequency and will usually have little effect. Another difference in performance arises from the fact that the effect of gyro torquing scale factors and misalignments depends on the angular velocity of the platform with respect to the inertial frame. Therefore these errors are negligible for the space stable platform but not for other systems. Britting (1971) discusses some of these phenomena in more detail and the interested reader is directed to this book for further reference.

It should be noted that (2) with the forcing function as given above describes the behaviour of an unaided inertial system. Such a system shows instability in the height channel and has to be aided by external height information. This will be discussed in the next chapter.

Although the analysis in terms of forced oscillations gives a good understanding of the error behaviour of an unaided inertial system, the case usually encountered in survey applications is that of a system updated at regular time intervals. The effects of these updates on system oscillations will now be studied. In a first step, a general discussion of some undesirable effects of an oscillating system will be given, while in a second step state feedback will be discussed.

3. Oscillations of Aided Inertial Systems

Assume that the state vector x describes the errors of the INS. It thus gives the deviations from the true trajectory. Such a system is nonlinear, unstable, and coupled.

The error equations are intrinsically nonlinear because the dynamics matrix is dependent on position and velocity. The equations are linearized using a reference trajectory, thus yielding the linear systems (1) and (2). The amplitudes of the oscillations are dependent on the initial errors and indicate how well the reference trajectory approximates the true trajectory. Large amplitudes will amplify the effect of the nonlinearities, and iterations may be necessary for an accurate solution. The objective is therefore to approximate the true trajectory as closely as possible.

Three-dimensional inertial systems are unstable because of the height channel. This does not only result in height errors growing exponentially with time, it also leads to large nonlinear effects. As the different channels are coupled, large height errors will affect the determination of the horizontal position adversely.

The differential equations are coupled due to the Coriolis term. If the matrices A, B, and C in equation (2) were diagonal, the oscillations in the different coordinates would be decoupled and could be treated individually. Decoupling makes the analysis of the dynamic behaviour - which in the presence of coupling requires the manipulation of a ninth-order polynomial - much easier. Coupling may be due to the choice of coordinates. In that case it can be removed by a principal axis transformation. However, the principal coordinates obtained may not give a simple interpretation of the error behaviour. Also, this type of decoupling is only possible if the matrix B is a linear combination of A and C (Natke, 1983), which is not the case for the three axes inertial navigation system.

A possible remedy for these unwanted characteristics is the use of feedback. In surveying applications the goal is to estimate the best approximation to the true trajectory via the state vector of the dynamical system. This holds for both the nonlinear and the linearized system. The dynamical behaviour of the inertial survey system cannot be changed because it is determined by physical laws. In contrast to control theory where the goal is to change the dynamical behaviour, it is the objective in survey applications to estimate the difference between true trajectory and reference trajectory. Feedback is a way to reduce this difference. This can be achieved by deriving an input u that changes the reference trajectory. Such an input can for example be a torque on the platform.

Even though the goal is different in control theory, the formulas are the same. For a linear time-invariant system a control law of the following type can be considered

$$u(t) = r(t) - K\,x(t) \tag{3}$$

where r(t) is a reference input and K is a real constant feedback gain matrix determined from given specifications.

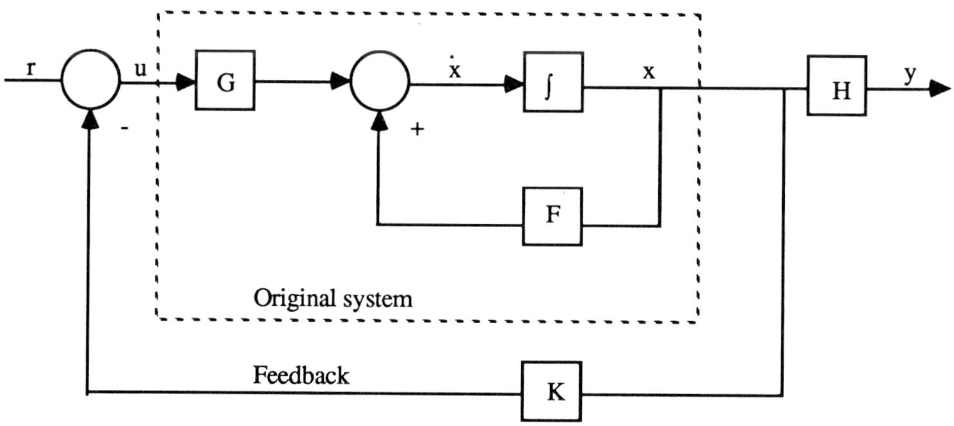

Fig.1: Blockdiagram of a feedback system

Combining (1) and (3) yields

$$\dot{x}(t) = (F - GK)x(t) + Gr(t) \tag{4}$$

which shows how the dynamics matrix is affected by feedback.

As the control law depends on the state x this is called state feedback. A control law depending on the output y would describe output feedback. However, state feedback is the more general of the two models; if a certain behaviour can be achieved by output feedback, it can always be achieved by state feedback, but not vice versa, see e.g. (Chen, 1970).

How a certain undesired behaviour can be reduced or eliminated by feedback will be shown for the three effects discussed previously, namely nonlinearity, instability, and coupling. The solutions are damping of large oscillations, stabilization of the system of differential equations and decoupling.

Feedback can be used to reduce large oscillations. To illustrate the principle, the simple example of the single axis inertial navigator will be considered. The state vector consists in this case of a position, velocity, and tilt error and the system equations are of the form

$$\frac{d}{dt}\begin{pmatrix} \delta r \\ \delta v \\ \varepsilon \end{pmatrix} = \begin{pmatrix} 0 & 1 & 0 \\ 0 & 0 & -g \\ 0 & 1/r & 0 \end{pmatrix} \cdot \begin{pmatrix} \delta r \\ \delta v \\ \varepsilon \end{pmatrix} \tag{5}$$

For the unforced oscillation, the reference input is zero.

The applied feedback control can for example consist of a torque applied to the platform. This will create an additional change in tilt. Thus

$$G = \begin{pmatrix} 0 \\ 0 \\ 1 \end{pmatrix}, \quad u = (-k_1 \delta r - k_2 \delta v - k_3 \varepsilon) \tag{6}$$

and the closed-loop dynamics matrix

$$F_c = \begin{pmatrix} 0 & 1 & 0 \\ 0 & 0 & -g \\ -k_1 & \frac{1}{r} - k_2 & -k_3 \end{pmatrix} \tag{7}$$

is obtained.

This type of feedback will drive the state vector to zero if the eigenvalues of F_c have negative real parts. To avoid solving the characteristic polynomial

$$\lambda^3 + k_3\lambda^2 + g(\frac{1}{r} - k_2)\lambda - gk_1 = 0 \qquad (8)$$

the Routh-Hurwitz criterion can be applied to find conditions for the feedback coefficients. These are $k_1<0$, $k_3>0$, and $k_1/k_3 - k_2 > 1/r$; k_1 changes the equilibrium of the oscillations, k_2 changes its frequency and k_3 introduces a damping. This type of feedback is discussed in more detail in chapter 4.

As an example for the stabilization by feedback, the height channel of an inertial survey system is considered. Neglecting the coupling with other channels the system equation without feedback is

$$\begin{pmatrix} \delta\dot{h} \\ \delta\ddot{h} \end{pmatrix} = \begin{pmatrix} 0 & 1 \\ 2\omega_s^2 & 0 \end{pmatrix} \begin{pmatrix} \delta h \\ \delta\dot{h} \end{pmatrix}$$

and has the solution

$$\delta h(t) = \cosh(\sqrt{2}\,\omega_s t)\delta h_o + \frac{1}{\sqrt{2}\omega_s}\sinh(\sqrt{2}\omega_s t)\delta\dot{h}_o .$$

As the height error grows rapidly with time the system is definitely unstable.

If the height error δh can be measured, it can be fed back and the following closed loop system is obtained

$$\begin{pmatrix} \delta\dot{h} \\ \delta\ddot{h} \end{pmatrix} = \begin{pmatrix} 0 & 1 \\ 2\omega_s^2 - k_1 & 0 \end{pmatrix} \cdot \begin{pmatrix} \delta h \\ \delta\dot{h} \end{pmatrix}$$

For $k_1 > 2\omega_s^2$ the system will be stable and have the solution

$$\delta h(t) = \cos\sqrt{k_1 - 2\omega_s^2}\, t \cdot \delta h_o + \frac{1}{\sqrt{k_1 - 2\omega_s^2}} \cdot \sin\sqrt{k_1 - 2\omega_s^2}\, t \cdot \delta\dot{h}_o .$$

The resulting oscillation has a period which is determined by the choice of k_1.

For the determination of δh either external height information or predicted heights are used. The control is applied by using these heights for calculating the reference trajectory.

Decoupling of the states by feedback is possible as well. From (4) it can be seen that if a control can be applied to all states then GK is of the same dimension as F and each individual element of F can be changed. The feedback would be designed in such a way that the elements of F which constitute the coupling are brought to zero. A discussion of the INS-coupling is quite involved. It will not be treated here.

4. The Use of Update Measurements in a Feedback Loop

This chapter will discuss the effect of update measurements which are fed back to the system by way of the state vector and study their effect on the oscillations of a single axis inertial navigator. Three cases will be discussed. First, both the system and the update measurements are without noise. Second, there is system noise but not measurement noise. Finally, both system and measurements are noisy.

In the first case, it is assumed that the error free state vector is known. As has been shown in the previous chapter a feedback of the tilt error into the tilt channel leads to damping. For $k_3 = 2\omega_s$ we have the limiting case. A feedback of the velocity error into the tilt channel leads to a change in frequency. For $k_2 \rightarrow -\infty$, $\omega \rightarrow \infty$ is obtained and δv, $\delta\varepsilon \rightarrow 0$. This means that all oscillations have been eliminated. For $k_1 = 0$, there will still be a constant position error. This error can be driven to zero by choosing $k_1 < 0$. This will create a change in tilt until $\delta r = 0$. The faster the system is torqued, the faster the state vector will go to zero. For the feedback coefficients this means that k_1 and k_2 become smaller and k_3 gets closer to 2ω. Only a noise-free system with noise-free measurements can be torqued at any rate. These assumptions are, however, unrealistic.

In the second case, the F-matrix describes the system behaviour only approximately, i.e.

$$\begin{pmatrix} \delta\dot{r} \\ \delta\dot{v} \\ \dot{\varepsilon} \end{pmatrix} = \begin{pmatrix} 0 & 1 & 0 \\ 0 & 0 & -g \\ -k_1 & \frac{1}{r}-k_2 & -k_3 \end{pmatrix} \begin{pmatrix} \delta r \\ \delta v \\ \varepsilon \end{pmatrix} + \begin{pmatrix} 0 \\ w_v \\ w_\varepsilon \end{pmatrix}$$

and $\delta\dot{v}$ and $\dot{\varepsilon}$ now contain components w_v and w_ε which describe uncorrelated white noise with the power spectral densities

$$S_v = 4.10^{-8}\ m^2 s^{-3},\ S_\varepsilon = 3.10^{-15}\ rad^2 s^{-1}$$

The initial conditions are $\delta r_0=0$, $\delta v_0=0$, $\varepsilon_0=2''$. As before, the state vector of this system is given without error at each point in time.

When applying feedback the system becomes unstable if it is torqued too fast, i.e. if the choice of feedback coefficients is made too extreme. Since the choice of the coefficients is limited by this fact it takes a much longer time for the system to regain its equilibrium.

The best results were obtained for $k_1 = 0$, $k_2 = -1/r$, $k_3 = 2\omega$ (see Fig. 2). Even though the state vector can be reduced considerably in this case, it cannot be made zero. This means that feedback control has to be applied all the time.

In the third case, it will be assumed that the state vector is estimated from noisy update measurements. Thus, a noisy state vector is fed back to the system. Such an estimate can for example be obtained by a Kalman filter. The system equations are in this case

$$\dot{x} = Fx + Gu + w, \tag{9}$$

with x, F, G, u as in (1) and white system noise w.
The observation or output equation is

$$y = Hx + e \tag{10}$$

where y is the observation or output vector, H is the design matrix, and e is the observation error. The equation of the continuous Kalman filter is

$$\dot{\hat{x}} = (F - KH)\,\hat{x} + Ky + Gu \tag{11}$$

where K is the Kalman gain matrix.
The control now depends on the estimated state

$$u = -k^T\hat{x} \quad . \tag{12}$$

Substituting (10) and (12) into (9) and (11) yields

$$\begin{pmatrix} \dot{x} \\ \dot{\hat{x}} \end{pmatrix} = \begin{pmatrix} F & -Gk^T \\ KH & F-KH-Gk^T \end{pmatrix} \cdot \begin{pmatrix} x \\ \hat{x} \end{pmatrix} + \begin{pmatrix} w \\ Ke \end{pmatrix} \quad . \tag{13}$$

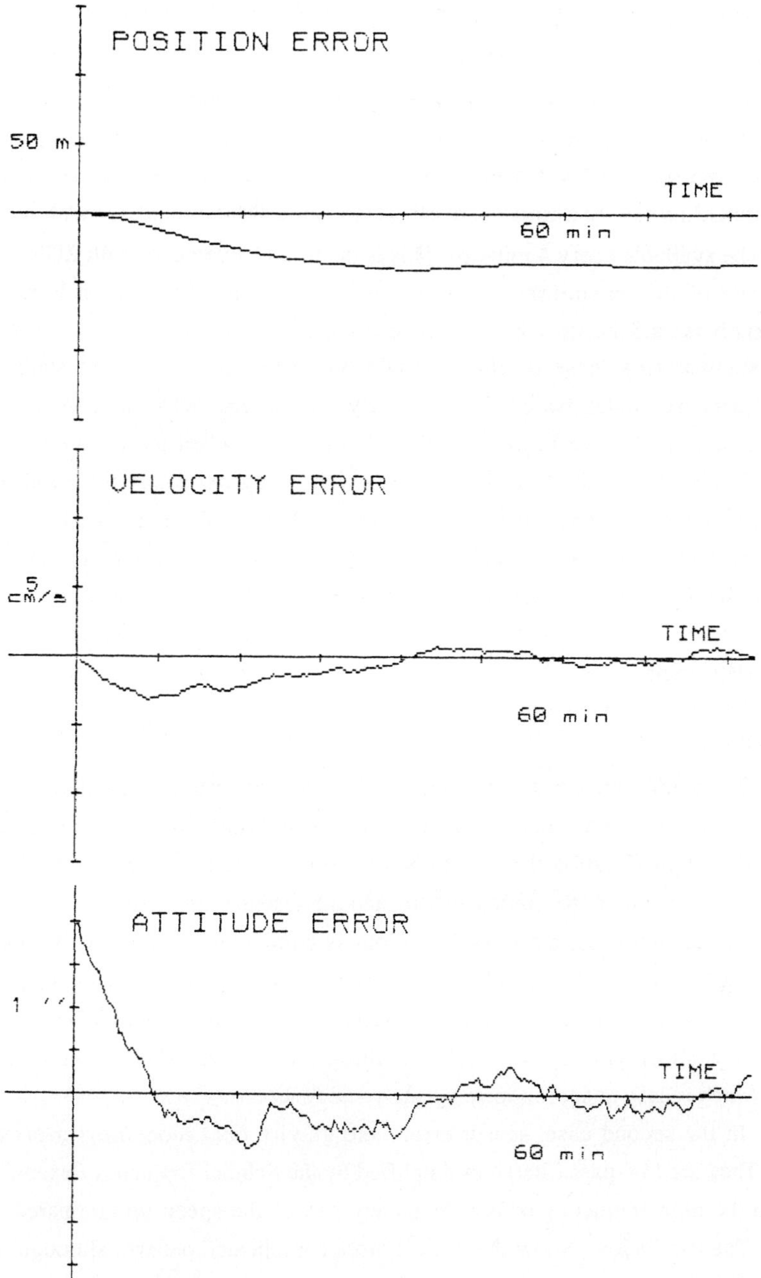

Fig.2: Behaviour of a noisy system with feedback

where the Kalman gain is chosen in such a way that $E(\hat{x} - x) = 0$ and the feedback coefficients k^T are chosen such that $x \to 0$ for $t \to \infty$ and thus also $\hat{x} \to 0$ as $t \to \infty$. Therefore feedback can be looked at as a mechanically or computationally realized iteration step. The performance of the system and the estimation error are shown for two different cases in Fig. 3 and Fig. 4. In the first case velocity updates with σ=1cm/s were assumed every 5 sec. This would be the case for an INS updated with GPS-velocity estimates. In the second case velocity updates with σ = 1 mm/s were assumed to be available every 4 minutes. This is the usual survey case with ZUPTs. In both cases the amplitude of the remaining oscillations are considerably larger than in the case without measurement noise and the settling time is much longer.

In summarizing these results it should be noted that even if the state vector can be determined error-free and is available continuously a linear feedback cannot bring the states to zero if there is system noise, see Fig.2. This is even more true when the update-measurements are noisy, see Fig. 3. During the first 1.5 hours - which is approximately the duration of a run for surveying applications - there will still be residual oscillations. Their amplitudes, however, will be much smaller than without feedback. The performance of the system is approximately the same if well-spaced update measurements of high accuracy are available. However, discontinuities in the system behaviour at update points are unavoidable. This leads to larger approximation errors when not at an update point.

5. Conclusions

By interpreting the erros in the output of an inertial system as oscillations about the true trajectory, an intuitive understanding of the error mechanism is possible, and interesting cross-connections to approximation theory can be obtained. Three cases have been discussed in this paper: the unforced system, the forced system, and the system with feedback.

In the first case, the error behaviour is completely dominated by the three natural frequencies of the system, namely the Schuler frequency, the Earth frequency, and the Foucault frequency. The system can be interpreted as oscillating in response to initial errors without taking into account plant or sensor errors. The resulting error curves show a well-defined periodic behaviour which is easy to approximate by analytical functions.

In the second case, sensor errors and gravity field modelling errors act as forcing functions. They are low-pass filtered or amplified by the Schuler frequency depending on whether they are in the high frequency or low frequency part of the spectrum compared to this natural frequency. The oscillations show therefore a more complicated pattern, although in most survey applications the natural frequencies are still dominant. In this case, the system mechanization plays a role and leads to distinct differences in the error behaviour.

In the third case, feedback is used to minimize the oscillations about the true trajectory.

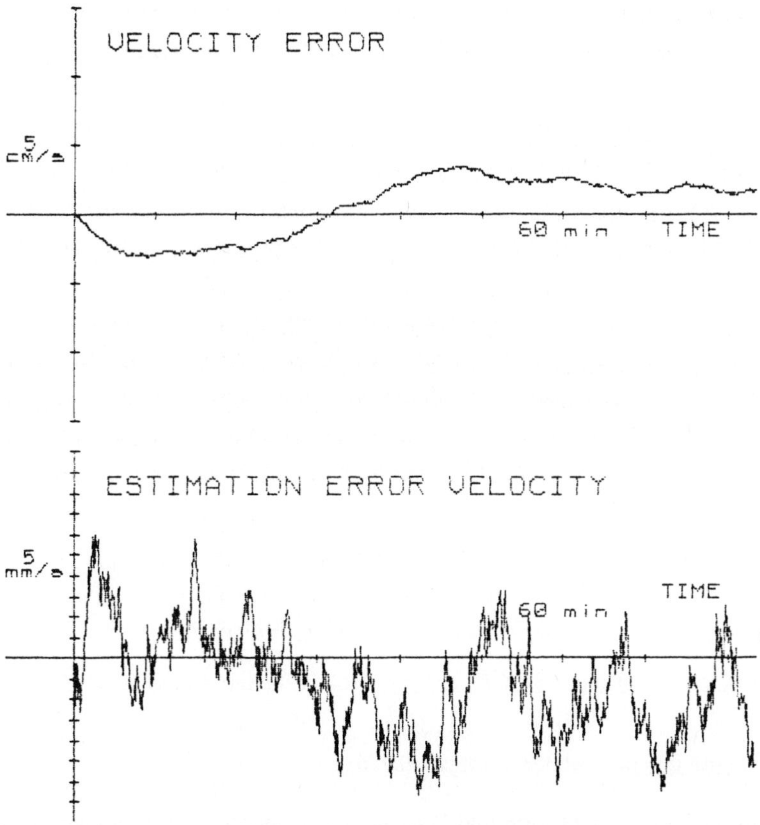

Fig.3: Velocity state and estimation error $x - \hat{x}$ in the case of system noise and continuous noisy velocity updates

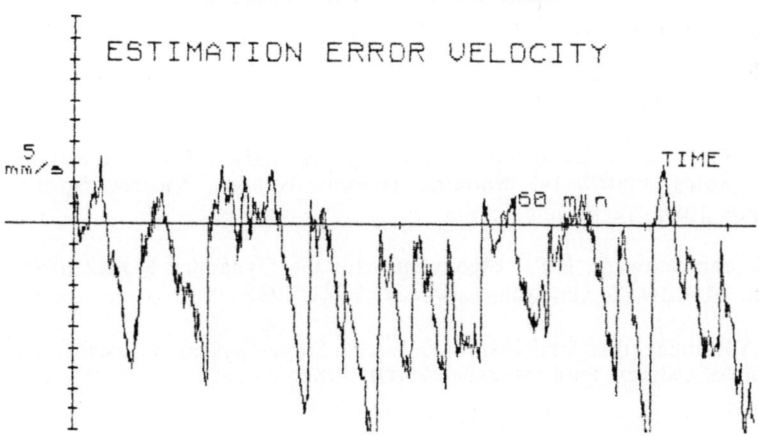

Fig.4: Estimation error in the case of system noise and well-spaced velocity updates

Theoretically, feedback can completely eliminate all oscillations; this means that the reference trajectory can be made to coincide with the true trajectory. This, however, requires a perfect system and a perfect knowledge of the state. In the real world, system noise is always present and the state vector is estimated, not known. In this case, state feedback will result in an approximation which shows small oscillations about the true trajectory during the initial period. With time these deviations become more and more stochastic. Obviously, the quality of the state vector estimation, i.e. in our case the quality and density of update measurements, is decisive for a good approximation.

When looking at these results from the point of view of approximation theory, feedback control can be seen as a computational or mechanical iteration step. It is applied in one form or other - especially for the height channel - in all existing survey systems in order to obtain a good reference trajectory in real time. The advantage of having a good first approximation is that effects of nonlinearities are small. This is paid for by discontinuities in the system dynamics at the update points.

References

Britting, K.R. Inertial Navigation Systems Analysis. Wiley-Interscience, New York, 1971.

Chen, C.T. Introduction to Linear System Theory. Holt, Rinehart and Winston Inc., New York, 1970.

Magnus, K. Schwingungen. Teubner, Stuttgart, 1976.

Natke, H.G. Einführung in die Theorie und Praxis der Zeitreihen - und Modalanalyse. Vieweg, Braunschweig, 1983.

Schmidt, G.T. Strapdown Inertial System - Theory and Applications. AGARD Lecture Series No. 95, 1978.

Schuler, M. Die Störung vou Pendel - und Kreiselapparaten durch die Beschleunigung des Fahrzeugs. Physikalische Zeitschrift, 24, 344-350, 1923.

Schwarz, K.P. The Error Model of Inertial Geodesy - A Study in Dynamic System Analysis. Lecture Notes, Fourth Int. Summer School, Admont, Austria; Springer-Verlag, Heidelberg, 1986.

Vassiliou, A.A. and Schwarz, K.P. Eigenvalues for the Dynamics Matrix used in Inertial Geodesy. Manuscripta Geodaetica, 10, 3, 213-221, 1985.

Wong, R.V.C. A Kalman Filter-Smoother for an Inertial Survey System of Local Level Type. The University of Calgary, Publication No. 20001, 1982.

Integration of INS and GPS Measurements

Alfred Kleusberg
Geodetic Research Laboratory, University of New Brunswick, Canada

Summary

The paper describes the basic features of INS and GPS absolute and relative positioning. The conceptual system differences are compiled and positioning results are classified according to their error characteristics. System integration advantages in navigation, positioning, and gravity field determination are discussed.

1. Introduction

Inertial Navigation Systems (INS) or Inertial Survey Systems (ISS) of various accuracy classes have been used extensively in the past for all types of navigation and for special tasks in surveying, e.g. densification of survey control networks. More recently, the satellite based Global Positioning System (GPS) has become available. Although this system is designed primarily for military navigation and remains in a prototype state as of today, its potential for precise kinematic and static surveying has been investigated thoroughly. In this paper, we will not attempt to describe either INS or GPS equipment, software, or results, and refer for this purpose to the related literature (e.g. K.P. Schwarz, 1986; D.E. Wells et al., 1986). Instead, we will restrict ourselves to a conceptual description of the two systems in order to explore the basic differences, and to compile a list of potential applications in navigation and surveying where system integration may provide advantages. The term "system integration" is used rather loosely here and will comprise all system combinations where INS and GPS measurements supplement or complement each other.

To characterize INS and GPS results, we will use in the following sections the phrases *high stability*, *medium stability*, and *poor stability* for position accuracies of decimetres, tens of metres, and hundreds of metres respectively. *Short term* and *long term* will stand for time intervals of minutes and hours respectively. *Absolute positioning* denotes the determination of coordinates in a global coordinate system, *relative positioning* means determination of coordinate differences between two stations in a global system.

2. Inertial Navigation System Measurements

The basic INS platform consists of three accelerometers mounted along orthogonal axes and three gyros defining a reference frame for the accelerometers. The INS may be mechanized in different ways (space stabilized, local level, strap-down) that will not be discussed here. The principal observations possible with this instrument are the accelerations along the orthogonal accelerometer frame axes, and the orientation of this frame. Therefore, the INS is basically an ***acceleration and orientation sensing system.*** Since these observations are available at a very high rate, the INS provides ***virtually continuous*** measurements. The INS does not require external communication or input during the measurement process; thus it is a ***self-contained system***.

The measured vector of accelerations, **a**, is expressed by the physical law

$$\mathbf{a}(t) = \frac{d^2}{dt^2}\mathbf{x}(t) + \mathbf{f}(t) - \mathbf{g}(\mathbf{x}(t)), \tag{1}$$

where

- **x** position vector in an earth-fixed reference system
- **f** expresses rotational accelerations of the accelerometer frame with respect to inertial space
- **g** gravity vector.

Obviously, if **f**, **g** and the initial state of the platform (absolute position, velocity and orientation) are known, the velocity and position changes relative to the initial state can be computed from eqn.(1) through integration

$$\frac{d}{dt}\mathbf{x}(t) = \frac{d}{dt}\mathbf{x}(t_0) + \int_{t_0}^{t} [\, \mathbf{a}(\tau) - \mathbf{f}(\tau) + \mathbf{g}(\mathbf{x}(\tau)) \,]\, d\tau \tag{2}$$

$$\mathbf{x}(t) = \mathbf{x}(t_0) + \int_{t_0}^{t} [\, \frac{d}{d\tau}\mathbf{x}(\tau) \,]\, d\tau. \tag{3}$$

This determination of ***change in absolute position*** finds its counterpart in ***inertial gravimetry***, if the platform position **x** and acceleration $d^2\mathbf{x}/dt^2$, and the apparent accelerations **f** are known (e.g. E.W. Grafarend, 1981). In this case we obtain from eqn. (1) for the gravitational accelerations

$$\mathbf{g}(\mathbf{x}(t)) = \frac{d^2}{dt^2}\mathbf{x}(t) + \mathbf{f}(t) - \mathbf{a}(t). \tag{4}$$

Various instrumental and initial state errors, and inaccuracies in the gravity field modelling employed in eqn. (2) complicate the processing of inertial data considerably and lead to relative position errors in the integration according to eqn. (3). For INS instrumentation presently available, these position errors are of the order of several centimetres after integration intervals of several minutes, and grow to hundreds of metres for integration intervals of several hours (e.g. D. Schröder et al., 1987). Thus the error behaviour of inertial positioning can be characterized by ***high short term stability*** and ***poor long term stability***. This unfavourable error situation is controlled in land surveys by frequently bringing the inertial platform to a complete rest. During these "Zero Velocity Updates", the platform velocity computed from eqn. (2) can be compared and reset to its known zero value. Using this observation procedure and appropriate error modelling, accuracies of a decimetre in position and several mGals in gravity over distances of tens of kilometres can be achieved (K.P. Schwarz, 1986). Obviously, the "Zero Velocity Update" technique is not feasible if the INS is continuously in motion like in airborne or marine applications.

3. Global Positioning System Measurements

GPS measurements used for navigation and kinematic surveying are pseudo-ranges P using either P-code or C/A-code (e.g. Wells et al., 1986) and/or carrier phases Φ. Presently available receivers provide these measurements at a rate between 0.05 Hz and 3 Hz. For the purpose of the following discussion we use the following simplified observation equations:

$$P(t) = ||\mathbf{x}^S(t) - \mathbf{x}(t)|| + b \tag{5}$$

$$\Phi(t) = ||\mathbf{x}^S(t) - \mathbf{x}(t)|| + b + A \tag{6}$$

where

$\mathbf{x}$ receiver position vector
$\mathbf{x}^S$ satellite position vector
b range bias due to clock errors
A carrier phase ambiguity, constant in time.

If pseudo-ranges P_i to at least four satellites S_i, i =1,2,3,4, are observed simultaneously at epoch t_k, eqn. (5) can be inverted to solve for the instantaneous receiver position

$$\mathbf{x}(t_k) = f(\mathbf{x}^{Si}(t_k), P_i(t_k)). \tag{7}$$

Thus the GPS receiver can be seen as a ***discrete epoch absolute position sensor***. Position accuracy is affected by measurement noise (few metres) and unmodelled short and long term systematic effects of the order of a few tens of metres. The propagation of these errors into the position solution can be characterized by a Dilution of Precision factor DOP > 1, expressing the geometry between the satellites and the receiver. Therefore, GPS pseudo-range absolute positioning can be said to have ***medium short term and long term stability***.

If carrier phases are observed simultaneously to at least four satellites for two subsequent epochs t_k and t_{k+1}, and the initial position of the receiver $\mathbf{x}(t_k)$ is known, we obtain from the difference of two eqns. (6) the ***change in absolute position***

$$\mathbf{x}(t_{k+1}) - \mathbf{x}(t_k) = f(\mathbf{x}(t_k), \mathbf{x}^{Si}(t_{k+1}), \mathbf{x}^{Si}(t_k), \Phi_i(t_{k+1}) - \Phi_i(t_k)). \tag{8}$$

These position differences are contaminated by carrier phase noise at the centimetre level and long term systematic effects of a few tens of metres, and therefore, show ***high short term stability*** and ***medium long term stability***.

In ***pseudo-range GPS relative positioning***, the position of a receiver at location $\mathbf{x}_2$ is determined relative to another receiver at a known location $\mathbf{x}_1$, from simultaneous pseudo range observations to at least four GPS satellites. In this case, the difference of two eqns. (5) is inverted to yield

$$\mathbf{x}_2(t_k) - \mathbf{x}_1(t_k) = f(\mathbf{x}_1(t_k), \mathbf{x}^{Si}(t_k), P_{2i}(t_k) - P_{1i}(t_k)). \tag{9}$$

Although most of the systematic effects are eliminated in this differencing process, the errors in these position differences remain dominated by metre-level measurement noise and can be characterized by ***medium short term and long term stability***.

Simultaneous GPS carrier phase observations at two locations $\mathbf{x}_1$ and $\mathbf{x}_2$ can be used to determine the change in relative position. Differencing the corresponding eqns. (8) for the two stations we obtain (cf. B. Remondi, 1985)

$$[\mathbf{x}_2(t_{k+1}) - \mathbf{x}_2(t_k)] - [\mathbf{x}_1(t_{k+1}) - \mathbf{x}_1(t_k)]$$

$$= f(\mathbf{x}_1(t_k), \mathbf{x}_2(t_k), \mathbf{x}^{Si}(t_{k+1}), \mathbf{x}^{Si}(t_k), \Phi_{1i}(t_{k+1}) - \Phi_{1i}(t_k), \Phi_{2i}(t_{k+1}) - \Phi_{2i}(t_k)). \qquad (10)$$

If one of the two stations involved remains static during the survey, eqn. (10) obviously determines the change in position of the second receiver referenced to the fixed monitor station. As carrier phase measurement noise is much lower than pseudo-range noise, and most systematic errors are removed in the observation differencing, the relative position errors are of the order of a few centimetres over distances of a few tens of kilometres. Therefore, ***changes in relative positions*** determined from GPS carrier phases exhibit ***high short term and long term stability***.

A GPS receiver becomes an ***orientation and position sensor*** if pseudo-ranges and carrier phases are measured simultaneously through three different antennae mounted on a common antenna base platform. For this type of instrument, the position is determined by means of eqn. (7) utilizing the pseudo-range observations of one of the three antennae. The carrier phase observations determine two linearly independent relative position vectors between the three antennae and therefore, yield the platform orientation. The orientation accuracy depends primarily on the phase stability of the antenna array and the degree of signal multipath interference, and amounts to about 1 degree for existing receivers (Starnav Corporation, 1986).

Since proper reception of external signals is required, the GPS receiver is a ***non self-contained system*** for positioning. This obviously restricts the applicability of GPS positioning to regions of unobstructed view to the GPS satellites. Moreover, even short interruptions caused by temporary shading of the received signal, etc., cause changes in the otherwise constant phase ambiguity terms A in eqn. (6). If not monitored by other means, these "cycle slips" render useless the carrier phase observations for positioning according to eqns. (8) and (10).

4. Potential for System Integration

Based on the discussions in the two previous sections the question of integration of GPS and INS measurements can be inspected more closely. To facilitate this task, we summarize the results of the previous sections in Table 1.

	INS	GPS Pseudo Range	GPS Carrier Phase
self-contained system	yes	no	no
gravity field sensitivity	yes	no	no
continuous measurements	yes	no	no
attitude sensing	yes	no	antenna array
Short term stability (absolute positioning)	N/A	medium	N/A
Long term stability (absolute positioning)	N/A	medium	N/A
Short term stability (relative positioning)	N/A	medium	N/A
Long term stability (relative positioning)	N/A	medium	N/A
Short term stability (change in absolute position)	high	N/A	high
Long term stability (change in absolute position)	poor	N/A	medium
Short term stability (change in relative position)	N/A	N/A	high
Long term stability (change in relative position)	N/A	N/A	high

Table 1: Comparison of INS and GPS characteristics

The first four rows of Table 1 show the substantial features of INS and GPS observations. In contrast to GPS, the INS is a self-contained system and provides virtually continuous measurements. These differences will prove of major importance in the system integration for absolute and relative positioning discussed in sections 4.1 and 4.2. In section 2, the gravity field sensitivity of INS was found to be a nuisance for inertial positioning. However, in combination with gravity field insensitive GPS measurements they may provide a tool for rapid gravity field mapping as shown in section 4.3. Possible use of the orientation sensitivity of INS and certain GPS receivers are discussed in section 4.4. The remaining part of Table 1 shows the capabilities of INS and GPS for positioning. Obviously, INS provides changes in absolute position only. GPS pseudo-ranges determine absolute and relative positions, whereas GPS carrier phases allow the determination of changes in absolute and relative positions. The positioning error characteristic in terms of stability in time is graphically visualized in Figure 1 and will be discussed in the following sections.

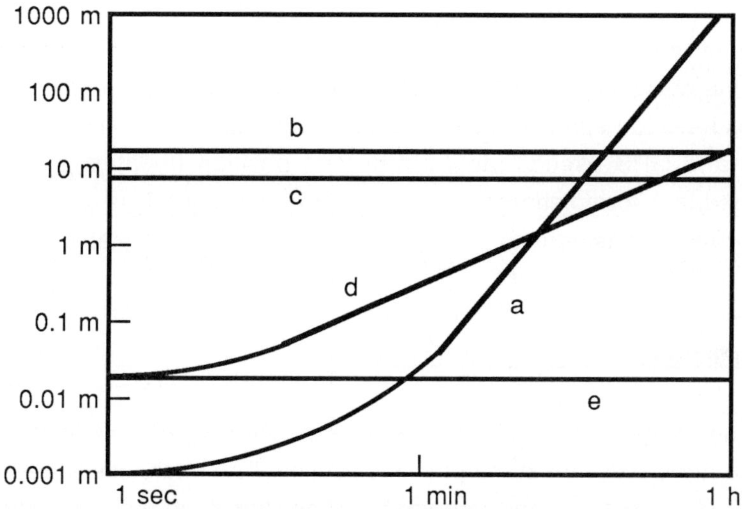

a: Change in absolute position, INS
b: Absolute position, GPS P-code pseudo-range
c: Relative position, GPS P-code pseudo-range
d: Change in absolute position, GPS carrier phase
e: Change in relative position, GPS carrier phase

Figure 1: INS and GPS accuracy stability characteristics

4.1 INS and GPS integration in absolute positioning

As can be seen in Table 1 and Figure 1, discrete epoch absolute positions determined from GPS pseudo-ranges exhibit a short and long term accuracy of several tens of metres. To achieve this level of accuracy, continuous reception of four GPS satellite signals in reasonable geometry, say DOP < 6, is required. Even for the full GPS satellite constellation, this condition will not be fulfilled for periods up to several minutes in certain areas (Wells et al., 1986). Additionally, in an operational environment temporary signal loss caused by shading has to be expected. During these periods of insufficiency of GPS positioning, the INS provides an adequate tool to maintain the required position accuracy, since INS is characterized by high short term stability in the determination of position changes, see Figure 1. In high dynamic applications, the low frequency discrete epoch GPS positions may not be sufficient to represent the trajectory of the receiver in the required resolution, and INS derived position changes can serve as an interpolator.

Typical applications of combined GPS and INS absolute positioning can be found in precise air and marine navigation. Since real time results are mandatory in navigation, the different systems have to be integrated in real time. Considering the INS as primary navigation system, and GPS pseudo-ranges providing position updates, a Kalman filter will be the adequate formulation for the system integration. This process will remove the pseudo-range measurement noise and retain the long term stability of GPS absolute positions.

4.2 INS and GPS integration in relative positioning

The reasoning for an integration of INS and relative positioning with GPS pseudo-ranges is identical to the discussion in the previous section. The main difference in the positioning results is the improved long term stability resulting from the elimination of systematic errors in the GPS pseudo-ranges, cf. Figure 1: Typical applications in this accuracy class can be found in inshore and river navigation, and hydrographic surveying.

As described in section 3, GPS carrier phase observations yield precise short and long term changes in relative position provided the initial relative position is known with a adequate accuracy. Figure 1 shows that comparable or better accuracies in position changes can be obtained from INS over very short time intervals only. Therefore, accuracy and stability gain is not a primary reason for system integration here. However, using GPS operationally for rapid kinematic surveys invariably will be hampered by more or less frequent signal loss caused by signal shading or interference. For obvious reasons, this problem is more pronounced in highly developed urban areas. Since GPS carrier phases become useless, if the period of signal loss is not bridged by other means (cf. section 3). According to Figure 1, the INS can maintain the required position stability for short time intervals and thus provide improved reliability in precise relative positioning. Additionally, the INS can serve as an interpolator in those applications requiring a continuous trajectory.

All of the above can be combined in an integrated system, if the INS is seen as the primary position sensor with frequent relative GPS measurements yielding position updates. Applications of the integrated system will include surveying, airborne photogrammetry, and close range precise navigation.

4.3 INS and GPS integration for gravity field determination

One of the principal differences between INS and GPS is the sensitivity of accelerometers to gravitational forces. Therefore, an integrated system can separate the gravity signal from the position information and operate as gravity vector mapping instrument in land applications (K.P. Schwarz, 1986). More advanced system could be used for the determination of area mean gravity anomalies from airborne gravimetry, if GPS carrier phase processing can be improved to yield mean accelerations for the platform of an INS (or a dynamic gravimeter) at the mGal level accuracy for integration intervals of several minutes (D.R. Bower and J.F. Halpenny, 1987).

4.4 INS and GPS integration in orientation determination

As mentioned in section 2, the INS yields implicitly the platform orientation referenced to the initial state by means of gyroscopes. Presently, the long term stability of high quality mechanical gyros utilized in precise INS exceeds by far the accuracy of a GPS orientation sensor as described in section 3. However, if low cost optical fibre ring gyros become available, their integration with improved GPS orientation sensors (L.E. Young et al., 1987) may prove advantageous. To make a fibre gyro useful for precise navigation, the large gyro drifts of up to 10 deg/h (W. Auch, 1985) will require frequent orientation updates. An integrated GPS - INS (- fibre gyro) sensor system could take advantage of the long term GPS orientation stability and the short term fibre gyro stability.

5. Conclusions

Of the four potential applications of an integrated GPS-INS system identified in the previous section, so far only only the approach described in section 4.1 has been thoroughly investigated and verified. For the other applications, only theoretical models exist at present. The development of operational instrumentation to be expected in the future will not only depend on technical feasibility and improved instrumentation, but also on the cost-benefit relation of integrated systems.

6. Acknowledgements

This paper was written while the author was on sabbatical leave at the University of Stuttgart with the Sonderforschungsbereich 228 "Hochgenaue Navigation". The hospitality of Prof. Dr. Erik Grafarend and discussions with the staff of the Department of Geodetic Science are gratefully acknowledged.

References

Auch W (1985): Optische Rotationssensoren. Technisches Messen 22, 199-207

Bower DR and Halpenny JF (1987): On the Feasibility of Airborne Gravity Measurements. Internal Report, Geophysics Division, Energy Mines and Resources Canada, Ottawa, Ontario

Boedecker G (1986): Inertial Gravimetry in a Test Network. in: Proceedings Third International Symposium Inertial Technology fo Surveying and Geodesy, University of Calgary, Calgary, Alberta, 161-172

Grafarend EW (1981): From Kinematical Geodesy to Inertial Positioning. Bull.Geod. 55, 286-299

Remondi B (1985): Performing Centimeter-Level Surveys in Seconds with GPS Carrier Phase: Initial Results. NOAA Technical Memorandum NOS NGS-43, Rockville, Md.

Schröder D, Wiegner S, Grafarend EW and Schaffrin B (1987): Simulation eines Geodätischen Inertialsystems. in: Schriftenreihe Studiengang Vermessungswesen, Universität der Bundeswehr München, Heft 22, 197-235

Schwarz, KP (1986): Geiod Profiles from an Integration of GPS-Satellites and Inertial Data. International Symposium on the Definition of the Geoid, Florence, Italy, May

Schwarz, KP, Editor (1986): Proceedings Third International Symposium Inertial Technology fo Surveying and Geodesy. University of Calgary, Calgary, Alberta

Starnav Corporation (1986): PAHRS Model 1 Product Announcement, Ottawa , Ontario

Wells DE, Beck N, Delikaraoglou D, Kleusberg A, Krakiwsky EJ, Lachapelle G, Langley RB, Nakiboglou M, Schwarz KP, Tranquilla JM and Vanicek P (1986): Guide to GPS Positioning. Canadian GPS Associates, Fredericton, N.B., Canada

Young LE, Wu SC and Dixon TH (1987): Decimeter GPS Positioning for Surface Element of Sea Floor Geodesy System. in: Proceedings International Symposium Marine Positioning, D.Reidel Publ. Comp.

B. SATELLITE TECHNIQUES

Adaptive Antennas for GPS

Samuel Bloch
Standard Elektrik Lorenz A.G., Stuttgart

Summary

The utilization of spread spectrum techniques by GPS enables the processing of satellite signals with a power level far below that of the normal background noise. The signal to noise gain provided by GPS correlation receivers is sufficient for operation in a benign electromagnetic environment. Thus, civil users generally need not be overly concerned about the noise performance of their receivers. A quite different situation applies for the military utilization of GPS. Here, users are expected to operate in the vicinity of hostile high-power "Jammers". The signals radiated by these noise sources may easily drive the SNR at the receiver front-end below an acceptable minimum, in many cases they will even force the receiver into saturation. This will result in a complete loss of the position-fix capability or at least lead to a considerably impaired performance.

In order to mitigate the jamming threat, military high performance GPS receivers will be equipped in the future with adaptive 'Null Steering' antennas. These antennas automatically adapt their pattern according to the actual jamming scenario. They generate deep nulls in the direction of all active jammers and thus selectively reduce the noise power passed to the receiver.

Although designed for military use, adaptive Null-Steering antennas might in special cases also be useful for civil applications. They will, for example, allow high precision geodetic GPS measurements in the proximity of strong interferers (e.g. L-Band radars, TACAN/DME transmitters, earth stations for satellite communication as well as harmonics of UHF transmitters etc.).

The paper gives an overview on various null-steering methods available for the GPS application.

1. The generation of selective nulls in the antenna pattern

Adaptive nulling antennas are comprised of a dedicated signal processor and the following main rf units:

- Antenna Array (referred to as CRPA = Controlled Reception Pattern Antenna)
- Adjustable Phase and Amplitude Weighting Elements
- Rf combiner

A simple example shall demonstrate the combined operation of the rf components.

Figure 1 depicts the reception of a jammer (denoted J) by a two-element array. The relative phase of the signals excited by the jammer in both branches of the array is given by:

$$\Delta\Phi = \frac{2\pi L}{\lambda} \sin\Theta \quad [\text{ rad.}]$$

L = array size [m]

λ = carrier wavelength [m]

Θ = angle of arrival with respect to broadside [radian]

In order to suppress the jammer, an appropriate phase delay needs to be introduced in one of the two array branches. This phase shift compensates $\Delta\Phi$ and thus causes both signals to arrive at the rf combiner with identical amplitude and phase. The subtraction performed by the rf combiner then perfectly cancels the jamming power. This result is comparable to steering an antenna null in the direction of the jammer.

Unlike jammers, wanted GPS signals S arriving from undisturbed directions will not be eliminated by the array, as the introduced phase shift $\Delta\Phi$ is 'mismatched' with respect to their angle of arrival. Consequently, these signals will be passed to the array output. The larger the angular separation between the GPS satellites and the interfering source the better the Signal/Jammer discrimination that can be provided by the nulling process (spatial filtering).

The array shown in figure 1 utilizes a voltage controlled phase shifter and an adjustable attenuator. (The attenuator is required for operation in a

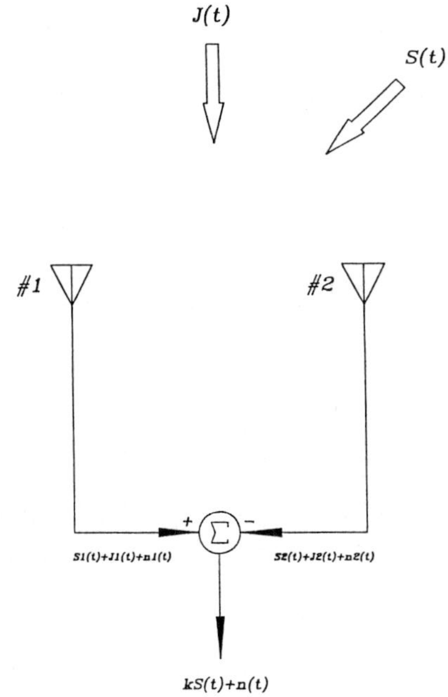

FIGURE 1: THE CONCEPT OF SELECTIVE NULL STEERING

multi jammer environment as discussed later). These two elements appropriately weight the signal in the variable array branch both in phase and amplitude.

In reality the implementation of adjustable high resolution phase shifters is rather cumbersome. For practical applications, therefore, an alternative signal weighting method is preferred. The weighting elements in this case exclusively deploy controlled attenuators as shown in Figure 2. To explain this weighting approach we assume that the amplitude and phase of a signal $X(t) = U \sin(\omega t)$ are to be modified such that $X(t)$ is transformed into a new Signal $Y(t) = A\, U \sin(\omega t+\Phi)$. This objective may be achieved by sub-dividing $X(t)$ into two components in phase quadrature. A hybrid used for this task provides the Inphase and Quadrature parts $X_I(t)$ and $X_Q(t)$. $X_I(t)$ and $X_Q(t)$ are weighted in amplitude by means of two parallel variable attenuators W_I and W_Q. The weighted signals are eventually superimposed to provide the combined result:

$$Y(t) = W_I X_I(t) + W_Q X_Q(t) = A\, U \sin(\omega t+\Phi)$$

$$\hat{U}\sin(\omega t) \;->\; A\hat{U}\sin(\omega t+\varphi)$$

$$A\hat{U}\sin(\omega t+\varphi) = \underbrace{A\cos(\varphi)}_{W_I}\;\underbrace{\hat{U}\sin(\omega t)}_{X_I(t)} + \underbrace{A\sin(\varphi)}_{W_Q}\;\underbrace{\hat{U}\cos(\omega t)}_{X_Q(t)}$$

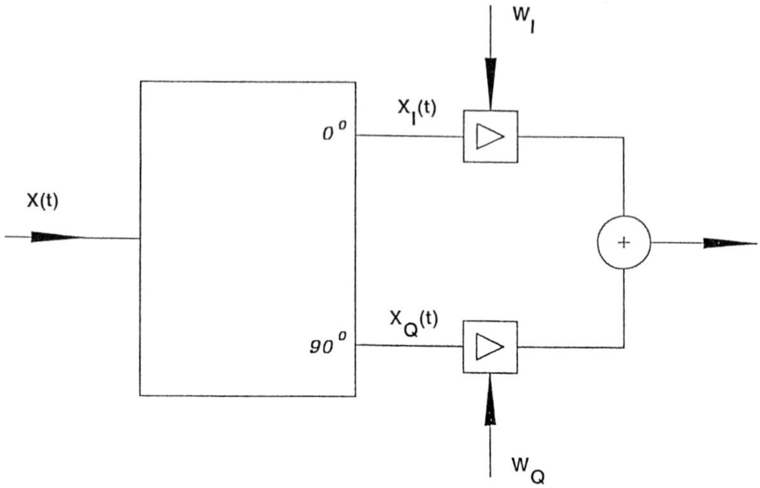

FIGURE 2: AMPLITUDE AND PHASE WEIGHTING

It can be shown that:

Amplitude weighting factor: $A = \sqrt{W_I^2 + W_Q^2}$

Phase shift: $\Phi = \tan^{-1}(W_Q/W_I)$

In the following discussions W_I and W_Q will be denoted by the complex coefficient **W**. The real and imaginary parts of **W** present W_I and W_Q respectively.

The two-element array shown in figure 1 can handle one single jammer only. For operation in a multi jammer scenario, a more complex antenna array with multiple weighting elements is required. The signals in various branches of this array need to be weighted both in phase <u>and</u> amplitude. Such an antenna configuration with N+1 array branches is shown in figure 3. This null steering antenna allows the suppression of up to N simultaneous jammers.
As indicated in figure 3, one of the array branches always remains unweighted. This branch is referred to as 'Reference Channel'. Its signals

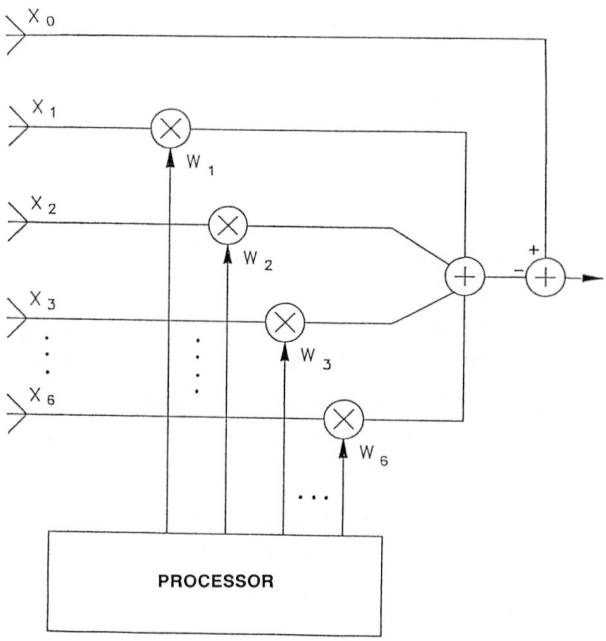

FIGURE 3: ADAPTIVE ANTENNA CONFIGURATION

are transferred directly to the combined array output. Each of the other N branches ('Auxiliary Channels') is equipped with a complex weighting element **W**. The signals in all auxiliary channels are weighted in phase and amplitude before being superimposed and eventually subtracted from the reference signal.

The auxiliary channels are inhibited during undisturbed GPS operation (no jammers). The array output in this case is established by the reference signal only. Thus, the nulling array behaves under beneign conditions like a conventional GPS antenna with almost hemispheric coverage.

The auxiliary channels become activated only when the GPS receiver is exposed to interferences. The auxiliary signals in this case get appropriately weighted and produce the required nulls in the antenna pattern. The number of nulls, their direction and shape are adapted to the actual jamming conditions.

Almost all GPS nulling antennas currently under development in the U.S.A, U.K. and Germany utilize circular array configurations. The CRPA reference element is usually positioned at the center while the auxiliary elements are located on a circle with a diameter of about 0.9 λ. Up to 6 auxiliary array elements are commonly used, they enable the nulling of maximum 6 simultaneous jammers.

2. Automatic Antena Adaptation

Having discussed the concept of selective null steering we now focus our attention on methods introduced for automatic antenna control.

A nulling antenna is expected to instantaneously react on changes in the jamming scenario. Therefore, the processor indicated in figure 3 is required to continuously monitor all received signals. It has to sense and analyze changes in the reception conditions and adaptively 'optimize' the setting of the antenna weights. For this task the processor is required to distinguish between wanted GPS signals and interferences. In the GPS case such discrimination can simply be based on power level observations. Due to the low power of the GPS signals (ca. -133 dBm) one can assume that the power measured at the antenna output almost entirely originates from interfering sources (namely thermal noise and jammers). This assumption leads to a rather simple adaptation algorithm. The processor attempts to minimize the signal power at the combined array output. This is equivalent to minimizing the reception of unwanted signals. The drawback of this simple algorithm is that it completely ignors all information pertaining to the wanted GPS satellite signals. For this reason the null steering operation may also affect the reception of wanted signals. (Remark! more sophisticated adaptation algorithms are applied for communications tasks. These algorithms maximize the Signal-to-Jammer ratio rather then merely minimizing the interference power at the antenna output. This is being achived by including data about the angle-of-arrival or alternatively about the spectral structure of the wanted signals. The adaptation process is then subjected to constraints that prevent an attenuation of the wanted signals. Unfortunately, such algorithms are inappropriate for the GPS application. Unlike communications, where usually a single signal is being received at a time, GPS receivers are expected to process transmissions of four, five or even more satellites simultaneously. Each satellite signal establishes another constraint, thus reducing the degrees-of-freedom available for jammer cancellation. This leads to an impaired multi-jammer capability of the antenna.)

The following vectorial equation expresses the average signal power at the array output:

$$P_{out} = E\{(x_o-\underline{W}^T\underline{X})^*(x_o-\underline{X}^T\underline{W})\} = E\{x_o^*x_o\}-E\{x_o^*\underline{X}^T\underline{W}\}-E\{\underline{W}^H\underline{X}^*x_o\}+\underline{W}^H E\{\underline{X}^*\underline{X}^T\}\underline{W}$$

E{ } = statistical expectation.
$()^T$ = Vector (or matrix) transpose.
$()^*$ = Complex conjugate vector (or matrix).
$()^H$ = Complex conjugate transpose.
N = Number of auxiliary array branches
$\underline{W}$ = $(W_1\ W_2\ W_3\ \dots\ W_N)^T$ vectorial antenna weights representation. All **W** are complex with real and imaginary parts denoting the inphase and quadrature components of the weighting elements.
$\underline{X}$ = $(X_1\ X_2\ X_3\ \dots\ X_N)^T$ vectorial representation of the signals at the N auxiliary array channels. All signals **X** are complex, with each element presenting both signal amplitude and phase.
x_o = Complex representation of the reference signal.

It can be shown, that the antenna weight vector leading to a minimized output power is given by the following expression:

$$\underline{W} = \underline{M}^{-1}\ \underline{P} \quad \textbf{(Wiener-Hopf condition)}$$

$\underline{M}$ = Covariance matrix of the auxiliary signals.
$\underline{P}$ = Cross correlation vector.
$()^{-1}$ = Matrix inverse.

$$\underline{M} = \begin{pmatrix} E\{X_1^*X_1\} & E\{X_1^*X_2\} & E\{X_1^*X_3\} & \dots & E\{X_1^*X_N\} \\ E\{X_2^*X_1\} & E\{X_2^*X_2\} & E\{X_2^*X_3\} & \dots & E\{X_2^*X_N\} \\ E\{X_3^*X_1\} & E\{X_3^*X_2\} & E\{X_3^*X_3\} & \dots & E\{X_3^*X_N\} \\ \cdot & \cdot & \cdot & \dots & \cdot \\ \cdot & \cdot & \cdot & \dots & \cdot \\ E\{X_N^*X_1\} & E\{X_N^*X_2\} & E\{X_N^*X_3\} & \dots & E\{X_N^*X_N\} \end{pmatrix} \qquad \underline{P} = \begin{pmatrix} E\{X_0^*X_1\} \\ E\{X_0^*X_2\} \\ E\{X_0^*X_3\} \\ \cdot \\ \cdot \\ E\{X_0^*X_N\} \end{pmatrix}$$

Different techniques are available for automatic antenna weight control. They all lead to an antenna setting which adequately approximates the optimum solution given by the Wiener-Hopf condition.
Three such methods are briefly discussed in the next paragraphs.

2.1 Sample Matrix Inversion (SMI)

This method provides a computational solution to the Wiener-Hopf equation. It utilizes periodic time sampling of the signals at the N+1 array elements. Each sample yields a new set of N+1 complex signal values. These data contain information about the amplitude and phase distribution across the array aperture at the sampling instance. By collecting enough time samples a mathematical processor is able to compute all elements of the covariance matrix and the cross correlation vector. The processor thus provides an avereged estimate of $\underline{M}$ and $\underline{P}$. It then computes the matrix inversion $\underline{M}^{-1}$ and multiplies the resulting matrix with the estimated cross-correlation vector. These methematical operations provide an approximation of the optimum weighting vector $\underline{W}$.
The main advantage of the SMI method is its fast reaction. The optimum weight solution may be determined within a few sampling periods. For an N element array about 2N time samples are needed to adequately adapt the antenna. Thus, without time limits set by the arithmetic processor, the SMI antenna adaptation may be completed within a few microseconds. The antenna control handles a wide dynamic range of jamming powers.
Unfortunately, the processing complexity of the SMI method is relatively high. Both the estimation of the covariance matrix and the determination of its inversed version set high mathematical load on the processor. Further, it can be shown that in many practical situations extremely high arithmetic precision is required in order to avoid numeric instability and excessive performance degradation. To solve this problem a novel method [2] has been suggested that allows the computation of the optimum antenna weights directly from the element data samples, without the need to evaluate $\underline{M}^{-1}$. This approach is less susceptible to problems caused by large spread in the received jamming power. It can be mechanized using dedicated Systolic Array Processors, thus providing high computational speed.

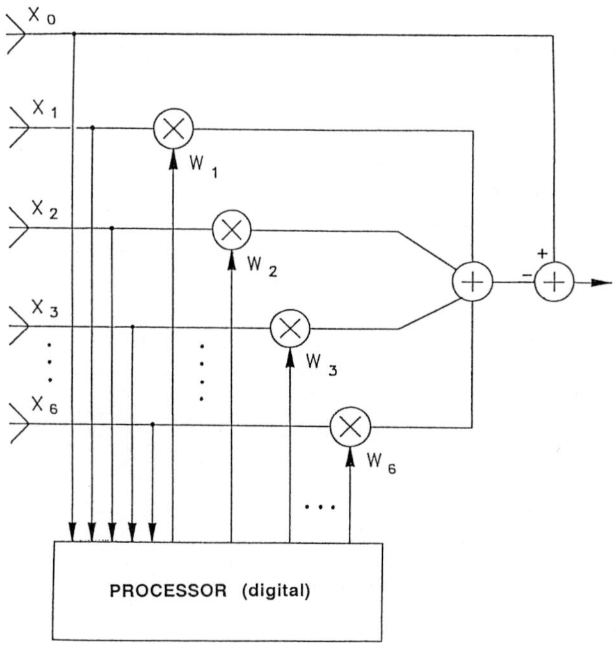

FIGURE 4: SAMPLE MATRIX INVERSION

A general drawback of the SMI method is that it is open loop. The entire information used for antenna adaptation is collected at the N+1 array inputs (s. figure 4). The processor does not get any feedback about the true performance at the combined array output. For this reason all components used for null steering need to be tightly specified and carefully calibrated. This strict requirement makes the implementation of SMI nulling antennas for GPS applications impractical.

2.2 Weight Perturbation (WP)

The Weight Perturbation antenna control leads iteratively to the power minimization at the combined array output. The processor dithers the weights in various array branches about their mean value. It perturbs the weight setting in small increments while simultaneousely monitoring the resulting variation in the output power. Doing this, the processor estimates by trial-and-error the path of steepest descent leading to the

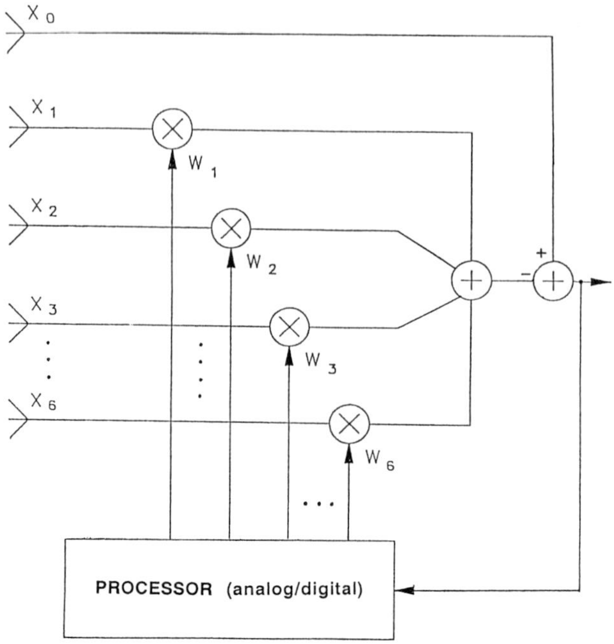

FIGURE 5: WEIGHT PERTURBATION

optimum setting. The magnitude of the introduced perturbation increments is adapted to the already achieved jammer cancellation.

The main advantage of this method is its simplicity. The Weight Perturbation approach leads to implementations which are based on microprocessor technology and therefore enables low cost realization. The closed-loop antenna control introduced here automatically corrects for nonlinearities and phase and amplitude differentials in various array channels. Hardware imperfections are compensated by feedback from the array output (s. figure 5). Thus, requirements pertaining to the rf components used are much less strict than in the case of the afore mentioned SMI approach.

A disadvantage of the WP method is that at least one of the array weights is always perturbed from its optimum setting. This inherent WP property gives rise to a so-called 'perturbation noise'. However, by using sufficiently small perturbation steps the added control noise can be kept adequately low.

The convergence time of the WP nulling procedure largely depends on the actual jamming conditions. It is very short when a single strong jammer is being received (a few milliseconds). Yet, the antenna reaction may considerably be slowed down in the case that several jammers are received simultaneously. This is especially true when the interfering sources are only slightly separated in space and/or when they significantly differ in power. The antenna response under such unfavourable conditions may become excessively slow (in the order of seconds).

2.3 Steepest Descent Correlation Loop (SD)

In this best known approach, the processor correlates the array output signal with each of the auxiliary input signals. All N cross-correlation results gained by this procedure are iteratively driven by the antenna control to zero. Thus, at steady state the combined array output and the individual auxiliary signals become decorrelated. This indicates that the antenna has generated spatial nulls in the direction of all interfering sources. The closed loop antenna adaptation leads to minimized output power along the path of steepest descent. At steady state the array approaches the optimum weighting solution given by the Wiener-Hopf condition. The SD method is tolerant to non linearities and deviations in the phase and amplitude response of the rf components used.
Figure 6 depicts the loop configuration utilized for antenna control in each of the N auxiliary branches. Both the Amplifier (Gain G) and the low-pass filter (Time Constant RC) determine the convergence behaviour of the control loop. It can be shown that for extremely strong jammers, the loop reaction becomes excessively sluggish or even unstable. The problem might be solved by reducing the amplifier gain factor G. This, however, simultaneously reduces the system sensitivity towards weak jammers. A fixed amplifier gain, therefore, confines the dynamic range of the control loop. This drawback may be mitigated by applying Automatic Gain Control. The AGC adjusts the amplifier factor G (and possibly also the time constant RC) according to the actual received interference power. The AGC function is readily implemented in the case of digitally controlled antennas. It, however, noticibly increases the hardware complexity when analog processing is applied.

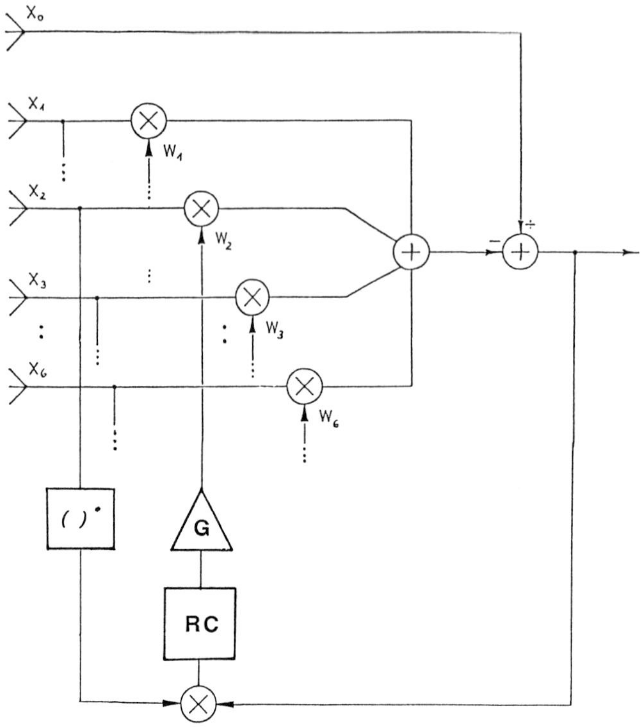

FIGURE 6: STEEPEST DESCENT CONTROL LOOP

When compared with the Weight Perturbation method, the Steepest Descent approach makes necessary extra couplers, rf amplifieres and transmission lines for the auxiliary input signal probes. It also requires a more sophisticated dedicated signal processor. Its performance, however, is superior to that of the Weight Perturbation antenna as it is less susceptible to perturbation noise and in general provides a faster reaction under unfavourable jamming conditions.

3. Discussion of the Nulling Performance

Figure 7 presents the response of a weight perturbation adaptive antenna to the reception of a single strong interfering source (40 dB above noise level). The jammer in this case is received from azimuth 15° and elevation

FIGURE 7: INTERFERENCE CANCELATION

(Single Strong Jammer 40 dB Above Noise)

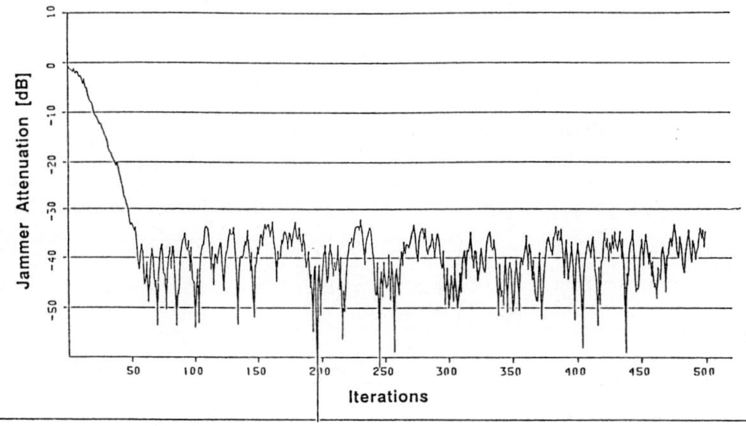

0°. The antenna nulling reduces the jammer power at the combined output by more than 32 dB. The jammer cancellation is performed within 50 iterations, this corresponds to a convergence time of 2 ms.

The antenna patterns shown in figure 9 pertain to the reception of three simultaneous interfering sources of moderate power (each 10 dB above noise level). The diagrams show the resulting antenna coverage at six different elevation angles. The results indicate the performance following an adaptation time of ca. 40 ms. Figure 9 demonstrates that two of the three jammers (Az_1=15° El_1=0° and Az_2=45° El_2=15°) become attenuated by the nulling process well bellow the noise level. The power level of the third jammer (Az_3=90° El_3=30°), although reduced by 6 dB, still remains 4 dB above noise. This indicates that the antenna convergence has not yet been completed. The relatively low power of the three jammers and their mutual interaction slow down the antenna adaptation as compared to the single jammer case.

As already discussed in paragraph 2, the jamming signals force the adaptive antenna to modify its patterns disregarding the characteristics of the wanted GPS signals. Figure 8 for example clearly demonstrates that

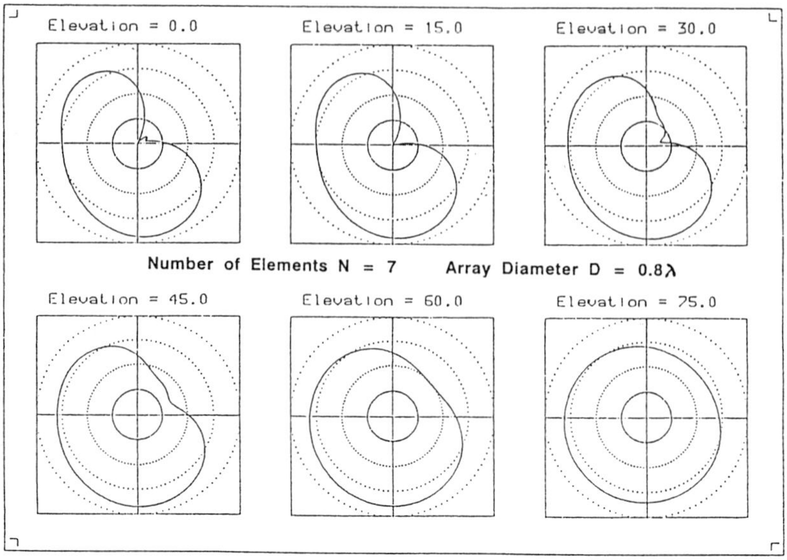

FIGURE 8:

ADAPTED ANTENNE PATTERNS AT 6 DIFFERENT ELEVATIONS
(3 Jammers 10 dB Above Noise, Convergence Time 40ms)

satellites received from azimuth sector 0° through 90° (at all elevations bellow 75°) will also be affected by the antenna nulls. The signals of these satellites may become marginal for a satisfactory GPS operation. In contrast, all signals arriving from other directions will be adequately received. They may in some cases even be enhanced by additional array gain. Here, the adaptive antenna retains reception of signals that otherwise would have become completely useless.

[1] IEEE Transactions on Antennas and Propagation, Special Issue on Adaptive Antennas, 1976, Vol. AP-24, No. 5.

[2] A Novel Algorithm and Architecture for Adaptive Digital Beamforming, C.R Ward P.J. Hargrave and J.G. McWhirter, IEEE Transactions Antennas and Propagation, Special Issue on Adaptive Processing Antenna Systems, March 1986, Vol. AP-34, No. 3.

Empirical Accuracy of Positions Computed from Airborne GPS Data

Peter Friess
Institut für Photogrammetrie, Universität Stuttgart

1. Introduction

The NAVSTAR Global Positioning System (GPS) is a satellite based navigation system designed especially for real-time threedimensional position and velocity determination. Real-time performance, however, is, except for navigation purposes, not necessarily required for the majority of the positioning tasks. Post-processing of the recorded raw GPS data is, therefore, the evaluation technique mainly applied, also by utilizing GPS in geodetic disciplines. Numerous investigations in static geodetic positioning (refs. 3, 14, 15) have meanwhile demonstrated a high accuracy potential of the GPS. Relative positioning accuracies in the order of centimetres are typical when using GPS carrier phase observations in the static mode.

In this presentation attention is concentrated solely on kinematic positioning, i.e. the determination of the trajectory of a moving object, whereby an aircraft is the subject of the investigations. Many airborne scientific applications such as laser bathymetry, airborne gravimetry, airborne laser profiling and aerial triangulation (refs. 1, 2, 6) would be facilitated if the trajectory of the aircraft could be determined within a certain degree of precision. GPS is an appropiate positioning system for these missions, as previous investigations in kinematic positioning (refs. 5, 8, 9, 11, 12, 16) have demonstrated.

At present, the accuracy characteristics of the positions computed from airborne GPS data are of main interest. For this reason a controlled photo flight with simultaneous GPS data registration over a test area in the Netherlands has been carried out. The Survey Department of Rijkswaterstaat Delft was in charge of the planning, preparation and realization of the project (ref. 4).

Paper submitted to Comm. III, 16th ISPRS Congress, Kyoto, 1988

The aim of the test is the investigation of the accuracy of kinematic positioning with GPS in high dynamic applications, such as airborne applications. In order to guarantee a proper control of the positions derived from the GPS observations, the test flight was performed in combination with a large scale aerial triangulation. The aerial triangulation provides the coordinates of the projection centres of the aerial camera, which can be used for comparison with the GPS antenna positions.

The evaluation of the test flight data has not yet been completed. Therefore, this paper presents only first results of the analyses of the carrier phase observations with respect to the accuracy and systematic effects of the positions computed therefrom.

2. Test description

The test flight itself took place on 10.06 and 12.06.1987. The complete flight comprises 16 parallel strips, with a length of approximately 4 km each. The average speed of the aircraft during this mission was about 240 km/h, thus the flight duration of each of the strips was approximately 60 sec.

The utilized GPS instrumentation consisted of two Sercel receivers : one NR52 receiver stationarily placed at a known reference point (Fig. 1) and one TR5SB receiver on board the aircraft. Both are 5 channel, L1 C/A-code receivers providing pseudorange and carrier phase observations and additionally the broadcast satellite navigation message.

During the entire flight duration pseudorange and carrier phase observations were carried out and recorded simultaneously with an observation rate of 0.6 sec with each receiver, at the reference point and on the aircraft, respectively. In addition, aerial photographs were taken along the 16 parallel strips approximately every 3 sec, which resulted in a total of 360 aerial photographs. The exact time of exposure of the individual aerial photographs was registered on the time scale of the GPS aircraft receiver, in order to establish the reference between the results of the GPS positioning and those of the aerial triangulation.

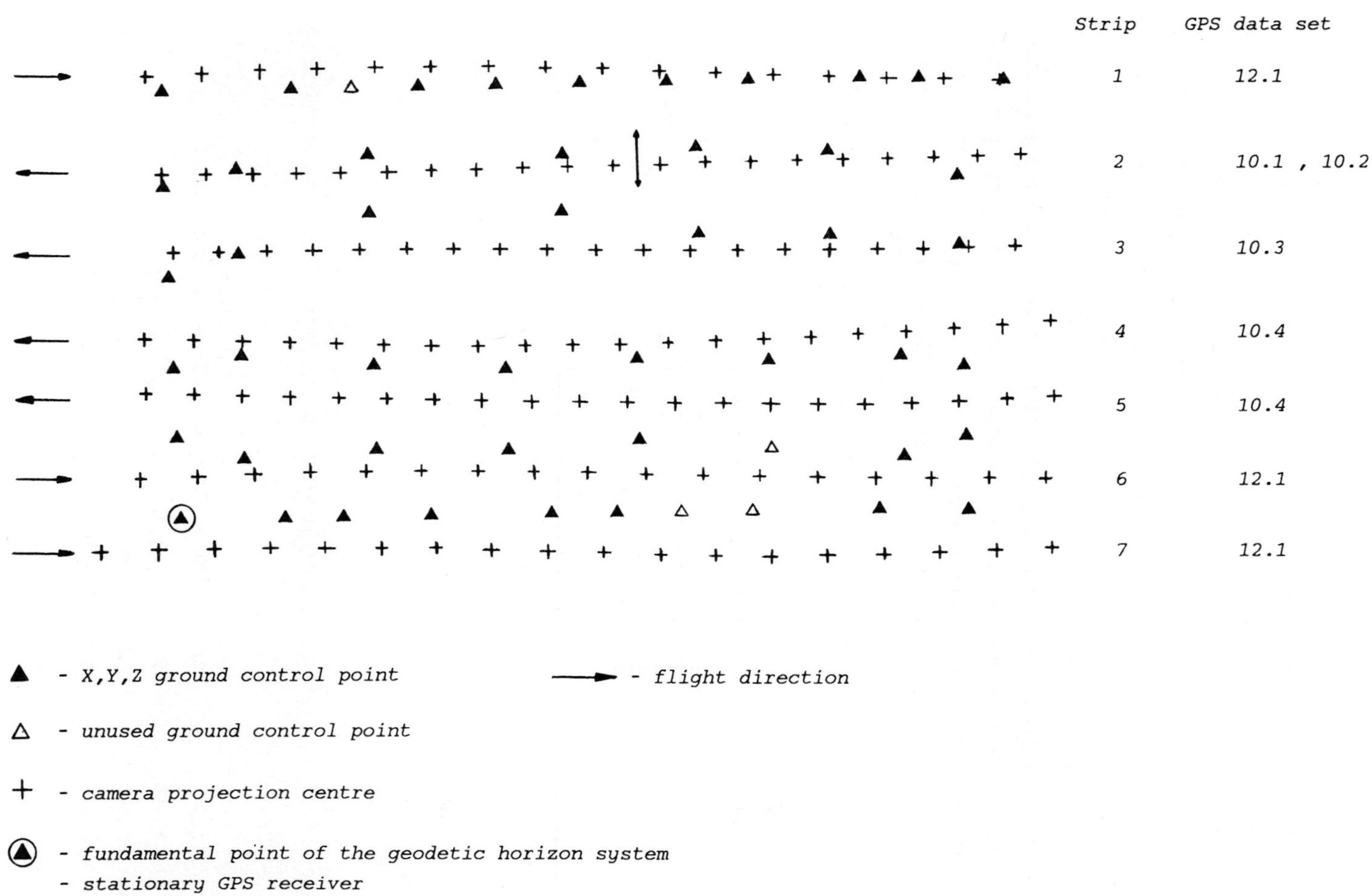

Fig. 1 : Test block configuration

2.1 GPS data

The preliminary investigations presented in this paper are based on a part of the block covering seven flight strips. The seven strips with a total of 130 photographs are chosen in such a way that the requirements with regard to a precise aerial triangulation and consequently for a precise determining of reference points for the GPS solution, are fulfilled. The aerial triangulation of these seven strips is described in the next chapter.

The GPS observations corresponding to the seven photo strips are subdivided into five independent data sets, due to data registration conditions. Each data set (tab. 1) contains pseudorange and carrier phase observations of both the stationary and the aircraft receiver. The observations were carried out at the L1 frequency simultaneously to five satellites (6,8,9,11,12), except for the first part of strip 2 (data set 10.1), where satellite 11 lost lock and thus only 4 satellites were tracked (tab. 1).

Date	Data Set	Strip	Satellites
10.06.87	10.1 10.2 10.3 10.4	2.1 2.2 3 4,5	6,8,9,12 6,8,9,11,12 6,8,9,11,12 6,8,9,11,12
12.06.87	12.1	7,6,1	6,8,9,11,12

Tab. 1 : GPS data sets

2.2. Aerial triangulation

The technical data of the photogrammetric block (fig. 1) can be summarized as follows :

block size	: 7 strips with a total of 130 aerial photographs
aerial camera	: Wild RC 10
focal length	: f = 213.67 mm
forward overlap	: p approx. 70 %
side overlap	: q approx. 50 - 60 %
flight altitude	: approx. 800 m
image scale	: M_B = 1 : 3800

number of points : 47 signalized X,Y,Z ground control points
236 signalized tie points
117 natural tie points
477 marked tie points

The coordinates of the ground control points were determined using both conventional geodetic methods and also GPS. The evaluation of the GPS observations has already been completed and the coordinates of the ground control points resulting therefrom were available with a precision of ± 2 to ± 3 cm with reference to the geocentric coordinate system WGS 84 (ref. 4). These coordinates were transformed into a local geodetic horizon system (ref. 13), with the location of the stationary receiver as the fundamental point (fig.1).

The aerial triangulation was then performed by the bundle method with additional parameters for compensation of systematic effects using the bundle adjustment programm PAT-BS. The results of major interest of this bundle adjustment are the coordinates of the camera projection centres, which are used for the testing of the GPS antenna coordinates.

Simulations with the above described photogrammetric block were performed in order to testify the accuracy of the projection centre coordinates. Hereby a standard deviation of the image coordinates of $\sigma_x = \sigma_y = \sigma_o = \pm 5$ µm (± 2 cm in the terrain) was taken into account, according to the σ_o achieved in the bundle block adjustment of the real data set. The simulations have shown that the accuracy of the coordinates of the camera projection centres can be indicated by ± 4 cm in X, Y and ± 2 cm in Z.

3. GPS data processing

The basic observation equations for pseudorange and carrier phase measurements can be written (refs. 7, 15, 17) as :

$$P_r^S = R_r^S + \delta R_r^S + c\cdot(\delta T_S - \delta T_r) + (\delta R_r^S)_{ion} + (\delta R_r^S)_{trop} + \varepsilon_P$$

$$-\lambda\phi_r^S = R_r^S + \delta R_r^S + c\cdot(\delta T_S - \delta T_r) + (\delta R_r^S)_{ion} + (\delta R_r^S)_{trop} + \lambda\cdot N_r^S + \lambda\cdot\varepsilon_\phi$$

whereby :

$R_r^S = \sqrt{(X_S - X_r)^2 + (Y_S - Y_r)^2 + (Z_S - Z_r)^2}$

P_r^S , pseudorange observation from receiver r to satellite S

ϕ_r^S , phase observation from receiver r to satellite S

X_r, Y_r, Z_r , unknown coordinates of the antenna phase centre of receiver r
X_S, Y_S, Z_S , coordinates of satellite S, calculated from the satellite broadcast ephemeris
δR_r^S , range error due to satellite ephemeris uncertainties
δT_S , clock error of satellite S
δT_r , clock error of receiver r
$(\delta R_r^S)_{ion}$, range error due to ionospheric refraction
$(\delta R_r^S)_{trop}$, range error due to tropospheric refraction
N_r^S , initial ambiguity parameter
c , speed of light
λ , wavelength of the satellite signal
ε_P , pseudorange observation error
ε_ϕ , phase observation error

These equations include all biases caused by certain systematic effects, which have to be removed by modelling or differencing to achieve unbiased positions. There are several possibilities for removing or rather reducing the biases using linear combinations of the original observations. The method applied in the present investigations is the following :

The satellite clock errors were computed from the clock parameters included in the satellite navigation message. The range errors due to satellite ephemeris uncertainties, ionospheric and tropospheric refraction were first of all neglected. This results in :

$$P_r^S = R_r^S - c\cdot\delta T_r + \varepsilon_P$$
$$-\lambda\cdot\phi_r^S = R_r^S - c\cdot\delta T_r + \lambda\cdot N_r^S + \lambda\cdot\varepsilon_\phi \qquad (2)$$

The initial ambiguity parameters N_r^S of the individual data sets of the stationary receiver were calculated using the known coordinates of the reference point. In contrast to this, the initial ambiguity parameters of the aircraft receiver data sets were computed with the help of initial positions derived from a pseudorange solution. The resulting observation equation for the corrected carrier phase observations can thus be written as :

$$-\lambda\phi_r^S = R_r^S - c\cdot\delta T_r + \lambda\varepsilon_\phi \qquad (3)$$

This observation equation was then used as a basis for the determining of single epoch positions of the receiver antennae at the reference point and on the aircraft. At each epoch the four unknowns, the three antenna coordinates and the receiver clock error, of both the stationary and the aircraft GPS receiver were calculated by independent least squares solutions. Carrier phase observations to five satellites (except for strip 2.1) for each receiver and each epoch were available. The known coordinates of the reference point were taken as approximate values of the unknown coordinates of both the stationary and the aircraft receiver antennae. The approximate value of the receiver clock error was assumed to be zero.

As a result of this process one obtains for each observation epoch i, i.e. every 0.6 sec, the coordinates of the stationary receiver antenna and of the moving antenna in the geocentric coordinate system WGS 84. The variance-covariance matrices of all epochs were also obtained :

$$\underline{X}_{ref}(t_i)\ ;\ \underline{Q}_{ref}(t_i) \quad \text{and} \quad \underline{X}_{air}(t_i)\ ;\ \underline{Q}_{air}(t_i) \qquad (4)$$

where :

$\underline{X}_{ref}(t_i)$, position vector of the GPS antenna at the reference point at observation epoch t_i in the WGS 84

$\underline{X}_{air}(t_i)$, position vector of the GPS antenna on the aircraft at observation epoch t_i in the WGS 84

$\underline{Q}_{ref}(t_i)$, corresponding variance-covariance matrices
$\underline{Q}_{air}(t_i)$

These single epoch positions are influenced by the above neglected systematic effects. Assuming that the systematic effects on the positions of both receiver antennae are identical or rather similar, these effects can be reduced by calculating relative positions of the aircraft GPS antenna with respect to the stationary GPS antenna.

$$\begin{aligned} &\underline{X}_{relair}(t_i) = \underline{X}_{ref} + [\ \underline{X}_{air}(t_i) - \underline{X}_{ref}(t_i)\ ] \\ * \quad &\underline{Q}_{relair}(t_i) = \underline{Q}_{ref}(t_i) + \underline{Q}_{air}(t_i) \end{aligned} \qquad (5)$$

where :

$\underline{X}_{ref}$,nominal position vector of the stationary receiver antenna

$\underline{X}_{relair}(t_i)$,relative position vector of the GPS aircraft antenna in the WGS 84

$\underline{Q}_{relair}(t_i)$,corresponding variance-covariance matrices

* The single epoch positions of the stationary and the aircraft receiver are assumed to be independent.

The relative positions $\underline{X}_{relair}(t_i)$ of the GPS aircraft antenna (existing every 0.6 sec) were interpolated onto the times of exposure of the individual aerial photographs and then transformed into the local geodetic horizon system, in order to achieve GPS aircraft antenna positions which are comparable to the camera projection centres obtained from the aerial triangulation.

4. Analyses of the GPS positions

The relative positions of the GPS aircraft antenna interpolated onto the exposure times of the aerial photographs are the subject of the following considerations. As mentioned above, an independent control of these GPS antenna positions is given by the comparison with the positions of the camera projection centres. Therefore, a reduction of the antenna centre positions onto the camera projection centres would, as a matter of fact, be necessary, considering the spatial separation of the antenna phase and camera projection centre.

In this presentation, however, the coordinate offsets, i.e. the differences dx, dy, dz between the GPS antenna coordinates and the projection centre coordinates are computed and analysed. In addition, the spatial distances ds between the antenna phase centre and the projection centre at the individual exposure times are calculated. The r.m.s. values of the coordinate offsets and the spatial distances ds with respect to their mean values are presented in table 2.

At a first glance, the results summarized in table 2 indicate a position accuracy of better than ± 10 cm for kinematic positioning with GPS in high dynamic (240 km/h) applications. The poor accuracy obtained in the first part of strip 2 occurred due to the fact that satellite 11 lost lock which results in a very unfavourable satellite geometry. The PDOP, a criterium for the assessment of the satellite geometry (PDOP= $\sqrt{Q_{xx} + Q_{yy} + Q_{zz}}$; position dilution of precision) was about 40 during the observation period of strip 2.1, in comparison to the PDOP values of the other observation epochs which were about 3.5 .

GPS data sets	Strip	rms dX [m]	rms DY [m]	rms DZ [m]	rms DS [m]
10.1	2.1	1.970	5.818	1.168	6.238
10.2	2.2	0.072	0.047	0.050	0.047
10.3	3	0.065	0.098	0.049	0.053
10.4	4 5	0.061 0.056	0.126 0.098	0.048 0.052	0.117 0.088
12.1	7 6 1	0.113 0.063 0.062	0.084 0.073 0.079	0.069 0.069 0.053	0.085 0.073 0.074
quad. mean *		0.073	0.089	0.056	0.080

* without data set 10.1 (strip 2.1)

Tab. 2 : Accuracy of the coordinate offsets between the camera projection centres and the GPS antenna centres

The position accuracy really obtainable in GPS kinematic positioning is even better than it appears at the first glance, because the coordinate offsets are influenced by at least two effects. On the one hand, the coordinate offsets are not constant, due to the inclination variations of the aircraft between consecutive positions. This effect is not taken into account in this presentation. But its influence is obvious if one compares the r.m.s. values of the coordinate offsets with the corresponding r.m.s. values of the distances. The distances ds are independent of the aircraft inclinations and are for this reason more appropriate as criteria for the position accuracy.

On the other hand, the coordinate offsets include all unmodelled, and in spite of relative positioning remaining systematic effects of the GPS observations. These effects also cause variations of the coordinate offsets. The method of linear regression was therefore applied to analyse the coordinate offsets with regard to a linear time dependent drift. The results of this investigation are listed in table 3. The drift values are given in mm/sec. It is obvious that the unfavourable satellite geometry during the observations of strip 2.1 causes an extreme drift in the coordinates. The drift of the coordinate offsets of the other strips varies between 0 and 6 mm/sec, which result for the worst case in 36 cm/strip.

GPS data sets	Strip	drift DX [mm/sec]	drift DY [mm/sec]	drift DZ [mm/sec]
10.1	2.1	- 237	+ 700	-141
10.2	2.2	+ 6	+ 1	+ 4
10.3	3	+ 2	+ 5	+ 2
10.4	4 5	+ 0 + 0	+ 6 + 4	- 2 - 1
12.1	7 6 1	- 4 - 1 - 1	+ 3 + 1 + 0	- 3 - 3 - 2

Tab. 3 : Linear drifts of the coordinate offsets between the camera projection centres and the GPS antenna centres

The r.m.s. values of the coordinate offsets calculated after removing the linear drifts are presented in table 4. These results demonstrate that, when using a refined model for processing the GPS carrier phase observations, which considers all systematic influences in such a way that no drift in the coordinates remains, a kinematic position accuracy of ± 5 cm or even better can be obtained. If one recalls that the check positions, the projection centres of the aerial camera, were calculated with a precision of ± 4 cm in X,Y and ± 2 cm in Z, a considerable part of the r.m.s. values of the drift corrected coordinate offsets (tab. 4) would thus arise from the accuracy of the coordinates of the camera projection centres. This fact is confirmed by the results of ARI-modelling (autoregressive integrated processes) and variance component estimation (ref. 10). The estimated standard deviations of the relative GPS antenna coordinates are summarized in (tab. 5) varying between ± 0.7 cm and 3.8 ± cm.

The accuracy analyses of the GPS aircraft positions has not yet been completed, thus not all questions of interest could be investigated up to now. Of interest are, for example, the correlations of the coordinates of consecutive positions, especially for the further utilization of the GPS positions in the evaluation processes of the above mentioned airborne scientific applications. Preliminary results of such analyses (ref. 10) are summarized in table 7. They indicate that the correlations of the coordinates of consecutive positions are negligible. The cross-correlation of the GPS antenna coordinates, derived from the variance-covariance matrices of the GPS solution are summarized in table 6. The variation of the crosscorrelation coefficients are due to the continuous changes of the satellite geometry.

GPS data sets	Strip	rms dX [m]	rms DY [m]	rms DZ [m]	rms DS [m]
10.1	2.1	0.045	0.073	0.033	0.037
10.2	2.2	0.041	0.045	0.031	0.028
10.3	3	0.057	0.056	0.035	0.034
10.4	4	0.061	0.055	0.032	0.032
	5	0.056	0.057	0.045	0.031
12.1	7	0.066	0.045	0.042	0.036
	6	0.058	0.072	0.045	0.034
	1	0.055	0.078	0.045	0.039
quad. mean		0.055	0.061	0.039	0.034

Tab. 4 : Accuracy of the coordinate offsets between the camera projection centres and the GPS antenna centres after removing linear drifts

Strip	σ_X [cm]	σ_Y [cm]	σ_Z [cm]
1	2.7	1.1	3.8
2	-	-	-
3	2.0	0.9	2.6
4	2.4	0.9	2.7
5	2.0	1.3	2.4
6	1.4	1.1	1.4
7	1.2	0.7	1.0

Tab. 5 : Accuracy of the GPS antenna positions derived from variance component estimation

Strip	r_{XY}	r_{XZ}	r_{YZ}
1	0.01	0.58	0.20
2	-0.57	0.61	-0.70
3	-0.53	0.59	-0.65
4	-0.47	0.56	-0.51
5	-0.39	0.54	-0.36
6	-0.13	0.56	0.06
7	-0.23	0.55	-0.08

Tab. 6 : Cross-correlation coefficients of the GPS antenna coordinates

	X	Y	Z
r(1d)	0.46	0.69	0.52
r(2d)	-0.06	0.11	-0.01
r(3d)	0.06	-0.15	-0.10

Tab. 7 : Auto-correlation coefficients r of the GPS antenna coord. (d=0.6 sec)

5. Conclusions

The presented results of first analyses of the GPS test flight data have demonstrated the high accuracy potential of kinematic positioning with GPS in high dynamic applications. Although a simple processing technique was applied for the calculation of the positions from the GPS carrier phase observations, a coordinate accuracy of better than ± 10 cm was nevertheless directly reached.

The results indicate as well that the potential kinematic position accuracy of GPS is at the centimetre level. This accuracy level is attainable when using a refined processing technique for the GPS carrier phase observations e. g. double differencing or triple differencing (refs. 15, 16,17), in order to reduce or even to eliminate the influence of the systematic effects of the GPS observations. Subsequent investigations will be devoted to these points.

Acknowledgements

The test flight was planned and carried out by the Survey Department of Rijkswaterstaat, in cooperation with KLM Aerocarto, Technical University Delft (Faculty of Geodesy) and Sercel. Their support by generously supplying the entire test flight data is gratefully acknowledged.

6. References

/1/ ACKERMANN, F. (1984) : Utilization of Navigation Data for Aerial Triangulation, IAP, Vol. 25-A3b, Rio de Janeiro

/2/ ACKERMANN, F. (1986) : Use of Camera Orientation Data in Photogrammetry - a Review, ISPRS Comm. I Symposium, Stuttgart

/3/ BEUTLER, G., ROTHACHER, M. (1986) : Auswertung der 1984-Alaska-GPS-Kampagne, Schweizerische Zeitschrift für Vermessung, Photogrammetrie, Kulturtechnik, Heft 6, 213-218

/4/ BOSWINKEL, D., WITMER, R., v.d. VEGT, J.W. (1988) : Utilization of GPS in large scale photogrammetry, ISPRS Congress Comm. I, Kyoto

/5/ CANNON, M.E., SCHWARZ, K.P., WONG, R.V.C. (1986) : Kinematic Positioning with GPS - An Analysis of Road Tests, Fourth International Geodetic Symposium on Satellite Positioning, Austin, Texas

/6/ FRIESS, P. (1986) : A Simulation Study on the Improvement of Aerial Triangulation by Navigation Data. ISPRS Comm. III Symposium, Rovanimi

/7/ KING, R.W., MASTERS, E.G., RIZOS, C., STOLZ, A. (1985) : Surveying with GPS, Monograph No.9, School of Surveying University of New South Wales, Kensington, NSW, Australia

/8/ KLEUSBERG, A. (1986) : GPS Positioning Techniques for Moving Sensors, ISPRS Comm. I Symposium, Stuttgart

/9/ LACHAPELLE, G., LETHABY, J., CASEY, M. (1985) : Airborne Single Point and Differential GPS Navigation for Hydrographic Bathymetry, Hydrographic Journal 24, 1-8 ·

/10/ LINDENBERGER, J. (1988) : Modelling Orientation Parameters of Sensor Platforms, ISPRS Congress Comm. I, Kyoto

/11/ MADER, G.L. (1986) : Dynamic Positioning Using GPS Carrier Phase Measurements, manuscripta geodaetica Heft 11, 272-277

/12/ MADER, G.L., CARTER, W.E., DOUGLAS, B.C. (1986) : Decimeter Precision Aircraft Positioning Using GPS Carrier Phase Measurements, Geodetic Research and Development Laboratory, Rockville U.S.A.

/13/ NAGEL, E. (1976) : Die Bezugssysteme der Satellitengeodäsie, Dissertation, Publication of the German Geodetic Commission (DGK), Series C No. 223

/14/ Proceedings of the First International Symposium on Positioning with GPS (1985) Rockville, Md. April 15-19

/15/ REMONDI, B.W. (1984) : Using the Global Positioning System (GPS) Phase Observable for Relative Geodesy : Modeling, Processing and Results, Ph. D. Thesis, Center for Space Research, The University of Texas at Austin, CSR-84-2

/16/ REMONDI, B.W. (1985) : Performing Centimeter-Level Surveys in Seconds with GPS Carrier Phase : Initial Results, Global Positioning Papers published in Navigation Vol. III, The Institute of Navigation, Washington D.C.

/17/ WELLS D.E., Ed., (1986) : Guide to GPS positioning, Canadian GPS Associates Fredericton ,N.B.

Integrated Navigation System for Approach Guidance for Regional Air-Traffic Using GPS

Thomas Jacob
Institut für Flugführung, Technische Universität Braunschweig

The underlying project of this report has been sponsored by the Federal Minister of Research and Technology. The responsibility for the contents of this publication lies with the Institute for Flight Guidance and Control.

1. Introduction

With todays commercial aircraft missions, the various standard means of position finding and navigation systems are required. According to the flight phase, different systems are necessary, for instance, the taxiing on the taxiway can only happen in acceptable visual conditions with the aid of optical markings and lights. At the take-off process, the outside view information is supported by supplementary radio navigation , the localizer, for lateral aircraft piloting. In the area around the airport, the airplanes move on a straight-line connection from radio beacon to radio beacon (VOR/DME AND NDB) or in the long-distance navigation from point to point with the aid of inertia navigation and radio navigation help, on the long or medium frequency band. The landing approach and the final approach are led again along straigh t flight paths under the use of the radio navigation tools and the instrument landing system (ILS)./1/. With these navigation systems, todays management of flight traffic is safe and reliable, though also costly. Because all of the major commercial airports are equipped with instrument landing systems navigation and approach systems are available for take-off and landing, also for use under bad weather flight conditions. However, instrument landing systems are not found on most of the regional airfields, aircraft operation areas in the regional flight traffic (feeder service or, for example, business aircrafts) and on very few of the airfields of the Third World countries. Here, is frequently no ILS installed for reasons of cost, and thereby a landing under bad weather flight conditions is neither permissible nor possible.

The "Integrated Navigation System for Aircraft," which is being integrated in the research and flight attempt airplane (DORNIER DO 128) of the Institute for Flight Guidance and Control at the Technical University of Braunschweig, should distinctly improve the possibility of using the regional air traffic. Through the realization of the Satellite based, worldwide available position finding system (Global Positioning System), it appears possible that the guidance of an aircraft can be led from Terminal A, including runway-, take-off-, travel/flight-, landing- and taxiing-guidance to Terminal B from an Integrated Aircraft Navigation System alone. During the realization of such a system, the following question must be resolved: is the precision of the determined position, as well as the effect of errors in the flight guidance, tolerable?

2. Accuracy Requirements

According to the prevailing visibility conditions, the most exact standards are placed on the accuracy of the position of a location system for the landing approach. At the present time on commercial airports, the employed instrument landing system is specified by the International Civil Aviation Organisation (ICAO) in ANNEX 10 of the Accuracy Requirements, differentiated from the visibility conditions. It is differentiated in the following operation stages (CAT) (Figure 1):

Non Precision Approach (NPA); CAT I; CAT II; CAT III a,b,c

In order that a passenger service aircraft with a mass greater than 5.7t may carry out a landing with one of these operation stages, the prerequisites must be complied by these /3/ : the ground installations, the aircraft installations and the air crews. Since a license for an aircraft in the regional air traffic for landing in CAT II requires a considerable improvement of punctuality and reliability, only the following standards of a landing in CAT II (decision altitude 30m) are figured.

According to this, the ground installation of the instrumental landing system must reach the following precision at the runway threshold. (ILS reference point) (Figure 2) /2/

the horizontal alignment of the centerline at exactly +/- 4.5m (effective for a new system)

deviation through irregularities of the guide beam characteristic must be, with 95% probability, less than 3.45m ($2\ \sigma$ = 3.52m).

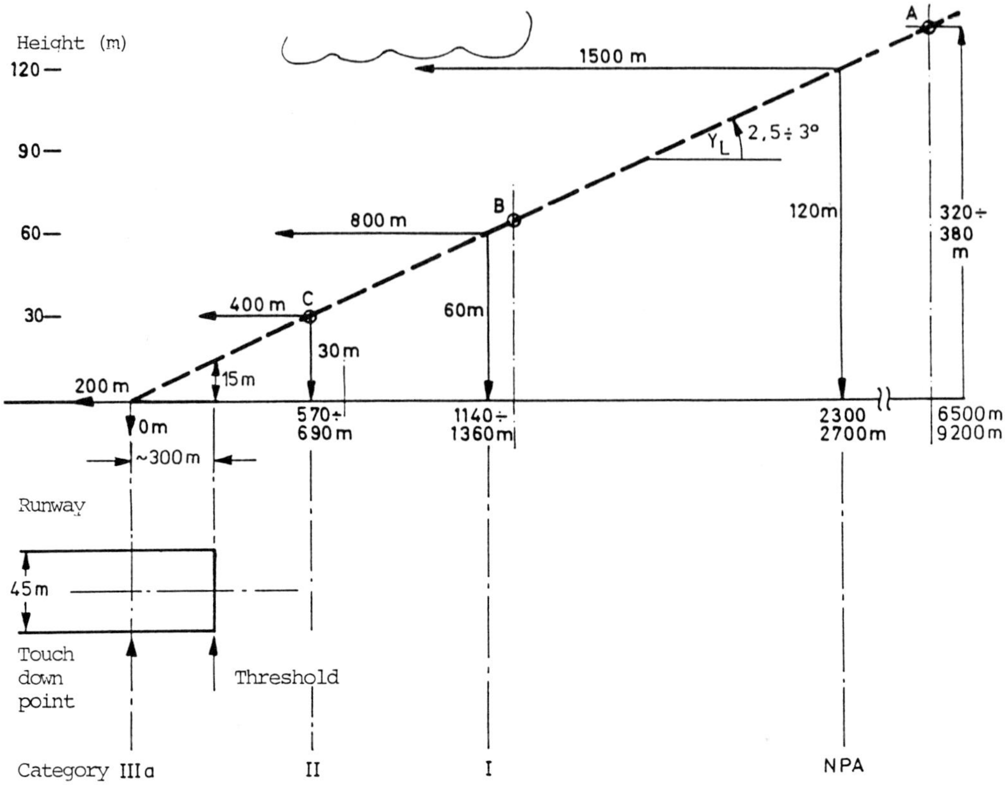

Figure 1: Categories and visual conditions in IFR

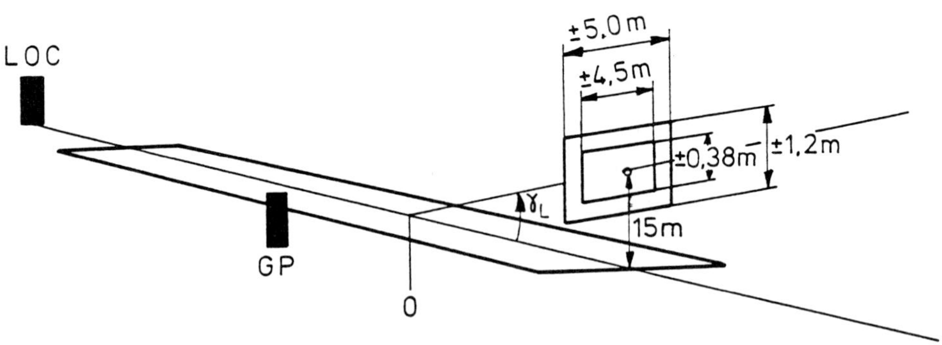

Figure 2: Tolerable misalignment of the ILS ground equipment (CAT II)

This results in the following maximum allowable deviations from the ideal centerline:

horizontal: $\Delta y = \pm 5.67m$ vertical : $\Delta z = \pm 0.38m$

(Prerequisite: distance antenna - threshold = 300m)

Irregularities in the glide path must remain, with 95% probability, less than:

$\Delta z = \pm 0.5m$ ($2\sigma = 0.51m$)

To these errors add the errors of the airplane receiver. The requirements for the guide beam characteristic are made in a way that the deviation of the entire system, consisting of the ground installation, airplane receiver and aircraft, should be as a consequence from the irregularities, less than:

horizontal: $\Delta y = \pm 5.0m$ vertical : $\Delta z = \pm 1.2m$

How extreme the requirements for the landing system are, is shown by measurements, which the DFVLR Braunschweig have performed in Frankfurt. As shown in figure 3, the standard deviation of the vertical deviation from the glide path, is determined from a number of actually carried out approaches. One recognizes, that the actual staggering from the nominal path increases drastically coming to the middlemarker, and arrives at the value of 8m. This makes the glaring contrast distinct between the value the ICAO has required, and that which actually has been flown.

If the "Integrated Navigation System for Aircraft" should be applied in a commercial airplane for landing approaches in CAT II conditions - without using an ILS - it must accomplish the position finding precision of an ILS. Beyond that it should be cheap, reliable, safe and compatible in relation to the ILS.

A Position finding system which is able to accomplish these demands is the GLOBAL POSITIONING SYSTEM (GPS) /4,5/, which is now developed. But this system will accomplish these demands just with the use of the differential technique and with extra support.

3. Error Characteristics of the GLOBAL POSITIONING SYSTEM by Aircraft Application

With the use of GPS in aircrafts for high precision landing approach guidance,

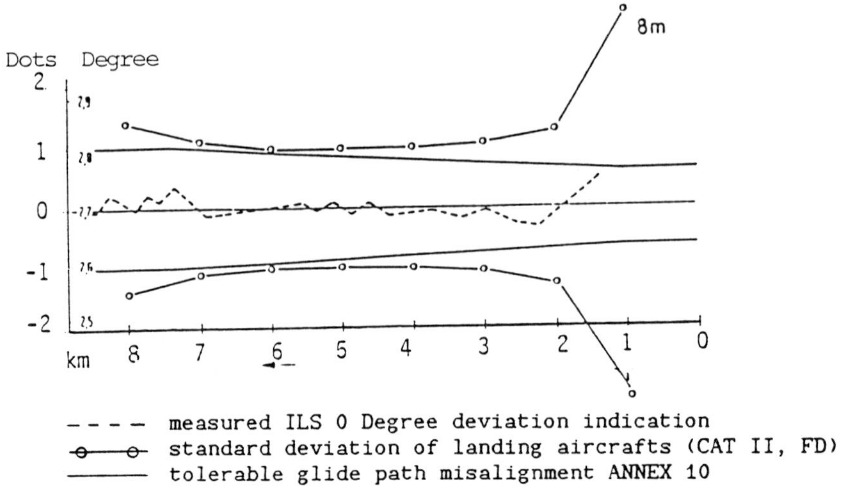

Figure 3: Measured deviation of transport aircrafts from the glide path

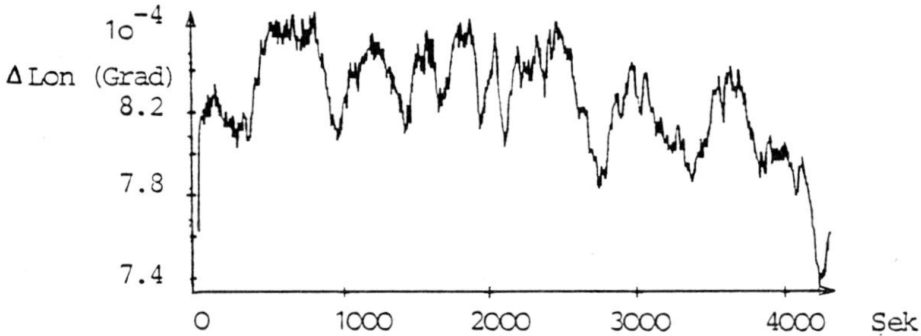

Figure 4: Deviation of GPS-Latitude (fixed antenna, V =0. m/s,
C/A Code 5 Channel Receiver)

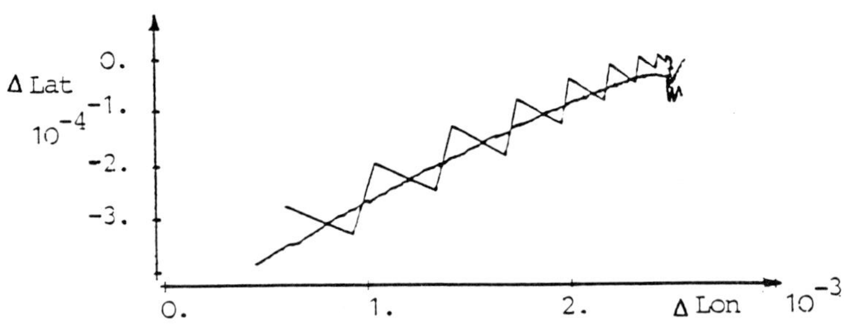

Figure 5: Ground Track while accelerating for take-off speed
C/A Code Multiplex Receiver / Integrated Navigation System

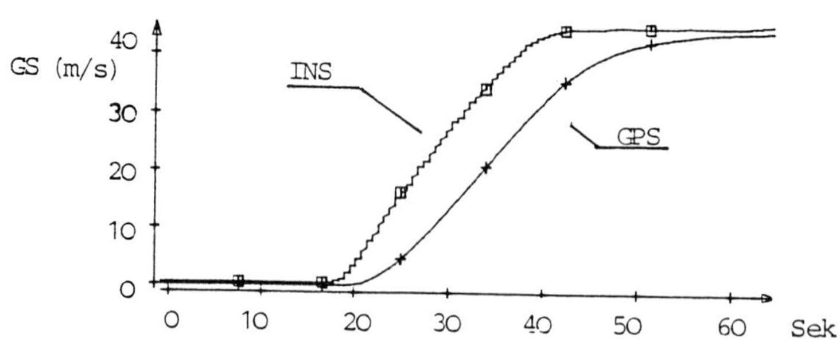

Figure 6: Comparison of ground speed of INS and GPS

the following errors will recognizably disturb:

a) the measured GPS - position has an offset against the real position
b) the measured GPS - position changes even with no aircraft movement of course the Satellite constallation change (GDOP, selective availibility)
c) the GPS - positions have a noise, the amplitude of the noise is receiver dependent; an oscillation is heterodyned (figure 4)
d) according to the receiver type there are dynamic errors during acceleration phases of the aircraft
e) shadow effects will increase position errors, or breakdowns in the GPS - position

Tests with industrially produced GPS-receivers have shown that with a favourable satellite constellation, the position failed with 15m against the known points in the local coordinate system. The positions determined by GPS, present themselves with a deviation, of 3m. With that, errors in the " stand alone" operation are greater than the errors which have been qualified for instrument landing systems for CAT II by the ICAO. Only through the use of different supporting measures, and how they would be conveyed into the integrated aircraft navigation system, can these bad error characteristics be positively influenced. The offset, as well as the time dependency of the offset, can be extensively eliminated through the differential GPS. Through these correction measures the dynamic error characteristics, however, are not improved.

How strong the GPS-position test data swings by acceleration is shown in figure 5. Here are applied the GPS test data during the acceleration phase by take-off of the twin-engine research airplane (DO 128) of the Institute of Flight Guidance and Control of the Technical University of Braunschweig. Here, with increasing acceleration, a lateral staggering of the GPS position from up to 7m is established, which is about half of the half-runway width of a regional airport.
A further error that likewise appears in dynamic stages of flight is distinctly recognizable by a comparison of the flight path speed between GPS and INS (Figure 6). Clearly, a lag error is registered up to 12m/s, during the take-off acceleration up to a speed of 43 m/s.
The ground track computed from the GPS during a circle over the airport is plotted in figure 7. During that flight test some full circles have been

flown. On the positions A and B on the figure 7 always one circle, on position C some holding circles, because the runway was not cleared for landing. Shadow effects from the circles over A and B appear to look like misshappened eggs; in these curves a bank angle of about $\Phi = 25°$ was introduced. Over C the curves were not able to be registered by GPS. After finishing the flight curves, the GPS receiver needed about 17 s to lock on all 4 satellites and give a valid position. In that time a regional commercial aircraft would have flown a horizontal distance of about 1 km and a vertical one of 50m during a landing approach. Those situations during a landing approach under bad weather conditions must stop the landing approach for security reasons, so a go around manoeuver has to be initiated.

The error characteristic shows distinctly that the position sensor GPS in "stand alone" operation would be without any use for high precision landing approach guidance up to weather condition CAT II, if the stationary position precision especially in differential mode would not be so extremely high.

4. Sensor Concept of the Integrated Flight Guidance System

In the integrated flight guidance system for landing approach guidance, the unacceptable error characteristics are clearly diminished. This system is in the base concept composed of two parts; a position finding system and a guidance generator (figure 8). The position finding system computes the aircraft position using a differential- GPS position as well as the currently used on-board sensor information. But because the pilot is not able to fly just by a position information, it is necessary to bring the position to a signal the pilot is used to. Also, to show some additional specified commands how to bring the aircraft on an ideal path to the target place. This includes a computation of the specified flight path as well as a glide path computation during the landing approach.

To improve the dynamic characteristics of the entire system, and to get enough information about the flight path during a breakdown of satellite information, the integrated flight guidance system is coupled with inertial sensors (acceleration and gyroscope sensors), as well as sensors for barometric and radar height. In this coupling different attributes of two systems are used. Acceleration and gyroscope sensors are working just short of high precision, but over a longer time period they show recognizable drifts, which would lead to intolerable errors in the navigation computation. On the other hand, the GPS has a complementary error characteristic; it has a high

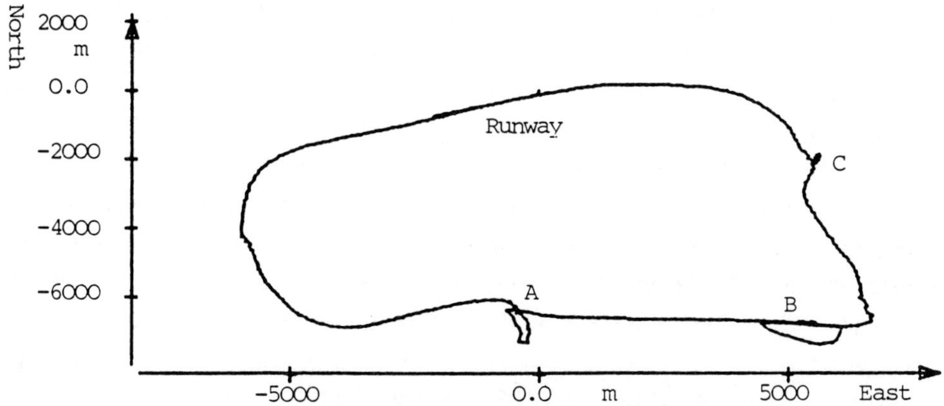

Figure 7: GPS ground track / flight test

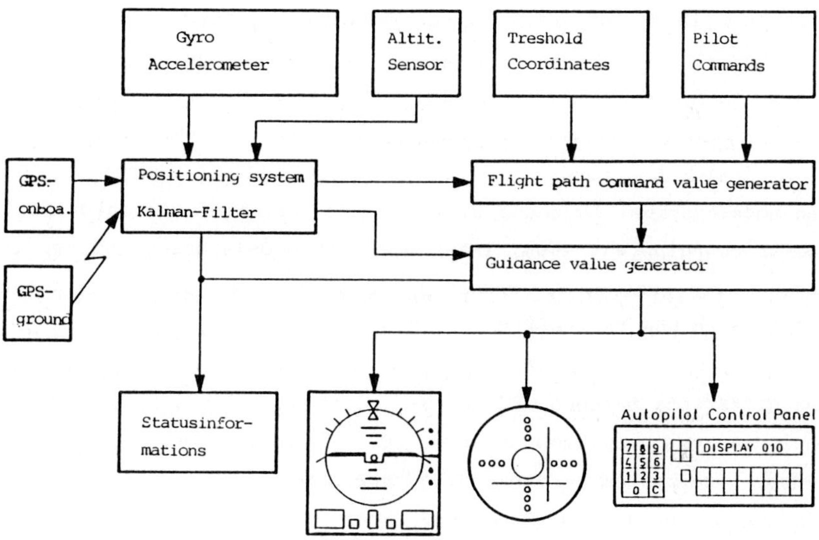

Figure 8: Integrated Navigation System for Aircraft

long- time precision but is not precise in the short term (dynamic errors, etc.) By use of the Kalman filter technique both signals can be optimally combined. With this Kalman filter technique one is able to estimate the system and sensor errors online; and as it were, to make a permanent calibration of the sensors during the flight. With that, even with a breakdown of the GPS signal, the filter algorithmn give for a certain time highly precise information about the groundspeed vector, the flight path and position angle. A main prerequisite for that is a nearly accurate mathematical model of the error characteristics of the inertial system.

5. General Construction of the used Kalman Filter

Figure 9 shows a general construction by using function blocks. The inertial sensors give, aircraft fixed, the measured acceleration $\underline{a}$ as well as the Euler angle (course angle Ψ, the pitch angle Θ and the roll angle Φ) of the aircraft, in an aircraft fixed coordinate system. The output of the GPS receiver after differential support with the ground receiver is equal to the position of the aircraft in the earth-fixed coordinate system (WGS 84), as well as, in a topocentric navigation coordinate system after a transformation. Inside the function block of the navigation equations, one transforms the accelerations by means of the Euler angles of the aircraft in the navigation coordinate system, integrates the acceleration and results the speed vector over ground. A further integration step gives the covered distance relative to the take-off position.

The difference between GPS and NAV-position results in an error signal which includes only position errors. Out of their dynamic characteristics the sensor- and system errors can be determined. That can just lead to realistic results, when the mathematical error characteristics, with differential equations described, correspond almost exactly with the real error characteristics. To describe the complicated sensor characteristics, (which are dependent on the aircraft movements) more easily - The computation expenditure should not be too great in order that an computation can be achieved in the limited time frame - two error models as well as two Kalman- filters has been developed. These two filters will be alternated, depending on the actual dynamic of the aircraft, for example at the beginning of a flight curve. Therefore, it is possible, even with small error models, to estimate extra errors of sensor measures, during a flight curve in relation to the

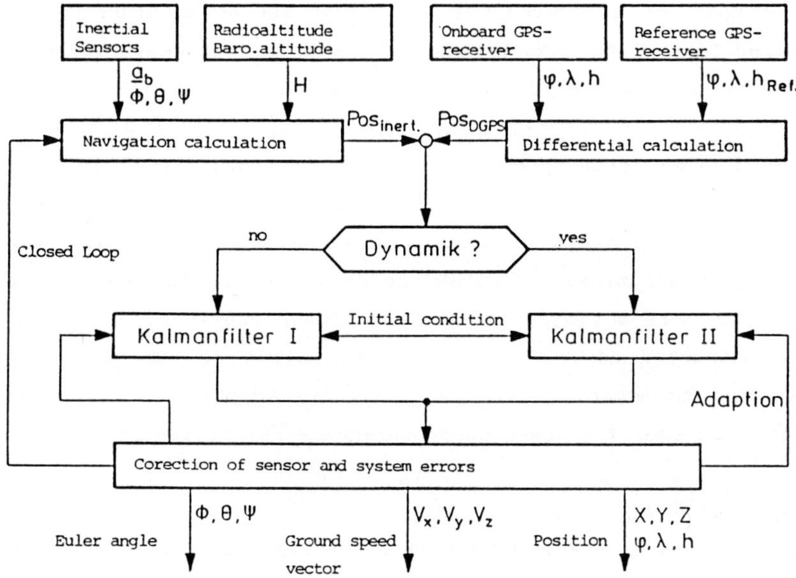

Figure 9: Position Finding Subsystem

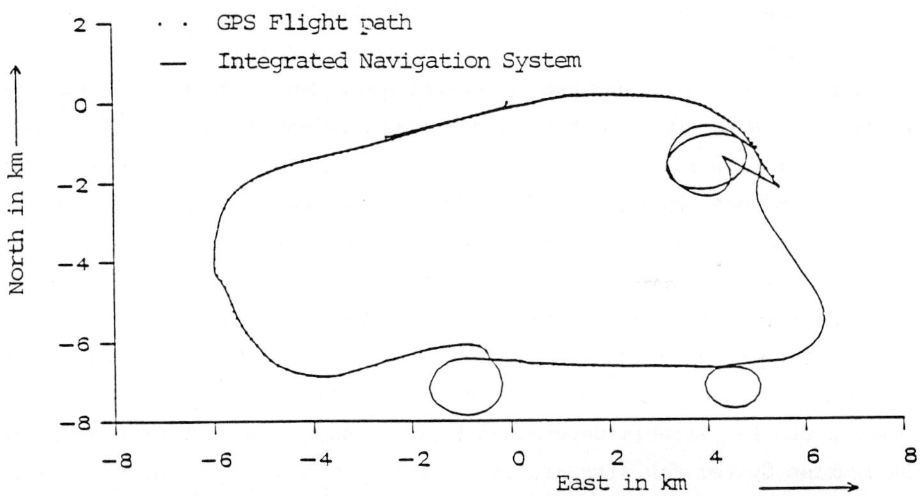

Figure 10: Ground track comparison of GPS and Integrated Navigation System

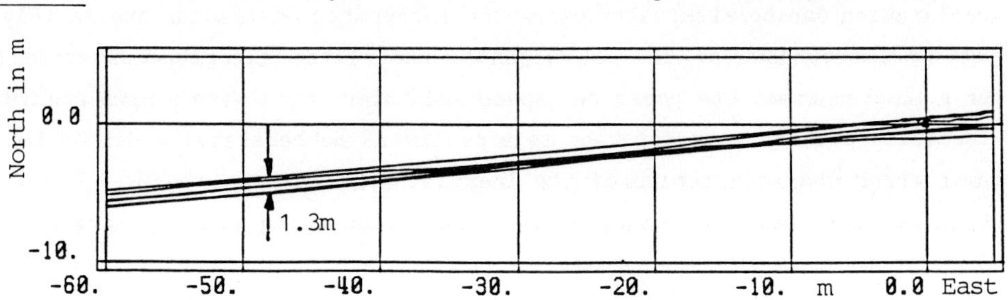

Figure 11: Ground track of the Integrated Navigation System for Aircrafts while taxiing, starting and landing (C/A Code Differential)

straight flight. As a result, the dynamic of the error characteristics can be described better. Now the navigation equations can be corrected, because the Kalman filters have estimated the errors.

Thus, the further transformation follows respectively. Integration is corrected with the calibration of the sensor signals, which is a decisive prerequisite, for if there were a breakdown of the GPS signals, the system errors would be delimited. Through the correction of the sensor and system errors one has here a closed loop Kalman filter submitted, which is supplementary constructed adaptiv for the computations of the performed flight tests. From a mathematical derivation of an adaptive, close loop Kalman filter, will be specified here under reference /6/ of the literature.

6. Flight-test Results

Figure 10 shows the ground track of the same flight as that shown in figure 7. The GPS- position and the position output of the position-finding part of the "Integrated Navigation System for Aircraft" are shown here. Based on the sensor errors which are determined by the system, the position-finding part is able to determine the flight curves at the points A and B "faultlessly". Striking is the leap in the course process in Position C. The cause for this is that the position-finding part of the "Integrated Navigation System for Aircraft", since the beginning of the flight curve at Point B, was no longer updated by GPS-measurements. Until the 3 full circles had been flown, as well as a course change of more than 90 degrees; a reliable GPS measurement had been carried out after 270 s. Based on the flight manoeuver and the long time without measurement update, the position of the system has drifted away from the GPS Position, so that by renewed updating, a leap to the GPS Position is taken. The position output during the remaining phase - the landing approach and landing can be reliably determined by the location's part of the "Integrated Navigation System for Aircraft".
Also the physically not possible swinging of the GPS position during the take-off accelaration can be eleminated using the integrated navigation system (Fig 5). Even by a breakdown of the GPS signals, the Filter algorithmens still give for a limited time, the position, speed and Euler angle with a high precision. A decisive prerequisite for that is a realistic mathematical model of the dynamic error characteristics of the inertial systems.
Which overall position accuracy can be achieved using the "Integrated Navi-

gation Systems for Aircraft"? In figure 11 the ground track of two take-offs, two landings and one back track are plotted in meters relative to the take-off position. During these aircraft movements the pilot tried to fly respectivly roll as exact as possible on the centerline. It can be recognized from the plots that the ground tracks deviates from each other less than 1.3m. This deviation is so small, that one cannot determine if the 1.3m is the position accuracy of the Integrated Navigation System or if the aircraft really deviated 1m from the centerline which is possible during landings.

7. Summary

The flight tests, which have been made with the "Integrated Navigation System for Aircraft" which has been developed at the Institute for Flight Guidance and Control of the Technical University of Braunschweig until now, have shown good results by combining two sensor systems with different, time dependable, signal qualities: the inertial sensors, with their good short- term charactertistics; and the GPS with good long-term characteristics. With the Kalman filter technique it is made possible, even in high dynamic flight phases, to determine a position of high precision and reliability. The horizontal position precision of the entire system which has been determined until now is better, than the precision of each system standing alone. A landing approach guidance under bad weather flight conditions up to CAT II is possible with the "Integrated Navigation System for Aircraft".

8.Literature

/1/ Avionics Navigation Systems
Kayton, M.; Fried, W. John Wiley & Sons 1969

/2/ Aeronautical Telecommunications
International Civil Aviation Organisation (ICAO) ANNEX 10

/3/ Richtlinien für den Allwetterflugbetrieb nach Betriebsstufe II
Nachrichten für Luftfahrer NfL I-350/72 Bundesanstalt für Flugsicherung

/4/ Principle of Operation of NAVSTAR and Szstemcharacteristics
Milliken, R.J; Zoller, C.J. AGARDograph AG 245

/5/ Differential Operation of NAVSTAR GPS
Kalafus, R.M. Journal of the Institute of Navigation Vol. 30, 1983

/6/ Techniques of the development of Error Models for Aided Strapdown Navigation Systems
Lechner, W. AGARDograph AG 256

GPS Receiver Development in the Space Research Centre Warsaw

Zdzisław J. Krysiński, Janusz B. Zieliński
Polish Academy of Sciences Space Research Centre, Warszawa, Poland

Summary

The current state of the design and the contruction of the receiver based on the GPS signals is presented. The interferometer concept has been drafted. Some of the receiver details are presented in the description of the prototype of the receiver.

1. Introduction

While considering the possibility of the use of the GPS signals for geodetic and geodynamical purposes we immediately encounter the problem of orbits. When we measure the short (< 50 km) baselines, the orbital information is necessary for processing of the phase differences data. In geodynamical solutions, when we have to do with long baselines of the length 100-1000 km, the orbital elements should be determined together with the station coordinates. Therefore in our first approach to the design and construction of an instrument for the GPS signal observation we concentrate on such a system which could be used for the determination of the orbits of the GPS satellites.

We also have to consider the fact that no decoded orbital information will be taken from the signal. Therefore, the instrument must be selfsupporting in area of the orbital data needs, when we plan the routine use of the GPS technique.

One method of the applying such a system is the accomodation of the technique known from the radioastronomy, i.e. the GPS signal is treated as a randomized radio signal. Then by means of the interferometry a time delay between the signals received in both ends of an interferometer is determined.

The time delay τ depends on the geometric variable $\Delta\varrho$, which is the difference of distances from the satellite to the ends of the baseline (see Fig. 1)

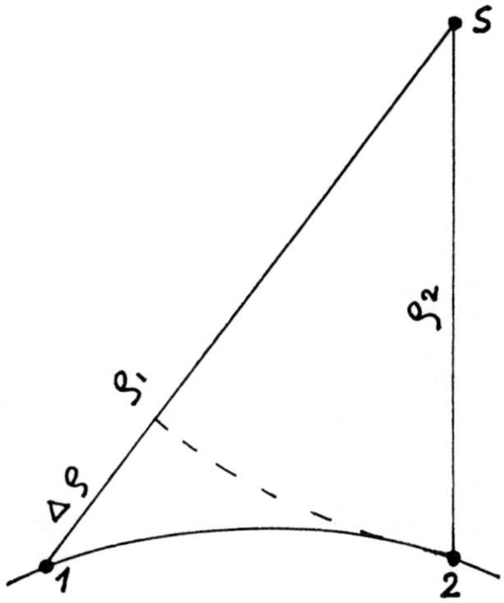

Fig. 1. The geometry of the interferometer

$$\tau = \Delta\varrho \cdot c^{-1} + \delta\tau + \delta\varrho = \frac{\Delta\Phi}{f} \qquad (1)$$

where:

$\Delta\varrho = \varrho_1 - \varrho_2$
$\delta\tau$ - synchronization correction
$\delta\varrho$ - signal propagation correction
$\Delta\Phi$ - phase difference
f - signal frequency

By this method we can also measure the time delay rate which is equal to the difference of relative velocities:

$$\dot{\tau} = \frac{\dot{\varrho}_1 - \dot{\varrho}_2}{c} + \delta\dot{\tau} + \delta\dot{\varrho} \qquad (2)$$

Since ϱ_1 and ϱ_2 are simple functions of the stations coordinates as well as the satellite positions, we can consider (1) and (2) as observational equations for the orbit determination.

2. Concept of the interferometer

The interferometer built in the Space Research Center is designed not for geodetic purposes only. This experimental construction allows to study the GPS technology and different solutions for the GPS data reduction and analysis.

Since no code is used in the interferometer there are codeless receivers applied. We use some of the VLBI ideas in the design of the receiver in order to use the bandwidth synthesis technique and also a wideband signal registration for its later correlation. We also pay attention to another method of the GPS signal preprocessing such as the signal conversion from a spread spectrum into a monochromatic form. So that we receive the possibly widest GPS spectrum, i.e. the total power distributed in the two 20 MHz-wide main lobes L1 and L2 (see Fig. 2). The signal is received by means of a wide-band antenna radiator and, after frequency separation, the signal from both L1 and L2 frequency bands is directed to the parallel arms of the signal transmission block. Then it is frequency converted and simultaneously filtered and amplified.

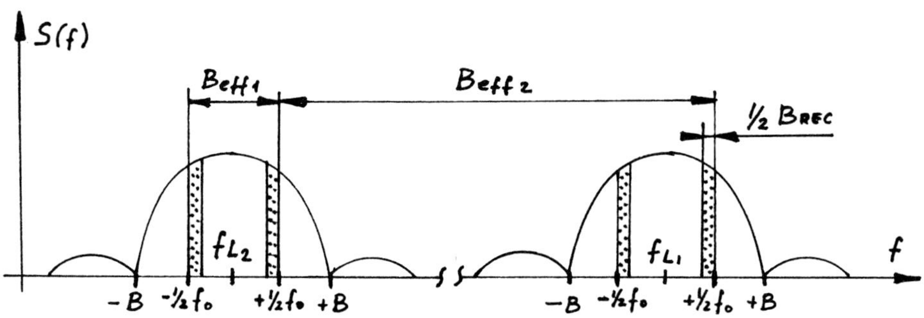

Fig. 2. P-code spectrum of the GPS signals:

f_o = 10.23 MHz; f_{L_1} = 154 f_o f_{L_2} = 120 f_o

B_{eff_1} = f_o B_{eff_2} = 34 f_o

B_{REC} - recorder bandwidth; B - the P-code bandwidth

We apply two methods for the GPS signal preprocessing without any codes. In the first case we use the single-sideband filtering (SSB), as in the VLBI. Similary as in the SERIES (2) separation of two frequency SSB channels is $\pm$ $0.5f_o$ with respect to the center frequencies of the L1 and L2 (where f_o is the GPS standard frequency). We also create two synthesized bands in order to apply the bandwidth synthesis technique: B_{eff_1} = f_o and B_{eff_2} = $34f_o$. It is easy to show that in this case we may remove the ambiguity of a fringe phase if only signal-to-noise ratio (SNR) is higher then 40. In this case we apply the correlation method of the data reduction and the signal is treated as a noise. We transform this signal into binary form and register before its off-line analysis.

In the second case we convert the wide-band and spread-spectrum signal into a monochromatic form through modulation removing from the last IF signal. The frequency of its is then compared to a standard frequency in order to determine a Doppler frequency value.

Heterodyne signals, necessary for the frequency conversion are synthesized from the local standard, as well as two signals used for the receiver calibration. The reference signal of the standard is also used in order to set a local time scale for a control and a registration system. The time scale is then synchronized to the UTC scale.

A capacity of a registration system of the interferometer depends on the amount of data necessary to set the SNR on the sufficient level. We meet hard criteria for the SNR when the VLBI method is used. It means that in this case we must use: a high gain antenna, a stable reference oscillator, a low noise front-end of the receiver and a wide-band recorder. These criteria became moderate when we profit by possibilities of the complex GPS spectrum. When we convert the spread-spectrum signal into the monochromatic form we can obtain the comparable SNR by use of unity gain antenna and fewer data.

3. The Receiver Prototype

The block diagram of the radio frequency (RF) and the intermediate frequency (IF) sections of the receiver is shown in Figure 3:

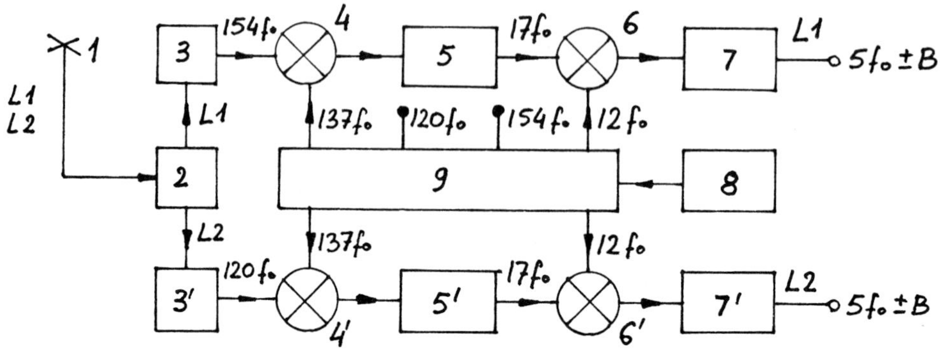

Fig. 3. BLock diagram of the RF and IF sections of the receiver:

1 - crossed dipoles; 2 - balun and directional filters; 3, 3' - parallel antenna amplifiers; 4, 4' and 6, 6' - the first and the second mixers; 5, 5' and 7, 7' - the first and the second IF amplifiers; 8 - the reference oscillator; 9 - the frequency synthesizer

The antenna with a paraboloidal mirror may be directed to the source by means of a small microcomputer and with angular accuracy better then 0,5° and resolution about 0.1°. The antenna can sense a signal from only one satellite in a particular period of the time. A radiator, consisting of two crossed dipols, and the balun are used in order to obtain circular polarization. They are placed in the focus of the antenna. The measured antenna gains are 23 dB for L1 and 26 dB for L2. Directional filters for L1 and L2 bands separations and a pair of parallel and low-noise antenna amplifiers are also close to the radiator. The amplifiers gains are equal to 35 dB and their noise temperatures are between 110 and 120 K. Semi-rigid cables connect the front-end of the receiver with the two parallel mixers of the IF section.

The signal from the bands L1 and L2 is then frequency converted and amplified and filtered. The total gain of each IF section is 75 dB. The bandwidth is about ± 7.5 MHz with respect to the center frequency of the band.

We use an oven controlled crystal frequency standard in the receiver. Its nominal frequency is equal to 5.115 MHz (i.e. the half of the GPS standard frequency f_o). A short-term (1 s) stability of the standard is better than $5 \cdot 10^{-12}$ and long-term (24 h) is better than $5 \cdot 10^{-10}$.

All heterodyne signals are synthesized in the common block. The first heterodyne frequency is $137f_o$ which is the arithmetic mean of the L1 ($154f_o$) and L2 ($120f_o$) frequencies. The first intermediate frequency is then $17f_o$. For the second step of conversion the heterodyne signal was chosen as $12f_o$, so the last IF signal has the center frequency equal to $5f_o$. We also synthesize the signals of 120 f_o and 154 f_o frequencies for calibration of the receiver.

As it was mentioned above, we preprocess the signal before its registration. When the SSB filtering is used the low -pass output filter limits the signal band up to 250 kHz. After binary conversion the output signal is registered by means of the standard video tape recorder. For the SSB filtering we synthesize a couple of pairs of heterodyne signals, reference, r, and ortogonal, i which are: $4.5f_{or,i}$, $5.5f_{or,i}$, and $0.0125f_{or,i}$. The sample rate is $0.05f_o$. When the Costas loop is used we synthesize signals which are $5f_{or,i}$. In the prototype we do not register the output signal from the loop.

Clock synchronization is performed by means of the signals transmitted by the braodcasting TV satellites. We use two receivers; the first one connected with the clock of the GPS receiver and the second one connected with the master clock referred to the UTC time scale. The resolution of this method is about 10 ns with very high repeatibility.

4. Conclusions

The aim of this paper is to present the current state of the design and engineering works on the prototype of the interferometer. All principal blocks of the both interferometer receivers are completed. The preceeding laboratory tests of the receivers parts confirmed the initial assumptions succesfully. The technological and field tests are beeing prepared and the software for the GPS data reduction and analysis is developed. The antenna sites for field tests are surveyed in the Astronomical Latitude Observatory. in Borowiec, near Poznań.

References

1. Zieliński J.B.: "Accuracy of GPS Surveying Technique and its Possible Application to the WEGENER Project", Delft Univ. of Technology, Rep. of the Dept. of Geodesy Nr. 85.3, Delft (1985)
2. MacDoran P.F.: "Satellite Emission Radio Interferometric Earth Surveying SERIES - GPS Geodetic System", Bull. Geod., 53, 117-128 (1979)
3. Spilker J.J. jr.: "GPS Signal Structure and Performance Characteristics", NAVIGATION, The Inst. of Nav., Washington D.C., 29-54, (1980)
4. Krysiński Z.J.: "RYSY - The Radio Interferometer for Surveying with the GPS", Scient. Intrumentation,Pol. Ac. of Sc., (1988)

The Monocular Electro-Optical Stereo Scanner (MEOSS) Satellite Experiment

Franz Lanzl
Institut für Optoelektronik der DFVLR, Oberpfaffenhofen

1. INTRODUCTION

In 1980 the Indian Space Research Organization ISRO offered to DFVLR to fly an experiment of 10-20 kg on a Stretched ROHINI Satellite (SROSS) of the 150 kg class.

The common understanding is to fly under the constraints of the SROSS mission (Table 1) an attractive low cost experiment provided by DFVLR.

Table 1: Nominal specifications for SROSS-MEOSS Mission
Mission Parameters

circular orbit	450 km
inclination	45.56° (48.20°)
orbital period	92.8 minutes
lifetime	6-12 months
launch	1988
telemetry	S-band
data rate	10.4 Mbit
payload weight	10 kg
payload power	25 Watt for 20 minutes/day

2. THE INSTRUMENT

The main characteristics of the proposed threefold stereoscopic CCD line scan camera are:

- a single lens (monocular) imaging system for reduction of weight and adjustment efforts,
- no moving mechanisms,
- passive cooling,
- three relatively long CCD lines of 3456 pixels in one common focal plane,
- one spectral band between 570 to 700 nm,

- 8-bit radiometric resolution,
- LANDSAT-MSS compatible geometrical resolution and
- correction for earth rotation by electronic shift.

The opto-mechanical design with a central, mechanically independent imaging unit consists of a space modification of a ZEISS Biogon lens and the focal plane containing three parallel CCD lines vertical to flight direction as shown in Fig. 1.

The central CCD line contains the optical axis and produces the nadir looking image, while the outer lines see the same parts of a scene at different angles and times (fore and aft looking images). A baffle shields the lens from stray-light. To reduce deforming effects from temperature differences the central case is fabricated from Titanium. The overall weigt of the camera is 6.5 kg. The instrument characteristics are summarized in table 2.

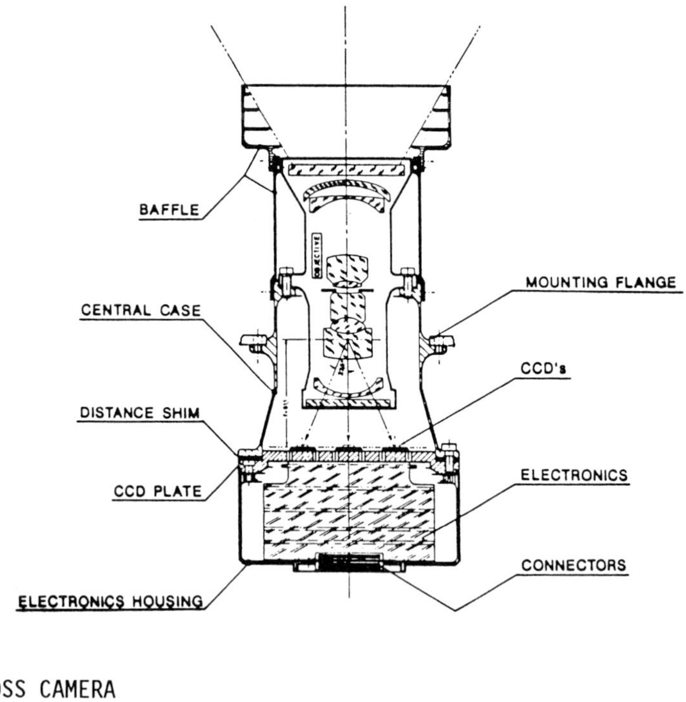

Fig.1 MEOSS CAMERA
The vital parts of the MEOSS camera are: space modified ZEISS BIOGON lens, 3 CCD's mounted in the focal plane, the electronics necessary to register and transform signals collected by the CCD's.

Table 2: Camera Data

optics	ZEISS Biogon
focal length	61.1 mm
distance of elements on CCD	Δy = 10.7 μm
number of elements on CCD	n = 3456
instantaneous field of view	IFOV = 0.01° square
distance of scanlines	d = 26.69 mm
convergence angle	β = ± 23.60°
scanline frequency	f = 131.1 Hz
spectral channel	570-720 nm

3. THE MISSION

The image data reception is performed by the ISRO-NRSA LANDSAT ground station near Hyderabad and by the DFVLR ground station Weilheim and is open to other ground stations within the mission constraints.

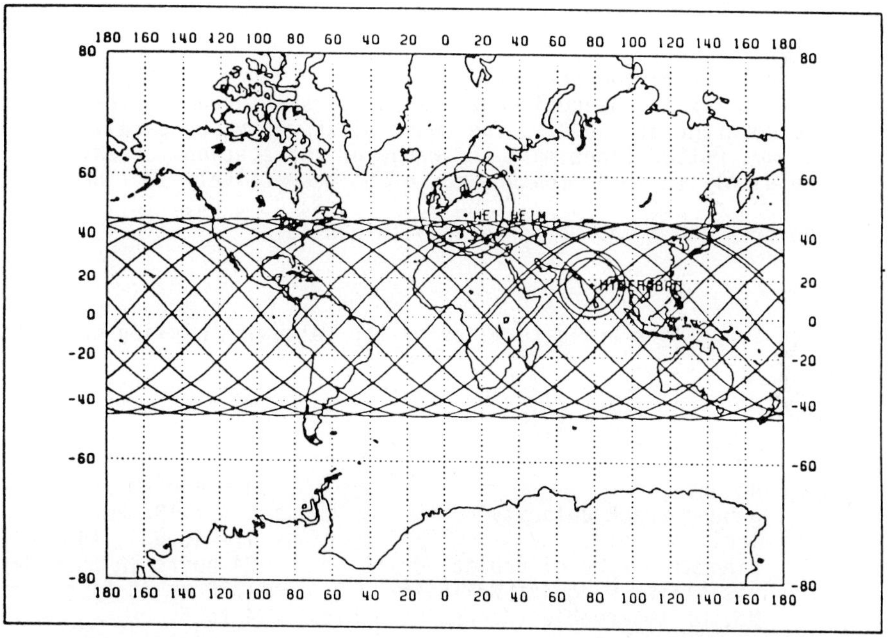

Fig. 2: The ground tracks for the stations Weilheim and Haderabad
The diagram displays the ground tracks for the MEOSS-SROSS-2 mission for 1 day. An altitude H = 450 km and an inclination of 45.6° is assumed. Shown are the visibility circles of Weilheim and Hyderabad for 5° and 10° satellite elevation. An orbit indlination of 48.2° is under consideration.

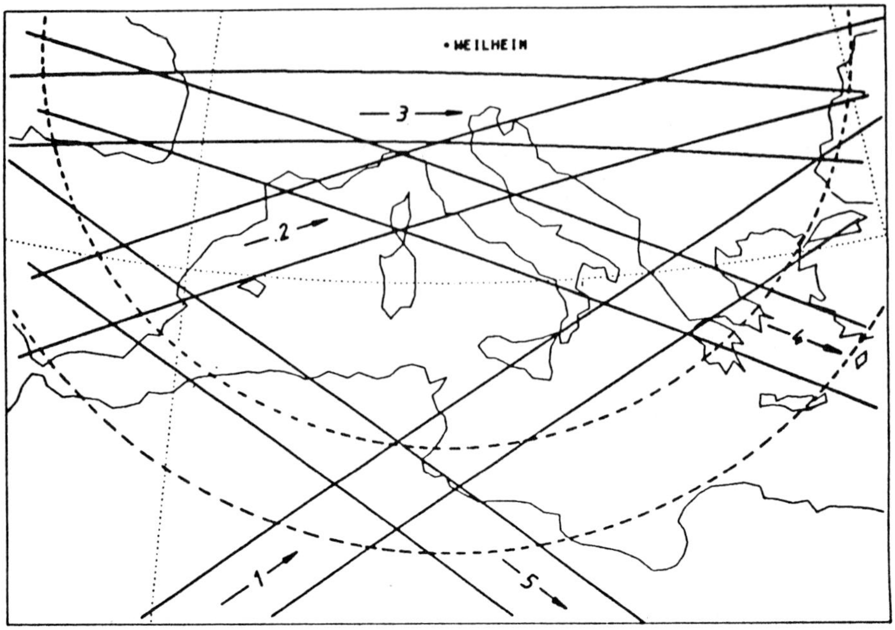

Fig. 3 Weilheim receiving station
Example for 5 consecutive orbits (numbers 1 through 5) respresenting the coverage of one day. The swath width marked by parallel lines is 255 km. Dotted lines mark the range of the station for an elevation of the satellite above the horizon of 5° respectively 10°. Because the pattern of passes is shifting 6.5° eastwards a day, full coverage of the area shown can be achieved within 6 days.

The received image data are determined by the following mission characteristics data:

Table 3: Orbit Data

height	h = 450 km
ground track velocity	V = 7.137 km/s
inclination	i = 45.60° (48.20°)
sinodal period of orbital drift	24 hours in 55.6 days
visibility range at Weilheim	1730 km at 5° elevation
No. of observable successive orbits	4 to 5

Table 4: Performance Data

radiometric resolution	8 bit
selectable gain factors	1,2 and 4
ground pixel along scanline	P = 78.8 m
inter scanline distance	P = 52.0 m
height resolution element	P = 54.7 m
swath width (central CCD)	SW = 255.8 km
baselength (nadir track)	BL = 396.1 and 2x198 km
corr. time interval (nadir track)	t = 58.1 and 2x29.5 sec
ground based convergence angle	ß° = ± 25.41° and 0°
effective base to height ratio	B:H = 0.950 and 2x0.452
active pixels per scanline	n = 3237
resolution for cross track wind	Δv = 1.12 and 2.24 m/s
height error for 1 m/s along track wind	Δh = 61.1 and 122.2 m

Image data evaluation is done in two steps

- preprocessing
  - radiometric calibration
  - earth rotation and curvature correction
  - orbit and attitude correction
- high-precision correction using
  - threefold stereo information redundancy
  - attitude modelling

The preprocessed data are available to all users. A high-precision correction scheme is also available and further dealt with in the next chapter under "Photogrammetric Image Correction and Evaluation".

4. SCIENTIFIC OBJECTIVES

The main goals of the MEOSS mission are
- generation of threefold stereoscopic line-scan imagery from space,
- investigation of procedures for image rectification,
- investigation of correlation accuracies over land and cloud fields

with the following derived products
- rectified line-scan images and stereo triplets at 1 : 1 000 000 scale,
- orthophotos and ortho-stereo triplets at 1 : 500 000 scale,

- heights and cross-track velocities of cloud fields.

MEOSS will be the first stereo scanner in space and offers the possibilities for

- basic investigations and tests of different evaluation methods for threefold stereo scan systems; this yielding experience for the plannung of future higher-resolution system,
- the incorporation of orbit and attitude data of the satellite for the evaluation of the stereo information.

The program scheme to achieve these goals is indicated in Fig. 4.

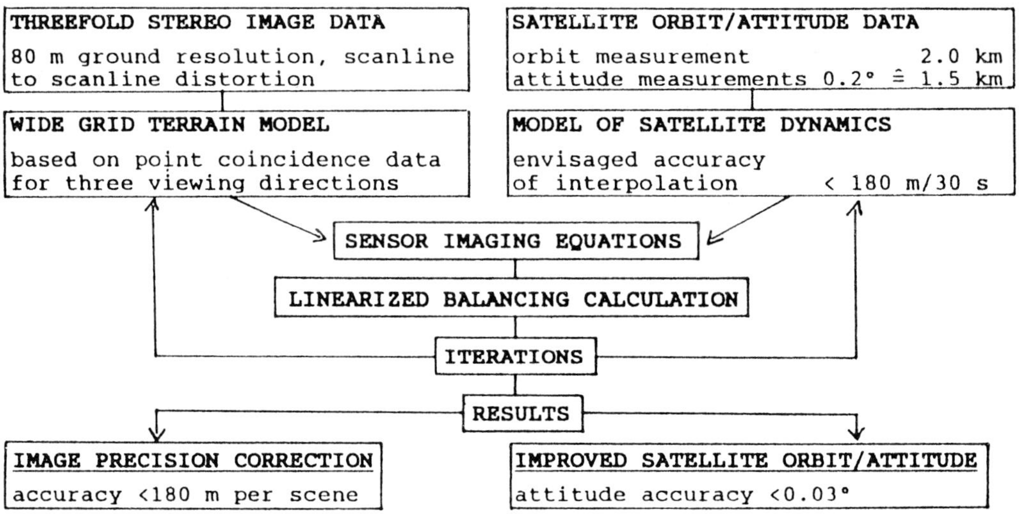

Fig. 4 MEOSS High Precision Correction Scheme

Geometric image correction will be based on ground control points and a large number of conjugate points, representing identical location in the three corresponding image strips. In addition, satellite orbit and attitude data are introduced. The result of these calculations involving photogrammetric collinearity equations are the position and orientation parameters of each recorded line, which can be applied to correct the image data.

In the following the type and magnitude of attitude variations of the satellite which cause a different orientation for each recorded line, are given in more detail. Fig. 5 shows the attitude-relevant aspects of the satellite schematically. Table 5 gives the attitude characteristics, tables 6 and 7 the resulting dynamic image distortions and the static image displacement. Table 8 summarizes the planned results of the indicated correction procedure.

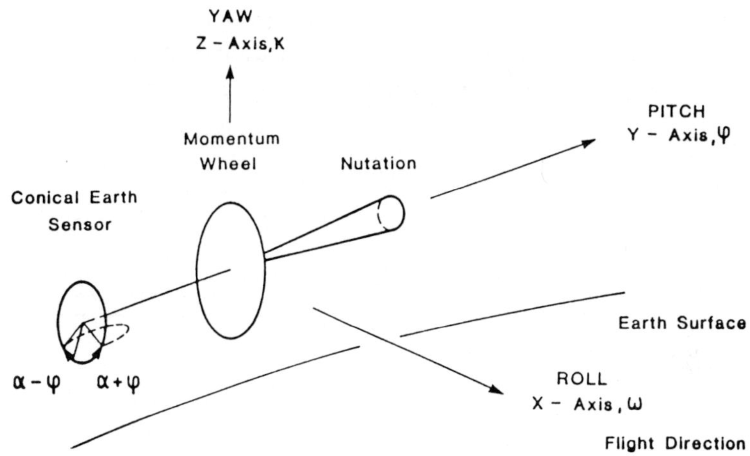

Fig. 5: Attitude Behavior of SROSS-02 Satellite

Table 5: Attitude Characteristics of SROSS-02 Satellite

	roll	pitch	yaw
Conical Scan Earth Sensor			
resolution	0.15°	0.15°	(0.4°)
Pitch Control System			
control range	---	+ 2.5° ± 0.5°	---
random drift	---	± 0.1°	---
drift periods	---	≥ 50 sec	---
drift frequencies	---	≤ $2 \cdot 10^{-2}$ Hz	---
Roll and Yaw Control System			
Control range	± 1.5°		± 3°
drift period	~ 5000 sec	---	~ 5000 sec
drift frequency	~ $2 \cdot 10^{-4}$ Hz	---	~ $2 \cdot 10^{-4}$ Hz
Nutation			
drift range	± 0.05°	---	± 0.03°
drift period	~ 12.5 sec	---	~ 12.5 sec
drift frequency	~ $8 \cdot 10^{-2}$ Hz	---	~ $8 \cdot 10^{-2}$ Hz

Table 6: Dynamic Image Distortion

Pitch random drift	
	± 0.1° ≙ ± 15 pixel
	shift along X-axis
	period > 50 sec; scene duration 29 sec
Nutation	
	± 0.05° ≙ ± 7 pixel
	shear between Y-axis and X/Z-axis
	period. ~ 12.5 sec, scene duration 29 sec

Table 7: Image Displacement

Tracking Error	
	0.4 - 2 km ≙ 8 - 40 pixel
	absolute position along X-axis
Earth Rotation	(latitude dependent, electronically compensated)
	± 8.6 km ≙ ± 110 pixel
	lateral displacement
	of fore and aft direction lines
Control system offset	
pitch	2.5° ± 0.5° ≙ 375 ± 75 pixel
	shift and scale change along X-axis
roll	± 1.5° ≙ ± 150 pixel
	shift and scale change along Y-axis
yaw	± 3° ≙ ± 450 pixel
	shear between X- and Y-axis

Table 8: Correction of Image Distortion and Image Displacement

1 MEOSS pixel

IFOV = 0,01° ≙ P_Y = 79 m; P_X = 52 m

1 MEOSS scene

swath width SW = 256 km ≙ 3237 pixel, SW : H = 0,566
bases (length) B = ± 198 km ≙ ± 3800 pixel, B : H = ± 0,452

Goal of correction:

better than 0,03° ≙ 3 pixel

Data of MEOSS will be of importance for the following fields of application:

A. Cartographic Applications

Suggested activities are:

- registration of stereo images to a standard grid system,
- derivation of orthophotos and ortho-stereo-triplets,
- resampling procedures,
- experimental derivation of topographic and orthophoto maps at scales 1:500 000 and eventually 1:250 000
- revision of topographical maps up to a scale 1:150 000.

B. Applications in Geology, Forestry, and Environmental Sciences

Suggested applications are:

- correlation of MEOSS stereo data with LANDSAT MSS, TM or SPOT multi-spectral data,
- geolocigal mapping (lithology and structures)

- studies on angular variations of radiances; discrimination between surface and atmospheric effects,
- forest inventory,
- land use inventory.

In performing these tasks two features of the MEOSS mission may be of special interest:

- varying illumination (approx. 26 min. time shift per day),
- nearly 90° difference of stereoscopic viewing direction for observation from ascending and descending orbit segment at low latitudes.

C. Meteorology

The following information could be derived:

1. From MEOSS data alone

   - cloud classification by three-dimensinal observation at high resolution,
   - cloud-snow discrimination,
   - discrimination of Cirrus against diffuse background as dense cloud fields, snow covered areas and turbity patterns in water bodies,
   - cross track cloud velocities with 1.2 m/s resolution,
   - estimate of sea state from sunglint, detectable at three differently inclined scan planes,
   - estimation of cloud field dynamics by stereo observation from successive orbits with 93 min. time gap.

2. From combination with other data sources

   - true heights of clouds and clouds displacement vectors by stereoscopy (along-track cloud displacement velocity required. 1 m/s error leads to 61.1 m error in height assigment)
   - optical density of clouds (radiation temperatures required).

Alternative Modeling of GPS Carrier Phases for Geodetic Network Analysis

Wolfgang Lindlohr
Geodätisches Institut, Universität Stuttgart

Summary

The observation equation for undifferenced GPS carrier beat phases is developed. The unknowns are divided into the desired cartesian ground station and satellite point coordinate parameters and different types of disturbing bias parameters. The classical Gauß-Markov Model and the Mixed Model which employs stochastic prior information are used as alternative mathematical models for parameter estimation and prediction, respectively. The applicability for static analysis of geodetic network is demonstrated in a six station local area network in the city of Berlin. Additionally, the concept of kinematic GPS relative positioning is described.

1. Introduction

For a long time, observations within geodetic network have only been made between the points themselves. With the coming up of single satellites and satellite systems, points along the path of a satellite motion have to be included into classical geodetic networks.

With the use of pseudo-range and especially carrier phase data from the Navstar Global Positioning System (GPS) the analysis of geodetic networks has to be reconsidered. The general concept of geodetic positioning is described in the books of R.W. King et al. (1985) and D.E. Wells et al. (1986), for example.

For high precision static GPS relative positioning the carrier phase observations are often processed in form of between-receivers and between-satellites double differenced quantities. A review of this techniques which intends to eliminate main parts of the disturbing influences and the achievable results and accuracies are given in G. Beutler et al. (1988). Alternatively, we have developed a model to investigate the original undifferenced carrier phases following the lines of W. Lindlohr and D.E. Wells (1985), B. Schaffrin and E.W. Grafarend (1986) and D.E. Wells et al. (1987). The corresponding processing program PUMA is described in full detail by W. Lindlohr (1988).

Besides the conventionally used Gauß-Markov Model, a Mixed Model which makes use of stochastic prior information for the parameters to be determined is applied as the mathematical model. This concept is very appropriate to treat the coordinates of geodetic network points no more as fixed but as quantities which need to be modified. The theoretical background and some numerical examples of this method is given in E.W. Grafarend and B. Schaffrin (1988).

2. The GPS carrier phase observation equation

The GPS carrier beat phase observable is defined by the difference of the satellite oscillator phase $\phi^s(t^s)$ at the actual time of transmission t^s and the receiver reference phase $\phi_r(t_r)$ at the actual time of reception t_r plus the initial epoch integer ambiguity N_r^s through

$$\Phi_r^s(t) := \phi^s(t^s) - \phi_r(t_r) + N_r^s \ . \tag{1}$$

Within the further developments all terms of eq. (1) have to be related to the nominal time instants of GPS signal reception t, as it is described by the basic ideas of B.W. Remondi (1984). Three Taylor series expansions for different quantities following W. Lindlohr and D.E. Wells (1985) and B. Schaffrin and E.W. Grafarend (1986) lead to the undifferenced GPS carrier beat phase observation equation

$$\begin{aligned}\Phi_r^s(t) = -\frac{f}{c}\left[1 - \frac{\dot\rho_r^s(t)}{c}\right] \rho_r^s(t) + \left[1 + \frac{\dot\rho_r^s(t)}{c}\right] \alpha_r(t) + \\ + \beta^s(t) + \gamma_r^s + (1+\delta_r) d_{trop} - d_{ion} + u_r^s(t)\end{aligned} \tag{2}$$

where

$$\rho_r^s(t) := \| \underset{\sim}{x}^s(t) - \underset{\sim}{x}_r \| \tag{3}$$

$$\dot\rho_r^s(t) := \| \underset{\sim}{x}^s(t) - \underset{\sim}{x}_r \|^{-1} \ \langle \underset{\sim}{x}^s(t) - \underset{\sim}{x}_r | \dot{\underset{\sim}{x}}^s(t) \rangle \tag{4}$$

$$\gamma_r^s := \phi^s(t_o) - \phi_r(t_o) + N_r^s \ . \tag{5}$$

In the case of single frequency observations, f denotes the L1 carrier frequency, c the speed of light in vacuum, d_{trop} and d_{ion} the tropospheric and ionospheric model correction terms, respectively, and $u_r^s(t)$ the measurement error. The range term $\rho_r^s(t)$ and the range rate term $\dot\rho_r^s(t)$ are expressed as a function of the static ground station position vector $\underset{\sim}{x}_r$, the satellite position vector $\underset{\sim}{x}^s(t)$ and the satellite velocity vector $\dot{\underset{\sim}{x}}^s(t)$. $\alpha_r(t)$ is a collection of the receiver-specific bias, $\beta^s(t)$ of the satellite-specific bias, γ_r^s the pair-specific bias and δ_r contains the tropospheric bias term according to a twofold hierarchical classification model. It has to be mentioned that the pair-specific bias parameters are non-integer quantities. Only after differencing the phase observations between-receivers and between-satellites, the initial epoch ambiguity biases exhibit their integer type nature.

3. Alternative mathematical models

A short review of the basic features of two alternative mathematical adjustment models, namely the classical Gauß-Markov Model of full rank and without restrictions and the Mixed Model is given. A detailed description of the mathematical background can be found in B. Schaffrin (1985).

3.1 The Gauß-Markov Model

The nonlinear functional model is described by the expectation of the undifferenced phase observations $\underset{\sim}{Y} := \Phi_r^s(t)$ and given by

$$E\{\underset{\sim}{Y}\} = A_1\underset{\sim}{\Xi} + A_2\underset{\sim}{\eta} \tag{6}$$

where the desired parameters are collected in the vector $\underset{\sim}{\Xi} := [\underset{\sim}{x}_r, \underset{\sim}{x}^s(t), \dot{\underset{\sim}{x}}^s(t)]$ containing the cartesian coordinates of the ground stations, the satellite positions and optionally the satellite velocities in the earth-centered earth-fixed reference frame. We have merged these unknowns into one group in order to be able to perform simultaneous relative positioning and orbit improvement. The vector $\underset{\sim}{\eta} := [\alpha_r(t), \beta^s(t), \gamma_r^s, \delta_r]$ is filled by nuisance parameters of type receiver-, satellite-, pair-specific and tropospheric bias.

The postulate of a Best Linear Unbiased Estimation (BLUE) leads after linearization at a fixed position $\underset{\sim}{\Xi}_o$, namely by introducing

$$\underset{\sim}{y} = A_1\underset{\sim}{\xi} + A_2\underset{\sim}{\eta} + \underset{\sim}{u} \tag{7}$$

where

$$\underset{\sim}{\Psi}_o := a_1(\underset{\sim}{\Xi}_o) \quad \underset{\sim}{y} := \underset{\sim}{Y} - \underset{\sim}{\Psi}_o \quad \underset{\sim}{\xi} := \underset{\sim}{\Xi} - \underset{\sim}{\Xi}_o \tag{8}$$

to the solution

$$\begin{bmatrix} \hat{\xi} \\ \hat{\eta} \end{bmatrix} = \begin{bmatrix} A_1^T P_y A_1 & A_1^T P_y A_2 \\ A_2^T P_y A_1 & A_2^T P_y A_2 \end{bmatrix}^{-1} \begin{bmatrix} A_1^T P_y y \\ A_2^T P_y y \end{bmatrix} . \tag{9}$$

3.2 The Mixed Model

An alternative approach which results in the so-called Mixed Model takes advantage of the fact that the approximate values being needed for linearization are quantities of stochastic prior information. A second linearization at a stochastic point $X_o := \Xi_o + o$, where o denotes the stochastic null vector, defines the Mixed Model

$$y = A_1 x + A_2 \eta + u \tag{10}$$

where

$$Y_o := a_1(X_o) \quad y := Y - Y_o \qquad x := \Xi - X_o = \xi - o . \tag{11}$$

A further decomposition of the a priori stochastic unknown vector x into

$$x := \kappa + e \quad ; \qquad \kappa := \kappa_o - X_o \tag{12}$$

contains the differential bias vector of prior information κ being a derivate of the vector of prior information κ_o and its error e which constitutes the basis of the variance-covariance matrix of prior information $C_\kappa := P_\kappa^{-1}$. In summary, the triple of prior information is given by (X_o, κ, C_κ).

The stochastic unknowns x are predicted using two alternative methods: the Best inhomogeneously LInear Prediction (inhom-BLIP) or the Best homogeneously Linear Unbiased Prediction (hom-BLUP) while the fixed unknowns η are again estimated by BLUE according to

$$\begin{bmatrix} \tilde{x} \\ \hat{\eta} \end{bmatrix} = \begin{bmatrix} N_{11}+P_\kappa & N_{12} \\ N_{21} & N_{22} \end{bmatrix}^{-1} \begin{bmatrix} A_1^T P_y y + P_\kappa \kappa \\ A_2^T P_y y \end{bmatrix} \quad \text{inhom-BLIP} \tag{13}$$

$$\begin{bmatrix} \tilde{x} \\ \hat{\eta} \end{bmatrix} = \begin{bmatrix} N_{11}+P_\kappa & N_{12} \\ N_{21} & N_{22} \end{bmatrix}^{-1} \begin{bmatrix} A_1^T P_y y + aP_\kappa \kappa \\ A_2^T P_y y \end{bmatrix} \quad \text{hom-BLUP} \tag{14}$$

$$N_{ij} := A_i^T P_y A_j \; , \qquad a := \frac{\kappa^T P_\kappa (N+P_\kappa)^{-1} A^T P_y y}{\kappa^T P_\kappa (N+P_\kappa)^{-1} N \kappa} . \tag{15}$$

If stochastic prior information is used for all parameters, our Mixed Model can be reduced to a pure Random Effects Model, cf. B. Schaffrin (1986), for example.

4. Results from a GPS network example: Berlin

4.1 The network

From November 2nd to November 6th, 1987 a GPS survey took place in the city area of Berlin. During the first two days five WM101 single frequency GPS-receivers of the Wild-Magnavox Satellite Survey Company have simultaneously been operated, while during the other three days even seven receivers had been available. The Institute of Geodesy and Photogrammetry, Technical University of Berlin, the Institute of Geodesy, University FAF München, the Institute of Surveying Engineering, Technical University of Braunschweig, and the Department of Geodetic Science, Stuttgart University, have taken part in this project.

The whole project network consists of a total of 19 stations. On November 4th the following 6 stations, which constitutes our local area GPS network, are illustrated in Figure 1 in a local horizontal plane. The baseline lengths between ground stations vary from about 6.2 km at minimum to about 31.3 km at maximum.

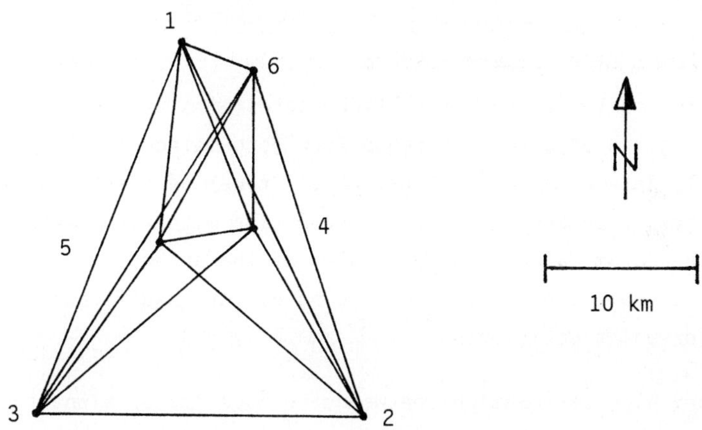

Figure 1: Part of the Berlin GPS local area network.

For demonstrating the capabilities and the performance of PUMA we have created a cycle slip free, limited data set which consists of simultaneous observations from the 6 ground stations to 5 GPS-satellites for the time span 4:01 to 4:33 UTC. Considering that the compressed measurements are recorded on magnetic tape in full minute intervals we arrive at

$$n = R \cdot S \cdot T = 6 \cdot 5 \cdot 33 = 990 \tag{16}$$

undifferenced phase observations which are used for the results computed by PUMA. For the most general case we would end up with a maximum of

$$\begin{array}{lllll} m = & 3 \cdot R & = 3 \cdot 6 & = 18 & [\underset{\sim}{x}_r] \\ & + 3 \cdot S \cdot T & = 3 \cdot 5 \cdot 33 & = +495 & [\underset{\sim}{x}^s(t)] \\ & + R \cdot T & = 6 \cdot 33 & = +198 & [\alpha_r(t)] \\ & + S \cdot T & = 5 \cdot 33 & = +165 & [\beta^s(t)] \\ & + R \cdot S & = 6 \cdot 5 & = +30 & [\gamma_r^s] \\ & + R & = 6 & = +6 & [\delta_r] \\ & - (R + S + T - 1) & & = -43 & \text{bias design defect} \\ & - 6 & & = -6 & \text{datum defect} \\ \hline & & & = 863 & \end{array} \tag{17}$$

estimable parameters resulting from a redundancy design investigation. A complete overview regarding GPS network design is given by D.E. Wells et al. (1987)

4.2 Static GPS relative positioning

Although the cartesian ground station coordinates are the parameters which are estimated in the adjustment process from both mathematical models, they are very weakly determined quantities. As all types of biased range observables, the GPS carrier phases are rather insensitive with respect to the translational part of the datum defect. The precision of the ground station coordinates is of the order of some tens of a meter, being by far insufficient for static GPS absolute or point positioning.

Due to very high correlations between the coordinates along their axes, especially in local networks, derived quantities as inter-station coordinates differences and baseline lengths, can be estimated or predicted with sufficient internal accuracy. Both types of GPS relative positioning measures, coordinate differences as well as baseline lengths offer a precision which is about three orders of magnitude higher than this for the coordinates itself.

The results of the baseline lengths for all possible combinations between network stations which are derived from the BLUE estimation, the inhom-BLIP prediction and the hom-BLUP prediction are summarized in Table 1. The third column of the table contains some approximate values to which the correction values given in columns 4, 6 and 8 have to be added. Their estimated or predicted mean errors are given in columns 5, 7 and 9. The differences between baseline lengths determined within alternative model formulations and prediction methods can be found in columns 10, 11 and 12 of the same table.

In general the differences between the BLUE estimation and the inhom-BLIP prediction are obviously the smallest ones. They are usually in the order of some mm and reach 1.2 cm at maximum. Especially for the three longer baselines the discrepancies between BLUE and hom-BLUP as well as inhom-BLIP and hom-BLUP are of the order of a few cm and reach their maximal values on the 25.5 km baseline with 4.0 cm or 3.7 cm, respectively. It indicates that under this circumstances the rather arbitrarily chosen stochastic prior information does not fully agree with the actual observation data set.

As it is theoretically known from the structure of the predicted variance-covariance matrix the precision in the case of inhom-BLIP is always better than this which can be gained from the hom-BLUP prediction. In our local network case the predicted mean errors from the inhom-BLIP prediction show in average an internal accuracy which is about 30 % higher.

4.3 Simultaneous and pure cartesian orbit improvement

The outcome of the treatment of an epochwise cartesian orbit improvement per satellite is demonstrated by two configurations related to the local area Berlin network. The first example investigates a simultaneous ground station and satellite orbit improvement. Besides one ground station which is held fixed for all other ground stations and all satellite orbit points an a priori mean error of 20 m is introduced, i.e. $P_\kappa = 2.5 \cdot 10^{-3} I$. The cartesian coordinate corrections for an arbitrarily chosen GPS satellite, i.e. PRN 12, are given in two minute intervals in Table 2(a) in the earth-centered earth-fixed reference frame. All coordinate corrections are below the 10 cm level. Their variation in all three of the coordinate axes show a rather random structure superimposed by some short wave pattern.

from	to	Approximate values [m]	BLUE (a) [m]	BLUE (a) [cm]	inhom-BLIP (b) [m]	inhom-BLIP (b) [cm]	hom-BLUP (c) [m]	hom-BLUP (c) [cm]	(a)-(b) [cm]	(b)-(c) [cm]	(c)-(a) [cm]
(1)	(2)	(3)	(4)	(5)	(6)	(7)	(8)	(9)	(10)	(11)	(12)
1	2	31320 +	1.582	± 3.1	1.591	± 3.2	1.583	± 4.2	-0.9	+0.8	+0.1
1	3	28520 +	1.878	± 5.5	1.894	± 6.1	1.865	± 7.1	-1.6	+2.9	-1.3
1	4	15520 +	7.281	± 2.7	7.290	± 2.7	7.288	± 3.7	-0.9	+0.2	+0.7
1	5	15590 +	7.263	± 4.0	7.275	± 4.3	7.262	± 5.2	-1.2	+1.3	-0.1
1	6	7040 +	6.723	± 5.9	6.720	± 6.5	6.728	± 7.6	+0.3	-0.8	+0.5
2	3	25530 +	3.749	± 8.5	3.746	± 8.6	3.709	± 11.3	+0.3	+3.7	-4.0
2	4	16380 +	3.548	± 4.2	3.549	± 4.3	3.541	± 5.7	-0.1	+0.8	-0.7
2	5	20460 +	1.672	± 5.8	1.671	± 5.9	1.671	± 7.8	+0.1	0.0	-0.1
2	6	25560 +	1.792	± 2.8	1.792	± 2.8	1.785	± 3.8	0.0	+0.7	-0.7
3	4	19110 +	6.071	± 8.4	6.072	± 8.6	6.041	± 11.2	-0.1	+3.1	-3.0
3	5	14160 +	4.003	± 7.5	4.004	± 7.7	3.981	± 10.1	-0.1	+2.3	-2.2
3	6	28350 +	2.606	± 7.3	2.599	± 7.4	2.578	± 9.7	+0.7	+2.1	-2.8
4	5	6230 +	9.048	± 8.8	9.047	± 7.4	9.050	± 11.7	+0.1	-0.3	+0.2
4	6	11380 +	7.743	± 4.5	7.737	± 4.5	7.739	± 6.0	+0.6	-0.2	-0.4
5	6	14320 +	2.851	± 7.0	2.843	± 7.2	2.843	± 9.4	+0.8	0.0	-0.8

Table 1: Estimated and predicted baseline lengths: (a) BLUE, (b) inhom-BLIP, (c) hom-BLUP.

	t	$d\tilde{x}_1^s(t)$	$d\tilde{x}_2^s(t)$	$d\tilde{x}_3^s(t)$
(a)	1	0.069	0.038	0.066
	3	-0.033	-0.040	-0.051
	5	0.011	-0.018	0.005
	7	0.035	0.001	0.042
	9	-0.063	-0.047	-0.056
	11	-0.025	-0.006	-0.004
	13	-0.064	-0.023	-0.061
	15	-0.024	-0.012	-0.035
	17	0.064	0.030	0.018
	19	0.058	0.021	0.016
	21	0.060	0.039	0.052
	23	0.094	0.078	0.099
	25	-0.057	-0.035	-0.039
	27	-0.081	-0.037	-0.031
	29	-0.051	-0.009	-0.003
	31	-0.029	-0.005	-0.006
	33	0.041	0.029	0.002
(b)	1	-0.503	-2.322	1.274
	3	-0.405	-1.975	1.120
	5	-0.330	-1.651	0.979
	7	-0.255	-1.333	0.819
	9	-0.186	-1.023	0.654
	11	-0.143	-0.732	0.497
	13	-0.086	-0.448	0.322
	15	-0.032	-0.172	0.136
	17	0.017	0.081	-0.049
	19	0.042	0.314	-0.220
	21	0.058	0.551	-0.399
	23	0.082	0.765	-0.585
	25	0.086	0.969	-0.761
	27	0.095	1.153	-0.942
	29	0.112	1.300	-1.110
	31	0.103	1.463	-1.284
	33	0.085	1.605	-1.442

Table 2: Predicted cartesian satellite coordinate corrections [m] for simultaneous (a) and pure (b) orbit improvement.

A pure cartesian orbit improvement is analysed in the second example. In this case all ground stations are held fixed and 31.6 m, i.e. $P_\kappa = 1 \cdot 10^{-3} I$, has been introduced as a priori mean error for the satellite points. For the same satellite the cartesian coordinate corrections are given in Table 2(b) and reach a maximum of approximately 2.5 m. In contradiction to the previous case, the correction terms show a strong systematic behaviour indicating that the satellite motion indeed follows some trend.

In order to enhance the clearness of the orbit improvement technique the estimated or predicted cartesian satellite coordinates are post-fitted to some other parameters describing the orbit. For example, a set of six Keplerian elements is estimated using the cartesian coordinates as pseudo-observations and the general error propagation law. In principle, the number of satellite points through which a set of Keplerian elements is fitted may vary between three at minimum and all points per arc at maximum. Instead of using Keplerian elements an alternative geometrical parametrization is given by three translational shifts and three tilts of the whole satellite arc. The methods of the translational orbit relaxation is well-known from Transit Doppler data processing, cf. J. Kouba and J.D. Boal (1975).

5. Kinematic GPS relative positioning

In order to approach navigational applications we switch from static to kinematic GPS relative positioning. The concept of kinematic positioning is to measure more relative positions in much shorter intervals of time. In principle, a kinematic network survey can be considered to be a decomposition of a lot of short period static relative position determinations.

Due to the fundamental ideas of B.W. Remondi (1985) the basic observable for kinematic surveys is the double difference observation. Once we apply the mechanism of differencing between-receivers ("Δ") and between-satellites ("∇") and neglect the atmospheric influences, we come up with a simplified observation equation for double differences

$$\nabla\Delta\Phi_r^s(t) := -\frac{f}{c}\,\nabla\Delta\rho_r^s(t) + \nabla\Delta N_r^s + \nabla\Delta u_r^s(t) \ . \tag{18}$$

In kinematic surveying the stationary antenna permanently occupies a known and well-determined network point. Then the initial relative coordinates of the roving antenna can be precisely determined by moving the stationary antenna to to the initial location of the roving antenna while, at the same time, moving the roving antenna from its initial position to the location of the stationary antenna. This procedure is referred to as "antenna swapping" and results in an elimination of the double difference integer ambiguities $\nabla\Delta N_r^s$. Afterwards the relative coordinates of all stations in the local network can precisely be determined in the kinematic mode. For some further ideas related to this concept we refer to C.A. Quirion (1987) and B. Hofmann-Wellenhof and B.W. Remondi (1988).

An alternative method for kinematic positioning using carrier phase data has been proposed by C.C. Goad (1988). At the beginning of a kinematic survey two known and well-determined stations have to be occupied for some time as long as the initial integer values of the pair-specific biases can be determined and stored for further processing. Afterwards the procedure continues in the same way as in the aforementioned method.

6. Conclusions

From the GPS carrier beat phase observations information for the analysis of geodetic networks can be gained. For certain reasons like ambiguity resolution as well as cycle slip detection and correction this type of observation can get support from the GPS pseudo-range observation type. Instead of applying between-receivers and/or between-satellite and/or between-epochs differencing techniques, we developed a model for analysing the original undifferenced GPS carrier beat phases.

These observations are investigated within two alternative mathematical models: the classical Gauß-Markov Model and a generalization of it, the Mixed Model. The Gauß-Markov Model leads to a best linear unbiased estimation (BLUE) of the parameters involved. The unknowns are divided into the desired parameters which contain the cartesian ground station and satellite point coordinates and on the other side the receiver-, the satellite- and the pair-specific as well as the tropospheric bias parameters. Two alternative prediction methods, the best inhomogeneously linear prediction (inhom-BLIP) and the best homogeneously linear unbiased prediction (hom-BLUP) allow to determine the desired parameters within the Mixed Model. The matrix of mean prediction errors of inhom-BLIP is minimal in the respectively class of predictors and the main characteristic of hom-BLUP is that it is robust against erroneous or wrong stochastic prior information.

A local area network which consists of six ground stations is situated in the city of Berlin. A limited undifferenced phase data set simultaneously observing five GPS satellites over a period of 33 minutes in full minute intervals is analysed in the network processing mode. For baselines of up to 30 km in length an internal relative positioning accuracy of 3 to 5 ppm can be obtained. In addition to this static GPS positioning technique the basic features of a kinematic mode GPS survey are generally described.

7. References

Beutler, G., I. Bauersima, W. Gurtner, M. Rothacher and T. Schildknecht (1988): Static positioning with the Global Positioning System (GPS): state of the art. Presented Paper, International GPS-Workshop, Darmstadt, April.

Goad, C.C. (1988): Investigation of an alternative method of processing Global Positioning System survey data collected in kinematic mode. Presented Paper, International GPS-Workshop, Darmstadt, April.

Grafarend, E.W. and B. Schaffrin (1988): Von der statischen zur dynamischen Auffassung geodätischer Netze. Zeitschrift für Vermessungswesen, Vol. 112, No. 2, 79-103.

Hofmann-Wellenhof, B. and B.W. Remondi (1988): The antenna exchange: one aspect of high-precision GPS kinematic survey. Presented Paper, International GPS-Workshop, Darmstadt, April.

King, R.W., E.G. Masters, C. Rizos, A. Stolz and J. Collins (1985): Surveying with GPS. Monograph No. 9, School of Surveying, University of New South Wales, Kensington.

Kouba, J. and J.D. Boal (1975): Program GEODOP. Geodetic Survey of Canada, Surveys and Mapping Branch, Department of Energy, Mines and Resources, Ottawa.

Lindlohr, W. and D.E. Wells (1985): GPS design using undifferenced carrier beat phase observations. Manuscripta Geodaetica, Vol. 10, No. 4, 255-295.

Lindlohr, W. (1988): PUMA: Processing of Undifferenced GPS carrier beat phase Measurements and Adjustment computations. Technical Report No. 5, Institute of Geodesy, University of Stuttgart, in preparation.

Quirion, C.A. (1987): The Global Positioning System in kinematic mode: formulation, analysis, and use. Department of Civil Engineering, University of Maine at Orono, Report No. 70.

Remondi, B.W. (1984): Using the Global Positioning System (GPS) phase observable for relative geodesy: modeling, processing, and results. CSR-84-2, Center for Space Research, The University of Texas at Austin.

Remondi, B.W. (1985): Performing centimeter-level surveys in seconds with GPS carrier phase: initial results. NOAA Technical Memorandum NOS NGS-43, Rockville.

Schaffrin, B. (1985): Das geodätische Datum mit stochastischer Vorinformation. Deutsche Geodätische Kommission, C-313, München.

Schaffrin, B. (1986): New estimation/prediction techniques for the determination of crustal deformations in the presence of prior geophysical information. Tectonophysics, Vol. 130, 361-367.

Schaffrin, B. and E.W. Grafarend (1986): Generating classes of equivalent linear models by nuisance parameter elimination: applications to GPS observations. Manuscripta Geodaetica, Vol. 11, No. 4, 262-271.

Wells, D.E., N. Beck, D. Delikaraoglou, A. Kleusberg, E.J. Krakiwsky, G. Lachapelle, R.B. Langley, M. Nakiboglu, K.-P. Schwarz, J.M. Tranquilla and P. Vanicek (1986): Guide to GPS Positioning. Canadian GPS Associates, Fredericton, N.B.

Wells, D.E., W. Lindlohr, B. Schaffrin and E.W. Grafarend (1987): GPS design: undifferenced carrier beat phase observations and the fundamental differencing theorem. Department of Surveying Engineering, Technical Report No. 116, University of New Brunswick, Fredericton, N.B.

Frontend Design for High Accurate Positioning with GPS

Guenther Schneck
Institut für Navigation, Universität Stuttgart

Summary

Within the next few years the Global Positioning System (GPS) will assert itself as a world wide navigational system. Today thanks to new technology, accuracies down to the nearest cm can be attained. Special attention should be drawn to the part of the receiver actually responsible for the reception of the satellite signals and commonly known as the frontend. The antenna phase centre also plays an important role by creating the spacial connection between electromagnetic waves and the geodesic circuit and at the same remaining invariant regardless of the direction of the received signals. Moreover weak signals must be amplified without causing additional noise and in such a way as to keep the thus resulting position uncertainty at a minimum. This is the task of the low noise pre-amplifier (LNA). The most significant problems that occur in the design of a frontend will be discussed.

GPS-Frontend

Fig.1 shows the block diagram of one possible solution of the GPS frontend. The antenna has to receive the satellite signals and is the geometric point to which the measurement is related. A bandpass filter rejects interference by adjacent signals and the multi-stage low noise pre-amplifier takes the received signals onto a level, that will be necessary to get the signal to the receiver without further disturbancies.

Antennas for GPS

It is important for high accurate positioning with GPS to get a form of antenna that meets most of the following requirements:

- hemispherical antenna pattern with half beam bandwidth of more than 70°
- right hand circular polarization
- accurate defined and stable phase centre

- low sensivity against multipath signals
- possibility of the reception of both GPS-frequencies
- handsome dimensions and weight
- high mechanical stability

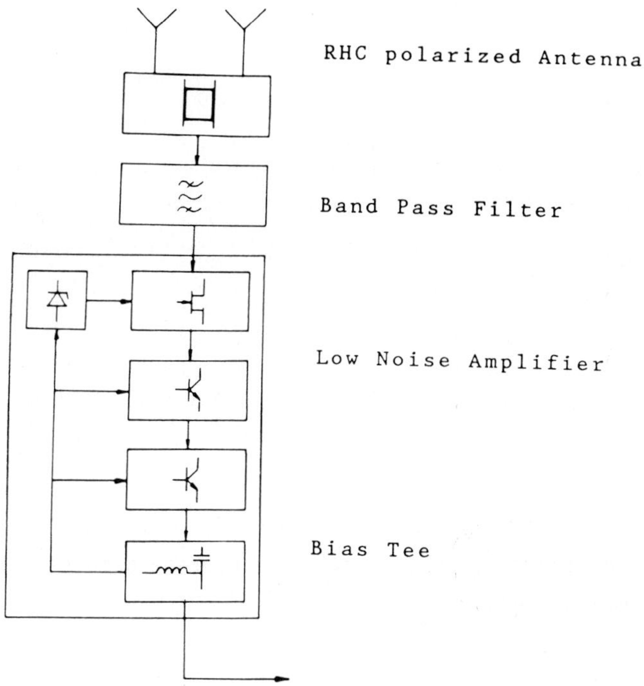

Fig.1 Block diagram

A few examples of suitable GPS-antennas will show their advantages, but also their disadvantages:

Dipole-Antenna
- simple form of antenna
- easy to manufacture
- moderate dimensions
- large bandwidth
- as a pair of crossed dipoles able to be polarized circular
- easy variation of desired antenna pattern

Helix-Antenna
- circular polarization
- easy to manufacture
- bad elypticity at wide half beam bandwidths

Double Ended Helix
- hemispherical antenna pattern
- circular polarization
- difficult to calculate and manufacture

Microstrip-Antenna
- very easy to manufacture
- very flat dimension
- possible to get circular polarization
- hemispherical antenna pattern
- narrow bandwidth
- low efficiency

Two of those antennas, the microstrip-patch-antenna and the circular polarized version of the dipole, the pair of crossed dipoles, have been optimized for GPS-measurements.

Pair of crossed dipoles

The dipole was taken because it is well described in literature and the results can easily be predicted. The two dipole-elements are formed as biconical-dipoles to gain a large bandwidth that can cover both frequencies. The disadvantage is, that an input filter has to be added. The antenna can be etched on a copper clad printed circuit board. Fig.2 shows the layout.

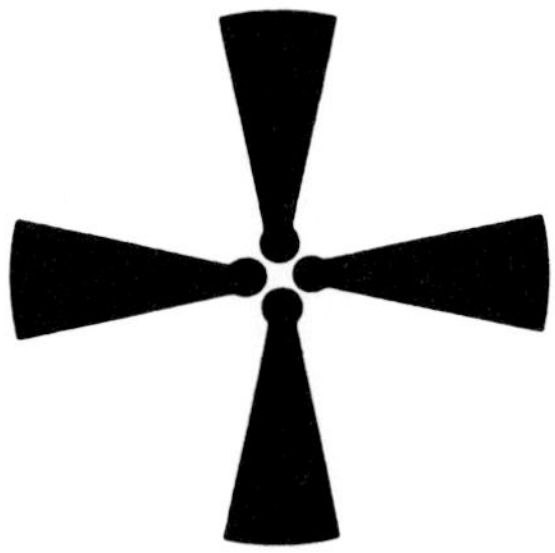

Fig.2 Layout of crossed dipoles (reduced scale)

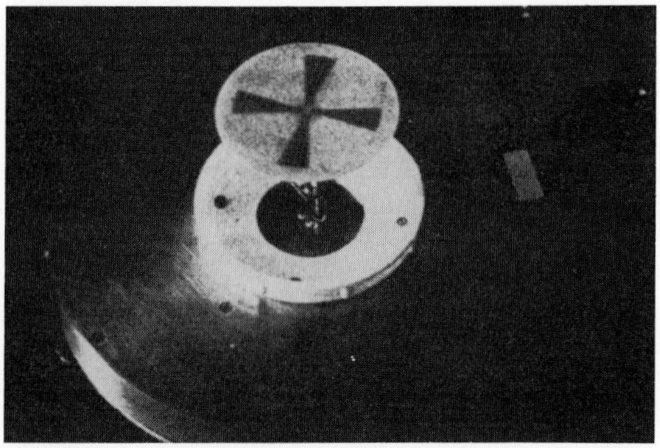

Fig.3 GPS-frontend

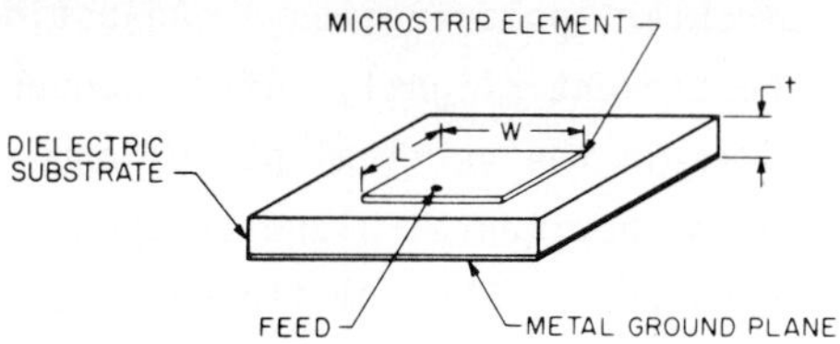

Fig.4 Basic microstrip-element

Both dipoles are fed with quadrature phase to produce circular polarization. The dipoles are positioned in front of a well conducting reflector. The half beam bandwidth of the antenna pattern can be varied by changing the distance between reflector and radiating elements. The reflector also serves as a case for the pre-amplifier and the input filter. The complete antenna is shown in fig.3.

Microstrip Antenna

The microstrip-antenna convicts by its flatness which is useful in many applications. It consists of a rectangular element that is photoetched from one side of a teflon-fibre printed-circuit-board. The element can be fed with a coaxial feed. Fig.4 shows the basic microstrip-element.

The length L is a critical dimension and slightly less than a half wavelength in the dielectric substate.

$$L \approx 0.49\ \lambda_d$$

The input impedance R_{in} is a function of the feed location.

$$R_{in} \simeq \frac{[120\lambda_0]^2 + \left[\frac{377\ t}{\sqrt{\epsilon_r}\ W}\right]^2 \left[\frac{\tan^2\beta l + \tan^4\beta l}{1 + \tan^2\beta l}\right]}{240\ W\ \lambda_0\ [1 + \tan^2\beta l]}$$

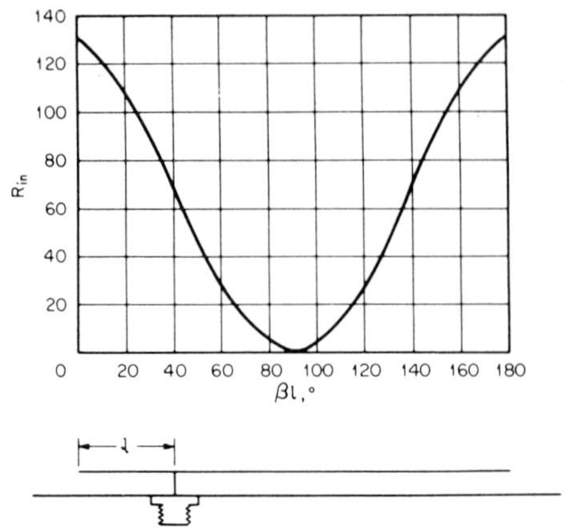

Fig.5 Input impedance as a function of feed location

The impedance can be matched to 50Ω by varying the distance from the edge of the element to the feed location *l* as shown in fig.5.

A microstrip-element that is square with *L* and *W* and equal to $\lambda_d/2$ will have two modes of radiation, horizontal and vertical. If these two modes are excited 90° out of phase the element will radiate in circular polarization. Adjusting the element slightly off resonance by +Δ in the vertical plane and -Δ in the horizontal plane by about 1% of $\lambda_d/2$ results in an increase of mode suspectance. The admittance in both planes is given by

$$Y_V = G + jB \qquad ; \qquad Y_H = G - jB$$

When $B = G$, the feed applies a uniform voltage to conjugate impedances. The two resulting modes of radiation are excited with equal power and 90° out of phase.

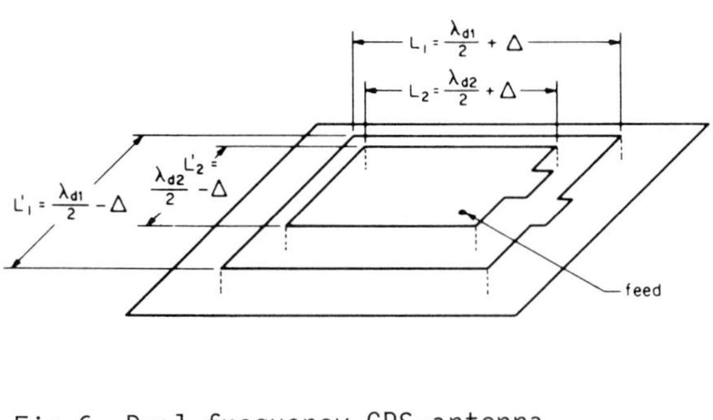

Fig.6 Dual frequency GPS-antenna

The bandwidth of the microstrip-antenna is proportional to the thickness of the dielectric substrate and therefor usually narrow. The advantage is, that no input filter will be required. If both GPS-frequencies should be received, it is possible to stack two elements fed in series, because off resonance the microstrip-element looks like a short circuit. This allows the resonant element to work independently. Fig.6 shows the dual frequency GPS-antenna.

Measuring Antenna Patterns

To measure the behavior of amplitude and phase of the antennas, a computer-controlled measurement-equipment was designed and installed. Its resolution is 3° in the elevation plane. Fig.7 shows the schematic structure.

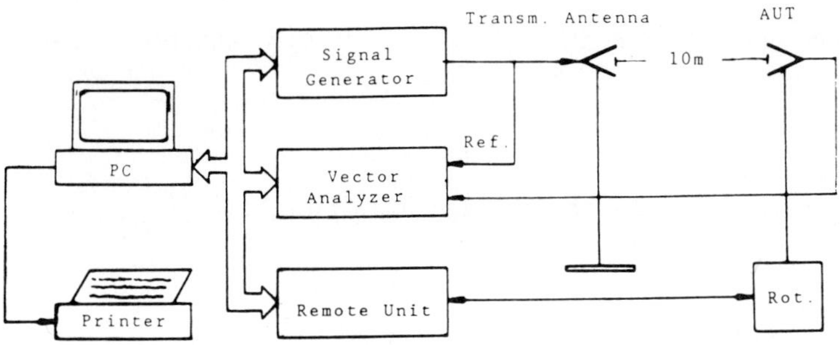

Fig.7 Schematic structure of measurement-equipment

Figs.8 and 9 show the measured results of the crossed dipoles and the microstrip-antenna. The amplitude pattern of the dipole gains proper values with a smooth decrease at lower elevation angles. This leads to less sensitivity against multipath signals. The phase error is less than ±10° at elevation angles of more than 20°. That equals to a position accuracy in the range of cm.

The microstrip-antenna has a wider amplitude pattern associated with increased phase errors. To get similar accuracies a software-supported correction should be used.

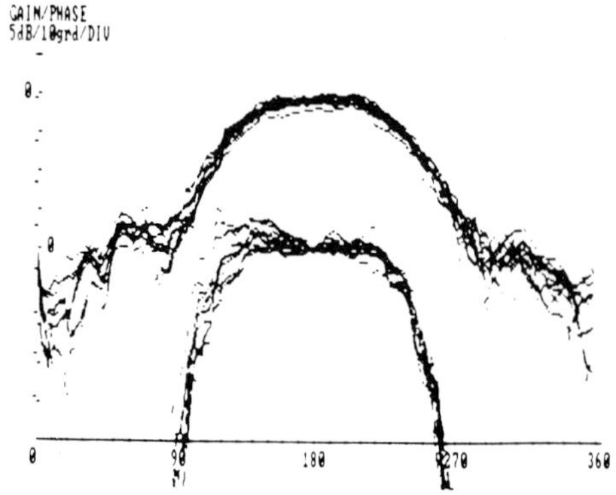

Fig.8 Antenna pattern of crossed dipoles

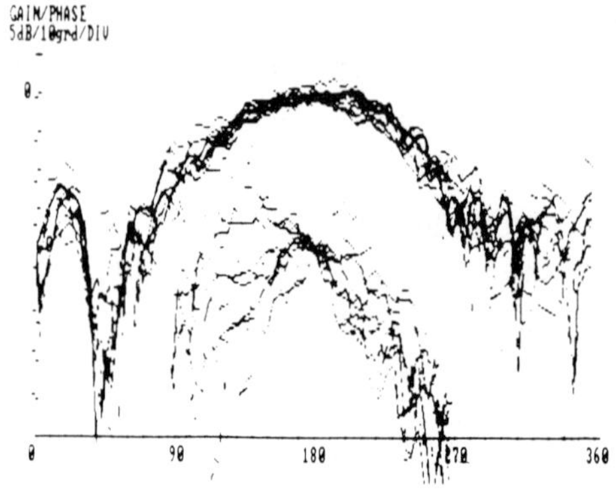

Fig.9 Antenne pattern of microstrip-patch

Low Noise Amplifier

The performance of the three stage amplifier is:

Gain	40 dB
Noise Figure	1.5 dB
Max. Output Level	14 dBm
Max. Input Level	-25 dBm

The maximum noise figure of the whole frontend is less than 2.5 dB. It is possible to separate the frontend from the receiver for up to 40 meters without affecting the signal quality itself while gaining a very high positioning accuracy.

Bibliography

[1] I.J. Bahl, P. Bhartia, "Microstrip Antennas", Second printing 1982, Artech House, Inc.

[2] R.C. Johnson, H. Jasik, "Antenna Engineering Handbook", Second Edition, McGraw-Hill Book Company

Combining GPS and Classical Geodetic Surveys for Crustal Deformation in the Imperial Valley, California

Richard A. Snay, Alice R. Drew
National Geodetic Survey, NOAA, Rockville, MD, USA

Summary

In 1986 the National Geodetic Survey (NGS) performed a GPS survey of a 38-station network in California's Imperial Valley. Here relative motion between the Pacific and North American plates has spawned several moderate earthquakes such as the M=7.1 event of 1940 and the M=6.6 event of 1979. Most stations occupied in 1986 had been positioned beforehand during one or more of the previously observed classical surveys. Triangulation was observed in 1934, 1939, 1941, and 1954-55; and triangulation/trilateration in 1967, 1975, 1979, and 1980. We supplemented the combined data set with prior information to estimate how individual stations have moved from survey epoch to survey epoch. The prior information has the form of quasi-observations that damp the amount of displacement that can occur at individual stations in the direction perpendicular to the plate boundary. The quasi-observations serve not only to remove datum defects encountered when estimating displacements from geodetic data; they also serve to better resolve attributes of the deformation that are hidden within the noise of the real observations. Estimated displacement vectors agree at the 0.1-m level with displacement vectors produced by relatively simple dislocation models.

1. Introduction

The significance of GPS technology differs from person to person. To the designer of geodetic surveys, GPS technology means that he not be concerned whether or not stations P and Q are intervisable. To the financer of geodetic surveys, GPS technology means that he need not dig so deep into his pocketbook. If we consider what GPS technology means to the data analyst who studies crustal motion, then we recognize that he now can work with more precise data than that to which he is accustomed and that he can readily obtain deformation in all 3 spatial dimensions. To elaborate, GPS technology provides the capability to obtain 3-D intersite vectors with approximately an 0.1 part-per-million precision. In comparison the precision of most triangulation/trilateration data is an order of magnitude worse. Moreover these classical data usually exist without adequate orientation and height information for crustal motion purposes. Because of GPS' greater precision and its 3-D feature, the data analyst will be able to characterize crustal deformation in terms of displacement vectors at individual stations as soon as a sufficiently long time base of GPS data becomes available. In contrast, when only classical horizontal data were available, the analyst generally characterized deformation in term of a spatially-averaged strain tensor. Displacement vectors are preferred because they are easier to interpret. Hence, rather than wait for GPS data bases of sufficient duration, we have become eager to generate such vectors with only the classical and GPS data presently available. This report describes a proposed technique for accomplishing just such a result.

The objective of estimating displacement vectors from classical horizontal data is ambitious in that observational noise exceeds the deformational signal in magnitude. For example, interseismic shear strain rates in California are generally less than 0.3 microradians per year [Savage, 1983], while for two triangulation surveys separated by 10 years, the accuracy of an observed angle change exceeds 0.6 microradians per year. To overcome this imbalance we propose the use of a technique whereby the geodetic observations are supplemented with prior geophysical information to estimate the displacement vectors. This prior information not only improves the signal-to-noise ratio, but also removes certain datum defects associated with estimating displacement vectors from classical horizontal data. This technique involves the creation of artificial observations, called quasi-observations, that mathematically embody the adopted prior information. These quasi-observations are assigned standard errors and entered with the geodetic observations into a least squares adjustment process to estimate values for selected parameters. In the process the quasi-observations are treated just like real observations. Here we describe a specific application of the technique to the geodetic data in California's Imperial Valley (Fig. 1). The data consists of 9 independent

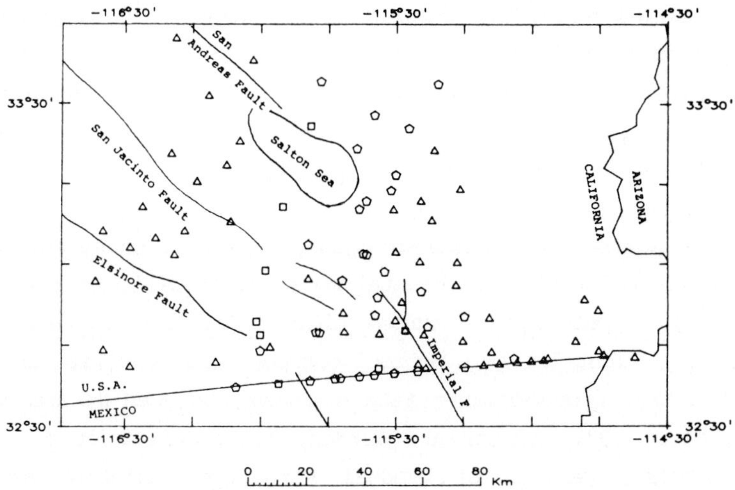

Fig. 1. Repeatedly observed geodetic stations in California's Imperial Valley. Map symbols specify the time interval between the earliest and latest observation at these stations: triangles = 10-20 years, squares = 20-30 years, pentagons = 30-52 years.

Table 1. Summary of geodetic surveys in Imperial Valley

Year of Survey	Stations Occupied	Observations			
		Directions	Azimuths	Distances	GPS Vectors
1934	80	460	0	4	0
1939	149	940	0	3	0
1941	66	570	0	4	0
1954-55	149	1342	0	18	0
1967	72	761	1	32	0
1975	23	74	3	114	0
1979	18	64	2	46	0
1980	16	134	2	89	0
1986	38	0	0	0	106
Totals	611	4345	8	310	106

surveys spanning the 1934-86 time interval (Table 1). The first 8 surveys were performed using triangulation/trilateration technology. GPS technology was used for the recent 1986 survey. The quasi-observations are derived from the hypothesis that displacements are oriented mostly parallel to the local boundary between the Pacific and North American plates--here assumed to equal N40°W. After describing the geodetic data and how the quasi-observations were combined with them to obtain the displacement vectors, this paper presents a preliminary evaluation of the technique.

2. Background

Since 1934 the NGS and its predecessor agency, the U.S. Coast and Geodetic Survey, have monitored horizontal crustal movement in California's Imperial Valley (Fig. 1). Deformation across this repeatedly surveyed region reflects ongoing right-lateral motion between the Pacific and North American plates that proceeds at a rate somewhere between 34 and 58 mm/yr [Prescott et al., 1987; Minster and Jordan, 1978; DeMets et al., 1987]. Within the network, periods of gradual strain accumulation have been punctuated by large, essentially instantaneous displacements associated with several moderate earthquakes including an M=7.1 event in 1940 and an M=6.6 event in 1979. During both of these events, patches of the Imperial fault slipped a meter or more.

Early geodetic surveys in the Imperial Valley (1934, 1939, 1941, and 1954-55) were performed in triangulation mode, that is, directions were observed from each of several geodetic stations to those stations that are visible from it. First-order directions have an accuracy of 0.6 arc-seconds (3.0 microradians) and second-order directions, 0.7 arc-seconds (3.5 microradians) [Gergen, 1975]. These surveys also contain a few astronomic azimuth observations for network orientation and a few taped distance observations for network scale.

Geodetic surveys performed in 1967, 1975, 1979 (prequake), and 1980 were performed in combined triangulation/trilateration mode, that is, several distance observations supplemented the direction observations. These distances were measured using single-frequency electronic-distance-measuring (EDM) instrumentation. The empirical formula

$$\sigma = (a^2 + b^2L^2)^{0.5} \tag{2.1}$$

is commonly used to express the standard error of a distance where a and b are specified parameters and L denotes line length. Typical values for NGS observations are a = 0.015 m and b = 10^{-6} [Gergen, 1975], but these values may vary greatly from survey to survey as technology has improved over time.

As previously mentioned, the 1986 survey was performed with GPS instrumentation. For each working day during the period from May 21 to June 2, as many as eight TI4100 instruments were deployed among network stations to receive simultaneously radio signals broadcasted by the then available constellation of GPS satellites. Most observing sessions spanned the 4 to 5 hour period of the day during which this constellation provided adequate geometric coverage over southern California. The GPS survey was designed so that most stations were occupied for at least two different observing sessions. In a Cartesian reference frame whose Z-axis parallels the Earth's pole of rotation, the components of individual vectors have a precision of about $10^{-6}L$, where L denotes vector length. This precision varies greatly from vector to vector, however, because of hardware problems that went undetected until after the survey was completed. Without such hardware problems and with the present capability to obtain more precise orbital information, we anticipate that future GPS surveys will be about an order of magnitude more precise. Neugebauer [1988] provides a more detailed description of the 1986 GPS survey and the procedures used to process and evaluate the corresponding data.

3. Estimating Station Displacements

Individual station displacements from survey epoch to survey epoch were estimated using a two step process.

In the first step we entered the combined data set into a least squares adjustment process in which the unknowns are horizontal coordinates of the stations at the various epochs. Thus a station P observed for n(P) epochs will have 2n(P) unknowns associated with it--its latitude and longitude for each of the n(P) epochs. To this adjustment we added quasi-observations which stipulate that the displacement of station P from epoch i to epoch i+1 (for i = 1,2,3,...,n(P)-1) is zero in the direction perpendicular to the plate boundary. In all we introduced $N = \sum_P [n(P)-1] = 269$ quasi-observations and assigned each a standard error of 0.06 m unless the adjustment revealed a conflict between a quasi-observation and the motion as implied by the real data. Such conflicts were found with several quasi-observations that correspond to coseismic displacements at stations near the Imperial fault, especially for the 1940 earthquake. (See fig. 2.) These coseismic displacements, that are distinctively nonparallel to the plate boundary, are thought to be real as they can be explained in terms of large spatial variations in slip that occurred on the Imperial fault during the 1940 earthquake. When such conflicts as these occurred we either discarded the corresponding quasi-observations or increased the corresponding standard errors by a factor of 10.

We also added "positional" constraints to the adjustment to obtain a nonsingular system of normal equations. A positional constraint specifies the latitude or longitude of a specific station at a specific epoch. The positional constraints are needed because the real observations (directions, distances, and azimuths) contain essentially no information about absolute position; the observations establish only relative positions

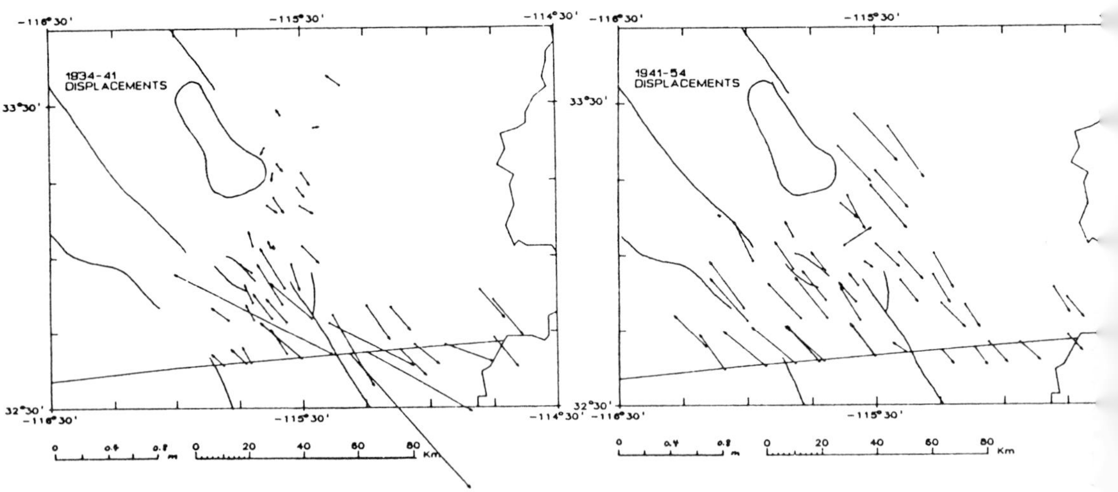

Fig. 2. Estimated displacements at selected geodetic stations.

among the stations. We added only the minimum number of positional constraints needed to compensate for the rank defect corresponding to fault-parallel position. A sample minimal set of positional constraints consists of specifying the latitude and longitude of one station during the first epoch and specifying the latitude of one station in each of the other epochs; the constrained station may differ from epoch to epoch. The fact that the set of positional constraints is minimal means that we have not biased the informational content of the real observations. The set of quasi-observations, however, is not minimal and thus they bias the informational content of the real data. This biasing is desirable, however, because of the low signal-to-noise ratio associated with the triangulation data and because network scale and orientation are poorly resolved by the few distances and azimuths that were observed (Table 1).

Whereas the first step is designed to quantify motion in the fault-perpendicular direction, the second step addresses motion in the fault-parallel direction. In this step we estimated a set of plate-parallel shifts, one for each data epoch, to be applied to the positional coordinates generated during the first step. These shifts are derived by a least squares process that assumes that if a station is sufficiently far from the line representing the plate boundary (we used 9 km as our criterion), then its velocity is constant and equal to half the plate rate where the plate rate is also an unknown in the least squares procedure. The positional coordinates generated during the first step serve as the "observations" for the least squares process of the second step. After we applied the derived shifts to obtain new horizontal coordinates for each survey epoch, we computed displacement vectors at the various stations from one survey epoch to another by differencing the new horizontal coordinates. The resulting displacements, consequently, are approximately balanced about the line that we chose to represent the plate boundary. Figure 2 presents the resulting displacements for the 1934-41 and the 1941-54 intervals.

4. Evaluation

For evaluating our results, we developed dislocation models to generate displacement fields that may be compared with the displacement vectors obtained from the geodetic data. For these models one assumes that the Earth's crust may be represented as a homogeneous, isotropic, elastic halfspace; and one specifies the location, size, and orientation of a few rectangles that represent certain geologic faults. Dislocation theory incorporates the mathematical relationships necessary for computing the displacement field that corresponds to a specific slip distribution on these rectangular surfaces [Chinnery, 1961].

For each of 3 intervals (1941-54, 1954-67, and 1967-86) we specified a set of 2 or 3 rectangular fault planes and then used a trial-and-error approach to find values for strike slip on these surfaces such that the corresponding displacement fields agreed closely with the geodetically-derived displacement vectors (Fig. 3). Table 2 identifies the adopted dislocation parameters. For each of these 3 intervals, the discrepancy vectors (difference between model and derived displacements) have an rms length of approximately 0.1 m. This rms discrepancy is quite acceptable given that the quasi-observations have an assigned standard error of 0.06 m. In addition to this anticipated 0.06-m uncertainty for boundary-perpendicular displacement, the 0.10-m rms also incorporates that uncertainty due to boundary-parallel displacement and that due to any inaccuracy associated with using dislocation models.

To evaluate the effect of introducing the quasi-observations, we generated 3 different solutions. For one solution we assigned a standard error of 0.06 m to each quasi-observation. For another, we assigned a standard error of 0.10 m, and for the third--1.00 m. Larger standard errors enable larger displacements to occur in the N50°E direction. Also as assigned standard errors increase, the corresponding solution better approximates the 'outer-coordinate' solution of Prescott [1981] in which the informational content of the real data remains unbiased. Figure 4 compares 1941-54 displacement vectors from the 1.00-m solution with those from the 0.06-m solution. Although these two sets of displacement vectors differ greatly, residuals of the real observations are statistically equivalent for the two solutions [Snay and Drew, 1988]. In other words, the real data are relatively insensitive to the standard error values assigned to the

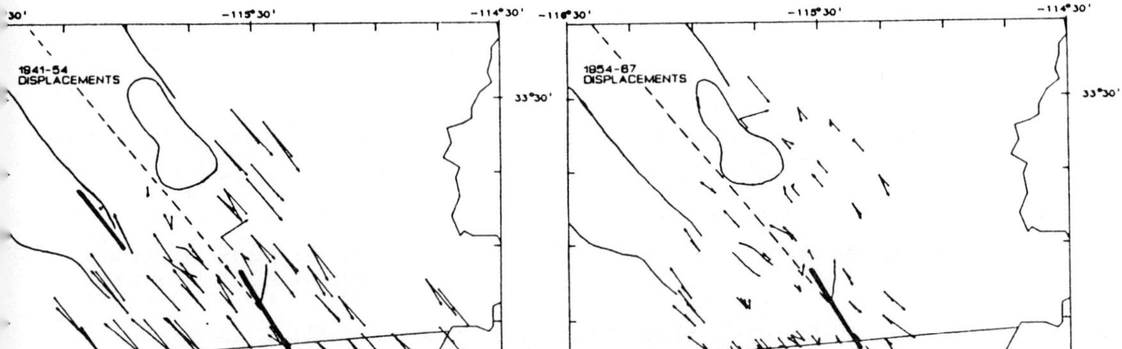

Fig. 3. Comparison of observed displacements (larger arrowheads) with modeled displacements at selected geodetic stations. Heavy lines locate the shallow dislocation surfaces. The dashed line locates the deep dislocation surface. Model parameters are specified in Table 2.

quasi-observations. The residuals for the quasi-observations provide a more sensitive measure of the appropriateness of the assigned standard errors. Statistics on these residuals indicate that a standard error of 0.04 m would be more appropriate [Snay and Drew, 1988]. Figure 5 supports this conclusion. We, however, did not generate a 0.04-m solution because the difference between the displacement vectors for the 0.06-m solution and those for the 0.10-m solution is virtually indistinguishable when plotted at the scale of Figure 4.

Table 2. Disloation surfaces for models of Figure 3.

Fault	Center	Length	Depth Range	Strike	Interval: slip (rate)
Imperial	32.796°N 15.464°W	40 km	0-10 km	N37°W	1941-54: 0.20m (15 mm/yr) 1954-67: 0.10m (8 mm/yr) 1967-86: 0.96m (*) Total: 1.26m (*)
plate boundary at depth	32.796°N 115.464°W	∞	10-∞ km	N40°W	1941-54: 0.80m (61 mm/yr) 1954-67: 0.27m (21 mm/yr) 1967-86: 0.85m (45 mm/yr) Total: 1.92m (43 mm/yr)
rupture surface for 1942 quake	33.080°N 116.120°W	30 km	0-10 km	N50°W	1941-54: 1.09m (*)

* No rate is given as the greater part of the estimated slip is thought to be coseismic.

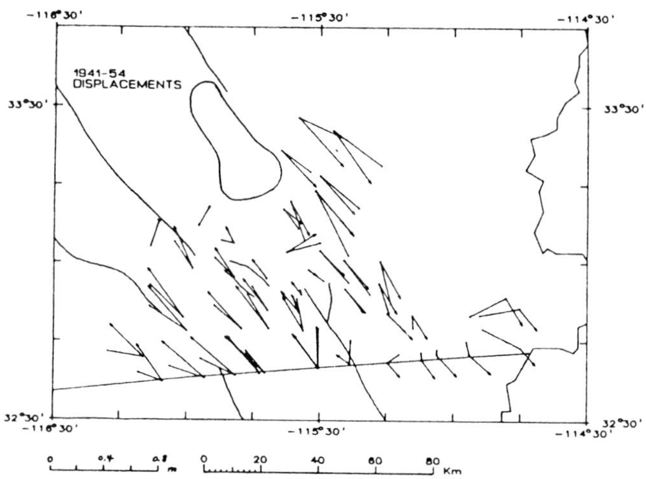

Fig. 4. Comparison of displacement vectors obtained from the 0.06-m solution (larger arrowheads) with those from the 1.00-m solution.

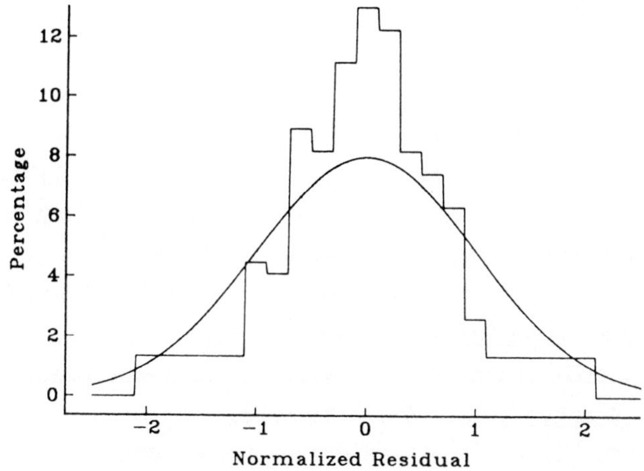

Fig. 5. The histogram of normalized residuals for quasi-observations in the 0.06-m solution. Also the probability density function for the standard normal distribution. The relatively small spread of the normalized residuals indicates that we could have assigned standard errors less than 0.06-m.

5. Conclusions

We have presented an example of how prior geophysical information may be used to better resolve crustal deformation by producing displacement vectors at individual stations. The prior information was introduced primarily to improve the poor signal-to-noise ratio associated with estimating crustal deformation from classical horizontal data. The prior information represents the hypothesis that displacements in the Imperial Valley mostly orient themselves parallel to the local plate boundary. Estimated displacement vectors agree at the 0.1-m level with displacement vector produced by relatively simple dislocation models.

References

Chinnery, M. A., 1961: The deformation of the ground around surface faults, Bull. Seismol. Soc. Amer., 51, 355-372.

DeMets, C., Gordon, R. G., Stein, S., and Argus, D. F., 1987: A revised estimate of Pacific-North America plate boundary zone tectonics. Geophys. Res. Lett., 14, 911-914.

Gergen, J. G., 1975: The new adjustment of the North American Datum: The observables, ACSM Bulletin, (Nov. issue), 9.

Minster, J. B., and Jordan, T. H., 1978: Present-day plate motions. J. Geophys. Res., 83, 5331-5354.

Neugebauer, H. C., 1988: The 1986 Salton trough GPS survey and data processing results. (Unpublished report) NOAA, Rockville, MD 20852.

Prescott, W. H., 1981: The determination of displacement fields from geodetic data along a strike slip fault. J. Geophys Res., 86, 6067-6072.

Prescott, W. H., Lisowski, M., and Savage, J. C., 1987: Velocity field along the San Andreas fault in southern California. EOS, Transactions, American Geophysical Union, 68, (abstract), 1506.

Savage, J. C., 1983: Strain accumulation in the western United States, Ann. Rev. Earth Planetary Sci., 11, 11-43.

Snay, R. A. and Drew, A. R., 1988: Supplementing geodetic data with prior information for crustal deformation in the Imperial Valley, California. (In preparation) NOAA, Rockville, MD 20852.

C. TOPOGRAPHIC APPLICTIONS

Digital Terrain Models of Forest Areas by Airborne Laser Profiling

Friedrich Ackermann
Institut für Photogrammetrie, Universität Stuttgart

Summary

The components of an airborne laser profiling system are briefly reviewed and the accuracy specifications discussed. A test flight is scheduled which is to demonstrate the operational features of the system. The data analysis is expected to allow conclusions about obtainable accuracy and the conditions and economy of practical application.

1. Review

1.1. Data acquisiton for establishing digital terrain models (DTM) has become a standard task of photogrammetry. Air survey methods are particularly suited because of easy coverage of large areas in combination with high accuracy performance which is, within a wide range, adaptable to respective specifications. Photogrammetric data acquisition of DTM's is predominantly done manually, by a human operator. This non-automated mode of operation is certainly tedious, although not critically slow. It has the advantage that the operator can intelligently capture morphological features (breaklines etc.) which are essential for efficient high-quality description of topography. Photogrammetric data capture is restricted to open terrain, although forest areas are adequately handled for DTM's of reduced requirements, equivalent to medium or small scale topographic contour mapping.

Modern technical means are used to increase the efficiency of photogrammetric DTM measurements by the semi-automated method of progressive sampling and via complete automation by digital image correlation.

1.2 There is a second approach to obtain DTM's in an efficient and highly automated mode which is based on recent progress in electronics, especially on laser and radar technology. It concerns direct airborne profiling or scanning of the terrain from an aircraft (or from space), in connection with a precise positioning system. Such systems are in operation already, see ref. 1-5. Their main feature is complete automation, realized by direct capture of digital terrain data at high rates and derivation of the DTM by subsequent data processing. Automatic extraction of morphological terrain features is feasible.

Particularly interesting and potentially most effective are scanning laser or radar systems - or alternately rigid array systems - which cover the terrain surface in a wide swath or strip. Very much simpler, but more readily available, are laser or radar profiling systems. They constitute an extension of the old airborne radar profile recorder (APR) of the 1950s and yield essentially vertical terrain profiles along the flight path. Dense area coverage requires many flight paths. Such systems are especially used for airborne bathymetry of shallow waters (ref.6) and in forest areas.

1.3 Within the scope of the SFB 228, project C1, we are specially interested in airborne laser profiling in forest areas in order to derive and investigate DTM's of such areas.

Contour mapping of high standards in forest areas has always been a serious problem, also and especially in well mapped countries. There exist no appropriate conventional survey methods, although certain special conditions may be successfully exploited, as for instance photogrammetric mapping of deciduous forests in winter time.

In the state of Baden-Württemberg, for example, the state survey authorities have the task to update the poor quality contour maps of the Black Forest (and other) areas to higher standards. The method presently applied is tacheometric field survey, which of course is tedious, expensive, and has a very slow rate of progress. The current cost figure is about DM 20 000,--/ km^2.

Airborne laser (or radar) profiling is practically the only feasible method which has a high potential for both accuracy and automation. Scanning modes are not applicable, as oblique rays would not reach the ground. The method of vertical airborne profiling has been successfully tested and applied in

Canada and in the United States. As far as we know it has not been applied, so far, in Europe. Therefore a test was scheduled within the SFB 228 with the aim of investigating the application under central European conditions, particularly in view of the high demands which are specified by the German 1 : 5000 base map, for instance.

Unfortunately, the scheduled hardware development (by IGN, Paris) has met unforeseen difficulties, which caused some delays. Therefore, no experimental results can be presented for the time being. The first controlled test flight is scheduled for summer 1988. It will use independent hardware units integrated provisionally for a test. This paper is limited to describing the system concept and to establishing some specifications.

1.4 The complete system of establishing a DTM in forest areas by airborne laser profiling is clearly subdivided in

(1) in-flight data acquisition, comprising the hardware system and its operation, and

(2) post-processing of data.

Additional points of attention are

(3) system calibration, and in case of a test

(4) analysis and verification of results.

The operational specifications depend on the accuracy requirements set up for the final DTM. It is not realistic to aim at the extremely rigid specifications of the German Base Map which ask for smallest contour intervals of 1 m in flat areas (and even 0.5 m for intermediate contour lines). The equivalent vertical accuracy of the DTM would be perhaps 20 - 30 cm (and 10 - 15 cm, respectively). Instead, for the time being, we aim at an average vertical accuracy of the final DTM of about 1-1,5 m, for non-accidented terrain. It is the equivalent to 5 m contour interval.

2. The hardware system for in-flight data acquisition.

2.1. The system of airborne terrain profiling is depicted in fig. 1. Its core is a distance measuring device, consisting of a pulsed laser mounted approximately vertical in the aircraft. The laser is capable of firing very

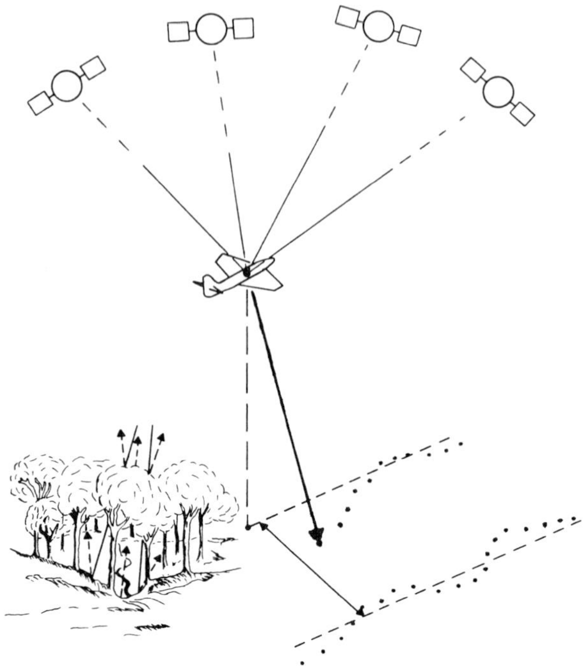

Fig. 1. Airborne Laser Profiling System

short pulses in quick sequence, the multiple return signals of each of which are recorded. The actually measured time delay gives the respective distance between the sensor and the point of reflection at the ground surface.

The narrow laser beam is slightly diverging to give typically a diameter of the footprint of about 1 m. When the wavefront of a pulse reaches the canopy of trees part of the energy is directly reflected. Other parts of the signal may penetrate deeper into the vegetation, giving partial reflections at intermediate levels. There is a lowest level of reflection which in most cases may refer to the ground surface (or a low layer of dense vegetation). The general result is that each emitted pulse may have one or several distinguishable return signals which are to be captured and recorded. Ideally, the electronic system would measure the travel time of an emitted and reflected pulse and output the converted distance digitally, for recording on a tape. It is essential that also the exact time marks of a system clock are recorded.

Available laser systems have a distance-resolution in the order of 10 cm. The measurement rate is usually up to 100 readings/sec, it can go up to 2000 readings/sec. If the laser is mounted in a light aircraft with a typical speed of about 80 m/sec, a measuring rate of about 40 readings/sec would be appropriate, which means a point every 2 m along the profile. At such rates there is no danger of misidentification of return pulses, the travel time being only 3 μsec, at 500 m flying altitude. For the scheduled test a Model 501 Laser Rangefinder of Optech Inc. will be used. Its maximum distance range is 500 m.

2.2 Whilst the laser system measures the approximately vertical distance between the sensor and the point of reflection on the ground surface, the position coodinates of the sensor must be known in order to derive the coordinates of the respective ground points. The position coordinates of the aircraft can only be determined by a high precision navigation system. The coordinates may be derived by post-processing, as they are not needed in real time, but the positioning is to be done in the kinematic mode.
The required positioning accuracy is critical in z; it should reach standard errors σ_z of < 0.5 m, preferably < 0.3 m. Moderate positioning accuracy in x and y of a few meters only would be sufficient in flat terrain. For accidented terrain, however, the horizontal positioning accuracy should be < 1 m, preferably < 0.5 m, as any horizontal displacement would cause an elevation error of similar magnitude. The frequency of GPS recordings should be about 10 Hz, although with a rate of 0.6 sec results have been obtained which did meet the above accuracy requirements. Such demands can be met by GPS phase-measurement in the kinematic mode, preferably with one airborne and one or two stationary receivers. As the GPS coordinates will be relative coordinates only, a subsequent transformation will be necessary, which will also consider the offset between GPS antenna and laser.

During the test flight 5-channel Sercel GPS receivers will be used with which very good results have been obtained.

2.3 The GPS positioning system and the laser distance measurement give eventually the position of the upper end point and the length of the distance vector. In order to obtain the coordinates of the lower end point of the distance vector its direction must be known.

As it cannot be assumed that the distance is vertical, its tilt has to be measured. Therefore an inertial platform is required which would record the tilt data, preferably all 3 attitude parameters, possibly at the same rate as the GPS recordings. The accuracy required is moderate for flat terrain, but in accidented terrain the tilt measurements should have standard errors of $\sigma_\omega = \sigma_\varphi < 10^{-3}$ = 60 mgon, which is equivalent to a pointing accuracy of 0,5 m at 500 m flying altitude.

2.4 All measuring units must be linked to a master clock, i.e. must record the same (or relatable) time signals. For purposes of calibration and for checking a small format spotting camera should be installed in addition, its axis being parallel or calibrated with regard to the laser beam within about 0,1 gon. It should take photographs at a rate of perhaps one /sec in order to be able to identify and interpolate the laser spots.

3. Distance between laser profiles, navigation.

3.1 The hardware system will give data which can be post-processed to xyz coordinates of the respective profile points along each flight profile. Within a profile there will be points about every 2 m. Some of them may not refer to the ground surface in forest areas. According to available experience it can be expected that at least every 5 or 10 m ground points will be obtained, see fig. 2. They can be identified by visual inspection of graphical plots. Automated methods will still have to be developed. The vertical accuracy of the ground points of the profiles would be < 1 m, hopefully reach 0.5 m. For the xy positioning of the points a few meters are sufficient, in flat or smooth terrain. In very steep terrain a positioning and pointing accuracy of 1 m, preferably 0.5 m is necessary, which can be obtained if the above mentioned specifications are kept. Within the profiles there is enough redundancy of points for applying filtering techniques and for assessing the precision.

3.2 The accuracy of the derived DTM will depend on the precision of the profile points only to a minor degree. The really decisive parameter will be the distance between profiles, i.e. the distance over which the interpolation of the DTM takes place. The permissible distance will depend on the required DTM accuracy, on the type of terrain, and, to a considerably

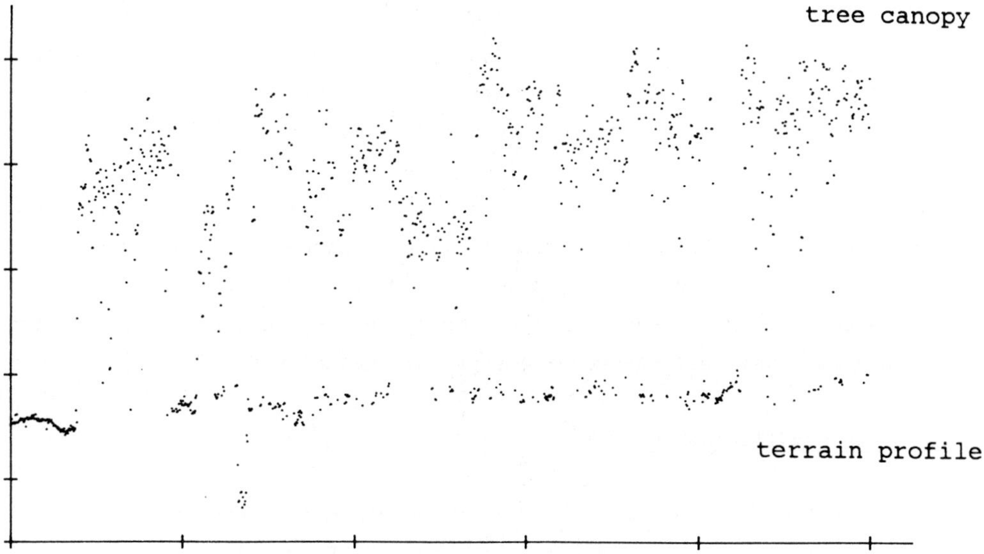

Fig. 2. Laser reflections over forest areas

lesser degree, on the accuracy of the profile points. Experience with DTM by ground surveys and photogrammetry indicates that for 1 m DTM accuracy the distance between profiles can be, respectively,

- -up to 100 m in very smooth terrain,
- -up to 50 m in rolling, hilly terrain,
- -less than 50 m in accidented terrain.

Those figures are rough estimates. The terrain classification is unsharp, and all terrain types must still be reasonably smooth. In forest areas it will not readily be possible to consider break lines in a DTM. The scheduled test flight will use profiles at 100 m distance. In a part of the area it will be attempted to fly profiles at 50 m distance.

3.3 This raises the problem of flight navigation. The area should be covered quite regularly with parallel profiles (plus some cross profiles), from which the DTM will be interpolated. In order to achieve uniform DTM accuracy one may permit lateral deviation of the actual flight lines from the flight plan of about 10 % of the distance. It means that the actual flight lines should follow the planned lines within 5 m or 10 m, respectively. Even if larger tolerances may be accepted it is evident that the specifications for the flight navigation are extreme. In fact, there is no navigation system ready, which would meet such specifications.

We expect, therefore, that the scheduled test will not succeed in providing the required regular distribution of profiles. Larger irregularities will have to be expected. They will,however, not disturb the test seriously. In future computer controlled GPS flight navigation systems will have to be used. They may just meet the requirements, at least for short strips.

4. Absolute positioning and DTM interpolation

4.1. The system described will give profile points in terms of xyz coordinates. However, the coordinates relate to a GPS coordinate system as no other coordinate system appears in the on-board data acquisition. It is essential, however, to relate the coordinates to a terrestrial coordinate system. This can be achieved in 2 ways:

The ideal solution will be, that GPS stationary receivers on given geodetic points will directly establish the relationship. It will have to be investigated, which accuracy can be reached over extended periods of flying. For the time being we have to rely on indirect orientation of GPS positioning data into the ground coordinate system via aerial photographs. The flying altitude of the laser profiling flight is too low to have it combined with taking aerial photographs. Instead, a separate photoflight is scheduled, by which wide-angle aerial photographs are taken. Photoscale can be about 1:12000 to 1:15000 which means a vertical measuring accuracy of terrain points of 0,2 m to 0,25 m. It is assumed, that the area contains enough open terrain in which identification and transfer of some laser points into the aerial photographs will be possible, with the help of the spotting camera. Special terrain features may also be identified directly from the laser profiles. Thus, it is expected that each laser profile can be separately oriented and transformed into the ground system. The transformation will also take care of the offset between laser and GPS receiver in the aircraft, whilst the secondary effects of attitude on the offset coordinate components may be neglected, for the time being, unless assessable by recorded attitude data. The aerial photographs themselves are oriented in the conventional way by ground control points and aerial triangulation.

4.2. As result of the direct data processing the xyz coordinates of the selected profile points will be available. They are treated as input observations for the interpolation of a regular rectangular or square DTM grid. There are DTM computer programs available which can handle the particular case which is characterized by high density of points within the profiles and empty areas between them. In the test the program SCOP will be used.

5. System Calibration.

The system performance depends largely on the internal calibration of the hardware units. It concerns in particular the alignment of the laser with the spotting camera, and the 0-position of the inertial platform. Also the offset between laser and GPS antenna should be determined. The calibration procedures will depend on the system configuration,especially if the laser, the camera and the inertial system are rigidly mounted and can be calibrated separately, previous to the flight.

Under the conditions of the scheduled test flight such a pre-flight calibration will probably not be possible. In this case in-flight calibration will have to be relied upon. It implies flying over terrain features which can be identified in the laser recordings and in the spotting camera, such as buildings, bridges, dams, football arenas etc. The special problem of separating position from tilt effects may be approached by deliberate pitch- and roll-manoeuvres. For flat and smooth terrain the calibration requirements are still moderate. It is one of the aims of the test flight to look into the calibration conditions under operational aspects in detail.

6. The scheduled test flight.

The test flight is scheduled for August 1988, after the hardware components will have been delivered and assembled in the experimental aircraft Do 228 of the Institut für Flugführung, Technische Universität Braunschweig. The test area is selected in a partially forested area near Stuttgart. The flying altitude will be about 500 m, and profiles will be flown 100 m apart. In part of the area 50 m distance between the profiles will be attempted, althougt the limitation of flight navigation will show up.

Aerial photographs will be taken separately, at photo scale 1:15 000. They will be used for the transformation of the laser profiles into the ground coordinate system. In addition, they will allow the accuracy assessment in open areas. Within forest areas the Landesvermessungsamt Baden-Württemberg will understake terrestrial measurements for direct checking of the results.

Supplement (October 1988)

Several experimental test flights were carried out 3 months after the presentation of the paper. The main experiment covered a test area of 6 km x 2,5 km, located near "Gammertingen", 80 km south of Stuttgart, flown between August 18 and 24, 1988. The area is largely covered with deciduous and coniferous woods at various stages of development. It also contains open and built-up areas. The topography is smooth, with some steep valleys. In addition, on September 9 some laser profiles were taken of a wetland area near Rotterdam and of an open pit mine near Cologne. Some preliminary results deserve interest and are briefly summarized:

1) At the Gammertingen area about 250 line km of laser profiles were flown, with 100 m and for half of the area with 50 m lateral spacing between the profiles. It had been anticipated that navigating such flight lines would be difficult. However, a digital counter was rigged-up for the pilot, displaying GPS coordinates in real time, in connection with a stationary GPS receiver, the data of which were telemetried to the flying aircraft. It was therefore possible to fly successfully parallel lines in east-west direction. The average deviations to the actual flight lines against the scheduled flight lines were less than 10 m.

2) The hardware equipment in the aircraft consisted of a GPS receiver in the aircraft recording data every 0.6 sec. on a portable GRID computer, a vertical laser operating at a frequency of 23 Hz, an inertial plat-

form recording tilt data also at 23 Hz frequency, and a small format (6 cm x 6 cm) camera. The camera failed partly. Unfortunately also the kinematic GPS positioning was effected by the fact that one of the satellites had been temporarily switched off. Also the time synchronization of the various pieces of equipment caused some diffculities.

3) The data processing is not yet completed. Preliminary results have been obtained, especially as the various transformations from the GPS reference system (WGS84) into the local state survey reference system have been completed.

   The laser profling was generally successful. It had been established previously, however, that the laser ranging did not operate reliably at the scheduled altitude of 500 m. The flying altitude therefore had been reduced to 350 m above ground. Focussing of the laser may not have been appropriate resulting in a footprint to large. The densitiy of laser profile points which refer to the terrain profiles in forest areas was generally satisfactory, see fig. 2. However, in areas covered with dense deciduous foliage and some young coniferous plantations very few laser points have penetrated to the ground. It is expected, that high frequencies might solve the problem.

   The spotting camera had failed partly. As it has a general control function only it might be replaced in future by video or CCD cameras.

   Visual analysis of laser profiles of open terrain and preliminary comparison with photogrammetric measurements have confirmed that the internal vertical precision of the laser points is better than 0.2 m, as expected.

4) Thus, the tests seem to confirm the operability of the laser profiling system, and the results seem to meet expectations. Evidently the system is capable of the expected accuracy. However, the experiments also point out some hardware difficulties which have to be investigated. They also confirm that a complete system calibration is necessary in order to make the system generally applicable and operational.

References.

(1) W.B. Krabill et al. (1984): Airborne Laser Topographic Mapping Results, Photogrammetric Engineering and Remote Sensing, Vol. 50, 1984, 685 - 694.

(2) H. Schreier et al. (1984): Calibrating an Airborne Laser Profiling System, Photogrammetric Engineering and Remote Sensing, Vol. 50, 1984, 1591 - 1598.

(3) R. Moreau, L.M.A. Jeudy (1986): Profilomètre Laser-Inertial (PLI): Un Essai Qualitatif en Terrain Boisé, The Canadian Surveyor, Vol. 40, 1986, 23 - 28.

(4) H. Arp, J.C. Griesbach, J.P. Burns (1982): Mapping in Tropical Forests: A New Approach Using Laser APR, Photogrammetric Engineering and Remote Sensing, Vol. 48, 1982, 91 - 100.

(5) J. Jepsky (1983): A Micro-processor Controlled Airborne Laser Profiling and Mapping System, ASP - ACSM Fall Convention, Salt Lake City, Utah, 19 - 22 Sept. 1982, 10 p.

(6) F.E. Hoge, R.N. Swift, E.G. Frederik (1980): Water Depth Measurements Using an Airborne Pulsed Neon Laser System, Applied Optics, Vol. 19, 1980, 871 - 883.

Advances in Radar Altimetry Techniques for Topographic Mapping

Hugh Griffiths
Dept. of Electronic Engineering, University College London, U.K.

Summary

This paper considers how radar altimetry techniques may be used to derive a global map of surface elevation. We consider firstly the requirements for topographic mapping, and then review a number of techniques to address these requirements. Finally we consider how these techniques may be used to form the basis of global topographic mapping missions.

1. Introduction

Satellite-borne radar altimeters have to date been designed for operation over ocean surfaces, and have achieved impressive results in determination of the marine geoid, as well as ocean wind, waves and swell, and sea ice. An early example of topographic mapping using altimeter data was the work of Zwally et al. in mapping the Greenland ice sheet [1]. This illustrated the power of the mapping by radar altimetry, but also showed the shortcomings of using a pulse-limited altimeter over rough and sloping surfaces. Subsequently Rapley et al. have examined a global 17-day orbit repeat dataset from the Seasat altimeter [2], and shown that a surprisingly high proportion of the non-ocean data are usable. It is necessary to analyse the echo waveforms themselves, since the telemetered elevation values are subject to tracker lag errors, and there is often much information on the nature of the surface contained in the echo shape. It is interesting to note that one of the most 'difficult' regions of the globe to track has proved to be the margins of the Greenland ice sheet, which was one of the first to be examined, and this may account for the slowness of the scientific community to appreciate the value of land altimeter data.

The ERS-1 mission will extend the results from the Seasat altimeter, using its 'ice mode', and orbiting to $\pm 82^{\circ}$ latitude. The ice mode uses a $\times 4$ coarser range window width, and a tracking algorithm that is more suited to non-ocean echo shapes. The ice mode will be of value not only over ice sheet surfaces, but also over land. The baseline mission will switch between ocean and ice modes under ground command, but an automatic mode-switching algorithm has been developed [3], and this may be used later in the mission.

2. Topographic mapping requirements

The Earth is at present mapped to a rather uneven degree. Although certain regions - in particular North America and Europe - are relatively well mapped, there are considerable variations in scale and accuracy. Other regions, especially those that are remote and inhospitable, are very poorly mapped; indeed the topography of the surface of the moon is better known than that of most of the Antarctic ice sheet.

In addition, there are inconsistencies in the datums used for different maps; for example, the United Nations lists eighteen separate ellipsoids in use in different countries. The best global Digital Terrain Map currently freely available is that due to NCAR, which provides maximum, minimum and modal height values on a 10 arcminute grid (equivalent to 18km x 18km at the equator, 12km x 18km at 50° latitude), with a quoted precision and accuracy of 30m. This has been produced by digitising existing contour maps, which themselves are subject to inconsistencies and errors.

A recent study [4] has attempted to identify and collate the scientific requirements for global topographic mapping. These have been split into:

(i) hydrology, vegetation and ecoscience
(ii) continental geophysics
(iii) geology and geomorphology
(iv) polar science

Within each of these a detailed list of requirements is given, and for each requirement, the spatial scale, height resolution, and repeat timescale has been identified. These cover a wide variety; for example, to provide a basic topographic map of the polar ice caps, a vertical resolution of the order of a few metres on a spatial scale of 100-300 metres is appropriate. Repeating the measurement at approximately decade intervals would allow simple mass balance studies to be carried out. In contrast, for studies of soil erosion, a vertical resolution as high as a few centimetres on a spatial scale of tens of metres, repeated annually, would be required, though not on a global scale.

Some additional requirements for topographic mapping include the correction of SAR imagery for the distortion caused by topography, and a requirement of radio engineers to assess whether a given microwave communication path may be line-of-sight or obstructed.

It is evident, therefore, that there are a large number of outstanding problems in several important scientific disciplines that require a global elevation map, and that the present state of knowledge of global topography is generally inadequate to satisfy these needs. There is therefore a strong case for a mission to address the problem. In particular, a single satellite-borne instrument providing global coverage would give a unified, consistent dataset.

There exist a number of satellite techniques that may be employed for topographic mapping. Optical stereo imaging (for example, the SPOT mission) is able to achieve surface elevation at a pixel size of approximately 10m and an elevation accuracy of about the same figure. However, the technique cannot be used when there is cloud cover or in darkness (which may be a significant problem in polar regions), and does not provide surface roughness or backscatter coefficient, both of which are also useful in the study of surface properties. More important, though, the cost of such data on a global scale is likely to be prohibitive. Laser altimetry represents another approach, and is capable of very high resolution, both spatially and in elevation. However, although significant advances are being made in lidar technology, the sort of laser lifetimes achievable are not yet at the point where a global mapping mission may be contemplated. A further problem, in common with optical stereoscopic mapping, is that of cloud cover.

Radar altimetry, on the other hand, is a relatively well-understood technique that gives all-weather coverage and can operate in darkness. The remainder of this paper looks at various radar altimeter concepts, starting with simple techniques which may be usable in the near future and moving to more ambitious ones which may take a decade or more before they can be realised, and considers how they can address the problem of global topographic mapping.

3. Pulse-limited altimeter techniques

It has already been mentioned that pulse-limited radar altimeters are able to achieve valuable results over non-ocean surfaces. The principal problem encountered by the Seasat altimeter over such surfaces was its inability to track surfaces whose elevation varies rapidly, and from which the echoes depart significantly from the standard 'ocean' shape. Two measures may be employed to address these problems:

Firstly, the width of the 'range window' may be increased to accommodate the more rapid variations of surface elevation. This can be achieved simply by reducing the bandwidth of the linearly frequency-modulated 'chirp' transmit pulse and the deramp local oscillator - all the other hardware in the altimeter remains unaltered. This is an elegant method of adapting the resolution of the instrument to the surface being tracked, and is the principle adopted for the ERS-1 ice mode, where the range window width is increased from 29m to 116m by reducing the chirp bandwidth from 330MHz to 82.5MHz. Although in the ERS-1 altimeter the different chirp pulse bandwidths are actually produced by two separate Surface Acoustic Wave (SAW) dispersive delay lines, a more attractive method for future instruments is to generate the chirp pulse by digital means [5]. This allows a range of pulse bandwidths to be generated by the same unit, simply by varying the control parameters of the digital circuitry. Digital pulse generation has an additional advantage, in that rather longer pulses can be generated than by SAW devices, reducing the peak transmit power requirement.

Secondly, a more robust tracking algorithm may be used, in the same way as that used by the ERS-1 ice mode [6]. Such an algorithm also controls the switching of the range window width. In the Advanced Terrain Tracking Altimeter (ATTA) system, which is currently the subject of ESA feasibility studies, six different range window widths are incorporated. The narrowest of these is for tracking ocean surfaces, and uses an ocean-specific tracking algorithm - the Suboptimal Maximum Likelihood Estimator (SMLE) developed for the ERS-1 altimeter. The widths of the other range windows increase successively by a factor ×4, to a maximum of 33km. The coarsest range window can accommodate the maximum global variation in topography, and so the altimeter will be able to maintain track over the entire surface of the Earth. This scheme also avoids the need for a dedicated acquisition sequence, since from switch-on the altimeter can start at its widest range window, and progressively reduce till it reaches the range window width appropriate to that particular surface type.

4. Beam-limited altimeter techniques

Although it is possible to achieve full global coverage with an adaptive pulse-limited altimeter, a fundamental problem is that over sloping surfaces the altimeter range measurement is made not to the nadir point, but to the point within the antenna footprint that is closest to the satellite. This point may be along- or across-track depending on the direction of the slope. Without a knowledge of the topography within the footprint the altimeter measurement must be assumed to be at nadir, and there is therefore an error in the surface elevation measurement, known as slope-induced error. This may be considered either as an error in the nadir range measurement, or an error in the surface position of a correct range measurement. The magnitudes of these errors can be quite high; for example, a 1° slope from a satellite altitude of 800km implies a nadir elevation error of 160m and a displacement of nearest point from nadir of 16km. Brenner et al. [7]

have demonstrated two ways in which slope-induced error may be minimised, either correcting the nadir elevation measurement, or the location of the surface point to which the correct range measurement was made. Both methods are iterative, and for three-dimensional correction rely on closely-spaced ground tracks, which in practice may not be available.

To overcome this problem it is necessary to use a much narrower antenna beam to fix the location on the surface of the altimeter measurement. This gives a beam-limited altimeter footprint. To achieve the sort of spatial resolution referred to in Section 2 requires either much larger antennas than are currently used, or higher frequencies. The table below summarises this requirement, and shows that even for modest spatial resolution very large antenna structures are required, and it will be necessary to use higher frequencies than 13.5GHz. A further consideration is that to determine the location on the surface of the measurement, the pointing angle of the antenna must be known to a fraction of the beamwidth.

spatial resolution	frequency 13.5GHz	35GHz	96GHz
1km	18m	6.8m	2.5m
300m	60m	23m	8.3m
100m	178m	68m	25m

Table 1. Antenna diameter required for a given spatial resolution, at frequencies of 13.5, 35 and 96GHz. Satellite altitude = 800km.

An altimeter with a single beam-limited beam, placed in a slowly-precessing orbit repeat cycle would completely sample the surface of the Earth in approximately 12 years; this could be reduced to a more realistic figure by using multiple feeds to provide a swath of, say, seven across-track beams. This would represent a relatively simple dedicated mission for global topographic mapping, but such a large spaceborne antenna would represent a major technological challenge.

5. Swath operation

To achieve more rapid global coverage than this latter scheme a wider swath is required. This introduces problems of its own, since non-normal incidence results in severe 'spreading' of the echo profile, even for very small off-nadir angles, and the only way to alleviate the problem is to use a very narrow antenna beam, which implies a large antenna structure. It is therefore much more difficult to extract surface elevation and roughness information to the same accuracy as a nadir-pointing beam. Furthermore, the measurement of surface elevation is extremely sensitive to any across-track antenna mispointing, so that if off-nadir surface elevation is to be properly recovered, any mispointing must be known to high accuracy. An analysis of these problems has been carried out [8], and the results are summarised in Table 2 below. The assumptions made in deriving these values are (i) a swath width of ±50km has been adopted; (ii) the height at nadir can be measured to 0.1m; and (iii) it is possible to track the echo profile to 1% of its extent in range.

rms height error (m)	required rms attitude knowledge (degrees)	required antenna beamwidth (degrees)	required antenna size (m)	
			f = 13.5GHz	f = 35GHz
10	8×10^{-3}	0.8	1.6	0.6
3	2.4×10^{-3}	0.24	5.2	2.0
1	8×10^{-4}	0.079	15.8	6.2

Table 2. Required attitude knowledge, antenna beamwidth, and antenna size (at 13.5 and 35GHz) as a function of rms off-nadir elevation error.

The problems of determining across-track mispointing are orders of magnitude more severe than those of present-day altimeters, and will require special techniques to address them. It is possible that methods using GPS may be of use here.

To obtain coverage up to ±50 beamwidths from boresight requires a phased array antenna. This may be achieved either with a single agile electronically-scanned beam, or with a set of fixed beams via a beamforming matrix, feeding multiple receivers. Such techniques, although used routinely in terrestrial radar systems, are unlikely to be cheap for spaceborne applications.

6. Synthetic aperture techniques

Since the high spatial resolution required for topographic mapping implies the use of large antenna structures, it is natural to consider whether the same resolution might be achieved using aperture synthesis techniques. This proves indeed to be the case, but can only provide improved resolution along-track, and if the same resolution is required across-track, this must be provided by an equivalent across-track real aperture, as described above.

This approach is similar in many ways to turning a conventional SAR on its side, but there are some important differences in the required resolutions, and in the ways in which resolution is achieved in each dimension.

Firstly, it is apparent that unfocused synthetic aperture processing may be used, since this provides adequate spatial resolution, and simplifies the required signal processing somewhat. Here, the synthetic aperture length is chosen such that the two-way phase excursion of the sequence of radar echoes is $\leq \pi/2$, and no compensation is provided in the processing for this phase excursion. The synthetic aperture processing is compatible with the 'full-deramp' form of pulse compression conventionally used for radar altimeters [9], so the high range resolution required can be achieved with relatively low processing bandwidths. In common with conventional SAR instruments, there is a constraint on the radar PRF to sample the Doppler shift unambiguously, and this results in a PRF rather higher than that conventionally used for radar altimeters (of the order of tens of kHz).

Secondly, there is the possibility of performing the processing on-board. The processing requirements are considerable, involving two stages of Fourier Transform with a 'corner-turning memory' in between. The first Fourier Transform forms the frequency-domain range bins, just as in a conventional altimeter, while the second forms a set of synthetic beams. The telemetry data rates from such an instrument are also considerable, especially if swath coverage is incorporated as well.

The techniques of a single scanned beam for swath coverage combined with aperture synthesis for along-track resolution are combined in the LORA instrument (Land and Ocean Radar Altimeter) to be proposed for the US Space Station Polar Platform. The purpose of this instrument is primarily global topographic mapping at a spatial resolution of 250m and a height resolution of 3m, though a five-beam interferometric mode of operation is to be incorporated to provide wide-swath ocean topography, at rather higher elevation resolution. The instrument operates at 35GHz, and uses an electronically-scanned beam from a 10m phased array to give a swath ±50km either side of the sub-satellite track. Equivalent along-track spatial resolution is provided by unfocused synthetic aperture processing.

The antenna system for the LORA instrument - a 10m x 1.2m 35GHz electronically-scanned phased array, represents a major engineering challenge, especially to operate over a 300MHz bandwidth for the interferometric oceans mode. The inclusion of this mode adds significantly to the value of the mission, and addresses a specific oceanographic problem. This is associated with the compromise that exists with single-beam altimeters between the spatial sampling (i.e. orbit track spacing) and temporal sampling (i.e. revisit time), such that at the equator a track separation of 150km can only be achieved at a revisit time of about 20 days, and the timescales of ocean mesoscale variability are considerably faster than this [10]. The use of the five-beam interferometric mode overcomes this, and should provide invaluable data on ocean mesoscale activity.

The global mapping task would be accomplished relatively early in the mission, and the possibility exists of subsequently mapping certain areas at higher resolution. A simple way of achieving this would be by allowing adaptive control of the chirp pulse bandwidth (in a similar manner to the Advanced Terrain Tracking Altimeter), using a digital chirp generator and adaptive mode-switching algorithm, to obtain better height resolution over those surfaces that permit it.

7. Conclusions

In conclusion, we have seen that there is a pressing requirement for global topographic mapping, from many different scientific disciplines. Although there are other candidate instruments, a radar altimeter appears best suited for all-weather global operation.

In the relatively near future, the ice mode of the ERS-1 altimeter should provide considerably better coverage than existing altimetry over non-ocean surfaces. Quite apart from the value of this in its own right, it should provide surface data that will be invaluable in the design of future instruments. In the slightly longer term, an adaptive pulse-limited altimeter mission would provide global coverage with only minimal increases in complexity over existing designs, though the height resolution of the dataset would depend on the range window width adopted, and the problems of correcting slope-induced error would remain.

In the longer term, a dedicated beam-limited altimeter system will be necessary to address the task. Such systems appear inevitably larger, more complex, and more costly than existing altimeter designs, and are likely to use larger antennas and higher frequencies. In particular, the LORA system poses some difficult (but exciting) engineering challenges, and promises to provide global surface elevation at 3m vertical resolution and 300m horizontal scale in a single consistent dataset.

8. Acknowledgements

I am indebted to my colleagues Dr Chris Rapley and Dr Duncan Wingham for many valuable discussions on this subject over the years. I would also like to acknowledge the use of information from workers at the Jet Propulsion Laboratory, in particular Dr Fuk Li, Dr Charles Elachi and Dr Tim Dixon.

9. References

[1] Zwally, H.J. et al., 1983, 'Surface elevation contours of Greenland and Antarctic ice sheets'; *J. Geophys. Res*, 88, No.C3, pp1589-1596.

[2] Rapley, C.G. et al., 1987, 'An exploratory study of inland water and land altimetry using Seasat data'; ESA Contract report 6483/85/NL/BI, ESTEC Publications Branch, Noordwijk, The Netherlands.

[3] Wingham, D.J., 1986, 'An automatic tracking mode switching algorithm for the ERS-1 altimeter'; Proc. *IGARSS-86* Symposium, Zurich, ESA SP-254, Vol.I, pp185-190.

[4] Burke, K. and Dixon, T.H., 1988 (in press), 'Report of the Topographic Science Working Group - a proposed experiment to determine global topography'; Lunar and Planetary Institute, Houston, USA.

[5] Griffiths, H.D. and Bradford, W.J., 1987, 'Digital generation of wideband FM waveforms for radar altimeters'; Proc. *RADAR-87* Conference, London; IEE Conf. Publ. 281, pp325-329.

[6] Wingham, D.J., Rapley, C.G. and Griffiths, H.D., 1986, 'New techniques in satellite altimeter tracking systems'; Proc. *IGARSS-86* Symposium, Zurich, ESA SP-254, Vol.III, pp1339-1344.

[7] Brenner, A. et al., 1983, 'Slope-induced errors in radar altimetry over continental ice sheets'; *J. Geophys. Res*, 88, No.C3, pp1617-23.

[8] Griffiths, H.D., Rapley, C.G., Milne, K. and Wingham, D.J.,1985, 'A study of antenna signal processing techniques for radar altimeters'; ESA Contract report 6001/84/NL/GM, ESTEC Publications Branch, Noordwijk, The Netherlands.

[9] Griffiths, H.D., 1988, 'Synthetic aperture processing for full-deramp radar altimeters'; *Electronics Letters*, 24, No.7, pp371-373.

[10] Monaldo, F.M., 1984, 'Prospects for improved measurements of the oceans and atmosphere with advanced radar altimeters'; URSI Commission F Symposium and Workshop, Israel.

A KU-Band Solid State Radar Altimeter for Topographic Application

Yong-Hoon Kim
Institut für Navigation, Universität Stuttgart

Summary

A Ku-band radar altimeter for a spacecraft or an aircraft is under development at INS. The main design features are the high accuracy, real time measurement, solid state integrated subsystems and a high degree of flexibility for versatile applications. A detailed description of the system - microwave subsystem, the microstrip scanning antenna and the signal processing - are given.

1. Introduction

Satellite altimetry is one of the newest space-borne remote sensing techniques. The measured data has already made a significant contribution to solid-earth observation. Altimeter measurements render possible a fine detailed global representation of the mean-sea-level surface. It is also a representation of the shape of the geoid, revealing ocean-floor features. Todays advanced radar altimeters such as SEASAT and ERS-1 illuminate only one point at nadir of the ocean's surface by their microwave beams. However, to obtain a global coverage of the earth's surface and to monitor the ocean's dynamics and the change of topography, a wide swath radar altimeter is required. As such a radar samples with high data rates the measured data should be processed very fast in real time which is not yet realized in the existing altimeter systems such as SEASAT and ERS-1.

A new type 13.74 GHz fully integrated radar altimeter with an electronically scanned antenna has been developed for use on an aircraft or a spacecraft where low power consumption, low mass and high realiability are required. This new radar altimeter consists of a **microwave subsystem unit**, a **frequency source unit**, a **digital pulse compression unit**, a **microstrip array antenna subsystem** and a **micro computer unit**. The microwave subsystem is realized in MIC (Microwave Integreated Circuit) technique. The "**digital pulse compres-**

sion" is performed by a new type component SAW (Surface Acoustic Wave) convolver which processes the range measurements in real time, because conventional techniques such as LFM (Linear Frequency Modulation) or "chirp" needs FFT (Fast Fourier Transformation) analyzers which showed long processing times for converting frequency signal into the time domain.
The design of scanning mechanism of a parabolic antenna, an interferometrical set up realized by two parabolic antennas and a phase scanning array antenna were studied to determine the peak gain - side lobe level performance. This paper will explain the system concept of an experimental radar which will gather digital topographical data over land- and sea surface. In addition this new type of radar altimeter will also be used for studying meteorological effects i.e. rain rate, cloud.

2. General system description

The system block diagram of the altimeter is shown in Fig. 1. A Phase-coded radar signal is generated by the PRC (Pseudo Noise Random Code) GN unit and this base band signal is modulated by the BPSK (Bi-Phase Shift Keying) modulator at 300 MHz. This signal is sharply filtered and up converted in the SSB mixer using a 13.45 GHz local oscillator . The signal centered at 13.74 GHz is then filtered and amplified to drive the power amplifier. This signal passes through the 1 watt GaAs FET power amplifier and is transmitted by 32*32 elements of the microstrip array antenna. This lightweight flat microstrip array antenna is electronically scanned by the microcomputer, controled in the cross track by +/-20 degrees. The scanning geometry is shown in Fig. 2.
The high speed PIN diode switch selects between the transmitted - and the received pulse. In the calibration mode, the PIN diode switch connects the transmitted pulse direct to the receiver through a high attenuator. The receiver consists of a band pass filter, a low noise amplifier and a down converter. The IF signal is centered at 300 MHz with 64 MHz bandwidth. The return signal is first pulse compressed with the PN coded reference signal in the SAW convolver. Because of the frequency doubling effect of the convolver, now the center frequency is 600 MHz and the bandwidth is 128 MHz. After video detection , the time delayed signal is evaluated by the TIC (Time Interval Counter). The pulse compressed signal has a impulse length of 40 nsec measured from zero to zero point and the signal-to-noise improvement is more than 30 dB without noise cancellation. The reflection

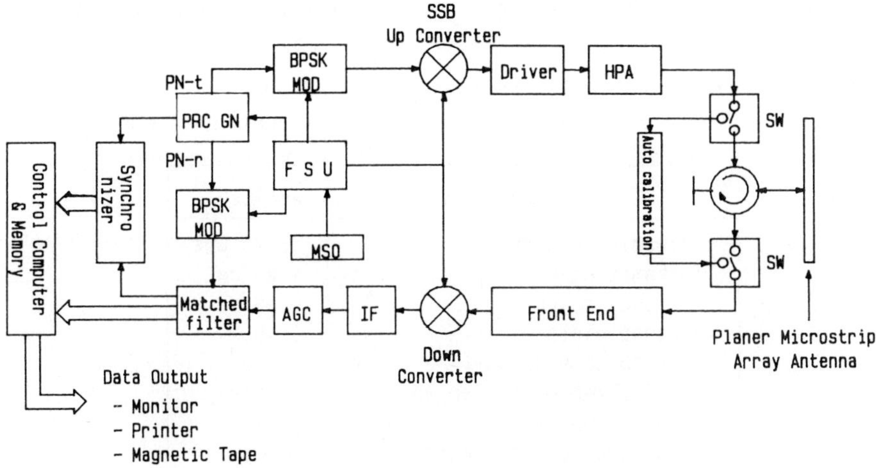

Fig. 1 Radar altimeter system block diagram

- Frequency : 13.74 GHz
- Antenna size : 0.48 * 0.48 m
- Real aperture footprint : 47 m
- Swath width : 870 m
- Number of dwells in swath : 36

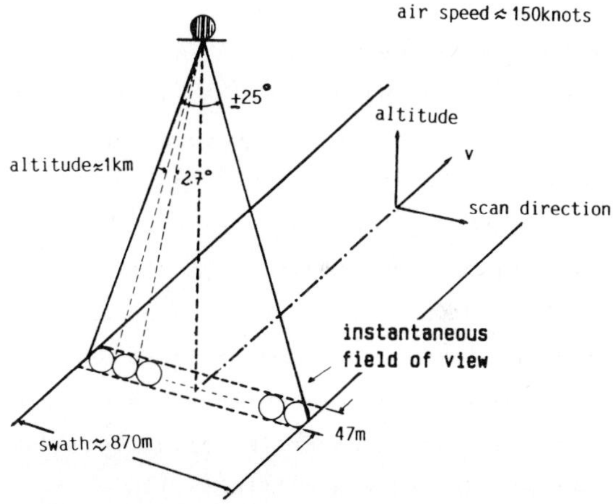

Fig. 2 Principal geometry of radar altimeter measurement

coefficient is measured using AGC. The measurement accuracy of the AGC is better than +/- 0.5 dB.
In table 1, the system design parameters are given.

Parameter		Values
Transmit frequency	F	13.74 GHz
Transmit power	Pt	1 W (cw)
Antenna length	l	0.48 m
Antenna width	w	0.48 m
Antenna loss (2-way)	La	-6 dB
Transmitted pulse length	Tp	16 usec
Compression ratio	N	1024
Compressed pulse length	Tc	40 nsec
Bandwidth	B	64 MHz
System noise figure	Nf	5 dB

Table 1. Radar altimeter system parameter

3. Microwave subsystem

The main components and performances are briefly described in this paragraph.

Up converter and driver

The SSB up-converter consists of a 180 degree rate-race 3 dB coupler and a 90 degree hybrid for IF. The suppression of the local frequency of 13.44 GHz and the lower side band of 13.14 GHz at the RF port is better than -20 dB. The driver module which consists of three amplifier stages has a gain of 21 dB. The output power is typically 13 dBm, which is required at the input port of the power amplifier.

High Power amplifier

The electrical performances of the Ku-band solid state GaAs FET amplifier are following:

- Amplifier gain of 19 dB
- Bandwidth of 1 GHz
- Gain variation of 0.5 dB
- The output power of the three stages is 1 watt (CW)
- The efficiency of each amplifier stage is more than 27 %

Front end

The input filter cancels the spurious signals of the out band at the input of the low noise amplifier (LNA). The LNA using GaAs FET is a typical noise figure of 2.5 dB. The LNA is matched for minimum noise figure, because a low loss microstrip circulator is used for a good matching at input and an output port. The total gain of the front end is 30 dB.

Down converter and IF section

The received signal is down converted in an SSB mixer with a local frequency of 13.44 GHz. The convertion loss of the mixer is typically 8 dB. The IF section consists of a preamplifier, high gain amplifiers, combline band pass filters, a AGC. The amplification of the IF section is more than 70 dB and the ripple of the gain is less than 0.5 dB in a pass band. The output power of the IF section is 10 dBm, which is needed at the SAW convolver input. The filter centered at 300 MHz has a 0.5 dB loss and a 100 MHz 1-dB band width.

4. Frequency source unit

The range accuracy depends on the instabilities of the transmitted frequency. To achieve a range accuracy less than 2cm, the frequency stability of the local oscillator must be better than $3*10^{-10}$ during 1 msec. To achieve this specification, a high stable frequency source has been developed using a dielectric resonator osciallator(DRO), a SAW resonator osciallator in Fig.3.

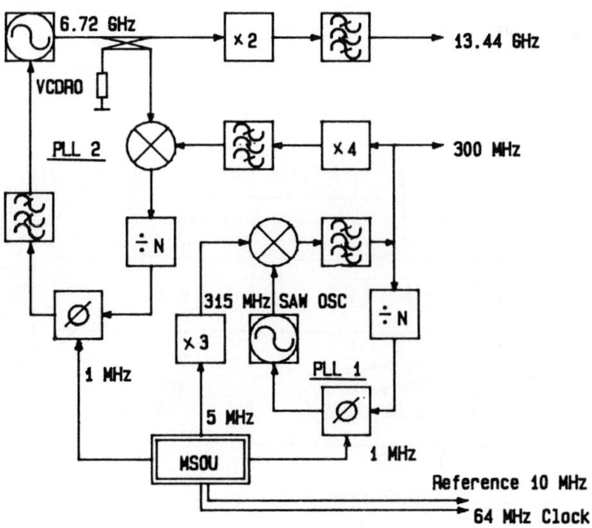

Fig. 3 A block diagram of the FSU

The MSOU (master stable oscillator unit) generates the reference signals of 1 MHz and 15 MHz for the PLLs and the 64 MHz TTL clock signal for the PRC GN. The 315 MHz SAW resonator oscillator is mixed with 15 MHz and is phase locked to the 1 MHz reference signal. The resulting 300 MHz reference signal is multiplied by four (i.e., 1.2 GHz) and filtered. The 6.72 GHz VCDRO (Voltage Controlled Dielectric Resonator Oscillator) is mixed with the 1.2 GHz and is further divided for phase locking to the 1 MHz. The GaAs FET multiplied the 6.72 GHz signal by two with a gain of 4 dB instead of a conversion loss. The phase noise of the VCDRO is -90 dBc at 10 KHz offset from carrier. The all harmonics lay below -40 dBc.

5. Digital phase coded signal generator

Digital matched filters which are e.g., transversal filters and the digital correlators can only work with fixed code sequences. The SAW convolver has the capability of signal processing continuously changing code patterns which must be fixed during one word period. In other word, SAW convolver is a programmable matched filter. However, it requires a reference signal which is the time reversed image of the transmitted signal. The transmitted signal might change its format e.g., code structure, code length, bit rate. To generate these variable format waveform signals, two EPROM components are used as shown in Fig. 4. In EPROM 2, PN code sequences of 127, 255, 511, 1023 bit length are loaded which creates codes with a cycle-time of 4, 8, 16, 32 usec, and 13 bit Backer code is stored.

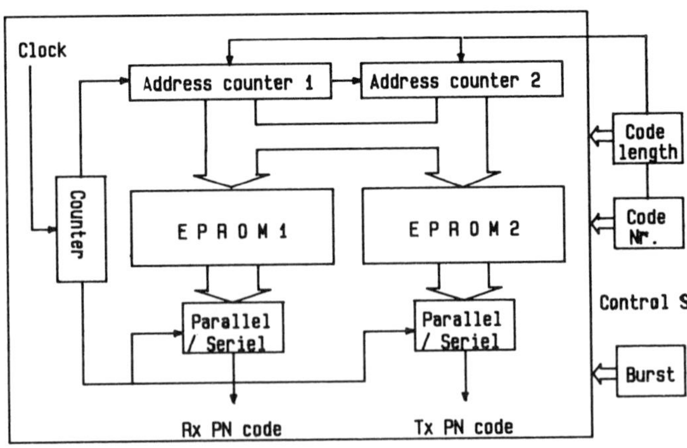

Fig. 4 Block diagram of the PRC generator

In EPROM 1, the time inveresed PN sequences of EPROM 2 are loaded. The counter divide the 64 MHz clock signal by eight. This signal drives address counter 1 and address counter 2. These counters address EPROM 1 and EPROM 2. The stored signal is read out by the parallel-serial shift register from EPROM 1 for the reference PN code sequence and from EPROM 2 for the transmit PN code. The control switch can select the length of the PN code and the code format. This variable format of the radar signal is especially usefull to different range measurement and to different range resolution. The burst control switch can select the transmit PN code sequence of a pulse which is a " Burst " of 1 to of 9.

6. Digital pulse compression using SAW convolver

The linear frequency modulation (LFM) or " Chirp " and the phase-coded pulse are the basic codings for pulse compression radar system. " Digital pulse compression " or " digital chirp " can also be obtained by the use of binary codes which are short compared to the duration of the total pulse. SAW technology has virtually dominated radar pulse compression realisations because the complicated pulse compressed radar signal generation and detection, which requires e.g., a special matched filter, can now be realized by very few components which in addition are very cost effective. SAW convolvers offer also significant advantage over the active correlators, because the correlation is independented on the critical alignment of the received signal with the locally generated reference. The time delayed received signal at 300 MHz is applied to a monolithic SAW convolver in approximate coincidence with a locally generated reference signal. The output signal is the convolution of the reference with the received signal provided that the two signals co-exist within the device processing time window of T as shown in Fig.5. The time delayed received signal and the reference signal are moving toward each other with a relative speed which is twice the SAW velocity, and causes a halving of the time scale at the output. Convolution of the two signals corresponds to a multiplication of the two spectra in the frequency domain. Therefore the time compression leads to a doubling of the frequencies in the output spectrum. Thus the output signal has twice the center frequency and the bandwidth of the input signal and the duration of the compressed pulse is halved. If the reference signal is the time replica of the received signal, then the output is the desired " Autocorrelation " of the received signal.

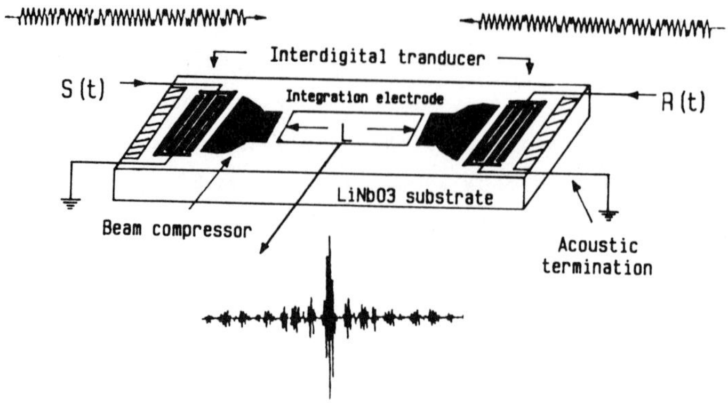

Fig. 5 A pulse compression principle of the SAW convolver

The convolution of two signals is defined by:

$$C(t) = S(t) * R(t) \quad (1)$$

where S(t) is a received signal

R(t) is a time inversed reference signal

After envelope detection

$$C(t) = \int_{-L/2}^{L/2} S(2t-\tau)\, R(\tau)\, d\tau \quad (2)$$

where $\tau = t - \frac{x}{v}$

Figure 6 shows the digital pulse compressed signal of 16 usec transmit pulse. The pulse compression ratio is 1024 and the compressed pulse length is 40 nsec which is wider than that the theorically expected 32 nsec. The right side of this compressed pulse is not symmetrical to the left side. This could be caused by matching problems at the input and output of SAW convolver.

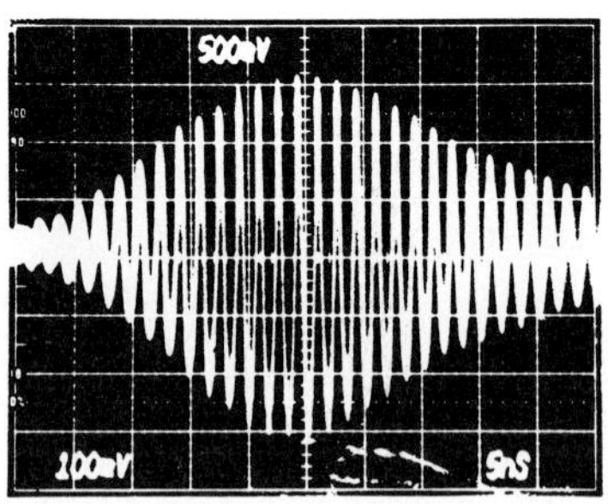

Fig. 6 The output of a digital pulse compression

7. Electronically scanned microstrip array antenna

The choice of a scanning antenna system is depend upon size, weight, width of total swath, feasibility and cost in space application. Therefore, we studied the scanning methods e.g., a scanning paraboloidal reflector, a phase scanning system and an interferometric scanning system for on radar altimeter application.

The beam shaped by a simple paraboloid reflector can be scanned over a limited angle by using a feed array which are displaced from the axis as shown in Fig. 7. This beam cannot be scanned too far without encountering serious deterioration of the antenna radiation pattern because of increasing coma and astigmatism. A paraboloid with f/D=0.5 (f= focal distance, D= antenna diameter) can be scanned +/- 6.5 beamwidths off axis before the gain is reduced to 80 percent of the maximum. This system already offers a better coverage, however the total field of view does not fulfill the requirements of the scanning radar altimeter. This technique may be used for an advanced beam limited radar altimeter of TOPEX or ERS. Comparing the return signal waveforms from the nadir footprint with return signals of the footprints lying within a ring concentric to the nadir as shown in Fig. 7, the height accuracy can improved especially at the high SWH (Significant Wave Height). The SNR will increase by the factor $\sqrt{7}$, the tracking area will be three times wider.

The technique of the interferometric scanning system has got a two times larger swath width with regard to the one reflector scanning method. This system ultilizes two scanning paraboloidal reflector antennas deployed on single hinged booms in a fixed measurement geometry. This array generates an interferometrical pattern over the desired total swath.

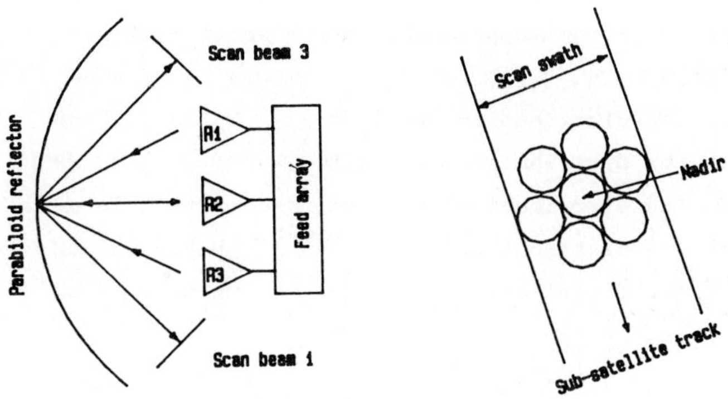

Fig.7 Scanning geometry of the parabolic antenna

In the phase scanning system the beam is positioned electronically by adjusting the phase differences between the elements of the array in a predetermined manner as shown in Fig. 8. A number of radiating elements are equally spaced along a straight line. Feeding these elements with in-phase currents of equal or tapered amplitude results in the propagation of a wave front parallel to the array, i.e., the formation of a broadside beam. This flexibility of the array antenna has lead to numerous applications such as electronical scanning and multiple beam generation.
One of the most important antenna parameter is the radiation pattern which can be expressed as a product of the parameters i.e., the spacing between the elements, the amplitude- , the phase distribution along the elements. A sidelobe suppression of the array antenna is dependent on these parameters mentioned above, and the tolerances in design and realisation . In Table 2, the highest sidelobe levels are compared for different amplitude distributions /1/. (k is the power difference between a maximum and a minimum at power distribution network). For the distributions factors k= 6, 8 and 10 dB, the cosine distribution represents the best performance but at high dynamic ranges e.g., k=16 dB the Dolph-Chebyshev distribution performs the lowest side lobe level. The deteriorative effect of the antenna pattern caused manufactoring tolerances is shown also in Table 3. The tolerance values are within the following boundaries:

- Intensity of peripheral array elements up to 10 %
- Intensity of central array elements up to 5 %
- Phase of array elements up to 7 degree (2 % of the wave length)
- Element position of array up to 2 %

The phase error contributes to the main error. Reading the errors the degradations of the sidelobe suppression in the worst case are:

With the distribution k=4 dB - up to 1.5 dB
With the distribution k=6 dB - up to 2.0 dB
With the distribution k=8 dB - up to 3.5 dB
With the distribution k=10 dB - up to 4.5 dB
With the distribution k=16 dB - up to 10.5 dB

The critical components of a microstrip array antenna are the radiation elements and feeding networks which must be designed and manufactured very properly. In order to obtain an optimum antenna design and the associated manufactoring data a microstrip array antenna CAD program was developed. The CAD program analysis and synthesis a microstrip array antenna considering the amplitude distributions and the errors. Fig. 8 shows a scanning performance of the 32*32 microstrip radiation elements.

- Center frequency	13.74 GHz
- 3 dB beam width	2.72 degrees
- Antenna gain	31.30 dB
- Size of array	48*48 cm
- Dolph-Chebyshev distribution	k=10 dB

The highest sidelobe level at scanning angle 0 degree is -26 dB and the second highest sidelobe level lay also under -30 dB. Increasing the scanning angle to 30 degrees there is no significant change in the half-power beam width and the gain. But at scanning angle 30 degrees the second highest sidelobe level increase drastically to -4.5 dB. This analysis show that, the optimum scanning angle of the pencil beam array antenna is 0 to 20 degrees.

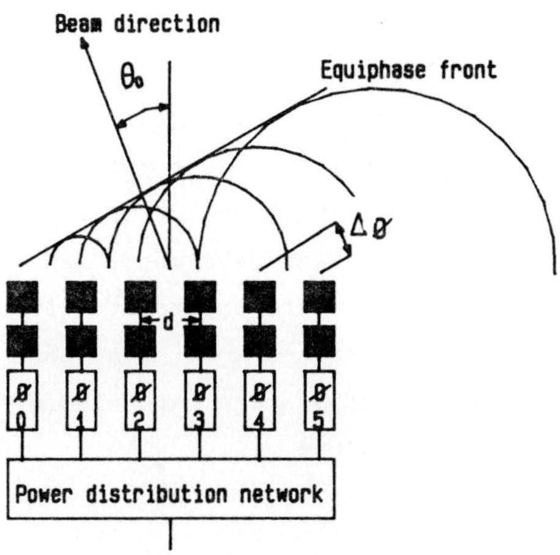

Fig. 8 Scanning concept of array antenna

Scanning angle:

$$\Delta\phi = \frac{2\pi}{\lambda} d \cdot \sin\theta_0$$

where

$\Delta\phi$ = phase defference between element

d = array distance

θ_0 = scanning angle

λ = wave length

Table 2

Imax / Imin k	distribution	Sidelobe level (dB)
4 dB	$Cos^a x$ a=0.85	-19.45
	$Cos^a x$ a=1.0	-18.64
	Dolph-Chebyshev a=1.5	-18.42
6 dB	$Cos^a x$ a=0.85	-23.26
	$Cos^a x$ a=1.0	-22.06
	Dolph-Chebyshev a=1.5	-21.73
8 dB	$Cos^a x$ a=0.85	-26.00
	Dolph-Chebyshev a=1.5	-25.43
	$Cos^a x$ a=1.0	-25.04
10 dB	$Cos^a x$ a=1.0	-28.44
	Dolph-Chebyshev a=1.5	-28.16
	$Cos^a x$ a=0.85	-25.76
16 dB	Dolph-Chebyshev a=2.0	-36.97
	Hamming	-33.77
	Dolph-Chebyshev a=1.5	-30.00

* 3 dB beam width is 2.4 deg.

The sidelobe level with amplitude distribution

Table 3

Distribution k=Imax/Imin	Random Error	Side lobe level (dB)
$Cos^a x$ a=0.85	Phase	-22.67
	Position	-23.15
k=6 dB	Intensity	-23.38
	Total	-22.75
$Cos^a x$ a=0.85	Phase	-25.84
	Position	-26.02
k=8 dB	Intensity	-25.29
	Total	-24.28
$Cos^a x$ a=1.0	Phase	-27.07
	Position	-28.41
k=10 dB	Intensity	-26.65
	Total	-25.43
Dolph-Chebyshev	Phase	-31.55
a=2.0	Position	-35.69
k=16 dB	Intensity	-32.38
	Total	-29.28

* 3 dB beam width is a 2.4 deg.

The sidelobe level with random errors

8. Conclusion

A concept of scanning radar altimeter was presented using " digital pulse compression " which was based on the SAW convolver, solid state microwave components and a microstrip array antenna. The measurement accuracy is better than 300 psec in the laboratory and the range resolution is 4.7 meters. Already several system components have been developed, manufactured and tested. Now they must be integrated, assembled to the radar altimeter. The microstrip array antenna will be manufactured and more broad-band signal processing (large time-bandwidth product) and a high stable, low noise receiver will be developed in the near future.

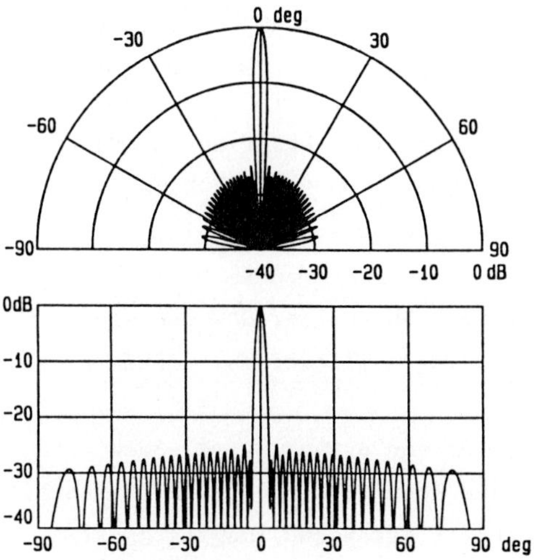

32 X 32 array, scanning angle = 0 deg.

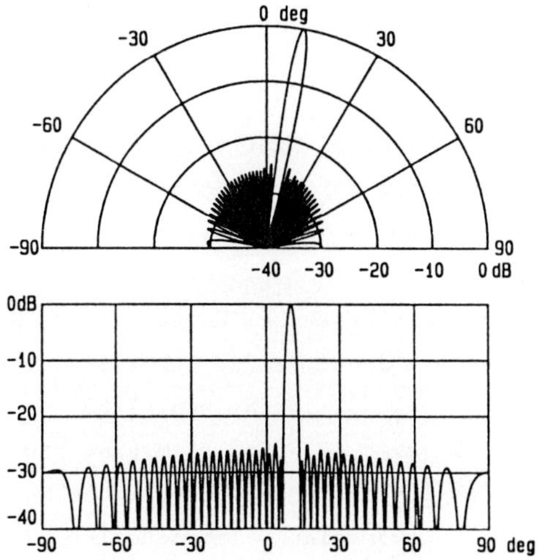

32 X 32 array, scanning angle = 10 deg.

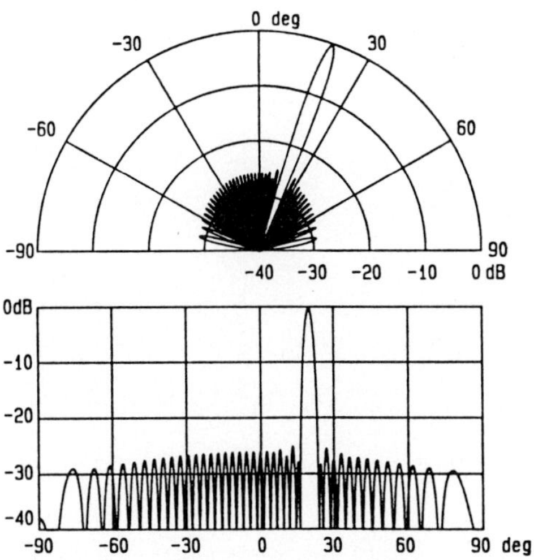

32 X 32 array, scanning angle = 20 deg.

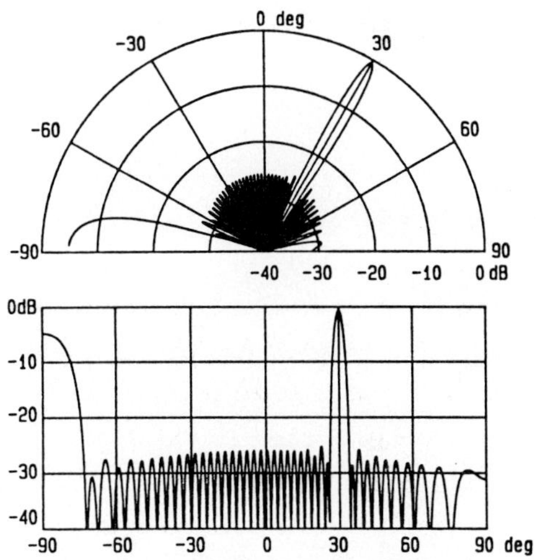

32 X 32 array, scanning angle = 30 deg.

Fig. 9 The radiation pattern of a microstrip array antenna

9. Reference

1. A. Nesic, D. Danilovic, M. Mikavica " Analysis of Distribution Functions Suitable for Printed Antenna Arrays and Radiation Pattern Dependence on Tolerances " Mikrowellen & military electroniec mag. vol.13, No.5, 1987.
2. A. Chatterjee, P. K. Das, L. B. Milstein " The use of SAW Convolver in Spread-Spectrum and Other Signal-Processing Applications " IEEE Trans. SU. vol. 32, No.5, sep. 1985
3. J.R.James, P.S.Hall " Microstrip antennas and arrays. Pt.2 - New array-design technique " Microwaves,Optics and Acoustics, vol.1, No.5 Sep.1977
4. S.A.Hovanesian " Introduction to Synthetic Array and Imaging Radars", Artech House, 1982.
5. Radar altimeter technical description (DRL Doc. RA-1-1), Selenia, 1983.

Quality Analysis of Platform Orientation Parameters for Airborne Laser Profiling Systems

Joachim Lindenberger
Institut für Photogrammetrie, Universität Stuttgart

Summary
Autoregressive integrated (ARI-) processes are introduced for modelling the orientation parameters of airborne sensor platforms. The ARI-model comprises a filtering algorithm and achieves several quality criteria. Empirical results are presented from applications of GPS and INS during aircraft flights.

1. Introduction

Airborne Laser Profiling Systems (ALPS) demonstrate a general development in recent technology, which may be characterized by the substitution of human operations by computer-controlled processes. The purpose of ALPS is the completely automatic production of a digital terrain model without manual measurements. The idea of ALPS can be briefly described as follows : a laser rangefinder mounted in an aircraft measures the distance to the ground surface. If the position and the attitude of the laser beam are known, the 3D-coordinates of the terrestrial laser footprint are determinable and can be directly introduced for generating a digital terrain model.

The ALPS is understood here as the whole system of several measuring, computing and registration devices which are necessary for the automatic DTM generation. Some properties of the most important devices shall be briefly sketched out :

- Laser rangefinder : measures the distances by reflection from virtually any remote surfaces. No special targets or retro-reflectors are required. The typical range measurement accuracy is about 0.2 m. High measurement rates (e.g. 100 Hz) are usual. The laserbeam is able to penetrate the canopy of trees with a rate between 10 an 70 percent depending on several factors such as deciduous or coniferous forests, winter or summer foliage, low-lying vegetation (Krabill et al 1984).

- Inertial Navigation System (INS) : provides the attitude data which are necessary for a highly accurate determination of the laser beam orientation. The original measurements on the gyros of the INS have to be evaluated by sophisticated algorithms in order to model several physical effects. Nevertheless, the INS must be supported by exterior control every few minutes during operation.
- Global Positioning System (GPS) : determines the 3D-position coordinates of a GPS antenna. The registration rate of most GPS receivers is about 1 Hz. The accuracy and feasibility of GPS measurements depend at present mainly on the satellite constellation which must be considered for the mission planning. Drift effects of the obtained coordinates are not expected to the same extent as with INS data and these effects are much easier to handle. For this reason it is desirable that the interferometric principle can be applied for attitude mesurements by the utilization of at least three GPS antennae on board (Hartl, Wehr 1986).
- Clock : A precise clock is necessary in order to give the reference in time for the data of the different devices measured with different rates. Under certain circumstances the internal clock of the GPS receiver can be utilized.
- Spotting camera : records permanently the surrounding of the laser footprint. It is recommended for visual examination of the results and for the identification of terrestrial control points. Usually a camera with conventional film is used, but tests using electronic video cameras or CCD cameras have also been carried out.
- Computer : samples and registers all measured data. Depending on the capacity of the computer, a partial on-board evaluation of the data is possible. At present most computational work must be done off-line after the mission.

The measuring devices of an ALPS provide very different types of data each referring to another coordinate system. The data-processing comprises the calibration and the specific evaluation of the measurments. The transformation of these results into a common terrestrial coordinate system requires reference measurements to control points which are, at present, not practicable without human intervention. Results of calibration tests are described by Schreier et al (1984).

Another problem concerns the different registration rates of the measuring devices. As the laser rangefinder works with the highest rate, the orientation parameters must be interpolated to obtain the orientation of every laser measurement. The interpolation is only efficient if the raw data are previously filtered and rectified in respect of gross errors.

The following part of this paper deals with the quality analysis of the measured or derived orientation parameters of an airborne platform. Every time-dependent orientation parameter, such as each position coordinate or each attitude angle, is considered to be a time-series which can be modelled by a stochastic process. A mathematical model describing the dynamic and stochastic properties of any time-dependent parameter will be presented. The evaluation of measured data takes full advantage of this model for the solution of a series of problems such as : stochastic description of the measurement process, filtering and smoothing of observations, detection of gross measurement errors, stochastic description of the filtered data. To summarize, the dynamic modelling of measured exterior orientation parameters provides filtered data with their stochastic properties for input into further evaluation of laser data.

The measured orientation parameters represent a trajectory of the platform, which is disturbed by the observation errors. As the observation errors are unknown, we are unable to reconstruct the true track of the platform. An approximation of the true movement will be found by the introduction of a model, which is built up here by the application of autoregressive integrated (ARI-) processes. The time-characteristic of each orientation parameter will be modelled by a specific representation of an ARI-process.

In chapter 2 the theoretical background of the time-series model will be outlined. Chapter 3 will present some applications of the theory : 1. modelling the inflight position parameters of an aircraft measured by the Global Positioning System GPS, and 2. modelling the inflight attitude parameters of an aircraft measured by an Inertial Navigation System INS.

2. Theoretical background of time-series modelling

2.1 Autoregressive integrated processes

Autoregressive integrated (ARI-) processes are a widely applied class of stochastic processes for various kinds of time-series (Haykin 1979, Kay and Marple 1981).

An autoregressive (AR-) process of order p describes a stationary time-series x_t by

$$x_t = - \sum_{i=1}^{p} a_i \, x_{t-i} + e_t \qquad V(e_t) = \sigma_e^2 \qquad (1)$$

The time-series considered in our applications are generally not stationary. These time-series have to be transformed into stationary time-series by the elimination of trends. A usual way is to take derivatives of the time-series. The autoregressive <u>integrated</u> process of order (p,d) is achieved if the d-th derivatives of the original time-series can be described by a stationary AR(p) process. Formally, the derivation can be expressed by an additional number d of process parameters a_i in eq.(1). Then the ARI(p,d) process fully describes the dynamic behaviour of any orientation parameter by a number of (p+d) process parameters a_i and by the variance σ_e^2 of the prediction errors.

2.2 The filtering algorithm

Two kinds of equations build up a Gauss-Markov model to estimate the filtered time-series x_t from the observed series y_t

$$E(y_t) = x_t \qquad V(y_t) = \sigma_n^2 \qquad (2)$$

$$E(e_t) = 0 = x_t + \sum_{i=1}^{p} a_i \, x_{t-i} \qquad V(e_t) = \sigma_e^2 \qquad (3)$$

Eq.(2) expresses the observation process with σ_n^2 being the variance of the observation noise. Eq.(3) presents the ARI model to which the unknown time-series x_t has to correspond.

The estimation of the filtered time-series x_t presumes the knowledge of the stochastic part of the Gauss-Markov model, i.e. the variances σ_n^2 and σ_e^2.

As the variances in general are a priori unknown, they must be estimated by applying the variance component estimation (VCE) technique. The formulation in the frequency domain published by Förstner (1984) is recommended for our application.

Together with the unknown ARI process paramters a_i we obtain a highly non-linear estimation problem which can be solved in an iterative computation scheme (Lindenberger 1987). The conditions to be fulfilled by the data for successfull solution of the GM model and the VCE are mentioned in the same paper (e.g. the variances σ_n^2 and σ_e^2 have to be of the same order of magnitude).

Gross errors in the observations would disturb the validity of the GM model in eq.(2). These gross errors are automatically detected and are taken into consideration by individual weights in eq(2). The weights are calculated following the robust estimation theory (Danish method, Krarup et al. 1980). On the other hand, a robust treatment of eq. (3) reduces the influences of edges and discontinuities in the data, which disturb the ARI model.

2.3 Capability of the algorithm

What are the main results from the algorithm ? Under the assumption that the true trajectory of the laser platform can be modelled by an ARI-process, we obtain a filtered data set x_t which is the most probable representation of the true track. Together with the estimated ARI process parameters, several demands of further evaluation of the laser data will be satisfied.

In addition the ARI-model yields important results, relevant for the analysis of the stochastic model. The estimated variance of the observation noise σ_n^2 describes the observation process. The variance of the prediction errors σ_e^2 gives a fidelity measure how well the the ARI-model is suited for describing the real physical process. The inversion of the normal equation system out of equations (2) and (3) provides accuracy criteria of the filtered data set, especially the variance of the filtered data σ_x^2 and the autocorrelation coefficients r(h). It is emphasised here that all stochastic results are obtained without any a priori information.

Any systematic effects in the time-series, such as drifts of the orientation parameters with the time, cannot be taken into consideration by the algorithm. For this reason the estimated accuracies must be understood not as absolute but as relative accuracies.

3. Applications

3.1 Position coordinates from NAVSTAR-GPS

The NAVSTAR Global Positioning System GPS enables the determination of the x,y,z position coordinates of one or more GPS antennae. In the case of a stationary GPS receiver the accuracy estimation of GPS measurements is relatively simple due to the redundancy of the observations. In contrast to this, the application of a GPS receiver in a moving vehicle such as an aircraft renders the accuracy estimation more difficult or rather impossible if a non redundant satellite constellation (i.e. 4 satellites or less) is available or if an unsuitable receiver is used. Then the suggested algorithm with ARI-processes can be utilized for stochastic investigations.

As an example some results of an accuracy study on real GPS data are presented in the following. The data for the analysis originate from a photogrammetric test flight with a Sercel GPS receiver TR5SB carried out by the Survey Departement of Rijkswaterstaat in the Netherlands on June 10th and 12th 1987 (D. Boswinkel, R.Witmer, J.W.v.d. Vegt 1988). From the primary phase measurements the position coordinates are calculated in the earth-fixed WGS system. After that the positions are transformed in a local horizon system. These coordinates are the subject of this analysis. Systematic effects which influence the determination of the coordinates are eliminated to a great extent in advance by simultaneous measurements with a second GPS receiver at a stationary reference point (Frieß 1988). Both receivers recorded with a registration rate of 1 measurement every 0.6 seconds. The data of the whole flight were divided according to 6 photo-strips with about 130 points each. Only the data within the strips are of interest. Within the strips the aircraft shows a very uniform kind of movement which can be modelled by a uniform kind of ARI-process beeing valid for the whole strip.

Each position coordinate is considered separatly by it's own ARI-process. The mean ARI process order was found to be (2,2) so that 4 ARI process parameters fully describe the model. The parameters of the stochastic model were estimated without any a priori information. In table 1 the estimated standard deviations of the noise-component in the GPS coordinates are listed. The main result is that the internal accuracy of the positions determined from a moving GPS receiver is about 0.01 m in the plane and less than 0.04 m in height. These results refer to this specific flight and depend mainly on the constellation of the satellites during the flight.

Table 1: Estimated noise σ_n [m] of GPS coordinates in horizon system

Strip	Date	X	Y	Z
1	12.06.87	0.003	0.005	0.047
3	10.06.87	0.005	0.007	0.034
4	10.06.87	0.013	0.007	0.036
5	10.06.87	0.012	0.010	0.031
6	12.06.87	0.006	0.005	0.037
7	12.06.87	0.006	0.005	0.028
rms of strips 3,4,5		0.011	0.008	0.034
rms of strips 1,6,7		0.005	0.005	0.038

These results can be compared with an estimation of the internal accuracy out of the least squares adjustment of the GPS positions. The LS adjustment determines the weight coefficient matrix Q and gives an estimate of the weight unit sigma naught, i.e. the GPS range measurement accuracy. In the case of a moving GPS antenna observing only 4 satellites the sigma naught value is in general not determinable because of the missing redundancy. The range measurement precision in static mode for the Sercel receiver is given from the manufactorer by about σ_0= 0.003 m. Out of different considerations the range measurement precision of the Sercel receiver in kinematic mode can be assumed reasonably as σ_0= 0.006 m. This value is twice as large as the precision obtained by a stationary receiver and depends mainly on the velocity of the aircraft. A confirmation of this value by the analysis of kinematic measurements to more than four satellites has not been carried out at the moment because of some difficulties with the handling of systematic drift effects. With this assumption the internal accuracies are calculated. In table 2 these results from least squares adjustment are set against the estimated internal accuracies obtained from the ARI-model. The accuracies in table 2 are not transformed in the horizon system, they refer to the WGS system. The comparision of the results show a very good agreement whereby the ARI-model has the great advantage that it does not require any a priori stochastic information (i.e. the knowledge of σ_0). Thus demonstrates that the ARI model is suited to be applied for modelling the dynamic characteristics of GPS coordinates during a photogrammetric flight and that the results are realistic.

Table 2: Comparision of estimated internal accuracies σ_n [m] from least squares adjustment with assumed range measurement accuracy of σ_0=0.006 m and from ARI-model. GPS coordinates in WGS system.

	LS-adjustment			ARI-model		
	X	Y	Z	X	Y	Z
rms (3,4,5)	0.022	0.009	0.017	0.021	0.010	0.026
rms (1,6,7)	0.020	0.009	0.020	0.018	0.010	0.021

Table 3: Accuracy criteria of filtered GPS data in horizon system

Estimated accuracies σ_x [m] of filtered coordinates			
	X	Y	Z
rms of strips 3,4,5	0.007	0.006	0.027
rms of strips 1,6,7	0.004	0.004	0.026
Correlation coefficients of filtered data d=0.6 sec			
	X	Y	Z
r(1d)	0.47	0.41	0.54
r(2d)	0.05	-0.02	-0.01
r(3d)	-0.08	-0.07	-0.11

All subsequent data-processing, for example the interpolation of orientation parameters to laser spot points, is based on the filtered GPS coordinates. These filtered data with their stochastic informations are achieved from the ARI-model. Table 3 summarizes the estimated accuracies of the filtered coordinates and their autocorrelation coefficients. It is an important result that in this example only the correlations from one to the next point (within 0.6 sec.) remain significant.

3.2 Attitude data from INS

Attitude data, i.e. the roll-, pitch- and yaw- angles of the aircraft body axis to the terrestrial fixed coordinate system can be directly measured by an inertial navigation system (INS) belonging to the equipment of every larger aircraft. The evaluation of INS measurements is in most cases very difficult and depends essentially on the physical model which describes the systematic influences of a large number of physical effects.

The analysis of INS data presented in this chapter examine only the registered attitude angles. Any systematic drift effects caused by remaining errors in the gyros of the INS are not taken into consideration. For this reason the filtered attitudes cannot be estimated as absolute values, but must be rectified for the systematic drifts. Then the estimated accuracies of the measured angles are understood as the precision of one reading of the gyro.

The attitude data are registered during an aircraft flight with a Litton LTN-72 Inertial Navigation System of a Falcon jet of the DFVLR. The registration rate was 10 Hz. The ARI process order was only (2,2). The analysis refer to the central part of the flight, not disturbed by take-off and landing maneuvers. During undisturbed parts of the flight the standard deviation of the ARI-prediction errors σ_e decrease below 0.0001 deg.

The estimated observation noise σ_n is in full accordance with other investigations from a stationary INS of the same type. In the stationary mode the precision of the attitude measurement is about four times higher than in the dynamic mode, which was expected in advance.

Table 4: INS attitude data from aircraft			
Estimated standard deviations in [deg]			
	YAW	PITCH	ROLL
observation noise σ_n	0.0023	0.0035	0.0030
ARI-model errors σ_e	0.0005	0.0013	0.0026
filtered data σ_x	0.0011	0.0013	0.0017
Correlation coefficients of filtered data d=0.1 sec			
r(1d)	0.86	0.79	0.80
r(2d)	0.57	0.40	0.41
r(3d)	0.29	0.09	0.08
r(4d)	0.09	-0.07	-0.09

4. Conclusions

This paper introduced autoregressive integrated processes for modelling the dynamic characteristics of the exterior orientation parameters of a platform for airborne laser profiling systems. The ARI-Model in combination with a variance component estimation enables the entire functional and stochastic description of the orientation parameters. The main advantages of this model are the easy handling, the low number of necessary ARI process parameters and the dispensation of any a priori information concerning the stochastic model. The joint treatment of dependent stochastic processes like the position coordinates X,Y,Z or the attitude angles roll, pitch, yaw in a vector process representation is the subject of further investigations.

The successful application of the ARI model to very different kinds of orientation parameters improves the strength of the concept. Comparisons to other methods for accuracy estimation demonstrate that the obtained results are realistic.

The analysis of the orientation parameters from aircraft flights with GPS and INS show a high accuracy of the parameters. The conclusions can be drawn that the accuracy of the 3D coordinate determination with airborne laser profiling systems depends mainly on the distance measurement accuracy of the laser rangefinder which is about 0.2 m. However, the accuracy of the generated digital terrain model does not depend primarily on the measurement accuracy. The accuracy of the DTM, that is the degree of accuracy with wich the real terrain is represented by the DTM, is essentially determined by the distance between the measured profiles.

5. References

- Boswinkel, D.; Witmer, R.; v.d. Vegt, J.W. (1988) : Utilisation of GPS in large scale photogrammetry; presented paper ISPRS Congress Kyoto Comm. I
- Förstner, W. (1984) : Determination of the additive noise variance of observed autoregressive processes using variance component estimation technique; Schriftenreihe Heft 10; Institut für Photogrammetrie Universität Stuttgart
- Friess, P. (1988) : Empirical accuracy of positions computed from airborne GPS data; pres. pap. ISPRS 16th Congress Kyoto Commission III
- Hartl, Ph.; Wehr, A. (1986) : Chancen der GPS Satellitennavigation für die Luftbildphotogrammetrie; BuL, Vol.54, Heft 6 1986, pp 221-240
- Haykin, S. ed. (1979) : Topics in applied physics vol 34 : Nonlinear methods of spectral analysis; Springer-Verlag 1979
- Kay, S.M.; Marple, S.L (1981) : Spectrum analysis - A modern perspective; proceedings of the IEEE, vol.69, no.11, Nov.1981
- Krabill, W.B.; Collins, J.G.; Link, L.E.; Swift, R.N.; Butler, M.L. (1984) : Airborne laser topographic mapping results; Photogrammetric Engineering and Remote Sensing, Vol. 50, June 1984, pp 685-694
- Krarup, T.; Juhl, J.; Kubik, K. (1980) : Götterdämmerung over least squares adjustment; pres. paper ISP Congress Hamburg Comm III, Int. Arch. of Photogr. vol. 23 B3, pp 369-378
- Lindenberger J. (1987) : Consideration of observation errors when modelling digital terrain profiles; Proceedings of the ISPRS WG III/3 Intern. Colloquium: Progress in terrain modelling, TU Denmark Lyngby, pp 227-237
- Schreier, H.; Lougheed, J.; Gibson, J.R.; Russell,J. (1984) : Calibrating an airborne laser profiling system; Photogrammetric Engineering and Remote Sensing, Vol. 50, November 1984, pp 1591-1598

Radar Altimetry for Land Applications

Harald Schüssler
Dornier System, Friedrichshafen

1. INTRODUCTION

An international programme for polar earth observations will be established by NASA /ESA/NASDA, starting in 1995. Among many other instruments, the first NASA mission (EOS-1) will possibly incorporate a sensor system for measuring global land topography. JPL (Pasadena, Cal.) together with the Institute of Navigation (University of Stuttgart) and Dornier System (Friedrichshafen) are currently conceiving such a sensor system, which will be a multimode radar altimeter for ocean and land applications.

The following paper covers system aspects related to land radar altimetry. Topographic mapping will be based on sampling theory for two-dimensional terrain functions.

2. SAMPLING OF TERRAIN

The radar maps the terrain around the satellite ground track as shown in fig. 1. For simplicity we start with one-dimensional terrain functions f(x). The altimeter will be moved along the x–axis, and the terrain will be sampled at regular intervals Δx. To satisfy the sampling theorem, we need a smoothed, i.e. bandlimited, terrain function $\overline{f}(x)$ prior to sampling:

$$\overline{f}(x) = f(x) * g(x) = \int_{-\infty}^{+\infty} f(\xi)\, g(x-\xi)\, d\xi$$

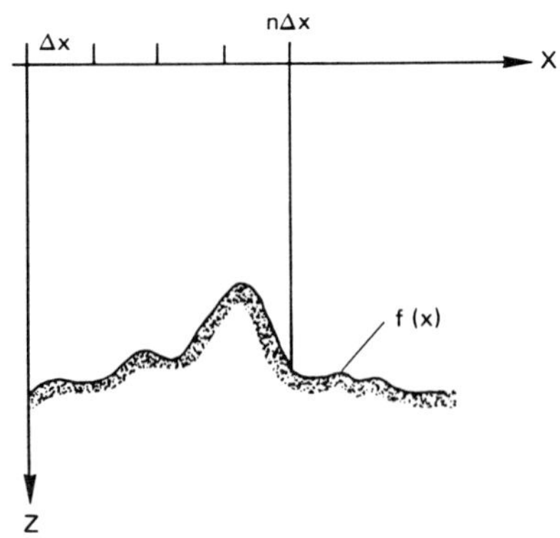

Fig. 1

The spatial filter should be ideal low-pass, viz.

$$g(x) = \frac{1}{\Delta x} \frac{\sin(\frac{\pi x}{\Delta x})}{(\frac{\pi x}{\Delta x})} = \frac{1}{\Delta x} \operatorname{sinc}\left(\frac{x}{\Delta x}\right)$$

The smoothed terrain function $\overline{f}(x)$ is the optimum bandlimited approximation of the real terrain $f(x)$, which minimizes the mean-square error [1]

$$e = \int_{-\infty}^{+\infty} |f(x) - \overline{f}(x)|^2 \, dx$$

Samples will be taken at regular intervals $x = n\,\Delta x$, thus

$$\overline{f}(n\,\Delta x) = \int_{-\infty}^{+\infty} f(\xi)\, g(n\,\Delta x - \xi)\, d\xi$$

The smoothed terrain may be recovered from these samples by

$$\overline{f}(x) = \sum_{n=-\infty}^{\infty} \overline{f}(n\,\Delta x) \operatorname{sinc}(\frac{x}{\Delta x} - n)$$

3. PROPOSED MEASUREMENT SYSTEM

Fig. 2 depicts the proposed measurement principle. The nadir-looking radar altimeter is moved along the x–axis to sampling points $(n\,\Delta x)$. The two-way antenna gain $g(x)$ is an even function $g(x) = g(-x)$ and normalized:

$$\int_{-\infty}^{+\infty} g(x)\, dx = 1$$

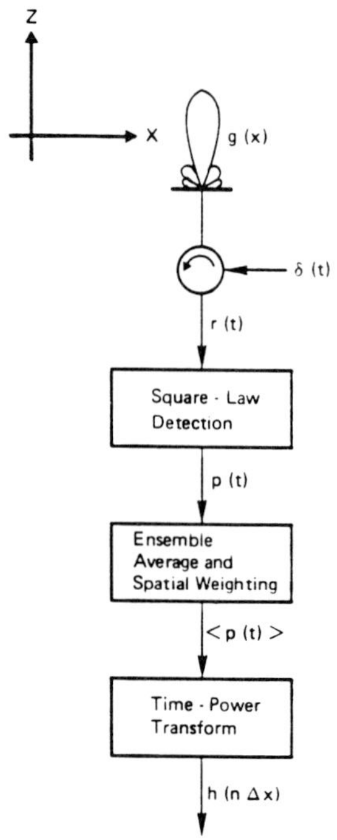

The radar emits short delta impulses $\delta(t)$, and the receiver includes a power meter measuring the power return $p(t)$. Area-extensive targets cause fading, since many randomly phased phasors add up in the receiver [2]. The ensemble average $\langle p(t) \rangle$ may be obtained by averaging many independent samples, which are taken from successive illuminations at slightly different orbit positions. The received delta impulses are time delayed in accordance with the terrain height $f(x)$. The receiver converts time delay to power by measuring the centroid of $\langle p(t) \rangle$.

Fig. 2

We consider antenna beamwidths less than one degree. Hence, we can (linearily) transform polar antenna coordinates to Cartesian space. The two-way antenna gain function, shifted to the sampling point, is $g(x - n\,\Delta x) = g(n\,\Delta x - x)$ The received power signal within dx, as illustrated in fig. 3, is given by

$$k(x)\;\sigma^0(x)\;g(n\,\Delta x - x)\;\delta(t - \tfrac{2}{c} f(x))\;dx$$

where
$\delta(t - \frac{2}{c} f(x))$ is the time-delayed transmit pulse,
$\sigma^0(x)$ is the backscattering coefficient, and
$k(x)$ accounts for propagation losses (free space, atmosphere).

Propagation losses are approximately constant within the antenna beamwith, i.e. $k(x) = k$.

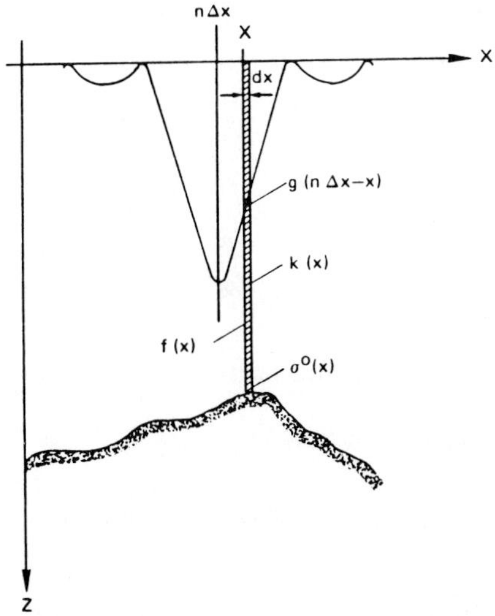

Fig. 3

The ensemble average of the power return within the beam is

$$\langle p(t)\rangle \ = \ k \int_{-\infty}^{+\infty} \sigma^0(x)\ g(n\ \Delta x - x)\ \delta(t - \tfrac{2}{c} f(x))\ dx$$

The centroid algorithm transforms delay-time into power:

$$h(n\ \Delta x) = \frac{c}{2}\ \frac{\int_0^\infty t\ \langle p(t)\rangle\ dt}{\int_0^\infty \langle p(t)\rangle\ dt}$$

Evaluation of above equation yields

$$h(n\ \Delta x) = \frac{\frac{c}{2}\ k \int_{-\infty}^{+\infty} \sigma^0(x)\ g(n\ \Delta x - x) \int_0^\infty \delta(t - \frac{2}{c} f(x))\ t\ dt\ dx}{k \int_{-\infty}^{+\infty} \sigma^0(x)\ g(n\ \Delta x - x) \int_0^\infty \delta(t - \frac{2}{c} f(x))\ dt\ dx}$$

Since $\int_{-\infty}^{+\infty} f(t)\ \delta(t - t_0)\ dt = f(t_0)$, and $\int_{-\infty}^{+\infty} \delta(t)\ dt = 1$ above equation becomes:

$$h(n\ \Delta x) = \frac{\int_{-\infty}^{+\infty} \sigma^0(x)\ f(x)\ g(n\ \Delta x - x)\ dx}{\int_{-\infty}^{+\infty} \sigma^0(x)\ g(n\ \Delta x - x)\ dx}$$

For flat terrain $f(x) = f_0$, the receiver output is $h(n\,\Delta x) = f_0$

Since both $f(x)$ and $g(n\,\Delta x - x)$ are positive functions, the mean value theorem of the integral calculus may be applied to the receiver output:

$$h(n\,\Delta x) = \frac{\sigma^0(x_1) \int_{-\infty}^{+\infty} f(x)\, g(n\,\Delta x - x)\, dx}{\sigma^0(x_2) \int_{-\infty}^{+\infty} g(n\,\Delta x - x)\, dx}$$

Recognizing that the two-way antenna pattern is normalized, we finally obtain

$$h(n\,\Delta x) = \overline{f}(n\,\Delta x) + \Delta\overline{f}(n\,\Delta x)$$

where $\overline{f}(n\,\Delta x)$ is the desired convolution of the sampling point, and

$$\Delta\overline{f}(n\,\Delta x) = \left(\frac{\sigma^0(x_1)}{\sigma^0(x_2)} - 1\right)\overline{f}(n\,\Delta x)$$

is the sampling error, which may be considered as a random variable. This error would vanish for homogeneous backscattering and/or flat terrain.

4. SPATIAL FILTERING WITH THE RADAR ANTENNA

Minimum antenna beamwidth is achieved with uniform aperture illumination. The two-way power gain is then given by [3]

$$S^2(u) = S_0^2 \operatorname{sinc}^4(a\,u)$$

where $u = x/\Delta x$. The factor a includes geometry, wavelength and antenna dimensions, and will be set to unity.

The desired impulse response of the spatial filter is $g(u) = \operatorname{sinc}(u)$. This response function can be approximated, as shown in fig. 4 and 5, by means of weighted superpositions of antenna illuminations from different sampling points, e.g.

$$g(u) \simeq C_{00} \operatorname{sinc}^4(u) + \sum_{i=0}^{10} C_i \left[\operatorname{sinc}^4(u - i - \tfrac{1}{2}) + \operatorname{sinc}^4(u + i + \tfrac{1}{2})\right]$$

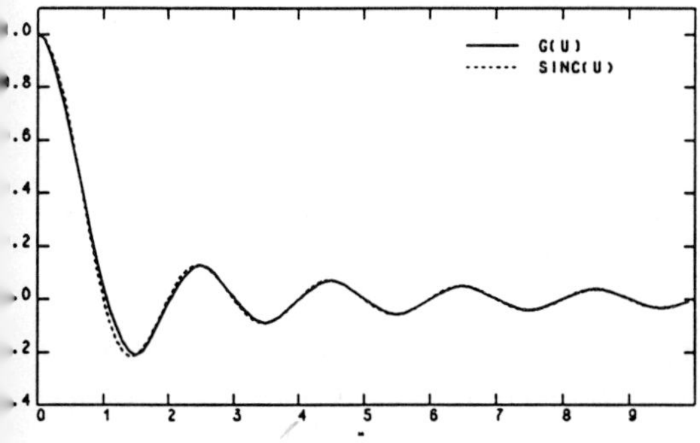

Fig. 4 Impulse Response

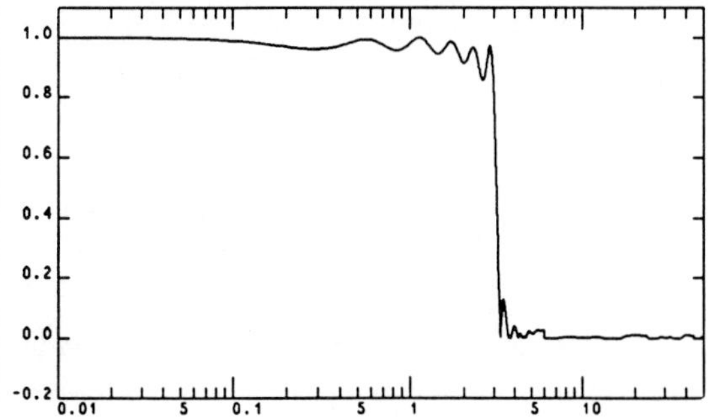

Fig. 5 Frequency Response

Weighting factors :		
	$C_{00} = +0.85$	$C_5 = -0.05788$
	$C_0 = +0.47747$	$C_6 = +0.04897$
	$C_1 = -0.21221$	$C_7 = -0.04244$
	$C_2 = +0.12732$	$C_8 = +0.03745$
	$C_3 = -0.09095$	$C_9 = -0.03351$
	$C_4 = +0.07074$	$C_{10} = +0.03032$

5. MAPPING WITH REAL APERTURE RADAR ALTIMETER

The proposed sampling/measurement scheme can be adopted to two-dimensional terrain functions $f(x,y)$ as well. Sampling nodes are arranged in a rectangular pattern in the (x,y)–plane. Spatial symmetry suggests $\Delta x = \Delta y = \Delta$. The smoothed terrain $\overline{f}(x,y)$ can be recovered from samples $\overline{f}(n\Delta, m\Delta)$ by [4]

$$\overline{f}(x,y) \;=\; \sum_{n=-\infty}^{\infty} \sum_{m=-\infty}^{\infty} \overline{f}(n\Delta, m\Delta)\ \mathrm{sinc}(\frac{x}{\Delta} - n)\ \mathrm{sinc}(\frac{y}{\Delta} - m)$$

Illumination of the (x,y)–plane is carried out with the antenna beam, scanned around the nadir in across-track direction. The one-way antenna half-power footprint diameter shall be $0.88\,\Delta$. Illuminations shall be spaced at $\Delta/2$ increments to permit suitable spatial filtering with weighted power returns around the sampling nodes.

However, antenna dimensions for spaceborne real aperture radar altimeters may become excessivley large. This is shown in the following example: Sampling interval $\Delta = 500$ m; Spacecraft altitude is 800 km, frequency is 35 Ghz, required antenna dimensions are 14 m x 14 m!

The antenna width can be significantly reduced with application of the SAR–principle, i.e. doppler filtering along the spacecraft ground track [5]. Across-track resolution is maintained with the antenna beam, see fig. 6

Since along-track resolution can be made extremely small, even with unfocused SAR–processing, the desired spatial filter function $g(u)$ can be easily approximated by suitable look weighting.

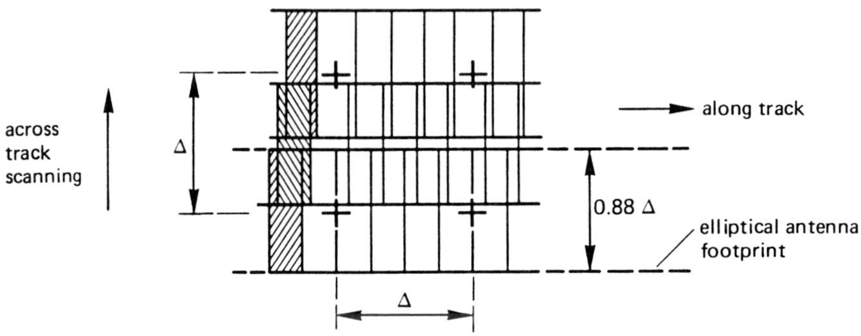

Fig. 6

Principal error sources, which must be considered for sensor design and performance analyses, are briefly discussed below.

Smoothing: Error between real terrain $f(x)$ and smoothed terrain $\overline{f}(x)$ is bounded by $\left|f(x) - \overline{f}(x)\right| \leq 2 \int_{\nu_c}^{\infty} F(\nu)\, dv$.
ν_c is filter cut-off frequency.

Aliasing: Will be negligible, if suitable spatial filtering is applied.

Pulse Length: Intrinsic range resolution produces quantization error of samples.

Fading: Can be reduced with look averaging.

Thermal Noise: Causes erroneous ensemble averages $\langle p(t) \rangle$ and, hence, erroneous samples $\overline{f}(n\,\Delta x)$. Permissible signal/noise ratio may be traded with number of independent looks [5].

Backscattering: Inhomogeneous backscattering (σ^0) produces erroneous ensemble averages $\langle p(t) \rangle$. This error cannot be avoided.

Localisation: High resolution topographic mapping requires precise knowledge of the sensor state vector (position, velocity, antenna orientation).

Further discussions on this subject together with detailed error analyses for special cases can be found in [5].

6. CONCLUSIONS

Topographic mapping with spaceborne (or airborne) sensor systems shall be based on sampling theory: The desired width of the measurement grid gives the upper limit of spatial frequencies to be recovered. The grid size determines the cut-off frequency of the spatial filter, which can be realized by means of the antenna pattern. Weighted superpositions of illuminations from different sampling points permit suitable approximations to the desired impulse response function.

Convolution in the spatial domain by utilizing the antenna power pattern implies, that the key measurement parameter, namely the terrain height, must also be translated into a power signal. This is realized with the power centroid algorithm.

Bibliography

[1] A. Papoulis *"Signal Analysis"*
McGraw–Hill Book Company

[2] F.T. Ulaby, R.K. MOORE, A.K. Fung
"Microwave Remote Sensing, Vol II"
Addison–Wesley Publishing Company

[3] F.T. Ulaby et al *"Microwave Remote Sensing, Vol I"*
Addison–Wesley Publishing Company

[4] A.W. Rihaczek *"Principles of High-Resolution Radar"*
Mark Resources, Inc.

[5] H. Schüssler *"Radar Altimeter mit konischer Keulenschwenkung"*
Bildmessung und Luftbildwesen 55(1987) Heft 4

D. IMAGE PROCESSING

Motion Estimation in Image Sequences

Norbert Diehl*, Hans Burkhardt
Arbeitsbereich Technische Informatik I, Technische Universität Hamburg-Harburg

Abstract

This contribution presents an efficient method to estimate the motion parameters of planar rigid objects in space. The problem is formulated as a model-based estimation problem which is solved by an iterative minimization strategy based on ordinary grey-scale images. Therefore no a priori knowledge as corresponding points or the displacement vector fields are required.

The advantages of the algorithm - a so called modified Newton-Raphson algorithm and its combination with Quasi-Newton methods - are the large region of stability, the high, image-bandwidth-adaptive convergence rate and the low numerical expense within each iteration step. Therefore the algorithm may be used for near realtime applications.

Furthermore the algorithm is relatively insensitive to additive noise because whole image areas are used. The variance of the motion parameters can be reduced further if the problem of motion estimation is formulated as maximum-likelihood estimation. Under a few assumptions the maximum-likelihood estimation can be interpreted as a generalized correlation, i.e. the correlation of the appropriately prefiltered images.

1. Introduction

The problem of motion estimation has become of great interest in several areas of image processing such as motion compensated image coding, remote sensing, robotics or autonomous vehicles [Huang 1981, Huang 1983, Nagel 1981, Nagel 1983, Nagel 1985, Musmann et al. 1985, Bähr 1985]. So far most algorithms for three-dimensional motion estimation use corresponding points [Roach und Aggarwal 1980, Huang und Tsai 1981, Tsai et al. 1982, Tsai and Huang 1984a, Tsai and Huang 1984b, Fang and Huang 1984, Longuet-Higgins 1981, Yen and Huang 1983] which have to be known a priori or they use the displacement vector field [Prazdny 1980, Adiv 1985, Kanatani 1986]. Both methods have a very high numerical complexity and are very sensitive to noise.

In this contribution the motion estimation problem is solved using model-based estimation techniques based on ordinary grey scale images. The advantage of these iterative algorithms is the combination of good performance features such as a large region of stability and a high, image-bandwidth-adaptive convergence rate with low numeric expense within each iteration step. Furthermore the algorithms are relatively insensitive to additive noise because whole areas of the

---

* Now with: AEG Forschungsinstitut, Sedanstraße 10, D-7900 Ulm, West-Germany

images are used. Similar algorithms may be found in [Spoer 1987, Netravali and Salz 1985, Lenz 1986].

The problem of motion estimation may be devided into two subproblems. First the scene has to be segmented into the different moving objects and the background where also effects of occlusion have to be taken into consideration. In a second step the motion parameters of the objects have to be estimated. This contribution deals with this second problem under the simplifying assumption of only one moving object. To achieve quite general results methods to estimate the parameters of arbitrary motion and therefore of general coordinate transforms are developed. Three-dimensional motion of a rigid planar patch is treated as a special example.

The parameter estimation algorithm is based on a simplified signal model. A camera looks at a rigid moving object in the three-dimensional space. Two grey-scale images $I_1(\mathbf{x})$ and $I_2(\mathbf{x})$ describe the motion of this object in front of a uniform background where no occlusion effects occur:

$$I_1(\mathbf{x}) = S(\mathbf{x}) + N_1(\mathbf{x}), \qquad I_2(\mathbf{x}) = S(\mathbf{x}') + N_2(\mathbf{x}). \tag{1.1}$$

$S(\mathbf{x})$ und $S(\mathbf{x}')$ are the projections of the moving object onto the image plane. $N_1(\mathbf{x})$ and $N_2(\mathbf{x})$ are additive noise terms. Thus equation 1.1 describes the connection between the projection of the object in position 1 ($I_1(\mathbf{x})$) and the same object in position 2 ($I_2(\mathbf{x})$) (cf. Fig. 1.1).

Generally the connection between the coordinates $\{\mathbf{x}\}$ and $\{\mathbf{x}'\}$ is given by a transform

$$\mathbf{x}' = \mathbf{h}(\mathbf{x}, \mathbf{T}), \tag{1.2}$$

with the motion vector $\mathbf{T}$. The transform $\mathbf{h}(\mathbf{x}, \mathbf{T})$ should be unique and invertible in the considered image region. For simplicity it should also be $\mathbf{x}' = \mathbf{h}(\mathbf{x}, \mathbf{T} = \mathbf{0}) = \mathbf{x}$ for $\mathbf{T} = \mathbf{0}$. If the image formation is described by central projection under the assumption of a rigid planar patch the coordinate transform is given by

$$x' = \frac{(1+a_1)\,x + a_2\,y + a_3}{a_7\,x + a_8\,y + 1}, \qquad y' = \frac{a_4\,x + (1+a_5)\,y + a_6}{a_7\,x + a_8\,y + 1}, \tag{1.3}$$

with the motion vector $\mathbf{T} = \mathbf{a} = (a_1, a_2, a_3, a_4, a_5, a_6, a_7, a_8)^T$. For parallel projection the coordinate transform may be described as

$$x' = (1+c_1)\,x + c_2\,y + c_3\,, \qquad y' = c_4\,x + (1+c_5)\,y + c_6\,, \tag{1.4}$$

with $\mathbf{T} = \mathbf{c} = (c_1, c_2, c_3, c_4, c_5, c_6)^T$.

The parameters of the coordinate transforms in the image plane are unique. But it is very hard to answer the question how to get the true three-dimensional motion parameters of the object. Using the central projection and two successive images there are normally two solutions [Tsai and Huang 1982]. Often one solution can be excluded using slight a priori knowledge such as a rough estimate of the object position at the beginning of motion. If three successive images are used the motion parameters are unique up to a common scale factor of the translation vector [Tsai and Huang 1984b].

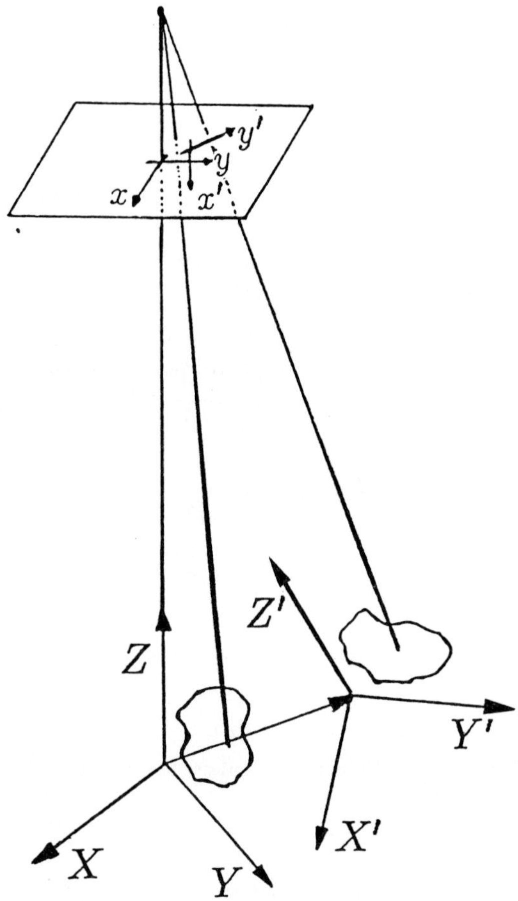

Fig. 1.1: Moving of an object in the three-dimensional space and central projection onto the image plane

If parallel projection is used the ambiguity is larger. There are usually four solutions up to the common scale factor [Lenz 1986]. These ambiguities mainly result from the fact that the parallel projection cannot describe the magnification of object parts coming closer and the reduction of those parts moving away.

2. Model-based parameter estimation

In this contribution the main aspect will be the efficient estimation of the parameter vector $\mathbf{T}$ of the coordinate transform $\mathbf{h}(\mathbf{x}, \mathbf{T})$. Therefore the idea of model-based parameter estimation will be used and a so called modified Newton-Raphson algorithm and its combinations with Quasi-Newton methods will be developed. To get an appropriate estimate $\widehat{\mathbf{T}}$ of the parameter vector $\mathbf{T}$ the motion is modelled by a model image $I_m(\mathbf{x}, \widehat{\mathbf{T}})$. This image results from $I_1(\mathbf{x})$ by the same transform $\mathbf{h}(\mathbf{x}, \mathbf{T})$ like $I_1(\mathbf{x})$ but now with the model parameter vector $\widehat{\mathbf{T}}$. In a second step the difference between $I_m(\mathbf{x}, \widehat{\mathbf{T}})$ an $I_2(\mathbf{x})$ is described by an appropriate error criterion $J\{e(\widehat{\mathbf{T}}\}$ which is minimized to get the optimal estimate $\widehat{\mathbf{T}} = \mathbf{T}^*$.

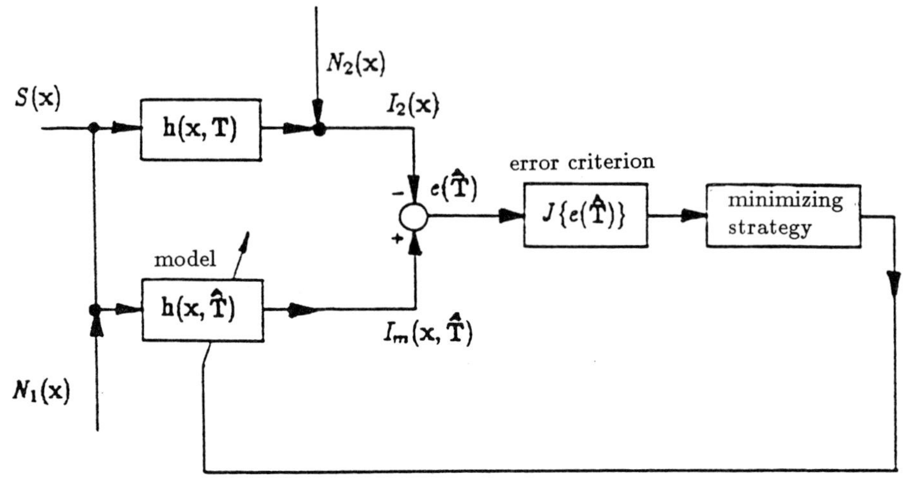

Fig. 2.1: Structure of model-based parameter estimation

2.1. The error criterion

The error criterion specifies the quality of the estimate. Therefore knowledge of the signal and noise statistics should be incorporated. The most general approach is given by a maximum likelihood estimate. At the moment the simpler error function

$$J\{e(\widehat{\mathbf{T}})\} = \frac{1}{2} E\{(I_m(\mathbf{x}, \widehat{\mathbf{T}}) - I_2(\mathbf{x}))^2\}, \tag{2.1}$$

i.e. the variance of the model error is chosen. The expectation value is realized by summing over the interesting image region. For stationary signals $J\{e(\widehat{\mathbf{T}})\}$ equals the negative cross correlation funktion of $I_m(\mathbf{x}, \widehat{\mathbf{T}})$ and $I_2(\mathbf{x})$ plus an additive constant.

This error criterion has to meet some requirements. First the contents of the images $I_1(\mathbf{x})$ and $I_2(\mathbf{x})$ must have a sufficiently high similarity such that the error function has a unique minimum. Second it is assumed that $J\{e(\widehat{\mathbf{T}})\}$ is at least twice differentiable at the optimum, i.e. the Hessian of the error function must exist. For ill conditioned images where the criterion has no dominant parabolic shape near the optimum appropriate filtering may be used.

2.2. The minimizing strategy

The minimizing strategy determines the velocity and robustness of the estimate. The region of stability, the convergence rate and the numerical complexity of the minimizing algorithm are very important and decide if near realtime applications are possible.

A global search for the optimal parameter vector $\mathbf{T}^*$ is unrealistic because it results in the calculation of a multidimensional cross-correlation function in conjunction with a tremendeous numerical complexity. Instead local, iterative methods are used, especially a so called modified Newton-Raphson algorithm and its combination with Quasi-Newton methods.

2.2.1. The modified Newton-Raphson algorithm

The modified Newton-Raphson algorithm originally has been developed for one-dimensional problems such as time delay estimation and stereoscopic range finding [Mesch 1982, Burkhardt and Moll 1978, Burkhardt and Moll 1979]. Then it was used to estimate simultaneously rotation and translation in image sequences [Burkhardt and Diehl 1986, Diehl and Burkhardt 1986] before it has been extended to general transforms $\mathbf{h}(\mathbf{x}, \mathbf{T})$ [Diehl 1987, Diehl 1988]. The modified Newton-Raphson algorithm differs from the normal Newton algorithm at one important point. Instead of using the Hessian $\mathbf{H}(\widehat{\mathbf{T}})$ at the actual iteration point $\widehat{\mathbf{T}} = \widehat{\mathbf{T}}_k$ the Hessian is always used at the optimum $\widehat{\mathbf{T}} = \mathbf{T}^*$. For pure translation ($\mathbf{T} = (d_1, d_2)^T$) the algorithm is given by the iteration

$$\widehat{\mathbf{T}}_{k+1} = \widehat{\mathbf{T}}_k - \mathbf{H}^{-1}(\mathbf{T}^*)\, \mathbf{g}(\widehat{\mathbf{T}}_k). \tag{2.2}$$

The special advantage of this algorithm is the possibility to compute the Hessian $\mathbf{H}(\mathbf{T}^*)$ only once before starting the iteration.

An interpretation will be given in the following if additive noise is neglected. Imagine that the error criterion equals the negative cross-correlation of $I_m(\mathbf{x}, \widehat{\mathbf{T}})$ and $I_2(\mathbf{x})$. At the optimum when $I_m(\mathbf{x}, \widehat{\mathbf{T}})$ is identical to $I_2(\mathbf{x})$ the cross-correlation equals the autocorrelation of $I_2(\mathbf{x})$. Therefore the Hessian can be calculated at the beginning as the curvature of the autocorrelation instead of the cross-correlation at the unknown optimum. Thus the numerical complexity is quite low. Only the gradient vector $\mathbf{g}(\widehat{\mathbf{T}}_k)$ has to be updated within the iteration.

If these results are extended to the general transform $\mathbf{h}(\mathbf{x}, \mathbf{T})$ a few small changes have to be made. Especially a moving coordinate system $\{\mathbf{x}_k\}$ and an additional motion vector $\widetilde{\mathbf{T}}$ have to be introduced [Diehl 1987, Diehl 1988]. This vector connects the coordinate system $\{\mathbf{x}_k\}$ of iteration step k with the coordinate system $\{\mathbf{x}_{k+1}\}$ of iteration step $k+1$:

$$\begin{aligned} \mathbf{x}_k &= \mathbf{h}(\mathbf{x}, \widehat{\mathbf{T}}_k) \\ \mathbf{x}_{k+1} &= \mathbf{h}(\mathbf{x}_k, \widetilde{\mathbf{T}}_{k+1}) = \mathbf{h}(\mathbf{h}(\mathbf{x}, \widehat{\mathbf{T}}_k), \widetilde{\mathbf{T}}_{k+1}) = \mathbf{h}(\mathbf{x}, \widehat{\mathbf{T}}_{k+1})\,. \end{aligned} \tag{2.3}$$

Thus the coordinate system $\{\mathbf{x}_{k+1}\}$ results directly from $\{\mathbf{x}\}$ by the vector $\widehat{\mathbf{T}}_{k+1}$ or indirectly from $\{\mathbf{x}_k\}$ by $\widetilde{\mathbf{T}}_{k+1}$. This is only possible if the set of transforms $\mathbf{h}(\mathbf{x}, \mathbf{T})$ is closed.

Now the model image $I_m(\mathbf{x}, \widehat{\mathbf{T}}_{k+1})$ is generated indirectly :

$$I_m(\mathbf{x}, \widehat{\mathbf{T}}_{k+1}) = I_1(\mathbf{x}_{k+1}) = I_1(\mathbf{h}(\mathbf{h}(\mathbf{x}, \widehat{\mathbf{T}}_k), \widetilde{\mathbf{T}}_{k+1})) \tag{2.4}$$

and the error criterion has to be minimized with respect to $\widetilde{\mathbf{T}}$. Then the new iteration is given by

$$\widetilde{\mathbf{T}}_{k+1} = -\,\widetilde{\mathbf{H}}^{-1}\,\widetilde{\mathbf{g}}(\widehat{\mathbf{T}}_k) \tag{2.5}$$

for the innovation at step $k+1$ and

$$\widehat{\mathbf{T}}_{k+1} = \mathbf{f}\,(\widehat{\mathbf{T}}_k, \widetilde{\mathbf{T}}_{k+1}) \tag{2.6}$$

connecting $\widetilde{\mathbf{T}}_{k+1}$ and $\widehat{\mathbf{T}}_k$ to the whole motion vector $\widehat{\mathbf{T}}_{k+1}$.

The Hessian $\widetilde{\mathbf{H}}$ at the optimum is given by

$$\widetilde{\mathbf{H}} = E\left\{\left(\frac{\partial \mathbf{h}(\mathbf{x},\mathbf{T})}{\partial \mathbf{T}}\right)^T \frac{\partial I_1(\mathbf{x})}{\partial \mathbf{x}} \left(\frac{\partial I_1(\mathbf{x})}{\partial \mathbf{x}}\right)^T \frac{\partial \mathbf{h}(\mathbf{x},\mathbf{T})}{\partial \mathbf{T}}\right\}\Bigg|_{\mathbf{T}=\mathbf{0}} \tag{2.7}$$

if additive noise is neglected. This expression does not explicitely depend on the actual motion vector. Therefore the Hessian actually can be calculated before starting the iteration. The gradient vector is given by

$$\begin{aligned}\widetilde{\mathbf{g}}(\widehat{\mathbf{T}}_k) = E\Bigg\{&\left(I_m(\mathbf{x},\widehat{\mathbf{T}}_k) - I_2(\mathbf{x})\right)\\ &\times \left(\left(\frac{\partial I_2(\mathbf{x})}{\partial \mathbf{x}}\right)^T \left(\frac{\partial \mathbf{x}_k}{\partial \mathbf{x}}\right)^{-1} \left(\frac{\partial \mathbf{x}_{k+1}}{\partial \widetilde{\mathbf{T}}_{k+1}}\right)\Bigg|_{\widetilde{\mathbf{T}}_{k+1}=\mathbf{0}}\right)^T\Bigg\},\end{aligned} \tag{2.8}$$

which has to be updated within each iteration step.

To get a better idea of the algorithm the formulas for a planar rigid patch using central projection are given in detail. With

$$\frac{\partial \mathbf{x}'}{\partial \mathbf{T}} = \begin{pmatrix} x & y & 1 & 0 & 0 & 0 & -x^2 & -xy \\ 0 & 0 & 0 & x & y & 1 & -xy & -y^2 \end{pmatrix}, \tag{2.9}$$

and

$$\begin{aligned}\frac{\partial I_1(\mathbf{x}')}{\partial \mathbf{T}}\Bigg|_{\mathbf{T}=\mathbf{0}} = \Bigg(&\frac{\partial I_1(\mathbf{x})}{\partial x}x, \frac{\partial I_1(\mathbf{x})}{\partial x}y, \frac{\partial I_1(\mathbf{x})}{\partial x}, \frac{\partial I_1(\mathbf{x})}{\partial y}x,\\ &\frac{\partial I_1(\mathbf{x})}{\partial y}y, \frac{\partial I_1(\mathbf{x})}{\partial y}, -\frac{\partial I_1(\mathbf{x})}{\partial x}x^2 - \frac{\partial I_1(\mathbf{x})}{\partial y}xy, -\frac{\partial I_1(\mathbf{x})}{\partial x}xy - \frac{\partial I_1(\mathbf{x})}{\partial y}y^2\Bigg)^T\end{aligned} \tag{2.10}$$

the Hessian can be calculated according to

$$\widetilde{\mathbf{H}} = E\left\{\left(\frac{\partial I_1(\mathbf{x}')}{\partial \mathbf{T}}\right)\left(\frac{\partial I_1(\mathbf{x}')}{\partial \mathbf{T}}\right)^T\right\}\Bigg|_{\mathbf{T}=\mathbf{0}}. \tag{2.11}$$

The gradient vector may be written in the simplified form

$$\widetilde{\mathbf{g}}(\widehat{\mathbf{T}}_k) = \mathbf{B}_k^{MNR}\,\mathbf{z}(\widehat{\mathbf{T}}_k) \tag{2.12}$$

with the simple weighting matrix $\mathbf{B}^{MNR}$ which is independent of the image content and the coordinates. The vector $\mathbf{z}(\widehat{\mathbf{T}}_k)$ is given by

$$\mathbf{z}(\widehat{\mathbf{T}}_k) = E \left\{ \begin{array}{c} \left(I_m(\mathbf{x},\widehat{\mathbf{T}}) - I_2(\mathbf{x})\right) \frac{\partial I_2(\mathbf{x})}{\partial x} x^2 \\ \left(I_m(\mathbf{x},\widehat{\mathbf{T}}) - I_2(\mathbf{x})\right) \frac{\partial I_2(\mathbf{x})}{\partial x} xy \\ \left(I_m(\mathbf{x},\widehat{\mathbf{T}}) - I_2(\mathbf{x})\right) \frac{\partial I_2(\mathbf{x})}{\partial x} x \\ \left(I_m(\mathbf{x},\widehat{\mathbf{T}}) - I_2(\mathbf{x})\right) \frac{\partial I_2(\mathbf{x})}{\partial x} y \\ \left(I_m(\mathbf{x},\widehat{\mathbf{T}}) - I_2(\mathbf{x})\right) \frac{\partial I_2(\mathbf{x})}{\partial x} \\ \left(I_m(\mathbf{x},\widehat{\mathbf{T}}) - I_2(\mathbf{x})\right) \frac{\partial I_1(\mathbf{x})}{\partial y} x^2 \\ \left(I_m(\mathbf{x},\widehat{\mathbf{T}}) - I_2(\mathbf{x})\right) \frac{\partial I_1(\mathbf{x})}{\partial y} xy \\ \left(I_m(\mathbf{x},\widehat{\mathbf{T}}) - I_2(\mathbf{x})\right) \frac{\partial I_1(\mathbf{x})}{\partial y} x \\ \left(I_m(\mathbf{x},\widehat{\mathbf{T}}) - I_2(\mathbf{x})\right) \frac{\partial I_1(\mathbf{x})}{\partial y} y \\ \left(I_m(\mathbf{x},\widehat{\mathbf{T}}) - I_2(\mathbf{x})\right) \frac{\partial I_1(\mathbf{x})}{\partial y} \end{array} \right\} . \tag{2.13}$$

To compute this vector the image difference between $I_m(\mathbf{x},\widehat{\mathbf{T}})$ and $I_2(\mathbf{x})$ has to be built and has to be multiplied by the weighted partial derivatives. Further details ar given in [Diehl 1988].

The transform

$$\widehat{\mathbf{T}}_{k+1} = \mathbf{f}\,(\widehat{\mathbf{T}}_k, \widetilde{\mathbf{T}}_{k+1}), \tag{2.14}$$

which connects $\widehat{\mathbf{T}}_k$ and $\widetilde{\mathbf{T}}_{k+1}$ is given by

$$\begin{pmatrix} \widehat{a}_1 & \widehat{a}_2 & \widehat{a}_3 \\ \widehat{a}_4 & \widehat{a}_5 & \widehat{a}_6 \end{pmatrix}_{k+1} = \begin{pmatrix} \widetilde{a}_1 & \widetilde{a}_2 & \widetilde{a}_3 \\ \widetilde{a}_4 & \widetilde{a}_5 & \widetilde{a}_6 \end{pmatrix}_{k+1} \begin{pmatrix} \widehat{a}_1 & \widehat{a}_2 & \widehat{a}_3 \\ \widehat{a}_4 & \widehat{a}_5 & \widehat{a}_6 \\ \widehat{a}_7 & \widehat{a}_8 & 1 \end{pmatrix}_k \tag{2.15}$$

and

$$(\widehat{a}_7,\, \widehat{a}_8)_{k+1} = (\widetilde{a}_7,\, \widetilde{a}_8,\, 1)_{k+1} \begin{pmatrix} \widehat{a}_1 & \widehat{a}_2 \\ \widehat{a}_4 & \widehat{a}_5 \\ \widehat{a}_7 & \widehat{a}_8 \end{pmatrix}_k . \tag{2.16}$$

Thus using these formulas and

$$\widetilde{\mathbf{T}}_{k+1} = -\,\widetilde{\mathbf{H}}^{-1}\,\widetilde{\mathbf{g}}(\widehat{\mathbf{T}}_k) \tag{2.17}$$

the modified Newton-Raphson algorithm is described completely.

The advantages of the modified Newton-Raphson algorithm are manifold. Because the Hessian is always used at the optimum the advantages of the gradient algorithm and those of the normal Newton algorithm are combined. The modified Newton-Raphson algorithm has the large region of stability like the gradient algorithm. Like the normal Newton algorithm it has good signal adaptive properties. The Hessian which can be interpreted as the curvature of the autocorrelation function of the images adjusts to the image bandwidth. Therefore the convergence rate of the

Euclidean error norm $||\widehat{\mathbf{T}}_k - \mathbf{T}^*||$ is at least of second order, independently of the chosen signals. Another important advantage is the low numerical complexity because of the ability to calculate the Hessian before starting the iteration. Therefore only the gradient vector has to be updated.

Nevertheless there are a few problems. If there is significant noise the Hessian calculated according to eq. 2.7 does not exactly represent the second derivatives of the error function. Therefore the convergence rate may be worse. Furthermore it is sometimes disadvantageous that the set of transforms $\mathbf{h}(\mathbf{x}, \mathbf{T})$ has to be closed. Quasi-Newton algorithms and especially their combination with the modified Newton-Raphson algorithm can succesfully deal with these problems.

2.2.2. Quasi-Newton methods

Quasi-Newton methods [Gill et al. 1981] try to build up the Hessian at the actual iteration point $\widehat{\mathbf{T}}_k$ only using values of the gradient vector which have already been calculated. Thus the curvature of the error criterion is approximated only by first derivatives. The iteration scheme is given by

$$\widehat{\mathbf{T}}_{k+1} = \widehat{\mathbf{T}}_k - \mathbf{C}_k^{-1}\, \mathbf{g}(\widehat{\mathbf{T}}_k)\ . \tag{2.18}$$

The Quasi-Newton matrix $\mathbf{C}_k$ converges to the true Hessian at the optimum if for example the BFGS-update [Gill et al. 1981] is used:

$$\mathbf{C}_{k+1} = \mathbf{C}_k + \frac{1}{\mathbf{g}(\widehat{\mathbf{T}}_k)^T \mathbf{s}_k} \mathbf{g}(\widehat{\mathbf{T}}_k)\mathbf{g}(\widehat{\mathbf{T}}_k)^T + \frac{1}{\mathbf{y}_k^T \mathbf{s}_k} \mathbf{y}_k \mathbf{y}_k^T\ . \tag{2.19}$$

$\mathbf{s}_k = \widehat{\mathbf{T}}_{k+1} - \widehat{\mathbf{T}}_k$ describes the difference of two succesive estimates and $\mathbf{y}_k = \mathbf{g}(\widehat{\mathbf{T}}_{k+1}) - \mathbf{g}(\widehat{\mathbf{T}}_k)$ the change in the gradient. The gradient vector is given by

$$\mathbf{g}(\widehat{\mathbf{T}}_k) = E\left\{ \left(I_m(\mathbf{x}, \widehat{\mathbf{T}}) - I_2(\mathbf{x})\right) \left(\left(\frac{\partial I_2(\mathbf{x})}{\partial \mathbf{x}}\right)^T \left(\frac{\partial \mathbf{x}'}{\partial \mathbf{x}}\right)^{-1} \left(\frac{\partial \mathbf{x}'}{\partial \widehat{\mathbf{T}}}\right)\right)^T \right\}\Bigg|_{\widehat{\mathbf{T}} = \widehat{\mathbf{T}}_k} \tag{2.20}$$

which again can be simplified for the case of central projection to

$$\mathbf{g}(\widehat{\mathbf{T}}_k) = \mathbf{B}_k^Q\, \mathbf{z}(\widehat{\mathbf{T}}_k) \tag{2.21}$$

with $\mathbf{z}(\widehat{\mathbf{T}}_k)$ of equation 2.13. The weighting matrix $\mathbf{B}^Q$ is similar to B^{MNR}. It is important to note that only the gradient vector has to be calculated whithin each iteration step. Therefore the numerical complexity - characterized by the number of operations on each pixel - is as low as for the gradient or modified Newton-Raphson algorithm. Because the Hessian is approximated the convergence rate only is superlinear. Furthermore the region of stability is quite low.

2.2.3. A combined algorithm

A very efficient algorithm can be realized if the modified Newton-Raphson and the Quasi-Newton algorithm are combined. The iteration is given by

$$\widehat{\mathbf{T}}_{k+1} = \widehat{\mathbf{T}}_k - \mathbf{M}_k^{-1}\, \mathbf{g}(\widehat{\mathbf{T}}_k)\ . \tag{2.22}$$

Here a matrix $\mathbf{M}$ is used as a combination of the Hessian $\widetilde{\mathbf{H}}$ at the optimum and a Quasi-Newton matrix $\mathbf{M}^{BFGS}$

$$\mathbf{M}_{k+1} = \frac{1}{2}\left(\widetilde{\mathbf{H}} + \mathbf{M}_{k+1}^{BFGS}\right). \tag{2.23}$$

The matrix $\mathbf{M}_{k+1}^{BFGS}$ results from $\mathbf{M}_k$ of the preceeding iteration step k by the BFGS-Quasi-Newton update. If the estimate $\widehat{\mathbf{T}}$ is far away from the true value $\mathbf{T}^*$ and therefore the Quasi-Newton method would diverge $\mathbf{M}_k^{BFGS} = \mathbf{0}$ is used. Thus the algorithm keeps stable and the innovation is twice the innovation of the modified Newton-Raphson algorithm, which improves the far away convergency. Near the optimum $\mathbf{M}_{k+1}^{BFGS}$ converges to the Hessian $\widetilde{\mathbf{H}}$ of the optimum such that the combined algorithm turns into the modified Newton-Raphson algorithm. An additional improvement can be reached if at the beginning only the translation parameters a_3 and a_6 are estimated under the assumption that the other parameters are zero. More deatails are given in [Diehl 1988].

Fig. 2.2: Picture of a woman digitized by 512×512 pixel

2.3. Tests using real image data

The algorithm has been tested with real image data. Therefore different images $I_1(\mathbf{x})$ have been transformed in the computer by a definite eight parametric motion vector $\mathbf{T} = \mathbf{a}$ to generate $I_2(\mathbf{x})$. Using two corresponding images $I_1(\mathbf{x})$ and $I_2(\mathbf{x})$ the motion vector $\mathbf{T}$ was estimated. Fig. 2.2, 2.3 and 2.4 show a typical experiment. The motion vector connecting $I_1(\mathbf{x})$ and $I_2(\mathbf{x})$ was $\mathbf{T} = (0.166, -1736, -5.0, 0.2409, -0.0152, -7.0, -0.00577, 0.0)^T$. This describes a rotation of 10° around the camera main axis, an inclination of 30° around the y-axis and a translation assuming that the object is very close to the camera. Fig 2.4 illustrates how a_1, a_2, a_3 and a_7 of the estimation vector $\widehat{\mathbf{T}}$ converge to the the true vector $\mathbf{T}$ using the combined algorithm. Further results are given in [Diehl 1988].

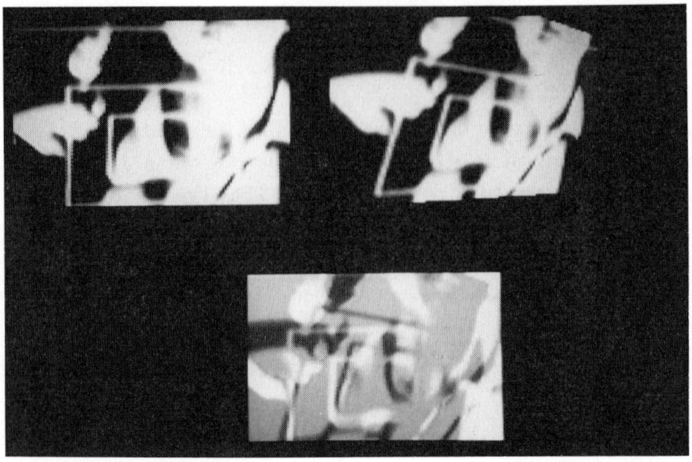

Fig. 2.3: Picture of a woman with $I_1(\mathbf{x})$, $I_2(\mathbf{x})$ and the difference image before starting the iteration

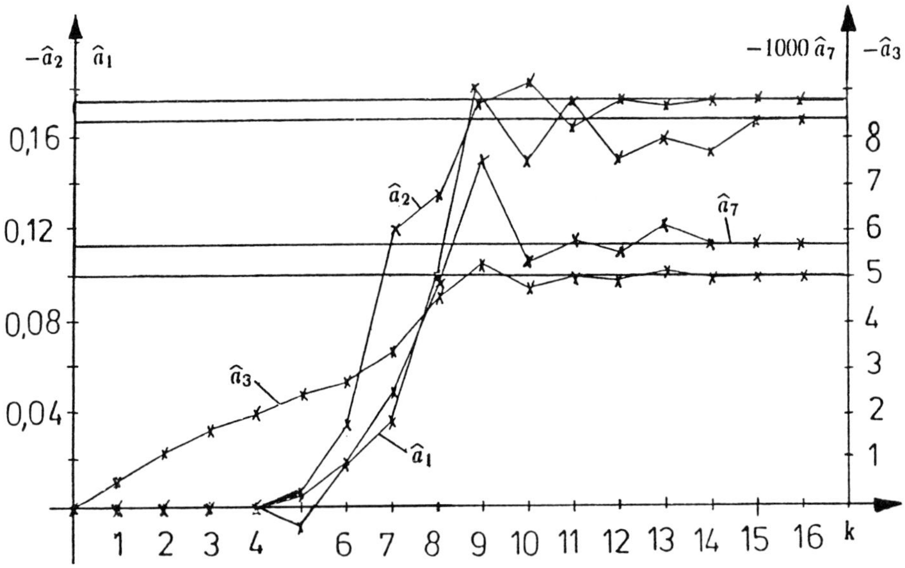

Fig. 2.4: The estimation values $\hat{a}_1$, $\hat{a}_2$, $\hat{a}_3$ and $\hat{a}_7$ using the combined method

3. The maximum-likelihood estimation of the motion parameters

3.1. The maximum-likelihood estimation as generalized correlation

So far the ad hoc error criterion $J\{e(\widehat{\mathbf{T}})\}$ of eq. 2.1, i.e. the variance of the model error was minimized. In this part the results of the maximum-likelihood estimation of the motion parameters will be given. To deal with this problem a few simplifying assumptions have to be made. The image $S(\mathbf{x})$ and the additive noise terms $N_1(\mathbf{x})$ and $N_2(\mathbf{x})$ are assumed to be stationary, zero mean gaussian processes which are not correlated to each other. Furthermore the coordinate transform $\mathbf{h}(\mathbf{x}, \mathbf{T})$ may be a linear transform, e.g. the affine transform $\mathbf{x}' = \mathbf{C}\,\mathbf{x} + \mathbf{d}$. Under these assumptions the likelihood-function $p(I_1(\mathbf{x}), I_2(\mathbf{x})|\mathbf{T})$ may be calculated. The value $\mathbf{T}$ which maximizes this function is called the maximum-likelihood estimate $\widehat{\mathbf{T}}_{ML}$.

It can be shown that the maximum-likelihood estimation can be interpreted as a generalized correlation near the optimum [Diehl 1988]. Similar results are known from time delay estimation [Knapp and Carter 1976] or pure translation [Beyer 1985]. Thus the estimate $\widehat{\mathbf{T}}_{ML}$ is the value which maximizes the cross-correlation function of the two images $\widehat{I}_1(\mathbf{x})$ and $\widehat{I}_2(\mathbf{x})$ which result from the original images $\widehat{I}_1(\mathbf{x})$ and $\widehat{I}_2(\mathbf{x})$ by linear filtering using the impulse response $w(\mathbf{x})$ calculated from $|W_{ML}(\mathbf{f})|^2$:

$$\widehat{I}_1(\mathbf{x}) = w_{ML}(\mathbf{x}) \otimes I_1(\mathbf{x}), \quad \widehat{I}_2(\mathbf{x}) = w_{ML}(\mathbf{x}) \otimes I_2(\mathbf{x}). \tag{3.1}$$

$\otimes$ denotes convolution. The prefilter is given by

$$|W_{ML}(\mathbf{f})|^2 = \frac{G_{S'S'}(\mathbf{f})}{G_{S'S'}(\mathbf{f})\left(G_{N_1'N_1'}(\mathbf{f}) + G_{N_2N_2}(\mathbf{f})\right) + G_{N_1'N_1'}(\mathbf{f})\, G_{N_2N_2}(\mathbf{f})} \tag{3.2}$$

$G_{S'S'}(\mathbf{f})$ is the power-spectral density of the image content $S'(\mathbf{x})$ of image $I_2(\mathbf{x})$ and $G_{N_2N_2}(\mathbf{f})$ the power-spectral density of the noise $N_2(\mathbf{x})$. $G_{N_1'N_1'}(\mathbf{f})$ denotes the power-spectral density of $N_1(\mathbf{x})$ which has been moved in direction of $I_2(\mathbf{x})$ under the assumption that the matrix $\mathbf{C}$ of the affine transform may be approximately known. If $N_1(\mathbf{x})$ and $N_2(\mathbf{x})$ have the same statistics $G_{N_1'N_1'}$ may be replaced by $G_{N_2N_2}$. Fig. 3.1 shows the structure of the generalized correlation.

The interpretation of the maximum-likelihood estimation as generalized correlation has an interesting impact. Because the error criterion of eq. 2.1 may be interpreted as a cross-correlation the algorithms developed in chapter 2 can be used directly for the maximum-likelihood estimation. Only the original images $I_1(\mathbf{x})$ und $I_2(\mathbf{x})$ have to be replaced by the prefiltered images $\widehat{I}_1(\mathbf{x})$ and $\widehat{I}_2(\mathbf{x})$.

3.2. Properties of the maximum-likelihood estimation

The maximum-likelihood estimate has some interesting properties. For high signal-to-noise ratios

$$G_{SS}(\mathbf{f}) \gg G_{NN}(\mathbf{f})$$

$|W_{ML}(\mathbf{f})|^2$ is given by

$$|W_{ML}(\mathbf{f})|^2 \approx \frac{1}{G_{N_1'N_1'}(\mathbf{f}) + G_{N_2N_2}(\mathbf{f})}\,. \tag{3.3}$$

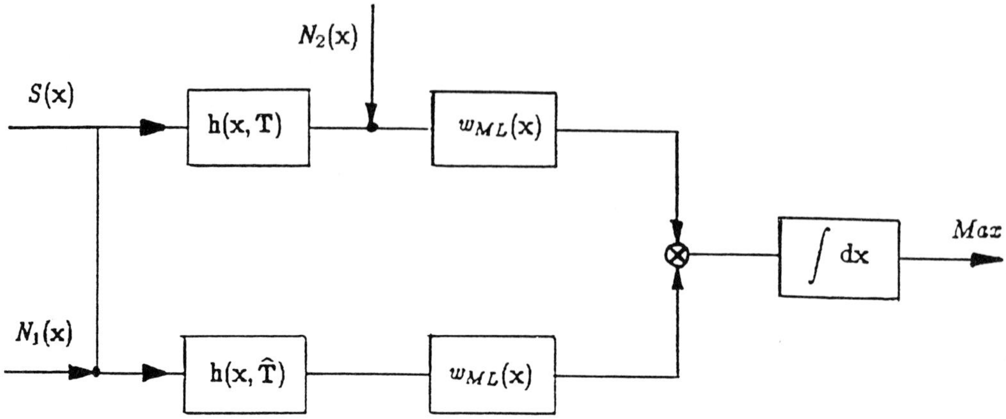

Abb. 3.1: The maximum-likelihood estimation as generalized correlation

If white noise, i.e.

$$G_{N_1'N_1'}(\mathbf{f}) = \sigma^2_{N_1} \qquad \text{and} \qquad G_{N_2N_2}(\mathbf{f}) = \sigma^2_{N_2} \tag{3.4}$$

is assumed, the prefilter is constant:

$$|W_{ML}(\mathbf{f})|^2 \approx \frac{1}{\sigma^2_{N_2} + \sigma^2_{N_1}} = const. \tag{3.5}$$

For this special case the generalized correlation is equal to the correlation of the original images. Thus the ad hoc error criterion of eq. 2.1 equals the maximum-likelihood criterion if stationary signals, white noise and a high signal-to-noise ratio are assumed.

Furthermore the maximum-likelihood estimate reaches the Cramer-Rao bound if the noise terms and the signal are not cross-correlated [van Trees 1976, Diehl 1988]. The Cramer-Rao bound which is the bound for the lowest variance of the estimation values can be expressed by the Hessian of the prefilterd signals:

$$var\{\mathbf{T}^*\} = \frac{1}{XY}\, E\{\widehat{\mathbf{H}}\}^{-1}\,. \tag{3.6}$$

3.3. Tests of the maximum-likelihood estimation

The maximum-likelihood estimation has been tested using real image data and different additive noise. Details are given in [Diehl 1988]. Fig. 3.2 and 3.3 illustrate the standard deviation of a translation in x-direction. If the signal-to-noise ratio is low and the noise is highly correlated a

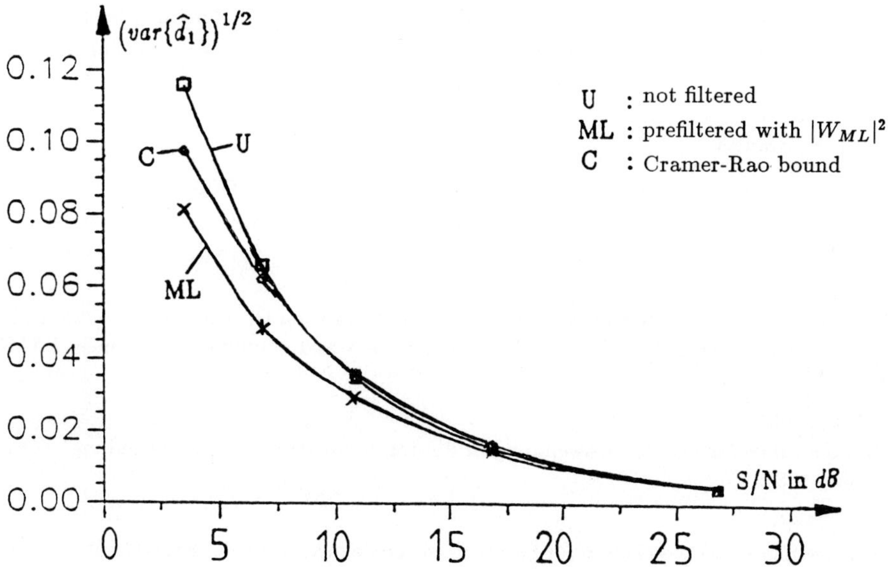

Abb. 3.2: Standard deviation of $\hat{d}_1$ using white noise

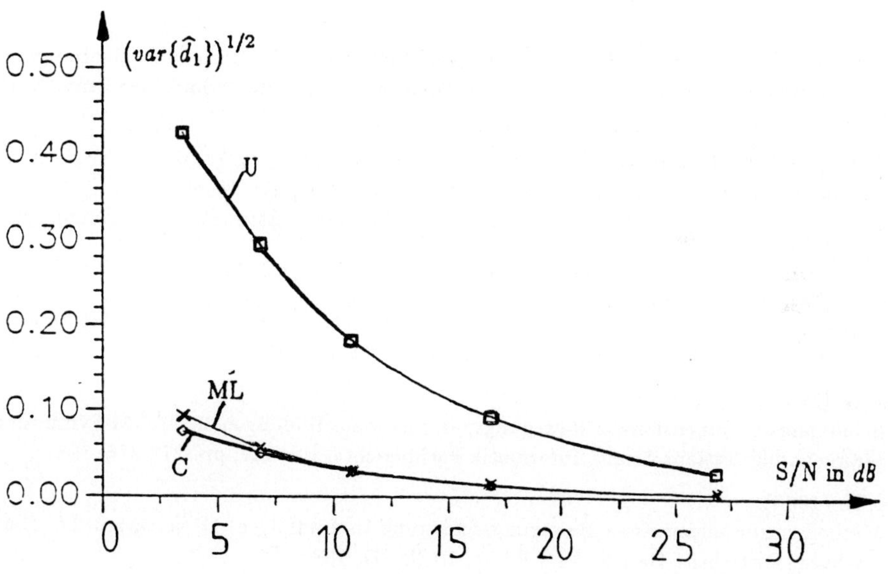

Abb. 3.3: Standard deviation of $\hat{d}_1$ using stronly correlated noise

significant reduction of the variance can be reached. For white noise and high signal-to-noise ratios there is according to eq. 3.5 no difference between the variance using the original or prefiltered images.

Acknowledgements

This work was supportet by the Deutsche Forschungsgemeinschaft (DFG).

References

Adiv, G. (1985):
Determining Three-Dimensional Motion and Structure from Optical Flow Generated by Several Moving Objects. IEEE Trans. Pattern Analysis and Machine Intelligence, PAMI-7, 4, pp. 384-401, 1985.

Bähr, H.-P. (1985):
Digitale Bildverarbeitung: Anwendung in Photogrammetrie und Fernerkundung. Wichmann Verlag, 1985.

Beyer, S. (1985):
Displacement-Schätzverfahren für Fernsehbildsignale mit minimaler Schätzfehlervarianz. Diss. Univ. Hannover, VDI-Fortschrittbericht: Reihe 10; Nr. 51, 1985.

Burkhardt, H.; Diehl, N. (1986):
Simultaneous Estimation of Rotation and Translation in Image Sequences. Proc. of the European Signal Processing Conference, EUSIPCO-86, Den Haag, pp. 821-824, 1986.

Burkhardt, H; Moll, H. (1978):
Ein modifiziertes Newton-Raphson-Verfahren zur modelladaptiven Identifikation von Laufzeiten. Mitteilungen aus dem Institut für Meß- und Regelungstechnik der Univ. Karlsruhe, 1978.

Burkhardt, H; Moll, H. (1979):
A modified Newton-Raphson-Search for the Model-Adaptive Identifications of Delays. In: Isermann, R. (ed.): Identification and System Parameter Estimation. Pergamon, pp. 1279-1286, 1979.

Diehl, N.; Burkhardt, H. (1986):
Planar Motion Estimation with a Fast Converging Algorithm. Proc. of the 8th Intern. Conf. on Pattern Recognition, 8-ICPR, Paris, pp. 1099-1102, 1986.

Diehl, N. (1987):
Schätzung dreidimensionaler Bewegungsparameter aus Bildfolgen. 9. DAGM-Symposium Mustererkennung, Braunschweig, Informatik Fachberichte, Bd. 149, pp. 272-276, 1987.

Diehl, N. (1988):
Methoden zur allgemeinen Bewegungsschätzung in Bildfolgen. Dissertation TU Hamburg-Harburg, Fortschritt-Bericht VDI Reihe, 10 Nr. 92, 1988.

Fang, J.-R.; Huang, T.S. (1984):
Solving Three-Dimensional Motion Equations: Uniqueness, Algorithms, and Numerical Results. Computer Vision, Graphics, Image Processing, 26, pp. 183-206, 1984.

Gill, P.E.; Murray, W.; Wright, M.H. (1981):
Practical Optimization. Academic Press, 1981.

Huang, T.S. (ed.) (1981):
Image Sequence Analysis. Springer, 1981

Huang, T.S. (ed.) (1983):
Image Sequence Processing and Dynamic Scene Analysis. Springer, 1983.

Kanatani, K.-I. (1986):
Structure and Motion from Optical Flow under Orthographic Projection. Computer Vision, Graphics, and Image Processing, 35, pp. 181-199, 1986.

Knapp, C.H.; Carter, G.C. (1976):
The Generalized Correlation Method for Estimation of Time Delay. IEEE Trans. Acoustics, Speech, and Signal Processing, ASSP-24, 4, pp. 320-327, 1976.

Lenz, R. (1986):
Ein Verfahren zur Schätzung der Parameter geometrischer Bildtransformationen. Diss. Techn. Univ. München, 1986.

Longuet-Higgins, H.C. (1981):
A Computer Algorithm for Reconstructing a Scene from Projections. Nature, 293, pp. 133-135, 1981.

Mesch, F. (1982):
Geschwindigkeits- und Durchflußmessung mit Korrelationsverfahren. Regelungstechnische Praxis 24, Heft 3, pp. 73-82, 1982.

Musmann, H.G.; Pirsch, P.; Grallert, H.J. (1985):
Advances in Picture Coding. Proceedings of the IEEE, 73, 4, pp. 523-548, 1985.

Nagel, H.H. (1981):
Image Sequence Analysis: What can we learn from Applications? In: Huang, T.S. (ed.): Image Sequence Analysis. Springer, pp. 19-228, 1981.

Nagel, H.H. (1983):
Overview on Image Sequence Analysis. In: Huang, T.S. (ed.): Image Sequence Processing and Dynamic Scene Analysis. Springer, pp. 2-39, 1983.

Nagel, H.H. (1985):
Analyse und Interpretation von Bildfolgen. Informatik Spektrum 8, pp. 178-200 and pp. 312-327, 1985.

Nagel, H.H. (1986):
Image Sequences - Ten (octal) Years - From Phenomenology towards a Theoretical Foundation. Proc. of the 8th Intern. Conf. on Pattern Recognition, pp. 1099-1102, 1986.

Netravali, A.N.; Salz, J. (1985):
Algorithms for Estimation of Three-Dimensional Motion. AT & T Technical Journal, 64, 2, pp. 335-346, 1985.

Prazdny, K. (1980):
Egomotion and Relative Depth Map from Optical Flow. Biological Cybernetics, pp. 87-102, 1980.

Roach, J.W.; Aggarwal, J.K. (1980):
Determining the Movements of Objects from a Sequence of Images. IEEE. Trans. Pattern Analysis and Machine Intelligence, PAMI-2, 6, pp. 554-562, 1980.

Spoer, P. (1987):
Schätzung der 3-dimensionalen Bewegungsvorgänge starrer, ebener Objekte in digitalen Fernsehbildfolgen mit Hilfe von Bewegungsparametern. Diss. Univ. Hannover, 1987.

Tsai, R.Y.; Huang, T.S. (1981):
Estimating Three-Dimensional Motion Parameters of a Rigid Planar Patch. IEEE Trans. Acoustics, Speech, and Signal Processing, ASSP-29, 6, pp. 1147-1152, 1981.

Tsai, R.Y.; Huang, T.S. (1982):
Estimating Three-Dimensional Motion Parameters of a Rigid Planar Patch, II: Singular Value Decomposition. IEEE Trans. Acoustics, Speech, and Signal Processing, ASSP-30, 4, pp. 525-534, 1982.

Tsai, R.Y.; Huang, T.S. (1984a):
Uniqueness and Estimation of Three-Dimensional Motion Parameters of Rigid Objects with Curved Surfaces. IEEE Trans. Pattern Analysis and Machine Intelligence, PAMI-6, 1, pp. 13-27, 1984.

Tsai, R.Y.; Huang, T.S. (1984b):
Estimating Three-Dimensional Motion Parameters of a Rigid Planar Patch, III: Finite Point Correspondence and the Three-View-Problem. IEEE Trans. Acoustics, Speech, and Signal Processing, ASSP-32, 2, pp. 213-220, 1984.

van Trees, H.L. (1968):
Detection, Estimation, and Modulation Theory - Part I: Detection, Estimation and Linear Modulation Theory. Wiley, 1968.

Yen, B.L.; Huang, T.S. (1983):
Determining 3-D Motion and Structure of a Rigid Body Using Straight Line Correspondence. In: Huang, T.S. (ed.): Image Sequence Processing and Dynamic Scene Analysis. Springer, pp. 365-394, 1983.

Precision of Geometric Features Derived from Image Sequences

Wolfgang Förstner

Institut für Photogrammetrie, Universität Stuttgart

Summary:

The paper discusses the accuracy potential of mono and stereo image sequences. Specifically treated are the effect of image blur onto the precision of image features, the precision obtainable for relative position, orientation and speed of the sensor platform with respect to a given coordinate frame or to other moving objects. The results can be used for designing a multisensor navigation system.

0. Introduction

Sensing and analysing changing environment is a basic requirement in navigation and is the reason for the development of vision based navigation systems. Here vision sensors are understood to be all types of sensors which produce a one, two or three dimensional geometric projection or image of the world. In contrast to "zero"dimensional sensors, which only determine the position or orientation at single points, possibly over time, they are not blind to changes in the environment, do not necessarily need a map and, in case they use maps, do not require them to accurately represent reality during the movement of the vehicle. These restrictions of sensors used e. g. in the Global Positioning System (GPS) or in Inertial Navigation Systems (INS) have their counterpart in the higher efficiency of these sensors for determining the orientation which in general is a magnitude higher than the efficiency of orientation determination of vision-type sensors, both with respect to accuracy and speed.

On the other hand vision-type sensors derive geometric properties of the observed scene out of the images in any case, either as a prerequitite for scene interpretation, e. g. to determine free space, or as a byproduct, e. g. for monitoring the own trajectory and/or the movement of other vehicles as a basis for the planning of collision free paths.

This paper wants to discuss which basic geometric properties can be derived from sequences of optical images, taken e. g. with a CCD-camera and what accuracies are obtainable solely from image sequences with the aim to get information what type of other sensors may be necessary to support vision based systems for high precision navigation. Specifically we want to discuss the effect of image blur onto the precision of image features, the precision obtainable for relative position, orientation and speed of the sensor platform with respect to a given coordinate frame or to other moving objects. Though the derivations are basic and general, the questions and examples are taken from the environment of a road vehicle.

1. On the Geometric Precision of Image Features

The derivation of the own trajectory or of the geometry of the observed scene has to be based on elementary geometric features which have to be found in a sequence of images. The precision of the derived geometric quantities in 3D depends on the accuracy with which the image features can be extracted and on the geometric relations between the objects and the trajectory. In this section we only deal with the precision of the extracted image features.

As we are interested in the potential of image sequence analysis for reconstruction 3D geometry we have to ask for the geometric precision inherent in the digital images. This precision is influenced by various factors, e. g. by

- the quality of the interior orientation of the camera, including principle point, focal length and lens distortion,
- the geometric quality of the sensor, mainly depending on the electronic properties of the camera and the frame grabber,
- the properties of the resolution of the sensor in terms of a modulation transfer function of simply the pixel distance,
- the electronic noise,
- the image blur induced by the non-zero integration time,

to name the main error sources.

Most of these effects can be reduced by proper calibration procedures and thus taken into account in the anlysis, especially the geometric properties of the camera. The resolution of images taken from moving sensors mainly depend on the image blur, which in this type of application because of the varying distance of the object points to the sensor is inhomogeneous and anisotropic.

The following analysis therefore neglects all effects except for the image content, including texture, blurr and noise. It is closely related to the theoretical investigations on the accuracy of image matching (cf. FÖRSTNER 1982, 1984). This is because selecting optimal points for image matching using cross correlation is identical to selecting optimal feature points, like corners or centres of circular features, for image analysis (cf. FÖRSTNER 1986, FÖRSTNER/GÜLCH 1987).

1.1 The geometric precision of image features

The precision of locating a known object, or determining a feature point in a digital image is given by:

$$\sigma_x = \frac{1}{\sqrt{n}} \cdot \frac{\sigma_n}{\sigma_{f_x}} \tag{1}$$

where n is the number of pixels involved

σ_n is the noise standard deviation and

σ_{f_x} is the standard deviation of the first derivative of the given or the true signal $f(x)$.

There are other versions of the same relation, which give rise to various interpretations and which are useful in our context. With the signal to noise ration SNR

$$\mathrm{SNR} = \sigma_f \ / \ \sigma_n \ , \tag{2}$$

the correlation coefficient ϱ between the measured and the given signal f and with the effective bandwidth b_x of the signal with power spectrum $P_f(u)$

$$b_x^2 = \int u^2 \ P_f(u) \ du \quad / \quad \int P_f(u) \ du \tag{3}$$

we obtain the relations:

$$\sigma_x = \frac{1}{\sqrt{n}} \cdot \frac{1}{\mathrm{SNR}} \cdot \frac{\sigma_f}{\sigma_{f_x}} = \frac{1}{\sqrt{n}} \cdot \sqrt{\frac{1-\varrho}{\varrho}} \cdot \frac{\sigma_f}{\sigma_{f_x}} \tag{1a,b}$$

$$\sigma_x = \frac{1}{\sqrt{n}} \cdot \frac{1}{\mathrm{SNR}} \cdot \frac{1}{2\pi b_x} = \frac{\sigma_n}{\sum_1^n f_{x_i}^2} \tag{1c,d}$$

If two images are matched the theoretical precision for the parallax can be determined from

$$\sigma_p = \sqrt{2} \cdot \sigma_x \quad . \qquad (4)$$

Examples:

A few examples want to demonstrate the usefulness of these relations, specifically how they compare to empirical findings.

a. Phase Measurements using narrow banded signals with average wavelength $l = 1 / b_x$:
   For GPS measurements the SNR of the recovered carrier is appr. 16 dB. This leads to

   $N = 1$, $SNR = 40$, $\Rightarrow \sigma_x = 1 / (40 * 2 * \pi) \approx 0.004\ l$

   and with a wave length of $l = 20$ cm we obtain a theoretical precision of 0.7 mm. This has to be compared with empirical findings for the accuracies of short distances in the range of a few mm.

b. Area based image matching using cross correlation or least squares matching:
   The power spectrum in aerial images can be approximated by $P(u) \cdot \exp(-a|u|)$. In good images $a = 0.2$ mm (cf. HELAVA 1976) which leads to a bandwidth of $b = \sqrt{2}/a = 7\ mm^{-1}$. Using windows of 12 x 12 pixels and assuming SNR = 5 this leads to a theoretical precision of the parallaxes of $\sigma_p = 0.54$ μmm which for a pixel size of dx = 20 μm is equivalent to 1/40 dx. The empirical findings range from 1/10 to 1/50 dx (cf. PERTL 1984, VOSSELMAN/FÖRSTNER 1988).

c. Extraction of feature points with an interest operator (cf. PADERES et al. 1984, FÖRSTNER/GÜLCH 1986):
   The extraction of feature points is usually based on small windows containing e. g. only 2 x 4 values. Under the same conditions as in the previous example (SNR = 5, $b = 7\ mm^{-1}$) this leads to a positional accuracy of appr. $\sigma_x = 1.6$ μm. It has to be compared to the empirical precision of 3 μm. (from HAHN/ FÖRSTNER 1988, rms differences of two independent height measurements 0.7 ‰ of the flying height).

d. Matching of two Digital Elevation Models (DEM):
   EBNER and MÜLLER (1986) report on using a DEM as vertical and horizontal control for aerial triangulation with three line scanner imagery. In their example the precision of the height is given by appr. 0.10 m. The standard deviation of the slopes of the hilly terrain can be approximated by 10 %. They use 100 control points which would lead to a theoretical precision of the horizontal position of the DEM of $\sigma_p = 0.14$ m (neglecting the geometric complex geometric relations). This has to be compared with their empirical finding of $\sigma_p = 0.20$ m.

Obviously the theoretical accuracies indicate the potential of the methods which could be reached if all other effects can be taken care of by proper calibration. ▪

1.2 The effect of image blur onto the geometric precision

We will now investigate the effect of image blur onto the precision of feature extraction using eq. (1c).

Let the unblurred and noiseless (onedimensional) image $f(x)$ be given with power spectrum $P_f(u) = P_f(0) \cdot \exp(-a|u|)$. The bandwidth of this image is (cf. 1.1b) $b = \sqrt{2}/a$. Blurring of $f(x)$ can be approximated by convolution with a box filter $h(x)$ with a width being the integration path δx. In order to derive the precision of feature extraction analytically in dependency of δx, we approximate $h(x)$ by a Gaussian with the same standard deviation $\sigma_h = \delta x/\sqrt{12}$. Its Fourier spectrum $H(u)$ also is a Gaussian with standard deviation $\sigma_H = 1/(2\pi\sigma_h) = \sqrt{3}/(\pi \cdot \delta x)$ (cf. CASTLEMAN 1979), which again can be approximated by an ideal low pass filter with upper frequency $u_o = 3/(\pi \cdot \delta x)$. We therefore assume that blurring with path δx can be approximated by an ideal low pass filter $h(x)$ with this upper frequency. The blurred image

$$g(x) = h(x) * f(x) \tag{5}$$

then has the effective bandwidth

$$b_g^2 = \int_0^{u_o} u^2 \, P_f(u) \, du \Big/ \int_0^{u_o} P_f(u) \, du \tag{6}$$

$$= \frac{1}{a^2} \; \frac{2 - (z^2 + 2z^2 + 2)\, e^{-z}}{1 - e^{-z}} \tag{7}$$

with $z = a\, u_o$, which for large values $\delta x \gg a$ leads to

$$b_g^2 = \frac{3}{\pi^2 \, \delta x^2} \; . \tag{8}$$

With eq. (1c) we finally obtain the precision of feature location under (strong enough) image blur δx

$$\boxed{\sigma_x(\delta x) = \frac{1}{\sqrt{12\, n}} \cdot \frac{1}{SNR} \cdot \delta x} \tag{9}$$

<u>Discussion:</u>

a. The precision is independent of the texture of the image measured by the parameter a. This is reasonable as only the low frequencies of the spectrum are relevant for the location after image blur.

b. The precision is proportional to the blurring path δx, which is to be expected.

c. The precision increases with the squareroot of the number of pixels. As in all versions of eq. 1 n is the number of pixels really contributing to the location, i. e. only those with gradient not zero.

d. For SNR = 1 and interpreting δx as pixelsize we obtain the theoretical precision of the location of a straight line of height n in a noiseless binary image (cf. FÖRSTNER 1986). It indicates that the standard deviation of the noise at an ideal binary edge can be approximated by the standard deviation of the signal. The derivation for this theretical result is based on the spirograph theory of DORST and DUIN (1984). This coincidence of relations derived on totally different theories is remarkable as such.

Example:
Let the pixel size be dx, the blur be δx = 10 dx, N = 10, and SNR = 3. Then the image feature (say an edge) in spite of image blur can be located with a precision of 1/3 dx. This may be e. g. achieved by local approximation of the signal by a polynomial of low degree (say 3) and determining the inflection point.▪

We are now ready to model the accuracy of feature location in blurred images. We express it as a standard deviation of the direction Φ from the sensor to the point in object space in dependency on

- the pixel size dx
- the focal length f
- the relative velocity v of the object with respect to the sensor
- the effective integration time t_I
- the distance s from the sensor to the object
- the angle α betwen the viewing direction and the velocity vector
- the assumed precision $k_1 \cdot dx$ of feature detection in the unblurred image. k_1 can be derived from eq. (1) and the pixel size: $k_1 = \sigma_x/dx$ and will lie in the range $0.02 < k_1 < 0.5$,
- the assumed precision $k_2 \cdot \delta x$ of feature detection in the blurred image. k_2 can be derived from eq. (9): $k_2 = 1/(\sqrt{12n} \cdot SNR)$ and will also lie in the range $0.02 < k_2 < 0.5$

We can reasonably approximate the precision of the direction Φ by

$$\sigma_\Phi^2 = \left[\frac{k_1 \ dx}{f} \right]^2 + \left[k_2 \cdot \frac{v \ t_I \ \sin\alpha}{s} \right]^2 \tag{10}$$

For v = 0 we obtain the theoretical precision in the unblurred image, while for large blur the second term is dominating.

A worst case analysis can be based on the following assumptions: dx = 20 µm, f = 20 mm, k_1 = 1/2, k_2 = 1/4 (in case image blur is taken into account), v = 20 m/sec and t_I = 1/100 sec. Neglecting the sin-term this leads to the model

$$\sigma_\Phi^2 \approx (0.031\ g)^2 + (0.05\ m\ /\ s\ [m])^2 \tag{11}$$

which has a structure commonly used for the variance of direction measurements in geodetic networks and which is used in the example in section 2.3.

Exploiting the potential of feature detection and using a longer focal length may result in standard deviations which are a factor 3 to 20 better.

2. Precision of Geometric Properties derived from Image Sequences

This section wants to discuss the precision of various geometric properties of fix and moving objects as they can be derived by tracking features in image sequences. Specifically we are interested in relative distances and velocities between the sensor and possibly moving objects, and in the time the sensor is expected to meet the object, e. g. when being overtaken by another vehicle. We will treat the case of mono and stereo sequences. Finally we are interested in the precision with which one is able to track the own position and orientation from continuous observation of targets. To simplify the derivations we restrict the discussion to a two dimensional environment. Due to the nature of the mensuration process, namely measuring the geoemtry of bundles of rays, i. e. angles, the results can easily be transferred to three dimensions.

The section is organized along three case studies

- estimating the distance to an object from a mono image sequence

- estimating the distance, the velocity and the meeting time from a stereo image sequence,
- estimating the own trajectory from image sequences.

The examples are based on two different sensor systems:

System A (low accuracy; focal length 20 mm, pixel size 20 μm, k_1 = 1/2) obtains directions with standard deviation of **1/2000 = 0.03 gon.**

System B (high accuracy; focal length 50 mm, pixel size 10 μm, k_1 = 1/4) obtains directions with standard deviation of **1/20000 = 0.003 gon,** thus with a 10 times higher precision than system A.

2.1 Estimating the distance to an object from a mono image sequence

Let us assume the target point P(x,y) is observed from a moving platform over a short path of length l_m measuring the direction α_i n times (cf. Fig. 1). The trajectory of the platform is assumed to be known. The object has an average distance of approximately s from the sensor and is seen under

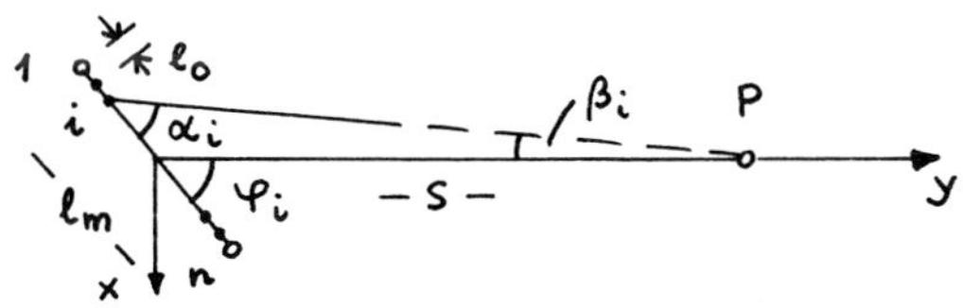

Fig. 1 Distance estimation from short image sequence

the aspect angle Φ. The parallactic angle $\beta_i = (\Phi - \alpha_i)$ can be written in dependency from the sequential number i of the directions as

$$\beta_i = (i - (n+1)/2) \cdot l_o \cdot \sin\Phi \;/\; s \qquad (12)$$

and can be used in the n observation equations for the overconstrained problem

$$(\Phi - \alpha_i) + e_i = dx \;/\; s + \beta_i \;/\; s \cdot dy \qquad w_i = 1 \qquad (13)$$

using the approximations $s_i \approx s$, $\sin\beta_i \approx \beta_i$, $\cos\beta_i \approx 1$. The directions are assumed to have equal weight $w_i = 1$. From the normal equation matrix

$$N = 1/s^2 \begin{bmatrix} n & 0 \\ 0 & \sum_{i=1}^{n} \beta_i^2 \end{bmatrix} \tag{14}$$

one obtains the standard deviations for the coordinates

$$\sigma_x = \frac{s}{\sqrt{n}} \sigma_\alpha \tag{15}$$

and

$$\sigma_y = \frac{s^2}{l_o \sin\Phi} \cdot \sqrt{\frac{12}{n \cdot (n^2-1)}} \cdot \sigma_\alpha = \frac{s^2}{l_m \sin\Phi} \cdot \sqrt{\frac{12 \cdot (n-1)}{n \cdot (n+1)}} \cdot \sigma_\alpha \tag{16}$$

which for large n can be approximated by

$$\sigma_y = \frac{s^2 \sqrt{12}}{l_m \sin\Phi} \cdot \frac{\sigma_\alpha}{\sqrt{n}} = \frac{s^2 \sqrt{12}}{l_o \sin\Phi} \cdot \frac{\sigma_\alpha}{n^{3/2}} \tag{17}$$

Discussion:

1. The x-coordinate just is the mean over all individual estimates, and therefore its precision increases with $\sqrt{n}$ and is proportional to the distance.
2. The precision of the distance decreases with the square of the distance, independent of the other parameters, which is known for the case of only two rays.
3. The precision of the distance depends on the angle Φ between the viewing direction and the direction of the movement. As to be expected, the distance cannot be estimated if the object is in the direction of the trajectory.
4. The precision of the distance can be related either to the path length l_m or to the distance l_o between two neigbouring sensor positions. In case the precision is related to a constant path length, the precision increases with $\sqrt{n}$. On the other hand, if the velocity v and the time intervall t_o between the observations is fixed, thus $l_o = v \cdot t_o$ is fixed, then the precision increases with $n^{3/2}$, thus an increased number of frames can be very effective.

Example:

Let a non moving object have a distance of approximately 100 m and be seen under an angle of $\Phi = 1/25$, thus lying 4 m aside the momentary driving direction. The directions of camera system A have a standard deviation of 1/2000 = 0.03 gon. Let the own velocity be $v = 20$ m/s = 72 km/h and $t_o = 1/25$ s, thus $l_o = 0.8$ m. Then the distance can be measured with a standard deviation of $86/\sqrt{n\cdot(n^2-1)}$ [m] with system A. If only two consecutive frames are used (n=2) the precision is appr. 35 m or 35 %. If the object is observed for 1/2 s, thus when using n=13 frames, the precision is 0.6 m or 0.6 %. With camera B already with 2 frames a relative precision for the distance of 3.5 % can be achieved. ▪

Though the own trajectory has been assumed to be known, this information is not difficult to obtain from image sequences over short time ranges (see below). However, a good relative accuracy for the distance obviously is obtainable only for point type objects lying far enough aside the momentary path, if a certain integration over time takes place or if the camera system yields a good enough precision. Aside from these restrictions no information on the relative speed to a moving object can be derived, as parallaxes due to distance and due to speed can not be separated.

2.2 Estimating geometric properties from stereo image sequence

The disadvantages of mono image sequences do not occur in stereo sequences. We want to develop the relations in several steps starting with the estimation of the distance from one image pair. With the base line b, lying perpendicular to the trajectory, and the parallactic angle δ the distance is given by

$$s = \frac{b}{2} \cot \frac{\delta}{2} \tag{18}$$

or for $b \ll s$ approximately

$$s = \frac{b}{\delta}. \tag{19}$$

The variance of the distance then is given by

$$\sigma_s^2 = \frac{1}{\delta^2} \left(s^2 \ \sigma_\delta^2 + \sigma_b^2\right) = \frac{s^4}{b^2} \ \sigma_\delta^2 + \frac{s^2}{b^2} \ \sigma_b^2 \tag{20}$$

which clearly shows that the uncertainty of the baseline is negligible for larger distances. The first part in eq. (19) can be obtained from eq. (15) with $n = 2$ and $\Phi = 100$ gon, $l_o = b$ and $\sigma_\delta = \sqrt{2}\ \sigma_\alpha$.

Example:

Let the following values be given $\sigma_b = 0.01$ m, $\sigma_\delta = \sqrt{2}/2000 =$ 0.02 gon (system A), b = 1 m. A distance of 10 m can be estimated with 0.12 m or 1 %, and a distance of 100 m will have a standard deviation of 7 m, leading to a relative accuracy of 7 %, which may be fully sufficient for certain tasks. These accuracies can be obtained instantaneously. With camera system B the relative accuracy of a distance of 300 m is still 21 %. ▪

A similar relation is obtained for the velocity derived from two image pairs. The difference s_m of two distances s_1 and s_2 with $\delta = \sqrt{\delta_1 \cdot \delta_2}$ and $\delta_m = \delta_1 - \delta_2$ approximated by

$$s_m = s_2 - s_1 = b\left(\frac{1}{\delta_2} - \frac{1}{\delta_1}\right) = \frac{b\ \delta_m}{\delta^2} = \frac{s\ \delta_m}{b} \tag{21}$$

and leads to the variance for the measured distance difference s_m

$$\sigma^2_{s_m} = \left[\frac{\delta_m}{\delta^2}\right]^2 \sigma^2_b + \left[\frac{b}{\delta^2}\right]^2 \sigma^2_{\delta_m} \tag{22}$$

the standard deviation of the velocity then is given by

$$\sigma_v = \sigma_{s_m} / t_m \tag{23}$$

Here t_m is the time difference between the two measurements.

<u>Example:</u>
Under the same conditions, as in the previous example the standard deviation of the distance difference for system A is σ_{sm} = 10 m, for system B σ_{sm} = 1 m. The influence of the uncertainty of the basis is negligible. Two consecutive frames cannot be used for estimating the velocity, as their standard deviation is σ_v = 250 m/s and 25 m/s resp. . Even if one uses two frames which are 1 s apart, the accuracy is still only 10 m/s = 36 km/h for an object which is 100 m far (system A). The situation, however, is much improved when using sytem B.▪

These relations directly can be used for estimating the time t until object and sensor meet. It is given by

$$t = s/v = \delta \cdot t_m / \delta_m \tag{24}$$

and has the relative accuracy

$$\frac{\sigma_t}{t} = \frac{\sigma_{\delta_m}}{\delta_m} \tag{25}$$

The time is independent of the baselength used in the stereo sequence. Obviously the bias in the baselength influences distance and velocity in the same manner, specifically the relative systematic error in the baselength is identical to the relative systematic erros in distance, distance difference and velocity.

<u>Example:</u>
In case the relative velocity would be 20 m/s and two frames with a delay of 1 s would be used for estimating the meeting time t, which is approximately 5 s, the relative accuracy would only be 50 % for system A, however 5 % for system B. ▪

We now want to investigate the effect of a larger number of stereo measurements onto the precision of distance, velocity and time. From the individual distances $l_i = b/\delta_i$ we can derive the average distance s_o and the velocity v from the overconstrained system

$$l_i + e_i = s_o + t_i \cdot v \,, \qquad \sigma_{l_i} = \sigma_l = b / \delta^2 \cdot \sigma_\delta \tag{26}$$

In a similar way as in section 2.1 we can build up the normal equation matrix and obtain the theoretical values for the standard deviation of the average distance and the speed:

$$\sigma_s = \frac{\sigma_l}{\sqrt{n}} = \frac{s^2}{b} \cdot \frac{\sigma_\delta}{\sqrt{n}} \tag{27}$$

and

$$\sigma_v = \frac{s^2}{t_o\ b} \cdot \sqrt{\frac{12}{n\cdot(n^2-1)}} \cdot \sigma_\delta = \frac{s^2}{t_m\ b} \cdot \sqrt{\frac{12\cdot(n-1)}{n\cdot(n+1)}} \cdot \sigma_\delta \tag{28}$$

which for large n can be approximated by

$$\sigma_v = \frac{s^2\ \sqrt{12}}{t_m\ b} \cdot \frac{\sigma_\delta}{\sqrt{n}} = \frac{s^2\ \sqrt{12}}{t_o\ b} \cdot \frac{\sigma_\delta}{n^{3/2}} \tag{29}$$

The precision for the time t = s/v until object and sensor meet is given by

$$\sigma_t^2 = \left[\frac{s^2}{b\ v}\,\sigma_\delta \right]^2 \cdot \left[\frac{1}{n} + \{\frac{t}{t_m}\}^2\ \frac{12\ n}{n^2-1} \right]^2 \tag{30}$$

For observation times $t_m < t$ and large n the 1/n term in the brackets, which represents the influence of the distance estimate onto the time, is negligible. This leads to the relations for the relative accuracy of the time estimate

$$\frac{\sigma_t}{t} = \frac{t}{t_m} \cdot \sqrt{\frac{12}{n}} \cdot \frac{\sigma_\delta}{\delta} = \frac{s^2}{v\ b} \cdot \sqrt{\frac{12}{n}} \cdot \sigma_\delta \tag{31}$$

It is interesting to note that the relative accuracy of the time t does not depend on the velocity nor on the basis, but only on the relative observation time t_m/t, the number of frames and the relative accuracy of the parallactic angle. Referring to a constant precision of the directions, only the parallactic angle $\delta = b / s$ is really influencing the relative precision.

Example:
If under the same conditions as in the last example 25 frames are taken for the estimation of the time t, it can be estimated with a relative accuracy of 25 % with system A and 2.5 % with system B, which is an increase of only a factor 2 compared to the result with only 2 frames. With camera system B a relative precision of 10 % would be reached for objects with a distance up to 200 m. ▪

The advantages of stereo sequences are obvious: They allow an instantaneous estimate of distances, achieve useful estimates for the absolute speed, when applied to fix objects or for the relative speed to other vehicles. In contrast to mono sequences they need no information on their own trajectory, epecially no orientation. On the other hand the results of this subsection can be used also in mono image sequences in case pairs of points with a constant distance between them are tracked. In this case the point pair can be treated as the base line and the parallactic angle δ is directly measured in one camera. As the relative error of the baselength usually is negligible or in the case of time estimates even of no influence, the discussed accuracies can also be achieved in mono sequences, again not requiring an absolute frame. Such point pairs could be the front or the rear lights of a car, which within a class of vehicles show only a limited variation in distance.

2.3 The precision of the trajectory from image sequences

In case the trajectory of the sensor is long compared to the distance to the object points, which connect the individual sensor positions, the accuracy of the trajectory follows the same rules as a strip of images in aerial triangulation.

The accuracy properties of strips have intensively investigated by ACKERMANN (1965). Of special interest was the dependency of the precision of the connecting tie points on the length of the strip. The result can be reduced to the following relations:

$$\sigma_X = k_X \cdot n^{3/2}, \quad \sigma_Y = k_Y \cdot n^{3/2}, \quad \sigma_Z = k_Z \cdot n^{3/2} \qquad (32)$$

where n is the number of images in the strip and the k_i are apropriate constants. These relations were found to hold for all types of calculation schemes, suggesting that the structure of the strips dominates all other effects, and differences in algorithm and control point density only influence the constants.

There are some differences of photogrammetric strips and image sequences taken from a vehicle:

- The viewing direction in image sequences usually is not 90° across the direction of the movement. This leads to much higher overlap in image sequences. In extreme cases single points may be trackable through hundrets of images. This locally may lead to extremely stable connections between the images especially with respect to the orientation parameters.
- The distances to the tie points in image sequences varies to a large extent. Smallest distances may be a few meters, long distances a few kilometers. This leads to a high inhomogeneity of the connections and makes realistic simulations difficult to achieve.
- in navigation one primarily is interested in the precision of the parameters for the sensor platform, as the positions of the object points can be determined with high enough accuracy.

The differences will be overruled by the strip-nature of the image sequence in the long run, thus, if only one sensor is used, leading to the same $n^{3/2}$ behaviour of the standard deviations of the positions of the sensor platform. On short distances the behavior may however be of totally different nature.

The orientation parameters were not investigated in this study. The principle of double summation which lead to the $n^{3/2}$ rule for the position can be transferred to obtain the the dependency of precision of the orientation parameters on the number of images. As a single summation takes place for the orienation angles, the standard deviations of them only increases with $\sqrt{n}$ (cf. FÖRSTNER 1983, p. 8).

Thus we can summarize the general structure of the precision of a long homogeneous trajectory from mono image sequences:

$$\sigma_i = k_i \cdot n^{3/2}, \quad i = x, y, z; \qquad \sigma_i = k_i \cdot n^{1/2}, \quad i = \Omega, \Phi, K \tag{33}$$

Investigations have to find out the constants k_i in dependency on the main parameters describing an image sequence.

Example:

In order to get an impression of the absolute accuracy of the trajectory derived from image sequences we simulated a short sequence. It consists of a point field with 100 points randomly distributed in a rectangle of 100 x 200 m². The trajectory consists of 50 points with a distance of 1 m each thus had a length of 50 m. It is situated in the centre along the major axis of the point field. The first two points are assumed to be known. All 100 points are assumed to be visible from each of the 50 points of the trajectory. This gives rize to 50 sets of direction. The orientation of the first two sets was also assumed to be known. The angular error is assumed to be 1/2000 = 0.03 gon with an additional a term 0.05/s[m] simulating the effect of image blur (cf. sect. 1.2). With an average of a distance of 40 m this leads to a standard deviation of the direction of 0.0013 = 0.085 gon. The 50 sets of bundles have been used for a joint estimation of all unknown parameters, including the coordinates of the tie points and the trajectory.

The result which reflects the last step of a rigorous Kalman-Filter is the following:

$$\sigma_x = 0.68 \text{ m}, \quad \sigma_y = 0.031 \text{ m}, \quad \sigma_\Omega \approx 0.008 \text{ gon}$$

The result is remarkable in two ways:

- The longitudinal error (x-direction) is only 1/2 m, thus 1 % of the travelled distance.
- The traversal error (y-direction) is even better. The trajectory is only off the starting direction by 0.08 % = 0.038 gon.
- The standard deviation of the orientaion angle has been estimated from the precision of the tie points and with $\sigma_\alpha/\sqrt{nt}$ very well can be predicted from the precision of the individual directions.

The result has been achieved with the camera system A including the effect of image blur. An increase of the accuracy can be expected if a more sophisticated feature detector is assumed. ▪

For stereo image sequences (cf. MATTHIES 1988) the geometry is much more stable in the direction of the trajectory. As each image pair has the same scale, which is fixed by the baseline, the longitudinal error results from a single summation, while the other error laws stay unchanged in structure. Thus the general structure for the precision of long homogeneous trajectories from image sequences is given by

$$\sigma_x = k_x \cdot n^{1/2}, \quad \sigma_i = k_i \cdot n^{3/2}, \quad i = y, z$$

$$\sigma_i = k_i \cdot n^{1/2}, \quad i = \Omega, \Phi, K \qquad (34)$$

Again the faktors k_i have to be determined in dependency of representative parameters of the image sequence.

Summarizing the study clearly demonstrated which precision can be obtained from image sequences. It is up to the designer of a system to integrate the geometric information coming from a visual sensor with other sensors which may support it. In any case the visual sensor will have its main application in the interpretation of the environment.

I want to thank Matthias Neureither for his support in the simulations.

References:

IAP	International Archives of Photogrammetry and Remote Sensing
IEEE T-PAMI	IEEE Transactions on Pattern Analysis and Machine Intelligence
SIPUS	Schriftenreihe des Instituts für Photogrammetrie der Universität Stuttgart

ACKERMANN F. (1965): Fehlertheoretische Untersuchungen über die Genauigkeit photogrammetrischer Streifentriangulationen, DGK C 87, München 1965

EBNER H., MÜLLER F. (1986): Processing of Digital Three Line Imagery Using a Generalized Model for Combined Point Determination, IAP 26-3/1, Rovaniemi, 1986, pp. 212-222

CASTLEMAN K. R. (1979): Digital Image Processing, Prentice Hall, 1979

DORST L., DUIN R. P. W. (1984): Spirograph Theory: A Framework for Calculations on Digitized Straight Lines, IEEE T-PAMI, Vol. 6, No. 5, 1984, pp. 632-639

FÖRSTNER W. (1982): On the Geometric Precision of Digital Correlation, IAP Vol. 24-3, Helsinki, 1982, pp. 176-189

FÖRSTNER W. (1983): On the Morphological Quality of Digital Elevation Models, Int. Colloquium on "Mathematical Aspects of Digital Elevation Models", Stockholm 1983

FÖRSTNER W. (1984): Quality Assessment of Object Location and Point Transfer Using Digital Image Correlation techniques, IAP 25-A3a, Rio de Janeiro, 1984, pp. 197-219

FÖRSTNER W. (1986): Prinzip und Leistungfähigkeit der Korrelation und Zuordnung Digitaler Bilder, SIPUS 11, Stuttgart 1986, S. 69-90

FÖRSTNER W., GÜLCH E. (1987): A Fast Operator for Detection and Location of Points, Corners and Circular Features, Intercommission Workshop on "Fast Processing of Photogrammetric Data", Interlaken 1987

HAHN M., FÖRSTNER W. (1988): On the Applicability of a Feature Based and a Least Squares Matching Algoritm for DEM Aquisition, IAP 27-3, Kyoto 1988

HELAVA U. V. (1976): Digital Correlation in Photogrammetric Instruments, IAP 23-2, Helsinki 1976

MATTHIES L. (1988): Depth from Stereo Image Sequences, Int. Workshop "High Precision Navigation", Stuttgart-Altensteig-Warth 1988

PADERES F. F., MIKHAIL E. M., FÖRSTNER W. (1984): Rectification of Single and Multipleof Satellite Scanner Imagery using Points and Edges as Control, NASA-Workshop on Mathematical Pattern Recognition and Image Analysis, Houston, 1984

PERTL A. (1984): Digital Image Correaltion with the Analytical Plotter Planicomp C 100, IAP 25-3, Rio de Janeiro 1984

VOSSELMAN G., FÖRSTNER W. (1988): The Precision of a Digital Camera, IAP 27-1, Kyoto, 1988

Towards Autonomous Navigation Utilizing Image Sequence Analysis

Ralf Kories, Nils Rehfeld, Georg Zimmermann
Fraunhofer Institut, Karlsruhe

Summary

Driving a vehicle under Central Europe's traffic conditions frequently consists of following a vehicle in front. This contribution addresses the problem of automatically detecting a leading vehicle in the process of starting off and to derive a model for further tracking. Very little knowledge about the appearance of the vehicle is utilized. Motion information is the main cue for selecting image structures belonging to the leading vehicle. Further an edge model is derived from two independent complementary motion estimation processes cross-validating each other. The suitability of the aggregated model for recognition and the precision of the motion estimation process is demonstrated by means of an image sequence consisting of 130 TV-images.

Introduction

High precision navigation can benefit a great deal from utilizing visual information. Most higher animals rely on vision for their orientation in their environment. Projects like PROMETHEUS within EUREKA as well as research activities within the ESPRIT program spawned a lot of interest in the computer vision community for vehicle navigation and guidance problems. Since navigation is a dynamic process, image *sequence* analysis in particular is a promising field.

Visual navigation is oriented at prominent landmarks. Unlike in other domains, like radio ranging, where a beacon is easily detectable, a visual landmark has to be recognized in a cluttered environment. This is a highly complex task in outdoor scenarios, in particular because landmarks generally provide very different appearances depending on the location they are seen from. In the context of driving in a convoy the vehicle in front temporarily serves as a "mobile landmark" that can be used to keep the vehicle on the road, to select a suitable speed etc.

This contribution addresses the problem of automatically detecting a leading vehicle in the process of starting off and to derive a model for further tracking. Very little knowledge about the appearance of the vehicle is utilized. Motion information is the main cue for selecting image structures belonging to the leading vehicle. Further an edge model is derived from two independent complementary motion estimation processes cross-validating each other.

Basic principle

The experiment we are conducting to demonstrate the capability of our method consists of the following situation. Two vehicles are standing in line on a minor service road surrounded by vegetation and buildings. The task of the second vehicle is to follow the one in front. In the first phase the cars are stationary. However, the imagery taken by the camera in the trailing car is not stable due to vibration caused by the running motor and compliance of the vehicle's suspension.

In the second phase the leading vehicle, a truck, starts up driving straight ahead. This causes the image of the truck to shrink i.e. the image flow is directed towards a point where the vehicle would disappear at the horizon should it maintain its straight path. Thus the displacement vector field describing the image motion in the entire image consists of two different regions. The background region shows only vectors caused by the motion of the camera, basically a small pan and tilt motion. The area on the vehicle shows vectors with a centrally oriented structure, and superimposed on to that, the component due to vibration of the camera. If it is possible to separate these two components, the location of the vehicle in the image can be derived. For further tracking the vehicle, not only its location in the image but a precise model is required. It undergoes incremental refinement as more motion information becomes available.

Underlying assumptions

Due to the kinematics in the experiment we have to accommodate two different motion models for the image area. The motion vectors caused by pan and tilt of the camera can be described with good accuracy as a constant flow field. This is particularly true for cameras with a small field of view.

$$\mathbf{V}(x,y) = (v_x, v_y) \tag{1}$$

The second model of the displacement vector field applies in the area of the receding truck. The truck we used in our experiment had a very flat back portion and therefore is very well approximated by a plane. Even though other vehicles occasionally show a prominent structure in depth, for distances that are larger than the extent of the vehicle, the planar approximation becomes valid. Due to the chosen camera orientation the optic axis is close to being perpendicular to the truck back surface. Under these circumstances the displacement vectors due to the motion of the receding truck can be expressed as

$$\mathbf{W}(x,y) = (a_x + b_x * x,\; a_y + b_y * y) \tag{2}$$

the components of the displacement vector field being linear functions of the image coordinates. Higher order functions and cross terms for more complicated constellations can be easily accommodated by the method we describe in the following.

Displacement vectors from tracking blobs

The method we apply to derive motion vectors is based on tracking features that are derived with the so-called monotonicity operator [1]. The monotonicity operation is invariant under monotone gray-value transformations (hence, the name of the operator). The invariance properties of the monotonicity operator guarantee that the operation is insensitive to changes of the camera gain, shading inhomogeneities of the sensor as well as against illumination changes. As such it is very suited to be applied to outdoor imagery for example taken from a vehicle. Tracking the bloblike features generated by the monotonicity operator from image to image allows the estimation of displacement vector fields. This method has been applied extensively to a variety of indoor and outdoor scenes and proved to be very robust [2].

Fig. 1 shows in its left portion an image from the sequence where the truck is still close to the following vehicle. The displacement vector field while the truck is just starting is shown in the right part of Fig. 1. Obvious is the effect of a slight camera tilt during the measurement time of 25 images (equivalent to 1 second).

Coarse vehicle segmentation

The purpose of the coarse vehicle segmentation is to separate a sub-population of the motion vectors which are very likely not to belong to the stationary background. The basic idea is the following. The distances between features belonging to the background remain constant under pan and tilt motion of the camera. This is due to the fact that the motion vector field is constant across the field of view. Vectors violating this rigidity constraint are either outliers due to malfunction of the feature extractor or they belong to a differently moving section of the image.

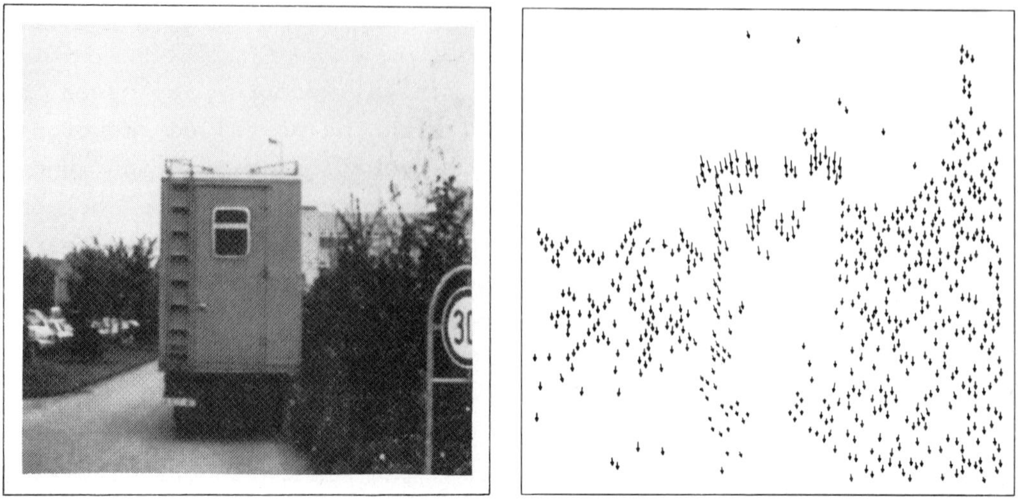

Fig. 1 left: First image of a sequence of 750 images taken by a camera mounted on a yet stationary car. Pan and tilt motion of the camera must be expected. The truck starts off during the image sequence. **Right:** Displacement vector field (using blobs) of the first second of the image sequence. The camera tilts, the truck starts to move.

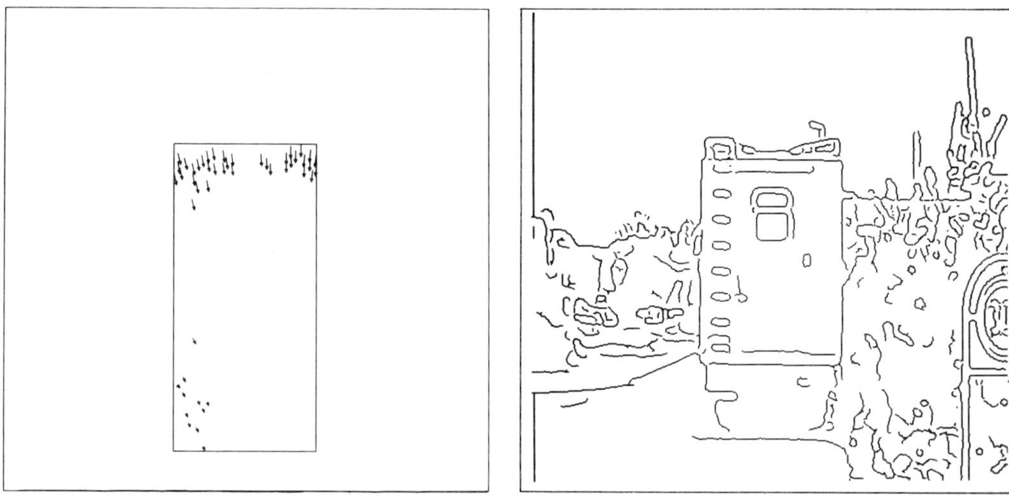

Fig. 2 left: Result of coarse vehicle segmentation using the non-homogeneous components of the displacement vectors from Fig. 1 right. **Right:** Contours extracted from the first image of the scene

Consider a matrix whose rows and columns are indexed by the features and the matrix coefficients are the distances between respective features. Such a distance matrix can be computed for every frame in the image sequence. Since the correspondences between consecutive frames are known from our motion estimation procedure, we can arrange rows and columns such that we obtain a sequence of distance matrixes of compatible structure. The differences between consecutive matrices represent the temporal change of feature distances. High differences along particular rows and columns indicate that these features changed their position relative to the rigid feature mesh. Such features are considered as not belonging to the stationary background. The threshold for discriminating between feature displacements caused by statistical errors of the monotonicity operator and a systematic drift can be easily determined beforehand. From earlier investigations we know the precision of our motion estimator applied to natural scenes [3].

The result of this procedure is a subset of vectors that are caused with high probability by a moving object in the field of view. A frame in Fig. 2 contains all the vectors selected from the field in Fig. 1 based on this hypothesis.

<u>Vehicle model from contour position verification</u>

Motion information alone will certainly not be sufficient to track the vehicle under all conditions as soon as the following vehicle is moving itself. We therefore try to obtain additional cues to re-recognize the leading vehicle using contours. Fig. 2 left shows the result of a precise contour extraction using an operator by Korn [4].

Since the coarse segmentation of the vehicle can not be expected to be perfect one always has to allow for contour elements within the selected frame that do not belong to the vehicle, such as trees etc. Occasionally shadows or reflections fall onto the vehicle creating contours that are not an integral part of its structure. The approach we use to generate and aggregate a model of the vehicle consists of incorporating motion information. We measure the displacement within the area of interest for a considerable duration (in our experiment up to 100 pictures). We are able to estimate a set of parameters describing this displacement field. This allows to predict edge locations in the following image provided the model is correct.

In order to verify and update our model we extract edges from the current image and compare them with the edge predictions using the motion field parameters. Areas where they are in good accordance are shown in the left acceptance map in Fig. 3. Since we required the edge positions to agree better than a pixel's diameter the derived edge model appears somewhat fuzzy.

The process described above verifies that the motion parameters are correctly estimated by the fact that predictions based on them are in good agreement with actual observations. However, a highly textured image will produce spurious agreements of predicted and observed edges even if the image area under observation does not undergo the supposed displacement. Consequently an additional test is implemented checking if contour elements are labeled with the proper displacement.

Vehicle model from contour displacement verification

We derive an alternate edge model from an independent and complementary image motion source. Edges by themselves do not provide enough information to obtain a reliable motion estimate. Contours, however, impose additional constraints and therefore yield good estimates. Contours are found at very different image locations than centers of monotonicity blobs. Also they undergo major changes in shape. Consequently a more sophisticated matching approach is needed.

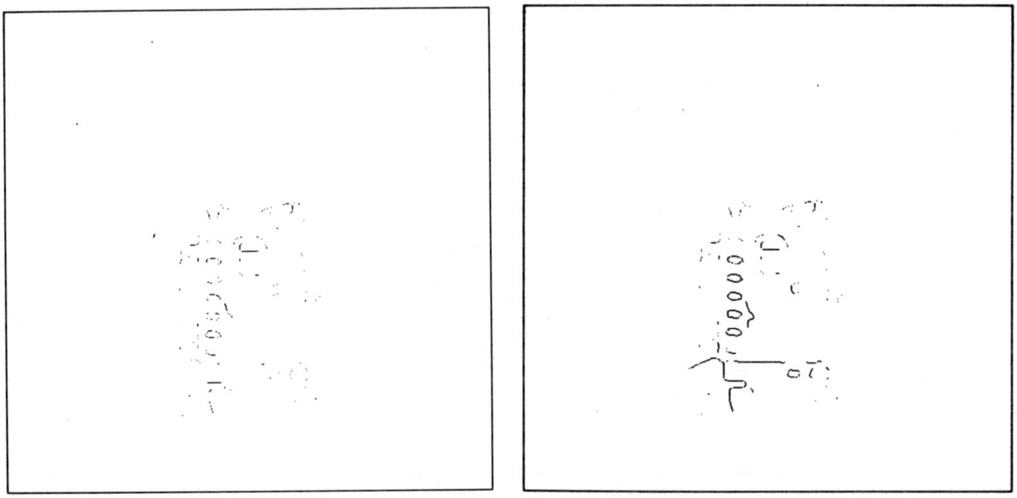

Fig. 3 left: Acceptance map obtained from the verification of contour position using blob displacement estimates. Each set pixel indicates that the contours predicted from the first image into the 25th (see Fig. 2) are in agreement with contours actually observed in the later image. **Right:** Acceptance map from contour displacement verification. Shown are edge points whose displacement estimated from elastic matching is in good accordance with the model derived from the vehicle segmentation based on blob displacements.

Fig. 4 left: Edge model of the truck after aggregation of the information of information from complementary motion information sources, i.e. blob displacement estimation and elastic contour matching. **Right:** This edge model projected into an image five seconds (125 images) later. Both the motion estimation process and the edge localization quality are very satisfactory.

The approach we have developed attempts to maximize a similarity function considering matches between single edgels. An appropriate design of the similarity function allows the application of a global optimization strategy based on a dynamic programming technique. The resulting elastic contour matching is (i) invariant under translation, (ii) selective in angular orientation, and (iii) shows elastic behavior in the presence of shape changes. This matching approach is more complicated than the matching of blobs. However, it allows a reliable contour displacement estimation over 25 and more images without recourse to intermediate images.

Each edge point is now associated with a displacement vector. The verification process compares the shift estimates edgel by edgel with the predictions by the model independently derived from blob displacements. The resulting map of pixels at edge locations is given in the right side of Fig. 3.

Incremental model aggregation and recognition

Each pixel which is set in any of the two acceptance maps has passed a test indicating that two independent and complementary motion estimation processes are in accordance, one based on contours, the other based on blobs. Since the requirements for passing the tests are rather stringent the acceptance maps have a somewhat fuzzy appearance. Now the two acceptance maps can safely be aggregated into a commonly generated edge model of the truck. Fig. 4 shows the result of such an operation. Each pixel in the model has been selected entirely based on motion criteria under the hypothesis expressed in Equation (2).

So far we did not utilize the fact that each image provides a new acceptance map. Since we have the displacement information from image to image available, the consecutive acceptance maps can be aggregated after motion compensation. An additional criterion is applied to the aggregated acceptance map: each edgel has to be verified n times until it is accepted. Contours with a sufficiently high "hit rate" become entirely part of the two dimensional contour model of the truck.

The temporal evolution of such a contour model is shown in Fig. 5. From the upper left corner down to the right hand side the number of images considered for computation increases. Obviously the left portion of the truck is favored. This is due to the higher number of features on this side of the vehicle. Even though the shrubs at the right hand side of the vehicle were moved by wind their edges are rejected due to incompatible motion. The incrementally aggregated contour model is sparse but highly reliable. None of the edges lies off the vehicle's image.

Fig. 4 illustrates in the right image the potential of the method for recognition tasks. Actually observed edges within the segmentation frame from the 25th image have been projected into an image 5 seconds (125 images!) later using the estimated motion parameters from the monotonicity operator. The predicted edge locations are superimposed onto image no. 125. It demonstrates that both the motion estimation process and the edge localization quality are excellent.

The same technique is applicable to the aggregated two-dimensional contour model. The edge elements selected automatically pixel by pixel based on motion criteria are transformed according to the observed motion. Matching between motion derived model edges and observed contours allows a precise localization of the vehicle and its recognition without recourse to human judgment in the entire process.

Discussion

We have demonstrated the capability of our method to extract a receding vehicle in a convoy situation generating an edge model based on motion criteria. We have shown that the method performs well in a realistic situation where the camera is not required to be absolutely fixed in space. The reliability of the method

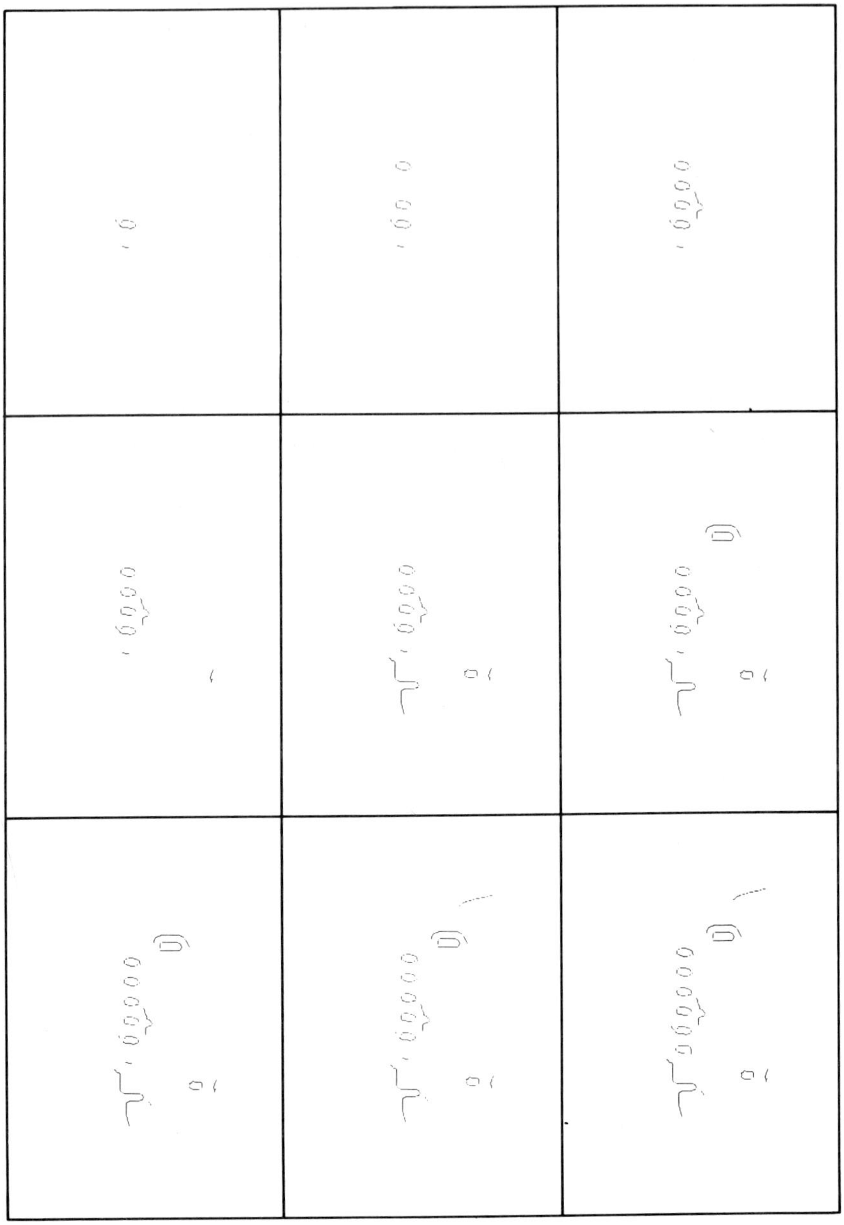

Fig. 5: Temporal evolution of the two dimensional aggregated contour model. The number of acceptance maps considered for aggregation increases from top left to bottom right. Shown is only the region around the truck's image. No contours are selected in peripheral image regions.

stems from the fact that two independent complementary motion processes are interacting cross-checking each other. These results support the assumption that image sequence analysis can offer a vital contribution to high precision navigation and vehicle guidance.

The two-dimensional vehicle model is quite simple. We intend to extend it towards a 2 1/2-D description. This will be necessary in order to allow for turns of the leading vehicle as they occur in real traffic situations. 3D data from stereo image sequences are now routinely collected in our laboratory. At the present time we are in the process of combining them with motion information [5].

References

[1] **R. Kories, G. Zimmermann**: Motion Detection in Image Sequences: An Evaluation of Feature Detectors. ICPR-84, pp.778-780.

[2] **R. Kories, G. Zimmermann**: A Versatile Method for the Estimation of Displacement Vector Fields from Image Sequences. IEEE Workshop on Motion: *Representation and Analysis*, May 7-9, 1986, Kiawah Island Resort, Charleston, South Carolina, pp. 101-106.

[3] **R. Kories, G. Hecker, G. Zimmermann**: On the Precision of a Feature Based Displacement Measurement. ICPR-86, pp. 1193-1196.

[4] **A. Korn**: Combination of Different Space Frequency Filters for Modeling Edges and Surfaces in Gray-value Pictures. SPIE-595 *Computer Vision for Robots*, Cannes, 1985, pp. 22-30.

[5] **N. Rehfeld**: Private communication.

Depth from Stereo Image Sequences

Larry Matthies
Computer Science Dept., Carnegie-Mellon University, Pittsburgh, USA

Abstract

Stereo vision is an attractive approach to depth sensing in many robotics applications. To date, most research in stereo has concentrated on the analysis of a single stereo image pair. However, many applications will involve moving robot systems that can acquire sequences of stereo pairs from successive robot positions. Such systems can achieve greatly improved stereo performance by appropriately controlling the motion of the cameras and by using depth information obtained from early images to guide the interpretation of later images. This requires a representation for the depth model at any point in time, methods for using the model to influence matching in subsequent images, and methods for controlling the motion of the cameras that take into account the degree of uncertainty in the depth model. In this paper, we propose a Bayesian approach to processing stereo image sequences that serves these requirements. The approach is based on representing, predicting, and updating depth and depth variance at every pixel in the image. We describe a vision system under development for a robot vehicle that incorporates this approach and summarize implementation results for parts of the system.[1]

1 Introduction

Stereo vision has a great deal of promise as a depth sensor for robotics. Typical modes of operation in many robot applications, such as navigation and manipulation, will involve moving robot systems that can or will acquire sequences of images from successive robot positions. Stereo systems for such robots can go beyond the static analysis of each stereo pair by using depth information from past images to guide the analysis of current images. Moreover, they can plan the motions of the robot to acquire image sequences that simplify the stereo correspondence problem.

Most stereo research to date has not taken this dynamic view of the problem, but instead has focused on how to constrain or disambiguate the search for correspondences in a single stereo image pair [2,3,11,13,14,23]. Often this has involved a reliance on restrictive or *ad hoc* assumptions about the nature of the application domain, such as that images contain primarily straight lines or that scenes contain only smooth variations in depth. More recent work has shown that using redundant images and conservative sensing strategies leads to more reliable depth measurement without the need for restrictive assumptions. Examples include trinocular stereo systems [21], systems that couple stereo with camera motion [7], and systems that obtain a sequence of closely-spaced images by moving a single camera [4,18,27]. While this work has led to improved depth estimation, it has not yet led to an efficient, general approach to obtaining reliable depth measurements from a moving robot platform.

[1]This research was sponsored by DARPA, monitored by the Air Force Avionics Lab under contract F33615-87-C-1499, by the ONR under contract N00014-81-K-503, and by a postgraduate fellowship from the FMC Corporation. The views and conclusions contained in this document are those of the authors and should not be interpreted as representing the official policies, either expressed or implied, of the funding agencies.

In this paper, we take three steps toward this goal. First, we introduce a Bayesian framework for processing stereo image sequences. This framework, which extends previous Bayesian methods for individual stereo pairs [9,14], is based on representing the depth at each pixel as a probability distribution. Second, we apply this framework in a system, currently under development, that uses fine camera motions to bootstrap reliable stereo correspondence. The system translates both cameras parallel to the image plane, estimates depth monocularly from image sequences acquired with each camera, and uses these depth estimates to constrain matching between the cameras. Finally, we show how to extend this system to process stereo image sequences obtained by a navigating robot vehicle. This involves using the depth map obtained from each stereo pair, together with estimates of the vehicle motion, to compute depth predictions that guide matching in the next stereo pair.

The next section presents the Bayesian formulation of the stereo problem. Following this, we describe our system for bootstrapping reliable stereo correspondence and show implementation results from the monocular component of the system. We then summarize the extension to stereo image sequences, together with the results of previous work with a visual position estimation system that can provide the necessary estimates of vehicle motion.

2 Bayesian Framework

We begin by representing the depth at each pixel as a probability distribution. This can express prior depth information available from known disparity limits, matching in previous images, other sensors, or map databases. We treat each pixel as independent of all others; this is similar to [9] but distinct from [14], where the emphasis was on correlations between depth at neighboring pixels. To compute the depth at a given pixel, we apply Bayes' rule to estimate the posterior probability, $P(d = d_i|\varepsilon_i)$, that the disparity at that pixel is d_i, given observed differences ε_i between the stereo pairs:

$$P(d = d_i|\varepsilon_i) = \frac{P(d = d_i)P(\varepsilon_i|d = d_i)}{\sum_i P(d = d_i)P(\varepsilon_i|d = d_i)}. \tag{1}$$

$P(d = d_i)$ is the prior distribution for the disparity at that pixel and is defined over the allowable range of d_i. Uniform and normal distributions are both useful models for the prior distribution. $P(\varepsilon_i|d = d_i)$ is the conditional distribution of the observation ε_i given that the disparity is d_i. For example, the observations may be intensity differences between the two images. We choose the disparity estimate to be the value of d_i that maximizes the posterior probability – that is, the MAP estimate.

A simple example of this approach is illustrated in Figure 1a. In this case, the prior depth distribution for pixel $I_l(x_0, y_0)$ in the left image is a uniform distribution[2] that confines possible matches to an unweighted search window in the right image. The observations for a disparity d_i in the search window are intensity differences

$$\varepsilon_{ijk} = I_l(x_0 + \delta x_j, y_0 + \delta y_k) - I_r(x_0 + \delta x_j + d_i, y_0 + \delta y_k)$$

for pixels in a region around x_0, y_0. If the noise in the images is white, the intensity differences will be normally distributed. Together with the uniform prior, this implies that the posterior distribution for disparities in the search window will have the form

$$P(d = d_i|\varepsilon_i) \propto \exp\left\{-\frac{1}{2\sigma^2}\sum_{jk}\varepsilon_{ijk}^2\right\}$$

where $\varepsilon_i = [\varepsilon_{i11}, \ldots, \varepsilon_{imn}]^T$ and σ^2 depends on the level of noise in the images. The MAP estimate of disparity is that which minimizes $\sum_{jk}\varepsilon_{ijk}^2$, the sum of the squared intensity errors. The variance of this

[2] determined from known minimum and maximum disparity limits, for example

estimate can be approximated in several ways; for example, by taking derivatives of the total squared intensity error [6], by fitting a curve to the total squared intensity error about its minimum [18], or by taking moments of the posterior distribution [9].

This example can be extended in a number of ways. The observations can be differences of linear functions of the images, such as band-pass filters, they can be taken over several color bands, or they can be defined for trinocular stereo. The procedure for two images can be extended to a sequential procedure for estimating depth from a series of images. In this case, the depth and variance estimates from the first two images become priors for matching in the third image. This leads to updated estimates which can be used for matching in the fourth image, and so on. Finally, the procedure can be extended to matching in a series of stereo pairs obtained from a moving robot. Given knowledge of the robot's motion, the depth estimates from the first stereo pair can be used to compute depth predictions that constrain matching in the next stereo pair. The last two of these examples are discussed in the rest of the paper.

Before moving on, we observe that by estimating depth and depth variance at every pixel in the image, this approach potentially supersedes edge-based approaches to stereo matching. Edge-based matching schemes attempt to focus on only the most distinctive points in the image – in other words, those that will have depth estimates with low variance. However, they sacrifice a large amount of the depth information potentially available in doing so, because it is difficult to design an edge-based matching scheme that works for all scales of intensity variation present in the image. By estimating depth at every pixel and explicitly representing the precision of the result, we can extract whatever depth information is available and weight it in later operations according to its quality.

3 Bootstrapping Stereo Correspondence

We are currently developing a vision system for a robot vehicle that employs this framework. The system focuses on two related issues: the use of exploratory camera motions to disambiguate stereo correspondence, and the use of depth from previous stereo pairs to constrain matching in the current pair.

The exploratory-motion aspect of the system is motivated by matching ambiguity in wide-baseline stereo. If little is known about depth before matching, search windows with wide stereo baselines are large and the potential for ambiguous matches is great. However, if we mount the cameras on a translation stage, we can move them in comparatively small steps to acquire image sequences from one or both cameras (Figure 1b). By appropriately controlling the amount of camera motion, we can guarantee that matching in these sequences is highly reliable. This will yield depth estimates that constrain subsequent matching between the cameras.

The overall system approach we are investigating uses exploratory camera motions to bootstrap reliable stereo correspondence, then uses the resulting depth map predictively to constrain matching in images acquired as the robot follows a task-oriented trajectory. We are currently implementing and evaluating the bootstrap phase of this approach. We describe the results of this work below and summarize our approach to the subsequent "steady state" phase in the following section.

3.1 Processing Lateral Camera Motion

The bootstrap phase can be realized by acquiring two or more images with one or both cameras. For simplicity and measurement sensitivity, the only camera motions we consider are translations parallel to the image plane and the image raster, so that image features flow along scanlines. We also consider only stereo configurations in which the optical axes of the cameras are parallel and epipolar lines are corresponding scanlines. Here, we will not consider the questions of how many images to take or how

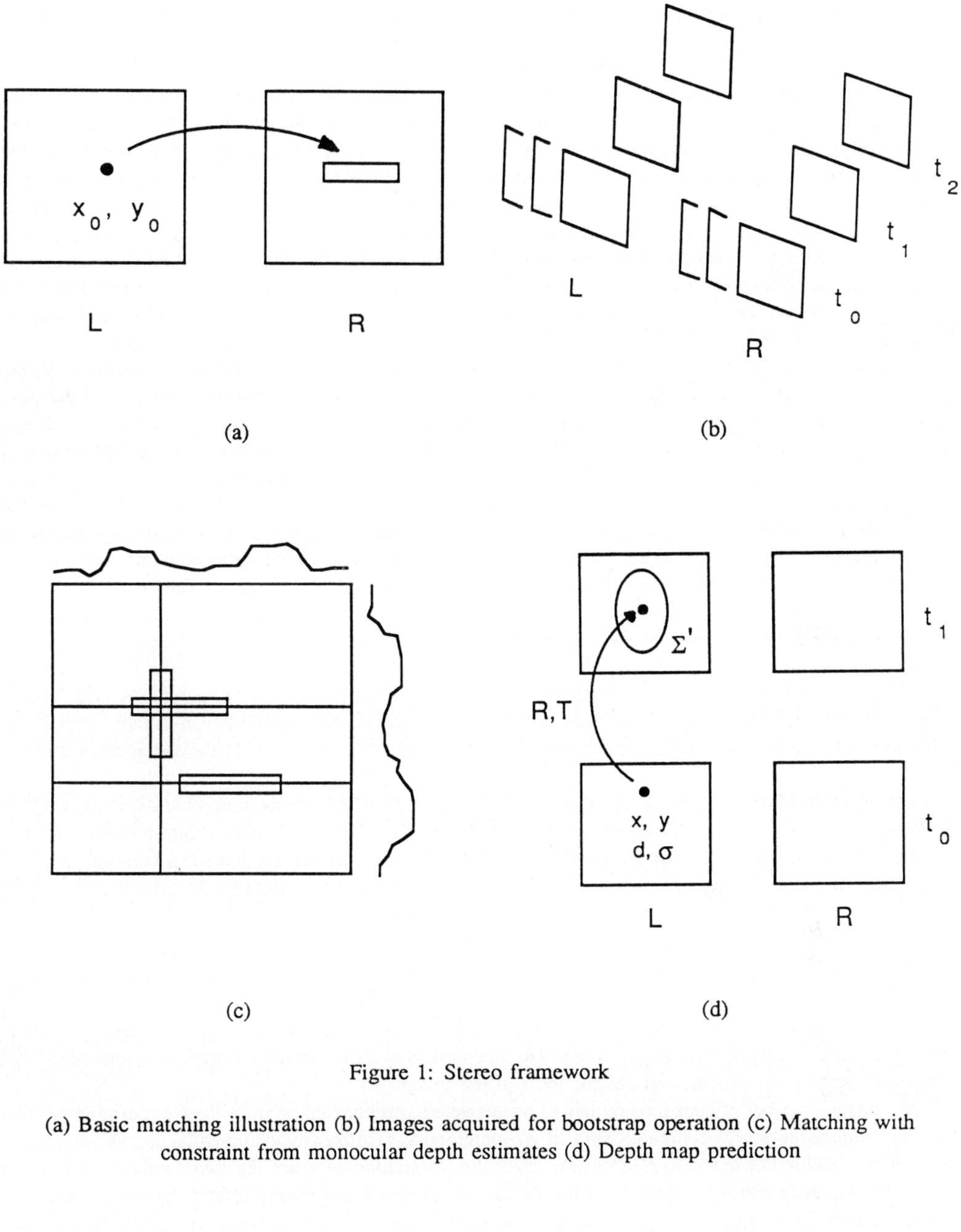

Figure 1: Stereo framework

(a) Basic matching illustration (b) Images acquired for bootstrap operation (c) Matching with constraint from monocular depth estimates (d) Depth map prediction

far to move the cameras between images. Instead, we will just ask how to process a sequence of two or more images from each camera and, for a particular procedure, how well it works.

If only two images are taken, equation (1) can be applied directly to match between the images and to estimate depth and variance. For more than two images, we require that they be processed in an on-line fashion; that is, that we match each new image with the previous image, update the depth map, and discard the previous image. It is possible to obtain a series of increasingly precise depth estimates in this way by increasing the distance between successive images and retaining only the results of matching between the most recent pair [27]. However, we wish to know if a sequential estimation scheme can be designed that will integrate depth measurements over the whole image sequence, so that the precision of the final estimate reflects matching between all images, not just the most recent two.

There are several ways to implement such a procedure. In [18], we developed an algorithm that treats the entire depth map as the state vector of a Kalman filter. The algorithm obtains depth measurements by matching between adjacent frames of the image sequence and estimating the disparity and disparity variance at every pixel. This is currently done with an intensity differencing approach, similar to that described in section 2, that computes disparity to sub-pixel precision. The estimation cycle of the filter operates as follows. Matching between the first two frames of the image sequence provides depth and depth variance estimates at each pixel in the second image. The known translation of the camera is then used to project each pixel into the third image of the sequence; this essentially projects the pixel grid onto the third image, warping it according to the depth at each pixel. This projection is resampled by linear interpolation at every pixel in the third image to generate depth predictions at each pixel. Matching is then performed between the second and third images to obtain new depth measurements that are used to update or correct the predictions. The update uses the standard Kalman filter update equation; in this case, it boils down to a weighted average between the new measurement and the predicted depth, with the weights given by the respective variances. The cycle then repeats with the next image.

We have evaluated this algorithm by deriving the theoretical error variance as a function of the number images processed and by comparing this with the actual variance in experiments with images of a flat poster. For constant camera translation between frames, our mathematical model implies that the error variance after n frames should be proportional to σ^2/n^2, where σ^2 is the variance of a single depth measurement. In the experiments, we computed the RMS relative error in depth estimates after processing each image. Figure 2a shows results for which RMS error was computed over all pixels in the image (top curve) and over only those pixels corresponding to distinct edges (bottom curve). Curves for the theoretical standard error are overlaid on both experimental curves, scaled to coincide with the first point on the experimental curves. The results show that depth estimates do improve, though at slightly lower than the expected rate.

We also tested the algorithm by using it to reconstruct depth from images of realistic, complicated model railroad scenes. Figure 2b shows an image of one of the scenes. Figures 2c-d show, respectively, a reconstructed depth map with depth coded as intensity (lighter is nearer) and a three-dimensional, perspective rendering of the scene created from the depth map and one of the original images. To obtain these results, the algorithm outlined above was augmented with a surface interpolation procedure that "filled-in" depth estimates in large, blank areas of the image.

Although a number of questions remain to be answered, our conclusion from these experiments is that we can quite successfully compute depth in an incremental fashion from monocular image sequences. A primary concern for the next stage of the algorithm (matching between the two cameras) is whether or not the variance estimates computed by the algorithm above accurately reflect the real variance of the estimation error. Our findings so far are that they are roughly proportional (this is supported, for example, by Figure 2a), but that the absolute scale of the variance estimates is not correct. This may be a calibration problem, but it requires more investigation.

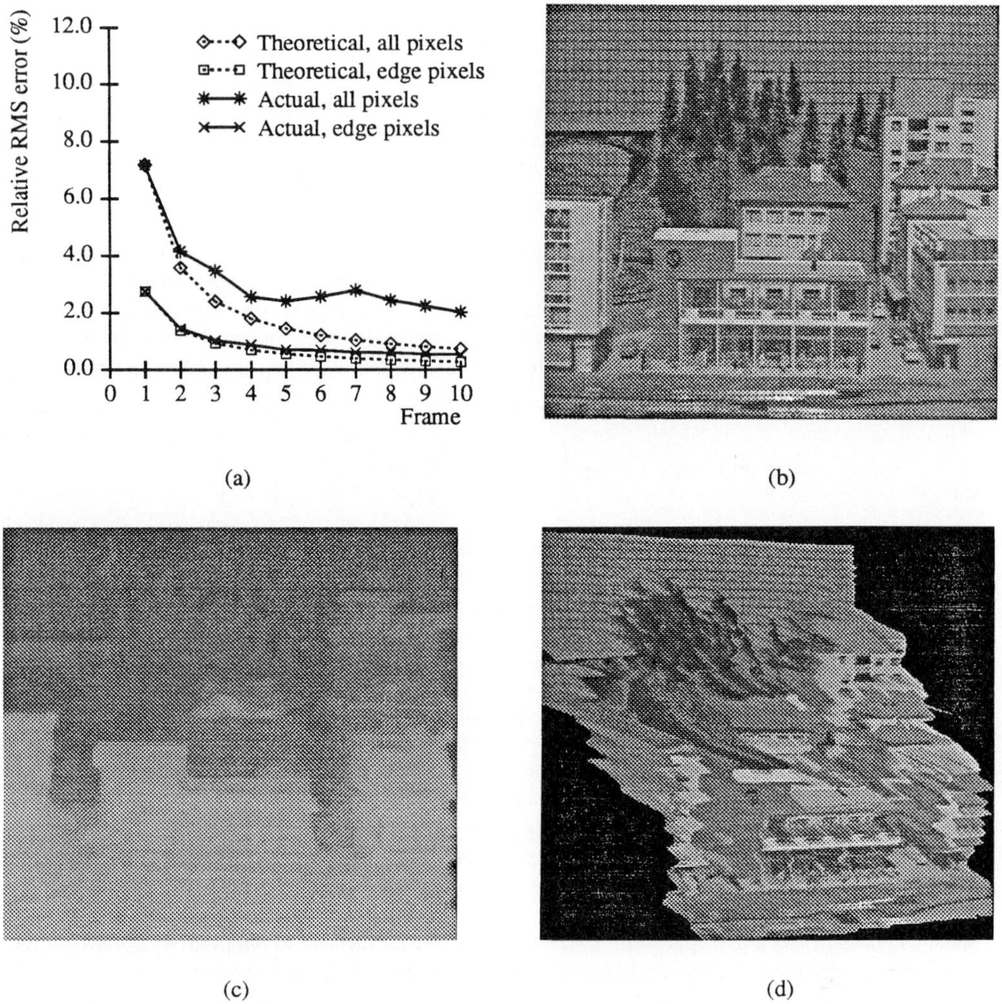

Figure 2: Depth estimation from an image sequence

(a) Error vs. number of images, poster experiment (b) Model train scene (c) Depth map for model train scene (d) Perspective reconstruction of model train scene

3.2 Stereo Matching

In the stereo matching stage of the bootstrap operation, we will use depth maps and variance estimates obtained from translating the cameras to guide matching between the two cameras. Using a depth map acquired with motion of only one camera is a straightforward application of the Bayesian formulation of equation (1), in which the depth estimates play the role of the prior distribution. Since this is less complicated than using depth maps acquired with both cameras, we will examine this case first. Implementation of the algorithms described in this section is currently in progress.

As discussed in section 2, our approach seeks the best match for each pixel independently of other pixels. With good prior depth estimates, this approach will be adequate; moreover, it will be considerably more efficient than approaches that attempt to satisfy global constraints on surface shape [2,3,13,14,23]. Since the monocular depth estimates model the estimation error as normally distributed, we can create a search window (as in Figure 1a) from a confidence interval for the mean of the prior distribution. Using equation (1), with the monocular depth estimates serving as priors, the posterior probability for disparities d_i in the search window will be

$$P(d = d_i|\varepsilon_i) \propto \exp\left\{-\frac{(d_i - d_p)^2}{2\sigma_p^2}\right\} \exp\left\{-\frac{1}{2\sigma^2}\sum_{jk} \varepsilon_{ijk}^2\right\} \tag{2}$$

where d_p, σ_p^2 are the mean and variance of the prior depth distribution, respectively. The MAP estimate of disparity determined by (2) is the value of d_i that minimizes

$$\frac{(d_i - d_p)^2}{\sigma_p^2} + \frac{1}{\sigma^2}\sum_{jk} \varepsilon_{ijk}^2 \tag{3}$$

In practice, however, minimizing (3) directly may not be the best implementation. For example, it may be better to evaluate just the second (likelihood) term within the search window and to use the first (prior) term only when necessary to select between multiple minima in the likelihood term. To our knowledge, this issue of how to use non-uniform priors in the matching process has not been addressed before, so the best way to do so is an open question. The stereo algorithm applies (3) to find a match for each pixel, with a confidence interval test on the value of the likelihood term to determine when no acceptable match exists.

When we use monocular depth estimates from both cameras to guide matching, they cannot both serve as priors in the above approach. Two equivalent possibilities are to treat one as a prior and one as an additional observation, or to treat both as observations and assume a non-informative prior. In either case, confidence intervals for the means of both monocular depth estimates can be used to determine which potentially matching pixels are actually consistent with each other. This is illustrated in Figure 1c, which shows the process of matching on a single pair of epipolar scanlines [13,23]. The intensity profile from the scanline in the left image is shown across the top of the figure and the profile from the right image runs down the right side of the figure. Intersections (*vertices*) of grid lines represent possible correspondences between the two scanlines. The rectangles represent search windows determined from the monocular depth estimates. Consistent match candidates are those vertices overlapped by search windows from both images. In this case, the stereo algorithm identifies "correct" matches as all consistent vertices for which

$$\frac{(d_i - d_l)^2}{\sigma_l^2} + \frac{(d_i - d_r)^2}{\sigma_r^2} + \frac{1}{\sigma^2}\sum_{jk} \varepsilon_{ijk}^2 \tag{4}$$

is minimal in the grid row and column containing the vertex. d_l, σ_l^2, d_r, σ_r^2 are the means and variances of the monocular depth estimates for the left and right images. Note that the considerations surrounding (3) apply here as well.

In summary, we augment existing approaches to stereo matching by using prior depth information obtained from camera motion to control search. The prior information can be used simply to define unweighted search windows, or it can be used to weight the likelihood function, as in (3) and (4). This is a new issue, so the most appropriate way to use the prior information remains to be seen. We expect that using monocular depth information from both cameras will provide significantly better results than from just one camera, because of the extra degree of redundancy in requiring consistency of the two sets of prior information. Once the stereo matching process is complete, we will compute updated estimates of the variance of the depth measurements. This is discussed in [15].

4 Processing Stereo Image Sequences

After the bootstrap operation, the robot follows a task-oriented trajectory and digitizes stereo pairs to monitor its environment. This is illustrated by images at times t1, t2, etc. in Figure 1b. For each new stereo pair, we would like to simplify the matching problem by taking advantage of depth information obtained from previous images. This can be done by projecting the existing world model (that is, the depth map from the previous stereo pair) onto the new stereo pair to create prior depth distributions for each pixel. To do this projection, we need an estimate of the camera motion between frames. In this section, we consider how to do the projection and how to obtain the motion estimate visually.

4.1 Generating Depth Predictions

Our approach involves projecting pixels forward from the previous depth map into the new images, as shown in Figure 1d, to predict the depth at each pixel in the new image. Given that the depth at pixel x, y in the previous map is d and that the robot's motion is given by a a vector of rotation angles $R = [\theta_x\ \theta_y\ \theta_z]^T$ and a translation vector $T = [t_x\ t_y\ t_z]$, it is straightforward to compute the image coordinates and depth for the corresponding 3-D point as seen in the new images. For brevity, we represent this transformation by the nonlinear function

$$[x'\, y'\, d']^T = f(x, y, d, R, T) \tag{5}$$

Since the depth and motion estimates d, R, T will be noisy, the projected image coordinates and depth estimate $[x' y' d']$ will be random variables. We can model their distribution as normal by approximating d, R, T as jointly normal, linearizing (5), and doing an approximate error propagation [20] to estimate the covariance Σ' of $[x'\, y'\, d']$. A confidence interval determined from the marginal distribution of the new image coordinates, $[x'\, y']$, will determine a window in the new image within which we expect the point to project (see Figure 1d). For each pixel x_i, y_i within this window, the conditional distribution $P(d'|x_i, y_i)$ yields a prior distribution for the depth at x_i, y_i. In this way, projecting each pixel from the previous depth map into the new images will result in a set of correlated depth predictions for each pixel. A simple, worst-case approach to resolving these is to turn each prediction into a confidence interval for the depth at that pixel and to use the smallest interval that encloses these as a search constraint for matching.

4.2 Visually Estimating Relative Position

The relative position estimte (R, T) required for the prediction process can be obtained from motion sensors on the robot or, if necessary, from open loop trajectory predictions. However, a number of factors make it desirable to obtain these estimates directly from the images, including simplicity of calibration, modularity, and potential compliances between the cameras and other onboard motion sensors. In previous work [17,16,22,25], we developed a system that does so by locating and tracking distinct

(a)

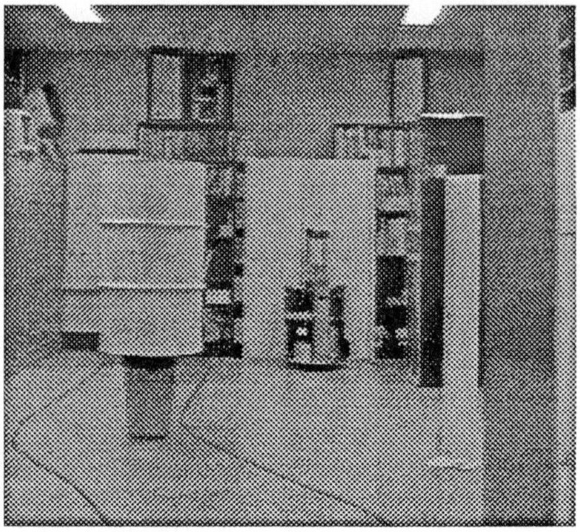

(b)

Figure 3: Motion estimation results

three-dimensional features of the environment. We review the main points of the system below. Related work is described in [1,10,26].

Visual position estimation requires

- obtaining a three-dimensional model of objects within view,
- constructing an image-processing system that can locate parts of the model in the current images,
- defining equations that relate the observed projection of the model to the unknown relative position of the current and previous images,
- solving these equations for the relative position.

In our system, the model consists of three-dimensional points in the environment of the robot. This model is constructed automatically from the first stereo pair by using an "interest operator" to locate distinctive points in one of the images and matching them in the other image. After the robot has moved and acquired a second stereo pair, the model points are located in the new images by a coarse-to-fine correlation procedure that uses approximate knowledge of the robot's motion to constrain the search. We then compute the relative position of the two stereo pairs, estimate the covariance of these parameters, and update covariance estimates for the three-dimensional model points. Details of these procedures are discussed in [17,25].

To test the quality of the position estimates, we used the system to monitor the global position of a robot vehicle as it drove across a room. Figure 3a shows the vehicle used for this experiment and Figure 3b shows one image of the room. Figure 4 shows the floorplan of the room with the actual and estimated robot positions for two experiments. The robot proceeded from left to right, digitizing images after every 10 cm of travel in Figure 4a and every 2 to 7 cm in Figure 4b. The dots in the figure show the actual robot position for every image; the diamonds show the position estimated by the vision system. The relative frame-to-frame estimates have been concatenated to show the global position.

Qualitatively, this system performs very well. The jogs in the estimated paths are due to incorrectly tracked model points that were not rejected by our existing outlier detection procedures. Further work will determine the extent to which this can be improved, the actual magnitude of the estimation errors, and whether the results are good enough to use in the prediction process above.

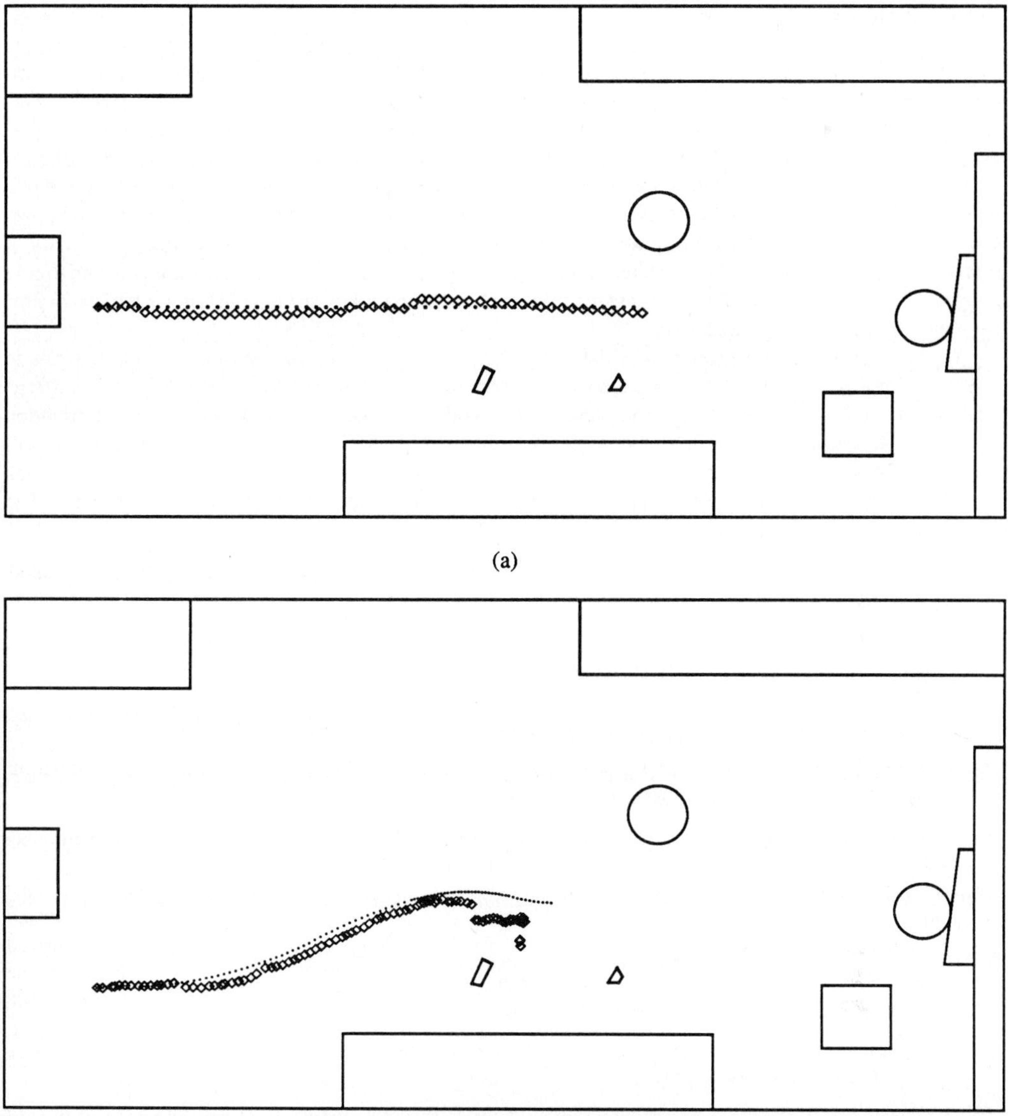

(a)

(b)

Figure 4: Motion estimation results

5 Conclusions

In this paper, we introduced a Bayesian approach to processing stereo image sequences that is based on representing the depth at each pixel as a probability distribution. This approach is appropriate for dynamic robot vision systems for several reasons. First, it allows us to express prior depth information available from a variety of sources, including known disparity limits, other range sensors, or map databases. Second, it provides a foundation for the use of exploratory camera motions to reliably initialize stereo fusion. This involves using fine camera translation to obtain coarse estimates of depth with each camera, then using the coarse estimates to guide matching between the cameras. Finally, it offers a way to use the results of stereo matching in previous images to constrain matching in future images. This involves projecting depth maps from each stereo pair forward to generate search constraints for matching in the next stereo pair.

We have demonstrated results from the fine translation portion of the exploratory motion scenario and are currently implementing the stereo fusion component. This will become a "stereo bootstrap" module for a robot vehicle. We plan to combine this with previous work on visual position estimation to extend the system to process stereo sequences obtained as the robot moves. Future research will determine the success of the predictive approach to processing stereo image sequences and will seek to make the system self-monitoring, so that the robot can use exploratory motions whenever needed to resolve ambiguous perceptions.

References

[1] N. Ayache and O.D. Faugeras. Building, registrating, and fusing visual maps. In *Proc. 1st Int'l Conf. on Computer Vision*, pages 73–82, IEEE, June 1987.

[2] H. H. Baker and T. O. Binford. Depth from edge and intensity based stereo. In *Proc. 7th Int'l Joint Conf. on Artifical Intelligence*, pages 631–636, August 1981.

[3] S. T. Barnard. Stochastic stereo matching over scale. In *Proc. DARPA Image Understanding Workshop*, pages 769–778, Morgan Kaufman Publishers, April 1988.

[4] R. C. Bolles, H. H. Baker, and D. H. Marimont. Epipolar-plane image analysis: an approach to determining structure from motion. *International Journal of Computer Vision*, 1:7–55, 1987.

[5] J. L. Crowley. *A Representation for Visual Information*. PhD thesis, Carnegie-Mellon University, November 1981.

[6] W. Forstner and A. Pertl. Photogrammetric standard methods and digital image matching techniques for high precision surface measurements. In E. S. Gelsema and L. N. Kanal, editors, *Pattern Recognition in Practice II*, pages 57–72, Elsevier Science Publishers, 1986.

[7] D. Geiger and A. Yuille. Stereopsis and eye-movement. In *Proc. 1st Int'l Conf. on Computer Vision*, pages 306–314, IEEE, June 1987.

[8] S. Geman and D. Geman. Stochastic relaxation, gibbs distribution, and the bayesian restoration of images. *IEEE Trans. Pattern Analysis and Machine Intelligence*, PAMI-6(6):721–741, November 1984.

[9] D. B. Gennery. *Modelling the Environment of an Exploring Vehicle by Means of Stereo Vision*. PhD thesis, Stanford University, June 1980.

[10] D. B. Gennery. Tracking known three-dimensional objects. In *Proc. of AAAI*, pages 13–17, Morgan Kaufman Publishers, 1982.

[11] W. E. L. Grimson. Computational experiments with a feature based stereo algorithm. *IEEE Trans. Pattern Analysis and Machine Intelligence*, PAMI-7(1):17–34, January 1985.

[12] M. Kass. A computational framework for the visual correspondence problem. In *Proc. 8th Int'l Joint Conf. on Artifical Intelligence*, pages 1043–1045, Karlsruhe, W. Germany, August 1983.

[13] D. Marr and T. Poggio. Cooperative computation of stereo disparity. *Science*, 194:283–287, 1976.

[14] J. Marroquin, S. Mitter, and T. Poggio. Probabilistic solution of ill-posed problems in computational vision. *Journal of the American Statistical Association*, 82(397):76–89, March 1987.

[15] L. H. Matthies. *A Bayesian Foundation for Motion Stereo*. PhD thesis, Carnegie-Mellon University, *in preparation.*

[16] L. H. Matthies and T. Kanade. The cycle of uncertainty and constraint in robot perception. In *Robotics Research 4*, MIT Press, *in press.*

[17] L. H. Matthies and S. A. Shafer. Error modeling in stereo navigation. *IEEE J. Robotics and Automation*, RA-3(3):239–248, June 1987.

[18] L. H. Matthies, R. Szeliski, and T. Kanade. *Kalman Filter-based Algorithms for Estimating Depth from Image Sequences*. Technical Report CMU-CS-87-185, Computer Science Department, Carnegie Mellon University, December 1987.

[19] P. S. Maybeck. *Stochastic Models, Estimation, and Control*. Volume 1, Academic Press, New York, NY, 1979.

[20] E.M. Mikhail. *Observations and Least Squares*. University Press of America, Lanham, MD, 1976.

[21] V. J. Milenkovic and T. Kanade. Trinocular vision using photometric and edge orientation constraints. In *Proc. DARPA Image Understanding Workshop*, Dec. 1985.

[22] H. P. Moravec. *Obstacle Avoidance and Navigation in the Real World by a Seeing Robot Rover*. PhD thesis, Stanford University, September 1980.

[23] Y. Ohta and T. Kanade. Stereo by intra- and inter-scanline search using dynamic programming. *IEEE Trans. Pattern Analysis and Machine Intelligence*, PAMI-7(2):139–154, March 1985.

[24] R. Szeliski. Regularization uses fractal priors. In *Proc. AAAI-87*, pages 749–754, Morgan Kaufman Publishers, July 1987.

[25] C. E. Thorpe. *Vision and Navigation for a Robot Rover*. PhD thesis, Carnegie-Mellon University, December 1984.

[26] H.-J. Wunsche. Detection and control of mobile robot motion by real-time computer vision. In *Proc. SPIE Conf. on Mobile Robots*, October 1986.

[27] G. Xu, S. Tsuji, and A. Minoru. Coarse-to-fine control strategy for matching motion stereo pairs. In *Proceedings of IJCAI*, pages 892–894, 1985.

Systolic Arrays for Block Matching Algorithms

Peter Pirsch, Thomas Komarek
Inst. für theoretische Nachrichtentechnik und Informationsverarbeitung, Universität Hannover

Summary

This paper discusses architectures for realization of block matching algorithms with emphasis on highly concurrent systolic array processors. A three step mapping methodology for systolic arrays known from the literature is applied to block matching algorithms. Examples of 2-dimensional and 1-dimensional systolic arrays are presented. The needed array size, the transistor count and the maximum frame rate for processing video telephone and TV signals have been estimated.

1. Introduction

Motion estimation is one of the key components of low bit rate video coding. It improves predictive, transform, and interpolative coding. Because of a less computational complexity mainly motion estimation based on translational models have been applied. The favorite schemes are block matching algorithms (BMAs) because of their simplicity for hardware realization.[1,2,3]

The BMAs determine one displacement vector for a block of NxN pels. A reference block from the actual frame is compared with several candidate blocks on search points within a search area of the previous frame. A matching criterion specifies the displacement vector. Because the mean of the absolute differences (MAD) matching criterion performs the displacement estimation satisfactory with the least computational complexity, architectures for this are discussed here.[1] The BMA using an MAD criterion is specified by

$$s(m,n) = \sum_{i=1}^{N} \sum_{k=1}^{N} |x(i,k)-y(i+m,k+n)|, \quad -p \leq m,n \leq p \tag{1}$$

$$u = \min_{(m,n)} \{s(m,n)\} \tag{2}$$

$$\mathbf{v} = (m,n)|_u . \tag{3}$$

The displacement vector **v** is determined by the least sum s(m,n) for all possible displacements (m,n) from -p to p along each dimension within the search area. The sums s(m,n) consist of the absolute differences between the reference block data x and the candidate block data y.

A video signal with L lines per frame and P pels per line results in $L*P/N^2$ displacement vectors. The number of operations per frame for the BMA using the MAD criterion is $q*(2+3*N^2)*L*P/N^2 \simeq 3*q*L*P$ where q is the number of search points for each displacement vector. Full search considers all possible search points $q = (2p+1)^2$. For full search the computational rate results in 900 mega operations per second (MOPS) for a video telephone standard (L*P=288*352) with a frame rate of f_F = 10 Hz and in 9000 MOPS for a CCIR video signal with f_F = 25 Hz (L*P=576*720). Search strategies like the 2-D search procedure[3] reduce the computational rate by a factor of 10. The disadvantage of search strategies is the increasing control overhead and the non regular data flow into the operational part.

Real time processing of the BMA at video rates can only be achieved by massively concurrent processing. Algorithms with high through-put rates and high computational rates can be provided by pipelining and parallel processing.[4] To reduce design costs VLSI architectures have to posess properties like regularity and modularity. For simplification of the design implementations without central overhead control is aimed. Systems with local communication permit high clock rates and thus high processing speeds.

General purpose multiprocessor systems consisting of programmable processors are one solution. Data is globally preloaded from a system of busses. In order to have fast data access local memories are assigned to each processor. Parallelism of the algorithm allows single instruction multiple data stream architectures (SIMD) where instructions are broadcasted from a host. To distribute processing task among several processors in a multiple instruction multiple data stream architecture (MIMD) each processor posesses an own control unit, program, and data. A configuration of eight digital signal processors (DSP) ADSP 2100 and a control processor MC 68020, which allow operation of frame rates until 10 Hz for the major codec operations including block matching with special search strategies, has been proposed by May[5] for a 64 kbit/s video telephone codec. Volkers and Wehberg[6] developed a programmable parallel processor configuration where each PE performs 100 MOPS. Because communication bottlenecks occur when multiple processors access shared data resources at the same time architectures with global communication display limitations in computation rate as the number of processors and the through-put rate on the common bus increases.

Regular and recursive algorithms like the BMA mentioned above permit parallel processing and pipelining in special purpose array architectures. Systolic arrays combine the necessary high data througput with a simple design. They consist of only localy connected non-programmable processing elements (PEs) which are working concurrently. Data is pipelined through the array controlled by a global clock. The use of local interconnections between the PEs permits high clock rates and thus a high processing speed. Modular and regular arrays can easyly be derived from algorithms like the BMA. Dyanisan, Baker and Salinas[7] have presented an 8x8-array of bit-serial PEs for the BMA where only the data flow direction in the network is globally controlled by select signals in order to perform search strategies. The bit-serial architectures permit high clock frequencies and a small pin count but the entire computational speed is less than that of bit parallel architectures. Schwerzel[8] has proposed a configuration of a tree structure of subtractors and adders to perform summation according to (1) for a 4x2 block. Architectures consisting of simple PEs for the basic operation of (1) have been proposed by Kraus and Wendt[9] as well as Roth, Tholin, and Wirén[10]. The first configuration performs full search for broadcast video (720x576 pels) up to 25 Hz frame rate on 15 chips[9] whereas the latter one requires 7 chips for a video telephone signal up to 15 Hz.[10]

The architectures described in the previous paragraph are specific solutions of systolic arrays. Because of the strong regularity, recursivness, and locality of full search BMA, systolic arrays will be recognized as the best candidate architectures. Under considerations of the specific video constraints in the following section a generalized design of systolic arrays for BMA will be presented. The mapping procedure provides several alternative array architectures which will be compared concerning performance and estimated chip size in order to specify the most suitable array. Several data streams have to be supplied in parallel to the systolic arrays. In order to support a one-chip solution with limited pin-count, the systolic array has to be combined with a local memory. In addition to the processing architecture the configuration and size of the local memories have to be taken into account because of its severe impact on chip size.

2. Mapping the Block Matching Algorithm onto Systolic Arrays

A systolic array for a given block matching algorithm can be derived by a methodology described by Kung.[4] The first step is the dependence graph (DG) design. A DG is a directed n-dimensional (n-D) graph where nodes represent the computations and arcs represent the data dependencies. The algorithm has to be converted to a form where each result is assigned to a unique variable (single assignment principle). Thus the algorithm is defined over an n-D index space. By a straightforward implementation to each node of the DG a processing element (PE) will be designated. In order to achieve a better processor utilization and a more efficient design a PE can execute the operations of multiple nodes. This mapping requires a specification of a time schedule and an assignment of the nodes to the PEs by projections. The projection leads to a signal flow graph (SFG) which expands over an (n-1)-D space. Because processing and signal flow in the SFG are assumed without any delay additional delay operators assigned to the edges are required for correct timing. Time scaling and delay transfer transform the SFG into a localized form where all edges are having at least one delay element. The systolic array is derived from the SFG by substituting the nodes by PEs, the edges by local interconnections, and the delay operators by pipeline registers. Because time scheduling and projection provide several solutions the most suitable one has to be selected.

In order to implement on a VLSI chip an array with local communications only architectures with a 2-D expansion or less can be derived from algorithms which have to be defined over 3-D or 2-D index spaces. The BMA, which is defined over a 4-D index space, is a piecewise regular algorithm with two computational hierarchies. The double sum s(m,n) has to be determined (1) prior to the minimum search min{s(m,n)} (2). Because both computational hierarchies have 2-D index spaces they can be mapped onto 2-D or 1-D systolic arrays by separate procedures.

To achieve more alternative systolic arrays the BMA has to be decomposed into parts with higher dimensions because time scheduling and projections give more different solutions. As an example the BMA is decomposed by fixing the index n that 2p+1 3-D sub-index spaces occur. If full search is applied intermediate minima are calculated for the sums $s_n(m)$ along a search line (4) in a search line level. On a proceeding superior level the final minimum is determined (6). The notation of the BMA specifies by a subscript the fixed indices during an operation. It should be recognized that all equations below consider shifted indices m and n in order to have positive numbers for i,k,m, and n.

$$u(n) = \min_{(m)} \{s_n(m)\}, \qquad 0 \leq m \leq 2p \tag{4}$$

$$v(n) = (m)|_{u(n)} \tag{5}$$

$$u = \min_{(n)} \{u(n)\}, \qquad 0 \leq n \leq 2p \tag{6}$$

$$v = (v(n),n)|_u . \tag{7}$$

For fixed n the notation of the canditate block data will be modified in

$$y(i+m,k+n) \rightarrow y_n(i+m,k). \tag{8a}$$

The double sum $s_n(m)$ has to be estimated sequentially by two additions. Starting with the index i first it follows from (1)

$$s_n(m,k) = \sum_{i=1}^{N} |x(i,k)-y_n(i+m,k)|, \quad 0 \leq m \leq 2p, \tag{9a}$$

$$s_n(m) = \sum_{k=1}^{N} s_n(m,k). \tag{9b}$$

The DG is a graphical representation of the single assignment form of the algorithm. Because input data x(i,k) and $y_n(i+m,k)$ are required at several nodes in the (i,m,k)-space they have to be broadcasted. To achieve a localized form of the DG input data is assigned to intermediate variables which are exchanged between neighbouring nodes. These intermediate variables are denoted by capital letters. Only the variables entering and leaving the DG remain in their origin form. For further simplification the indices of y_n are transformed

$$y_n(i+m,k) \rightarrow y_n(j,k) \tag{8b}$$

that all input data move on a lattice across the (i,j,k)-DG (Fig. 1). The index j can be seen as the extent of all candidate block data in a search area line. The DG for one search line of the BMA consists of three regular parts. The first one extends over a 3-D space filled with nodes of type AD in which the subtraction, the determination of the absolute values, and the addition over the index i like equation (9a) are executed. The nodes of type A are placed in a plane. They represent the addition to the final sums (9b). In the third regular part with nodes of type M the minimum search and the selection of the vector components according to equation (4) and (5) is performed. For simplification only the nodes on the surface of the DG and the nodes with valid results are displayed.

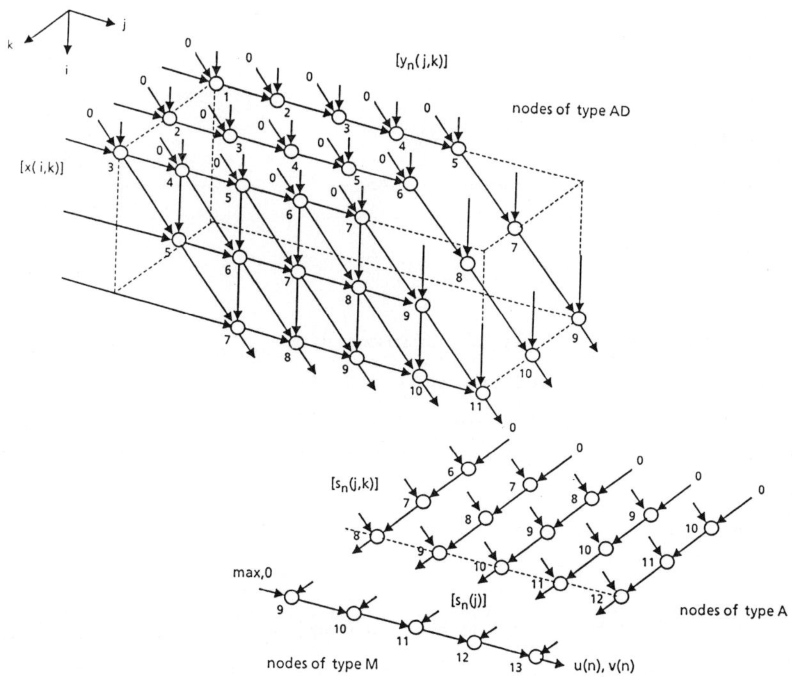

Fig. 1: Dependence graph for fixed index n with the computation nodes AD, A, and M of a BMA for N = 3 and p = 2. The timing is specified by the time schedule vector (1,1,1).

The exact operations of each node in the DG are specified below. Equation (9a) is transformed to (10), (9b) to (11), and (4) and (5) are transformed to (12). The relations below consider the single assignment principle and include the data transfer between the nodes. The recursion indices are denoted by exponents.

nodes AD:

$1 \le i,k \le N,\ 0 \le j \le 2p+N$:

$$S_n(j,k)^i = S_n(j-1,k)^{i-1} + |X(j,k)^i - Y_n(j,k)^i|, \qquad S_n(j-1,k)^0 = 0$$
$$X(j,k)^i = X(j-1,k)^i, \qquad X(0,k)^i = x(i,k)$$
$$Y_n(j,k)^i = Y_n(j,k)^{i-1}, \qquad Y_n(j,k)^0 = y_n(j,k)$$
$$s_n(j,k) = S_n(j,k)^N \qquad (10)$$

nodes A:

$1 \le k \le N,\ N \le j \le 2p{+}N$:

$$S_n(j)^k = S_n(j)^{k-1} + s_n(j,k), \qquad S_n(j)^0 = 0$$

$$s_n(j) = S_n(j)^N \tag{11}$$

nodes M:

$N \le j \le 2p{+}N$:

$$U_n^{\,j} = \min\{s_n(j),\ U_n^{\,j-1}\}, \qquad U_n^{\,N-1} = \max$$

$$\mathbf{V}_n^{\,j} = \begin{cases} j-N & \text{if } s_n(j) < U_n^{\,j-1} \\ \mathbf{V}_n^{\,j-1} & \text{otherwise} \end{cases}, \qquad \mathbf{V}_n^{\,N-1} = (0)$$

$$u(n) = U_n^{\,2p+N}$$

$$\mathbf{v}(n) = \mathbf{V}_n^{\,2p+N} \tag{12}$$

Using the schedule vector (1,1,1) minimal computation time can be achieved. The projection onto a plane perpendicular to the projection vector (0,0,1) leads to a 2-D SFG (Fig. 2). The final systolic array, denoted as AS1, can be derived from the SFG without any time scaling and delay transfer (Fig. 3). One displacement vector is determined in Nx(2p+1) steps considering a final minimum search (eq. 6 and 7) by an additional processor M.

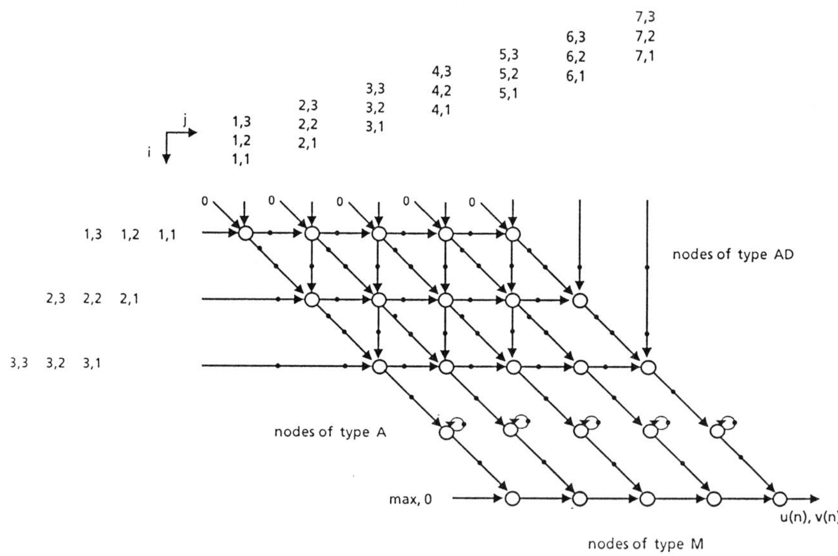

Fig. 2: Signal flow graph (SFG) with the computation nodes AD, A, and M derived from the DG in Fig. 1 by projection along (0,0,1).

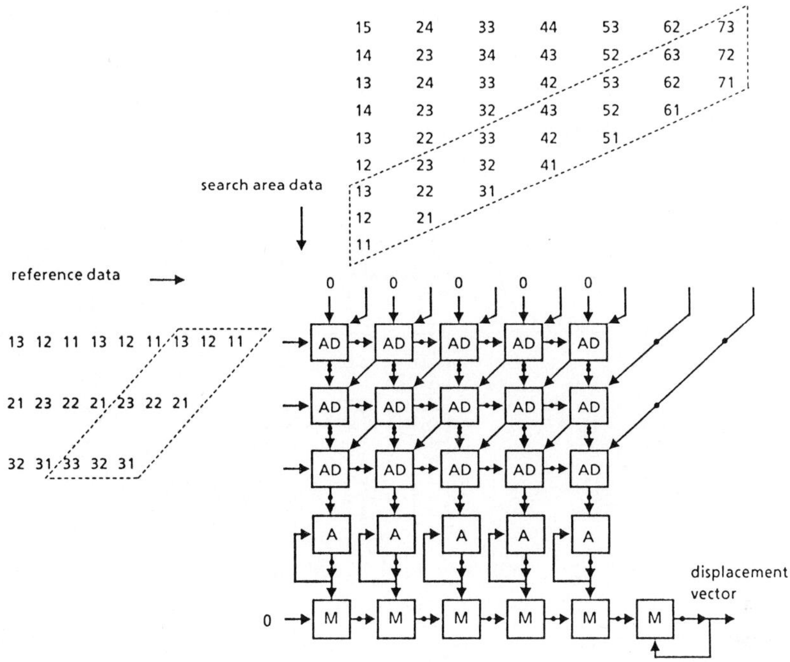

Fig. 3: Systolic architecture AS1 with processing elements AD, A, and M derived from the SFG and the indices of input data x(i,k) and y_n(j,k).

Alternative systolic arrays can be easily derived by modifications of DG description and different projections. The example AB1 in Fig. 4a was found by changing the sequence of summations (eq. (1)) and executing the same steps of development for a 3-D index space. The interconnections between the PEs M in the arrays AB1 and also AS1 permit in addition to full search only search strategies with equidistant search points. The smaller architecture AB2 in Fig. 4b was derived according to the decomposition of the 4-D index-space into two 2-D sub index spaces for a search area and a block level. Array AB2 allows all kinds of search strategies because only one search block is processed at a time.

3. Systolic Arrays for Block Matching

The needed frame rate f_F and the pel count L*P of one frame are the major constraints for real time computation. Because of pipelining the two computation hierarchies (determination of s(m,n) and minimum search) are executed sequentially. For this reason the computation time is determined by the number k of PEs AD. If $c = N^2/k$ additions are performed sequentially $L*P/N^2$ displacement vectors of a frame have to be estimated in the time

$$T_V = T * q * c * L*P/N^2 = q/k * T * L*P = r * T * L*P \leq 1/f_F \tag{13}$$

with the rate r as the average number of clock cycles T per pel

$$r = q * c * 1/N^2 \leq 1/(T*f_F) * 1/(L*P). \tag{14}$$

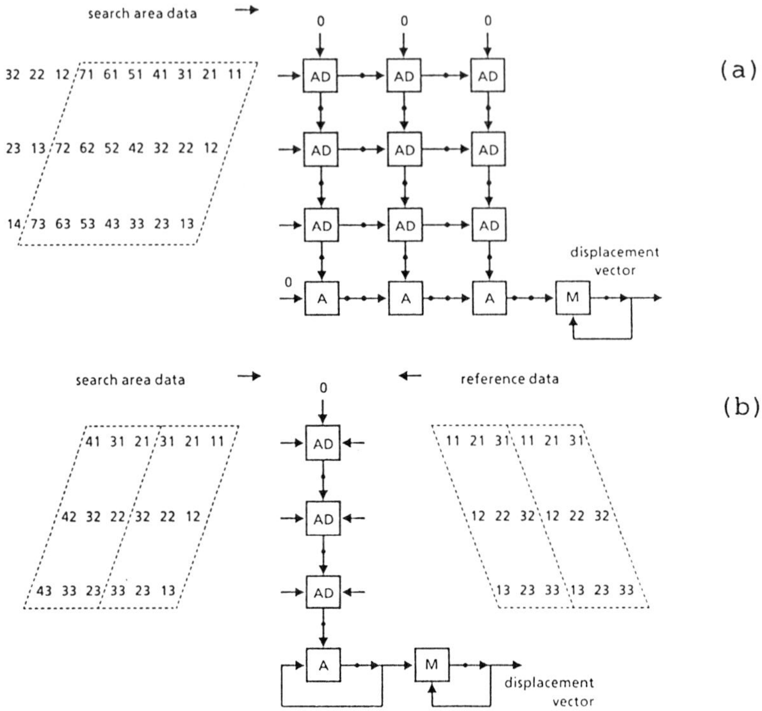

Fig. 4: Systolic architectures AB1 (a) and AB2 (b) for N = 3 and p = 2 with the indices of search area data $y_n(j,k)$. The reference block data $x(i,k)$ of architecture AB1 remain fixed in the PEs AD.

Table 1 shows the possible expansions of arrays, their rates r for full search, and the number c of parallel computations. The assumptions are made that all PEs are working concurrently. With N^2 PEs AD the computations of (1) can be done in parallel in one time step and with only one PE AD s(m,n) has to be determined sequential in N^2 time steps. If the number k of PEs AD is extended over the search area like in the architecture AS1 the necessary number of parallel computations q*c for one displacement vector decreases to $q*c = N*(2p+1)$. Finally with (2p+1)x(2p+1) PEs of type AD $q*c = N^2$ time steps are required for the determination of one displacement vector.

k	r	c	presented architecture
1	$(2p+1)^2$	N^2	
N	$(2p+1)^2/N$	N	AB2
2p+1	2p+1	$N^2/(2p+1)$	
NxN	$(2p+1)^2/N^2$	1	AB1
Nx(2p+1)	$(2p+1)/N$	$N/(2p+1)$	AS1
(2p+1)x(2p+1)	1	$N^2/(2p+1)^2$	

Table 1: Array size k, average number r of clock cycles per pel, average number c of parallel computations per vector for full search on $q = (2p+1)^2$ search points.

To support a one-chip solution a local memory has to be developed. It contains data which has to be used several times in a short period. Global clocked shift registers permit high speed processing with a minimum of circuitry. But their architecture preassumes a regular data flow. Search strategies with a conditional data flow require a local RAM circuitry. Its disadvantage is a higher transistor count and more sophisticated control but smaller rates r can be achieved.

architecture	AD	A	M	transistors	f_{Fmax} (CIF)	f_{Fmax} (CCIR)
AB1	8x8	8	1	70,000	87.3	26.7
AB2	8	1	1	50,000	10.9	3.3
AS1	17x8	17	17+1	120,000	185.2	56.6

Table 2: Number of PEs, transistor count for three arrays including local memory for the BMA with 8x8 blocks, 24x24 search window, and full search, maximum frame repetition frequencies f_{Fmax} for video telephone (CIF) and broadcast TV (CCIR) format assuming 40 MHz clock.

Table 2 presents the properties of three configurations consisting of systolic arrays and their local memories. Nearly 80 per cent of the transistor count which is estimated by a known cell library[11] results from local memories. The maximum frame repetition frequencies f_{Fmax} are assumed for a data flow without any dummy cycles.

Because search area data is flowing continously into the array AB1 (Fig. 4a), at the end of each search point line nonsensical results occur for N-1 clock cycles. These dummy cycles have to be taken in account for the total number of clock cycles for the determination of one displacement vector:

$$(q*c)' = (2p+1+N-1)*(2p+1).$$

The average number of clock cycles per pel rises to

$r' = (2p+N)*(2p+1)/N^2$.

For N = 8 and p = 8 the computational speed decreases about r/r' = 70,1 %.

Even the architecture AB2 is able to estimate the displacement vector field for the video telephone format in real time. The computational speed of the architectures AB1 and AS1 with a 2-D extent is sufficient for broadcast TV signals. For full custom design the needed number of transistors can be further reduced.

4. Conlusions

Because of the regularity of block matching with full search or regular search strategies systolic arrays are ideal candidates for VLSI realization. This paper describes the procedure of mapping the block matching algorithm onto systolic arrays and examples of 2-D and 1-D arrays are given. Because of processing speed and limited pin count such a block matching chip has to incorporate local memories for parallel data transfer to the operational part. Considering the state of the art CMOS technology real time processing of video telephone and TV signals is possible for a single chip realization. For the future new mapping procedures are requested which includes the design of local memories.

References:

[1] R. Srinivasan, K. R. Rao, "Predictive Coding Based on Efficient Motion Estimation," Proc. ICC'84, pp. 521-526, May 1984

[2] H. G. Musmann, P. Pirsch, H.-J. Grallert, "Advances in Picture Coding", Proc. IEEE, vol. 73, pp. 523-548, April 1985

[3] J. J. Jain, A. K. Jain, "Displacement Measurement and Its Application in Interframe Image Coding, IEEE Trans. on Comm., vol. COM-29, pp. 1799-1808, July 1981

[4] S. Y. Kung: VLSI Array Processors. Prentice Hall, Englewood Cliffs, 1988

[5] F. May, "Full Motion 64 kbit/s Video Codec with 8 DSP's," International Workshop on 64 kbit/s Coding of Moving Video, Hannover, FRG, June 1988

[6] H. Volkers, T. Wehberg, "Architektur eines Prozessors für die Echtzeit-Verarbeitung von Algorithmen der Bildcodierung," ITG-Diskussionssitzung, Königstein, FRG, Nov. 1987

[7] R. Dianysian, R. Baker, J. L. Salinas, "A VLSI Architecture For Template Matching and Motion Estimation, ISCAS'88, Helsinki, Finland, June 1988

[8] W. Schwerzel, "Prozessorarchitektur zur Bewegungsvektorsuche in der digitalen Bildanalyse," ITG-Fachberichte No. 98, Großintegration, pp. 221-225, vde-Verlag 1987

[9] G. Roth, P. Tholin, B. Wirén, "A VLSI for Motion Compensation", Picture Coding Symposium, Stockholm, Sweden, 1987

[10] J. Kraus, H. Wendt, "VLSI-Multiprozessor zur Bewegungsvektorsuche für Experimente am 64 kbit/s ISDN Bildtelefon", to be presented at the ITG conference "Mikroelektronik für die Informationstechnik", Berlin, FRG, Oct. 1988

[11] LSI Logic Coorporation: LSI-Logic Databook and Design Manual, 1988

Biography:

Peter Pirsch received the Dipl.-Ing. and Dr.-Ing. degrees from the University of Hannover (FRG), where he was first research assistant and later senior engineer. During 1979 to 1981 he was on leave working in the Visual Communications Research Dept., Bell Labs., Holmdel, NJ. From 1983 to 1986 he was head of the Digital Signal Processing Dept. at the SEL Research Ctr., Stuttgart, (FRG). He has been professor of electrical engineering at the University of Hannover since 1987. Dr. Pirsch is a senior member of the IEEE and a member of the ITG. He was recipient of the NTG paper award in 1982.

Thomas Komarek received the Dipl.-Ing. degree of electrical engineering from the University of Hannover (FRG) in 1987. He is now working as a research assistant at the Institut für Theoretische Nachrichtentechnik und Informationsverarbeitung at the University of Hannover. His major research interests are the development of architectures for high speed signal processing.

Modulation Transfer Function Obtained from Image Structures

Fang Lei, Hans Tiziani
Institut für Technische Optik, Universität Stuttgart

Summary

A new and relatively simple method for edge gradient analysis (EGA) has been developed to determine the modulation transfer function (MTF) of photogrammetrical systems. The MTF can be calculated as the ratio of the amplitude of those frequencies present in the original edge to the amplitude of the same frequency in the scanned edge image. Thus a differentiation of the edge scan - as usually implemented - is not necessary. In order to avoid a division by zero we multiplied the edge images by the so called Hanning function before performing the Fourier analysis. From the point of noise sensitivity the method using a Hanning window is advantageous in comparison to the edge differentiation method. The edge gradient analysis method and the grating pattern method are compared. Artificial edge, natural roof edge, and grating patterns were used to determine the quality of the photogrammetrical system. Good agreement of all MTF measurements was found. Furthermore, the resolution found from MTF curves agreed well with the resolution obtained from three-bar targets. Generally, the MTF curves obtained from patterns in the flight direction were lower than the MTF's perpendicular to the flight direction due to airplane movement. The influence of linear image motion and its compensation was examined and is discussed. Furthermore, the longitudinal motion of the object as defect of focus is studied analytically and experimentally for a close range.

1. Introduction

In photogrammetry the image quality of photographs is of great importance. The optical transfer function (OTF) and especially the modulation transfer function (MTF) are useful techniques for measuring the image quality of photographs. The MTF is the result of the contributions of the links in the image transfer forming chain, namely, the camera, the film, image motion (in systems without forward motion compensation, FMC), vibratins, and the atmosphere /1/.

In this paper, a new method for edge gradient analysis (EGA), using a Hanning window, is developed and discussed. A comparison is made with the other methods to measure the MTF of aerial survey lens systems from the image structures. The measured MTF curves are given showing the reliability of different methods. Furthermore, the influences of linear image motion and focusing defect on image quality is examined.

2. Methods for Determining the MTF from Image Structures

The edge gradient analysis (EGA) is a technique for obtaining the MTF from edge images recorded on photographic film, especially useful for evaluation of images not containing targets. During the past 20 years different calculation methods for EGA have been developed /2-6/. One method, based on the differentiation of the edge, has often been used for MTF determination. The OTF is determined from the Fourier transform of the spread function, obtained from differentiating the edge image. It can be called an edge differentiation method. We have developed a new and relatively simple method for EGA /6/. The method is based on the Fourier spectrum amplitude ratio between the real and an ideal edge image. It can be called an edge spectrum ratio method. From a purely mathematical analysis both methods are equivalent. However for practical applications they have to be treated differently.

In the edge spectrum ratio method, the MTF is obtained by dividing the spatial frequency spectrum of an ideal edge. From the definition of the MTF, it follows that

$$T(R) = \left|D(R)\right| = \left|\frac{k'(R)}{k(R)}\right| \tag{1}$$

where k(R), k'(R), and D(R) are the Fourier-transforms of an ideal edge, a scanned edge, and the line spread funktion respectively.

The division of the direct numerical Fourier-transforms of the edge in the image and object space leads to a problem because infinitely expanded edges must be limited by a multiplication with a rectangular function (window), leading to zero amplitudes at even harmonic frequencies in the spectrum of the ideal edge. To avoid a division by zero, the rectangular function must be replaced by another window function. We found that the so called Hanning function is a very suitable window function.

The Hanning function can be written as /7/

$$H(u) = \begin{cases} 1/2 - 1/2\cos(2\pi u/L) & 0 \leq u \leq L \\ 0 & u < 0,\ u > L \end{cases} \qquad (2)$$

with the length L of the limited edge.

A smoothed curve is obtained from the spectrum of an edge image multiplied by the Hanning function, which is in good agreement with the spectrum's envelope. Zero amplitudes are no longer present when using the modified edge spectrum ratio method. With a Hanning window the equation (1) can be rewritten as

$$T(R) = \left| \frac{k'_H(R)}{k_H(R)} \right| \qquad (3)$$

where $k'_H(R)$ and $k_H(R)$ are the Fourier-transforms of the Hanning function multiplied by a scanned edge and an ideal edge respectively.

Figure 1 shows schematically the procedure for the MTF determination. Before applying the Fourier-transformation, both scanned

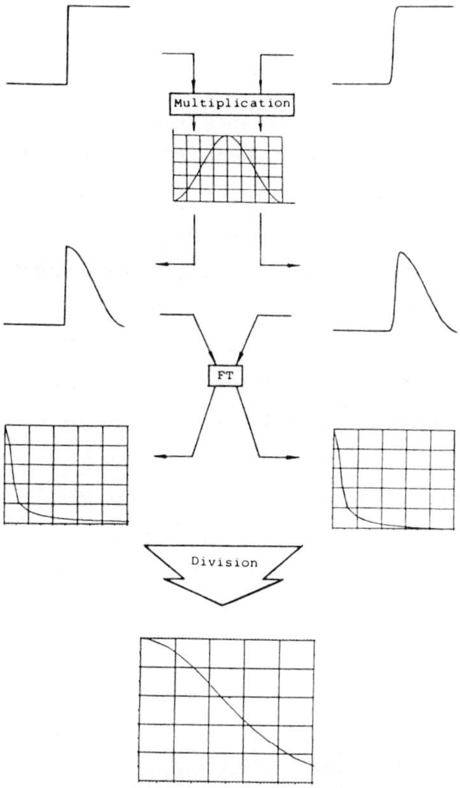

Fig.1 Diagram to calculate the MTF using the Hanning function

and ideal edges must be multiplied by the Hanning function. By the introduction of the Hanning function, noise on top of an edge image can also be suppressed. The edge spectrum ratio method has been compared with the edge differentiation method. From the point of noise sensitivity the edge spectrum ratio method using a Hanning window is advantageous.

For comparison with EGA, a variable spatial frequency grating pattern was used for the determination of the MTF. The contrast of the photographed pattern decreases due to diffraction and different disturbances with increasing spatial frequency. Using the convolution, the imaging process can be simulated with a computer. The intensity distribution of the grating pattern is convoluted with a computer simulated model of the spread function and the result should compare with that of the scanned image of the grating /8/. The MTF is calculated as the absolute value of the Fourier-transform of the chosen spread function.

3. Comparison between the MTF's Obtained from an Edge Image and a Grating Pattern

A pair of artificial edge and grating patterns was oriented in the flight direction, and another pair was oriented orthogonally to it. Two suitable natural roof edges were used to compare the artificial edge. The test pattern images and the natural edge images were scanned with a computer controlled microdensitometer and the measurements were evaluated by a computer.

Figure 2 shows the MTF's obtained from grating patterns and from artificial and natural edges both in the flight direction (a) and perpendicular to the flight direction (b). It should be noted that there is little difference in the MTF of a natural and an artificial edge. We found good agreement of the results obtained with different test patterns.

For an additional test of the MTF, we examined three-bar patterns with different spatial frequencies. They were put close to the grating and edge patterns in order to determine resolution limits. The spatial frequencies at the intersection of the MTF, for well-defined artificial edges, and the threshold curves compare well with the resolution limits obtained from the three-bar pattern in the flight direction and perpendicular to it, respec-

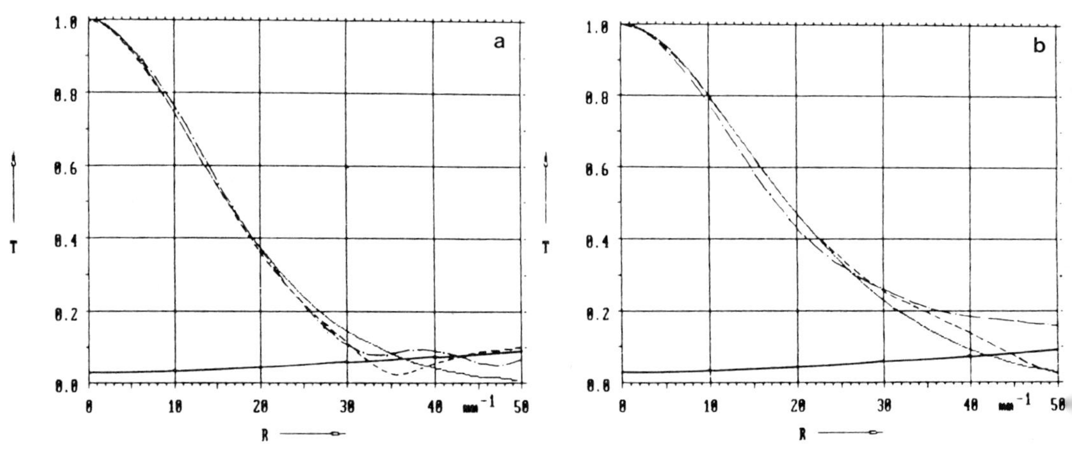

Fig.2 (a): MTF of the test patterns in flight direction. (b): MTF of the test patterns perpendicular to the flight direction.

———— MTF curve of a grating pattern.

-------- MTF curve of an artificial edge image.

—·—·—· MTF curve of a natural edge image.

———— Threshold curve SW.

tively. This comparison proves the reliability of the different methods for the determination of image quality.

4. Influence of Linear Image Motion on Image Quality

There are different types of image motions such as linear, parabolic, sinusoidal, and random /1/. This paper deals only with linear motion. The quality of airborne photographs can be improved by applying forward motion compensation (FMC). The photographs thus far referred to in this paper were obtained without using FMC, and this explains why in Figure 2 the MTF's of test patterns oriented in the flight direction are worse than those perpendicular to it. From the difference of the MTF's in the two directions, the image motion can be calculated. It will be compared with the theoretical image motion.

The image displacement during exposure time T is

$$a = vTf'/H \tag{4}$$

where v is the speed of the airplane, f' is the focal length of the objective, and H is the altitude of the airplane. The spread function corresponding to the movement is a rectangular function of width a. The corresponding MTF is its Fourier transform

$$MTF_B = a \frac{\sin \pi R a}{\pi R a} \tag{5}$$

where R is the spatial frequency and a is the image motion. This is a sinc-function. If the MTF is measured perpendicular to the flight direction and multiplied by this sinc-function, the result measured in the flight direction should be obtained providing the signal-to-noise ratio is not to bad. According to linear response theory, the MTF_F in flight direction can be written as

$$MTF_F = MTF_B \cdot MTF_P \tag{6}$$

where MTF_P is perpendicular to the flight direction and MTF_B is the MTF of the linear image motion. The image motion was therefore extracted from the measured MTF. The flight parameters were the following:

$v/H = 0.04342$ 1/sec

$T = 1/700$ sec

$f' = 303.64$ mm

From equation (4) the image motion is obtained as

$a = 19\,\mu m$

Figure 3 (a) shows the MTF of image motion with $a = 19\,\mu m$. In Figure 3 (b) MTF_F, MTF_P, and $MTF_B \cdot MTF_P$ are shown together. Good agreement was obtained between MTF_F and $MTF_B \cdot MTF_P$.

Two airborne photographs of the same scene, taken with and without FMC, were compared to investigate the linear image motion. Corresponding roof-edges were chosen and scanned with the microdensitometer. From the parameters of the photograph (v/H =

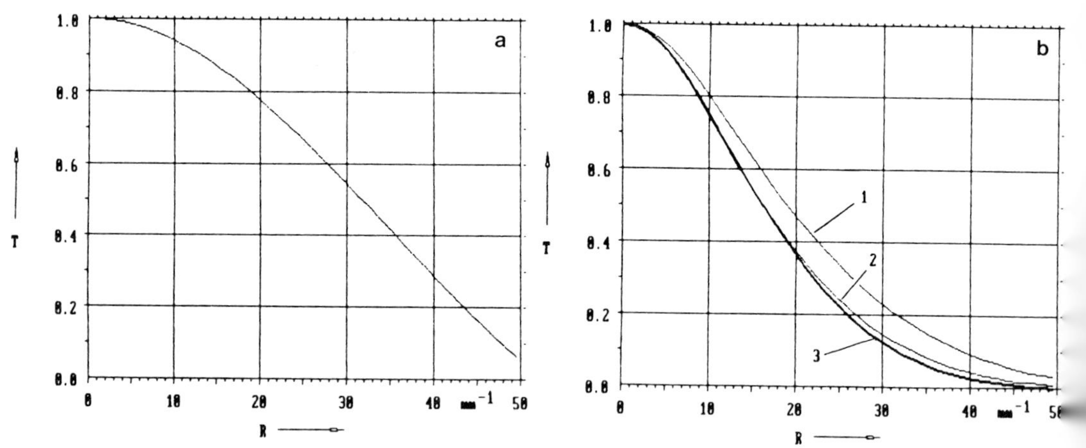

Fig.3 (a): MTF of an image motion with $a=19\,\mu m$. (b): MTF_F, MTF_P, and $MTF_B \cdot MTF_P$. 1 — MTF_P, 2 — MTF_F, 3 — $MTF_B \cdot MTF_P$.

0.05491 1/sec, T = 1/160 sec and f′ = 303.75 mm) the image motion was calculated according to equation (4) to be 104 μm as shown in Figure 4 (a). The MTF without FMC can be written as

$$MTF_O = MTF_B \cdot MTF_C \tag{7}$$

with MTF_B as the MTF of the linear image motion and MTF_C that of the airborne photograph with FMC. Figure 4 (b) shows the combined curves MTF_O, MTF_C, and $MTF_B \cdot MTF_C$. A major improvement of the image quality with FMC can be seen by comparing MTF_O and MTF_C. MTF_O compares well with $MTF_B \cdot MTF_C$. One should notice that the first minimum and maximum of the sinc-function can be seen regarding the MTF_O.

It is shown that the use of FMC makes substantially better image quality possible. Aerial survey cameras with FMC allow the use of high-resolution, but unfortunately low-sensitivity films, making long exposure times necessary. In order to test the image quality of airborne photographs with different film types, the same scene was photographed with films with different sensitivity, using a camera with FMC. These airborne photographs were

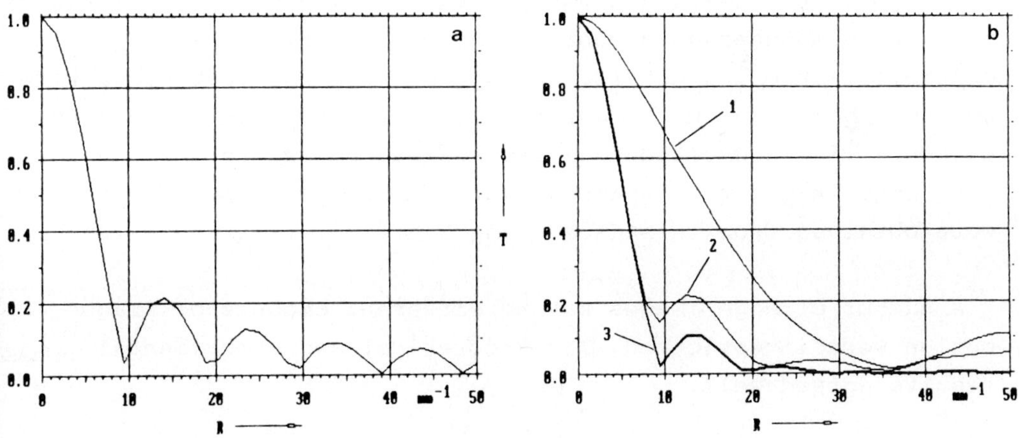

Fig.4 (a): MTF of an image motion with a=104 μm. (b): MTF_O, MTF_C, and $MTF_B \cdot MTF_C$. 1 — MTF_C, 2 — MTF_O, 3 — $MTF_B \cdot MTF_C$.

analysed by EGA. The measured results indicate that the fine-grain, high-resolution emulsion did not lead to the expected best image quality. We found that the airborne photographs taken with longer exposure times were affected by vibration and roll of the aircraft during exposure.

5. Study of the Transversal and Longitudinal Object Motion at a Close Range

In close-range photogrammetry, the object motion reduces the image quality of the photographs, especially when the object is very close to the camera. The linear transversal motion of the object leads to a linear image motion, as mentioned above. A longitudinal object displacement results in a defocusing of the image plane.

In a laboratory experiment the influence on the image quality of the transversal and longitudinal object motion at close range were studied. A test target consisted of four edges. A pair of the edge patterns was oriented in the direction of the transversal object motion in order to determine experimentally the MTF of the lateral image motion, and another pair was oriented orthogonally to it. The target was placed into a precision X-Y-coordinate stage mounted on an optical bench and driven by computer controlled stepping motors. Using a camera, the test edges were photographed with the linear transversal object motion during exposure with and without a longitudinal object shift. The resulting edge images were scanned with a microdensitometer. The MTF's were obtained from this data using EGA.

A number of edge images having different amounts of image motion were investigated. The theoretical and experimental results agreed well.

A longitudinal focal shift arises from a longitudinal object shift. On the basis of geometrical optics, the defocused image point consists of a circular patch of light of radius $\Delta z' \tan\alpha'$

where α' is the aperture angle in the image space and $\Delta z'$ the defocusing. The defocused MTF is obtained by the Fourier-transform of the image patch. This is given by a Bessel function of first kind and first order

$$MTF_{def} = 2J_1(Z)/Z \tag{8}$$

where $Z=2\pi R\Delta z'\tan\alpha'$. The first zero of (8) occurs when Z=3.83.

Figure 5(a) shows a defocused MTF curve calculated from the defocusing of $\Delta z'$=1.18 mm, and corresponding longitudinal object shift of 10 mm. Three curves are shown together in Figure 5(b). Curves measured with and without defocusing 1, 2 respectively, illustrate the effect of the defocusing on the MTF measurement; Curve 3 was obtained by multiplying the MTF values measured in the proper focus by the known values of the defocused MTF. Good agreement was found with the MTF curve measured from the defocused photograph. From a set of defocused

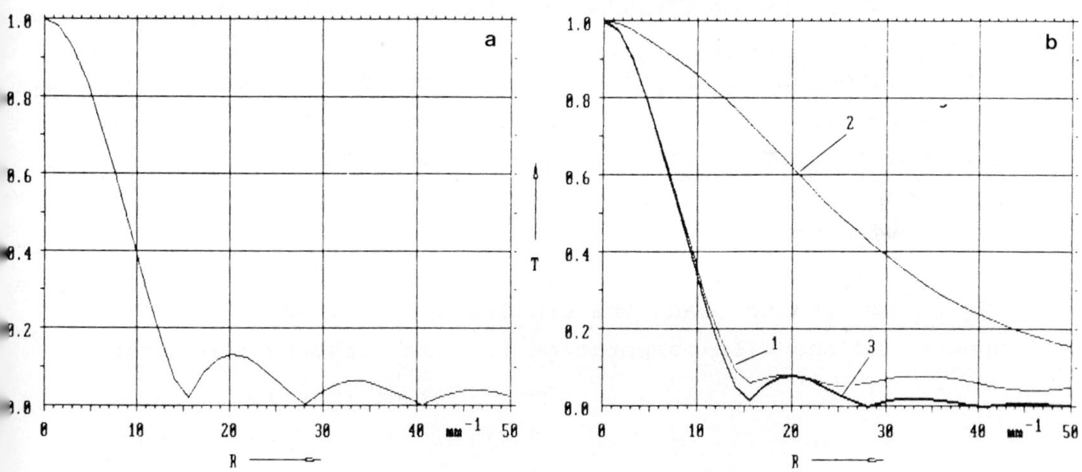

Fig.5 (a): Defocused MTF curve with $\Delta z'$=1.18 mm. (b): 1 — MTF curve measured with defocusing. 2 — MTF curve measured without defocusing. 3 — Calculated MTF curve with defocusing.

MTF's, we found good agreement with the calculated values even for large amounts of defocusing. It was found that geometrical optics holds for large defects of focus.

6. Conclusion

Different methods to determine the MTF from image structures were discussed and compared. For EGA a new and relatively simple method was developed, applying a Hanning function. Alternatively a grating pattern in the same photograph was examined. The MTF's of both methods were found to be in good agreement. The two techniques can be used complementary and increase the reliability of image quality determination. EGA however has, from a practical point of view, the advantage not to rely on artificial edges, but to work almost as well on natural roof-edges. The resolution obtained from measured MTF's were compared with the resolution from three-bar targets. Again a good agreement was found. Furthermore, the influence of linear image motion was examined and the improvement of image quality with FMC discussed. For photographs taken with high resolution films, FMC is found to be necessary. Due to longer exposure times additional disturbance are introduced and need to be considered like vibrations. In laboratory measurements the image motion and the defocus at close range were examined. Even for large amounts of defocusing, the measurements agreed well with the calculations.

Acknowledgement

We would like to thank the DFG(SFB 228) for the financial support and the WILD corporation for the airborne photographs.

References

/1/ Tiziani, H., 1977. Beurteilung der Bildqualität von Luftbildkammern: Vermessung, Photogrammetrie, Kulturtechnik. 4-77, pp. 137-146.

/2/ Scott, Frank, Scott, Roderic M., and Shack, Roland V., 1963. The Use of Edge Gradients in Determining Modulation-Transfer Functions: Phot. Sci. Eng., Vol.7, No.6, pp. 345-349.

/3/ Jones, Robert A., 1967. An Automated Technique for Deriving MTF's from Edge Traces: Phot. Sci. Eng., Vol.11, No.2, pp. 102-106.

/4/ Jones, Robert A., and Yeadon, Edward C., 1969. Determination of the Spread Function from Noisy Edge Scans: Phot. Sci. Eng., Vol.13, No.4, pp. 200-204.

/5/ Sievers, J., 1976. Die Kantenbildanalyse als Mittel zur Bestimmung von "Bildparametern". Beitrag für die Arbeitsgruppe "semantische Information" der Kommission E der OEEPE.

/6/ Lei, Fang and Tiziani, H.J., 1986. Eine neue Methode zur Bestimmung der MTF aus dem Kantenbild: Optik 74, No.1, pp. 17-21.

/7/ Brigham, E. O., 1974. The Fast Fourier Transform. Prentice-Hall, Inc. Englewood Cliffs, New Jersey.

/8/ Kölbl, O., 1985. Arbeitstagung: Kammerkalibrierung in der photogrammetrischen Praxis. Deutsche geodätische Kommission, Reihe B, Heft 275, S. 48-64. Verlag der Bayerischen Akademie der Wissenschaften, herausgegeben v. G. Kupfer u. W. Wester-Ebbinghaus.

Symbolic Image Descriptions for Relational Matching

George Vosselman
Institut für Photogrammetrie, Universität Stuttgart

Abstract

Relational matching can be used for searching corresponding objects in stereo images. It compares symbolic descriptions of two images. This paper presents a new method for extracting and representing symbolic image descriptions and discusses the suitability of several segmentation types for their use in relational matching.

0. Introduction

An important task in computer vision is to derive a 3D description of a scene by analyzing a series of images of this scene. In stereo vision, as well as in image sequence analysis, the first step in this derivation is the reconstruction of the recording geometry. The relative orientations between the images can be calculated if at least five pairs of corresponding points can be found in overlapping image pairs.

Not knowing anything about the orientation of the images, the method to solve this image matching problem can only use features which are more or less invariant with respect to changes in orientation.

One such method is relational matching. Although originally introduced for object recognition (Haralick and Shapiro, 1987), it also proved its applicability in matching stereo images (Boyer and Kak, 1988).

The main task of this method is to calculate the relative orientation between the images and to provide a rough 3D description of the scene which can be used for approximate values needed for more precise matching methods like the feature based matching algorithm (Förstner, 1986) which on its turn provides the approximate values for the most precise least squares matching (Pertl, 1984). Both the feature based and the least squares matching methods already have been developed within the project C2 of the SFB 228.

In relational matching, one systematically compares two symbolic image descriptions (socalled structural or relational descriptions) to find the best mapping from the primitives, i.e. the basic geometric elements, in one description to the primitives in the other description. A special feature of these descriptions is that they not only contain the extracted image primitives with their orientation invariant attributes, but also describe the spatial relationships among the objects. These relations play an important role in the matching process, since they are invariant against many transformations whereas simple matching methods like the optimization of a cross correlation coefficient between patches of the grey value images fail (fig 1.).

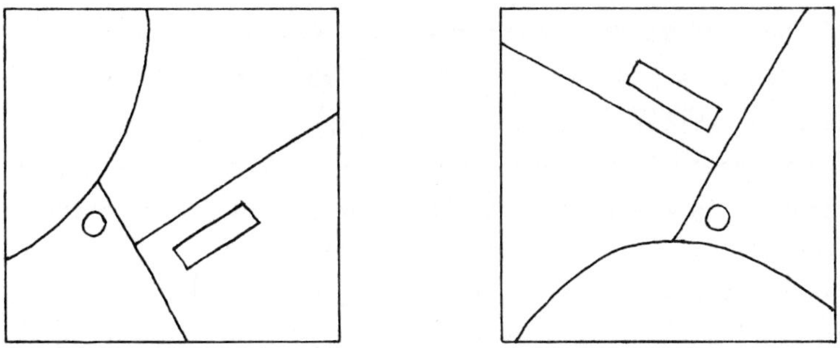

fig. 1 A matching problem with strongly rotated images

Classical concepts usually employ a region based method to obtain symbolic image descriptions. It will be shown in this paper that such region based descriptions can not be used for relational matching, since the extracted region boundaries often are very dependent on the illumination of the scene, i.e. they are very unstable. Therefore, an image description method has been developed which extracts a safer, i.e. a better matchable image description. Unlike classical concepts, this method is based on the extraction of lines.

After defining the primitives and describing a method to extract them from the digital images, the obtained descriptions of the primitives in a few test images are compared with those obtained by classical region segmentation algorithms such as region growing and histogram segmentation, in order to demonstrate the superiority of the new concept.

Finally, it is shown that the topology of the image can be represented well in a relational model and that the elementary relations among the primitives can be extracted easily from the relational model.

1. Extraction of primitives

In the relational matching process, primitives in the left image are matched to primitives in the right image. To maximize the success of the matching, those primitives should be extracted from a grey value image, which are likely to be extracted in the other image as well. This implies that the primitives should contain information on the objects in the recorded scene. In digital images this information is usually revealed by grey value gradients.

There are three elementary types of primitives one can extract:

Points are those locations in the image which show a significant change of the grey values in all directions. They often arise at corners of the lines but may also represent blobs.

Lines are sequences of eight-connected pixels which show a significant grey value gradient in one direction. The line pixels are connected in the direction perpendicular to the gradient direction.

Regions are the four-connected components of pixels which are bounded by points and lines, i.e. by pixels with strong gradients. For the matching, the regions of which the grey values of the pixels fulfil some homogeneity criterion are the most interesting since they are more likely to correspond to a surface element in the recorded scene than the other regions. However, these other regions are necessary to describe the topology of the primitives. Thus, all regions will be represented in the relational description.

1.1. Point extraction

Many interest operators have been developed for extracting points from digital images (Moravec, 1977, Dreschler, 1981). We use the operator first published in (Paderes et. al., 1984). In addition to a point strength and point roundness measure, it also provides the subpixel coordinates of the selected points.

Like most of the operators, this operator is gradient based. First simple row- and column-gradients are calculated. From these gradients the following matrix is computed:

$$N = \begin{pmatrix} \Sigma\, g_r^{\,2} & \Sigma\, g_r\, g_c \\ \Sigma\, g_r\, g_c & \Sigma\, g_c^{\,2} \end{pmatrix}$$

The summation is usually performed by a 4x4 or 6x6 box convolution (thus involving the grey values if the pixels in a 5x5 or 7x7 box). The trace of this matrix can be used as the interest value. The higher the trace, the more remarkable the point is. A threshold has to be used to select the pixels with strong gradients. After that a roundness criterion selects the points.

The roundness of the ellips described by this matrix can be assessed by:

$$q = \frac{4. * \det(N)}{tr(N)^2}$$

If the value of q (always between 0 and 1) is high, the ellipse is circular shaped, which implies that the gradient strength is about the same in all directions. This occurs at corners of the grey value edges and at blobs. A threshold in the range of 0.50 to 0.75 has been proven to be appropriate for the point selection in most applications.

The subpixel coordinates of the selected points can be estimated in a simple way, using the elements of the matrix N as weights. Full details on this interest operator can be found in (Förstner and Gülch, 1987).

1.2. Line extraction

Lines are eight-connected sequences of pixels with strong gradients. The selection of those pixels again can be done with the interest operator described above. But now, unlike in the point extraction procedure, there is no restriction on the roundness value, so that all pixels which have a strong gradient are candidate line pixels. The size of the operator for this application is usually 3x3.

The direction of the gradient can also be calculated from the elements of the matrix N:

$$phi = 1/2 * \arctan\left(\frac{\Sigma\, g_r\, g_c}{\Sigma\, g_c^{\,2} - \Sigma\, g_r^{\,2}}\right)$$

Two candidate line pixels can be linked if the gradient directions of the pixels are about the same and the normal to the gradient direction of the first pixel appr. points towards the second pixel. In this way, a graph of possible links can be constructed (see fig. 2). The costs of a link is a fixed value minus the gradient strength of the second pixel.

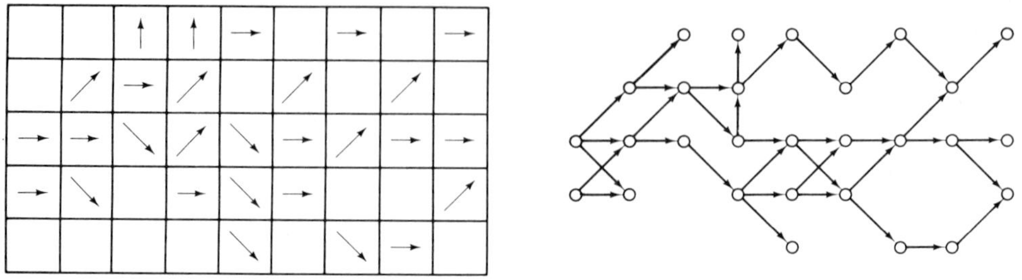

fig. 2 normals to the gradient directions and the constructed graph (adopted from Ballard and Brown, 1982)

The candidate pixel with the strongest gradient is assumed to be part of a line and is taken as the first point for the line following. Starting at this point, all paths of a certain length, i.e. a certain number of links, are evaluated. After selecting the combination of links with the lowest sum of link costs, the first part of that path, say half the path, is fixed and its pixels are classified as line pixels. Starting at the last fixed pixel, again all paths of a certain length are evaluated and the first part of the best path is fixed. Not fixing the best path over the complete length which has been evaluated, accomplishes that the selection of a pixel is influenced by the preceding pixels as well as by the following pixels. To avoid the extraction of many small parallel branches of a line, the pixels on the evaluated but not optimal paths may not be taken as start points for a new line.

The line following terminates when:

- there are no more candidate line pixels to continue, or
- the line following finds a pixel classified as a point primitive, in which case this point becomes the end node of the line, or
- the line following finds a pixel which is already classified as a line pixel. This may be a pixel of an old line, in which case the old line is split into two lines, or it can be a pixel of the line being followed, in which case a closed polygon has been found.

When the line following has terminated, the algorithm returns to the selected start pixel and traces the line in the opposite direction until one of the above conditions is matched. After finishing the second direction a new start pixel is selected and the algorithm starts to trace a new line. This process is repeated until there are no candidate line pixels left which can be used for a start point of a new line.

1.3. Region extraction

After the point and line pixels have been labeled, regions are the four-connected components of the remaining unlabeled pixels. This means that the regions are entirely surrounded by pixels with strong gradients. This definition allows the presence of other primitives (points, lines and other regions) within a region. For matching purposes, the regions with a homogeneous grey value distribution will be the most interesting regions.

1.4. Examples

Fig. 3 shows two images of a brick wall with a black dot, taken from different positions. The extracted primitives are shown in fig. 4. The points are represented by small circles. Note that not only the corners of the bricks, but also small speckles on the bricks are found by the interest operator. Most of the points are extracted in both images. The left images shows a few more points. These are mostly caused by speckles which are better visible in the left image.

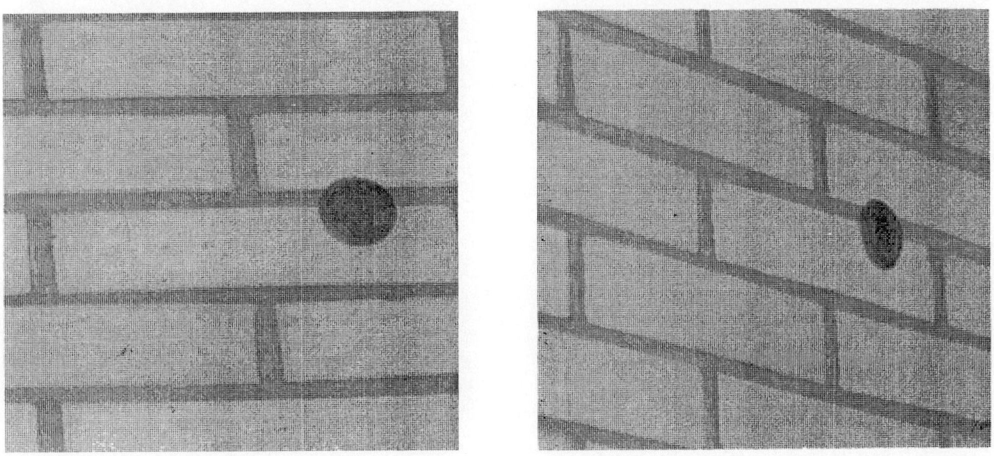

Fig. 3 Two images of a brick wall with a black dot

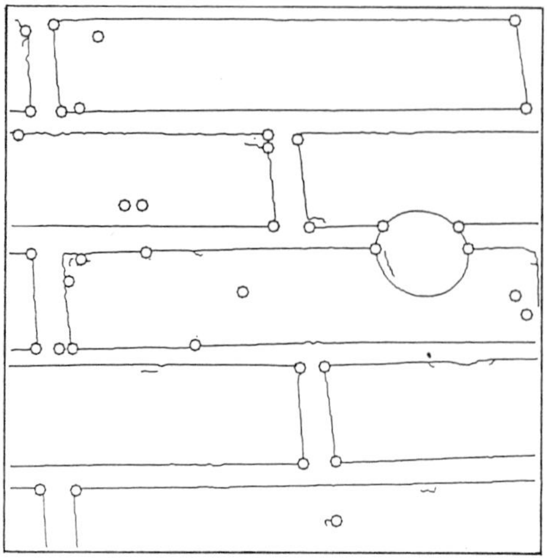

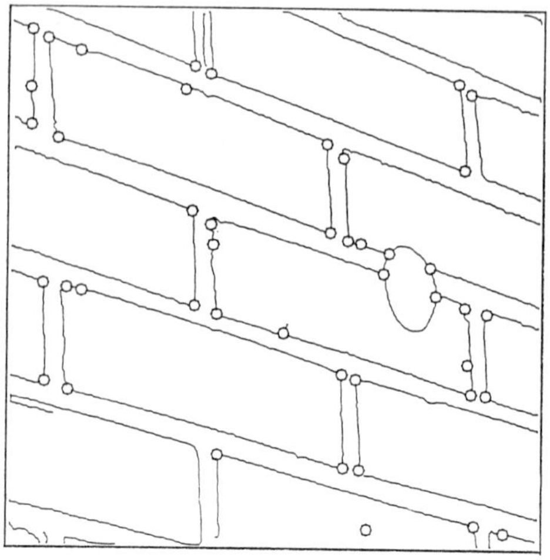

Fig. 4 Extracted primitives of the wall images

Fig. 5. shows two images of a part of a cup. The extracted primitives are shown in fig. 6. In the figure, also the regions have been displayed which have a low standard deviation of the grey values. The regions were omitted in fig. 4 since all bricks were found entirely surrounded by line primitives and all bricks and joints had low standard deviations of the grey values.

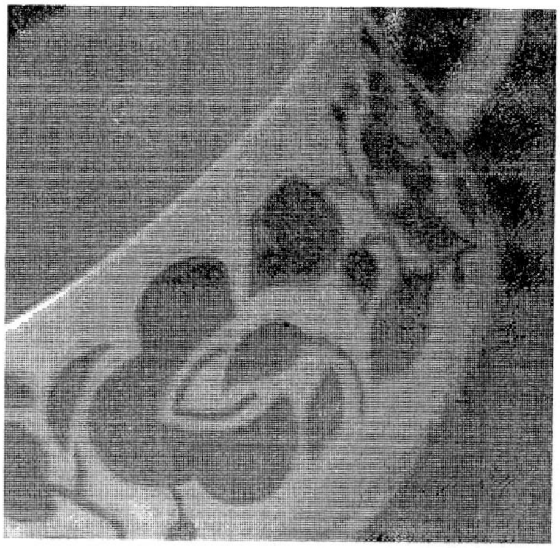

Fig. 5 Two images of a part of a cup

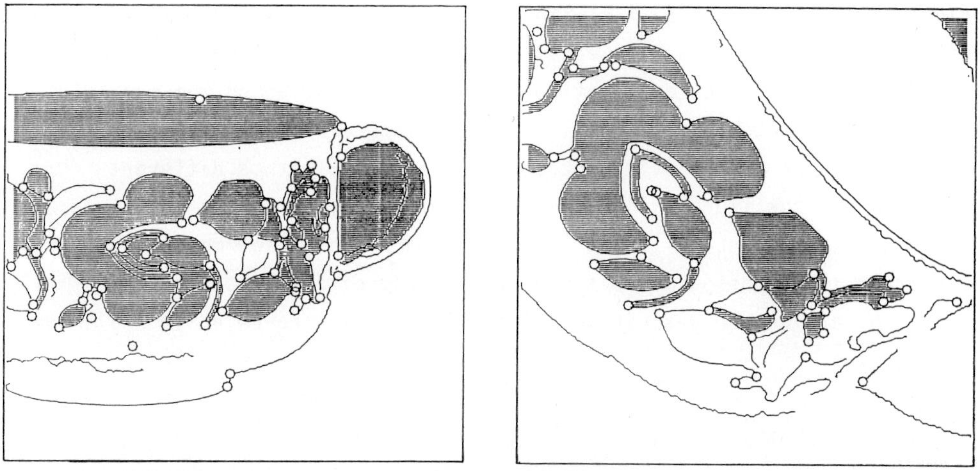

Fig. 6 Extracted primitives of the cup images

On the front part of the cup, the extracted primitives in the left and the right image correspond quite well. Near the ear, however, due to insufficient resolution and poor illumination, the line and region primitives will be hard to match. The point primitives of the right image, on the contrary, still can be found in the left image.

2. Line based segmentation versus region based segmentation

In the primitive extraction described above, the gradients of the grey values were used to find points and lines, whereas the regions were defined by the absence of pixels with strong gradients, i.e. the complement of the points and the lines. In regions based segmentation procedures, like region growing and histogram segmentation, the regions are extracted first. They are defined to be the four-connected components of pixels which fulfil some homogeneity criterion, usually a constant grey value. The lines then arise as the contours of the regions and the points are the junctions of the lines, i.e. the pixels where three or more regions come together.

2.1. Disadvantages of a region based segmentation

Region based segmentations often give a nice representation of the image, but they can not be used for matching purposes. In images where there is a slight change in the grey values over a larger part of the image (e.g. due to illumination differences), the boundaries of the extracted regions will be lines of pixels with the same grey values. Such lines usually will not coincide with edges of the recorded scene objects, i.e. the shape of these lines does not contain any information on the scene. They can not be matched to lines in another image of the scene, since in that image (taken from a different position and thus with a different illumination), the lines will be found at completely different positions.

A related disadvantage is that region based segmentation methods only can extract closed polygons, although under many circumstances open polygons give a better description of the information revealed in the image (see fig. 7.)

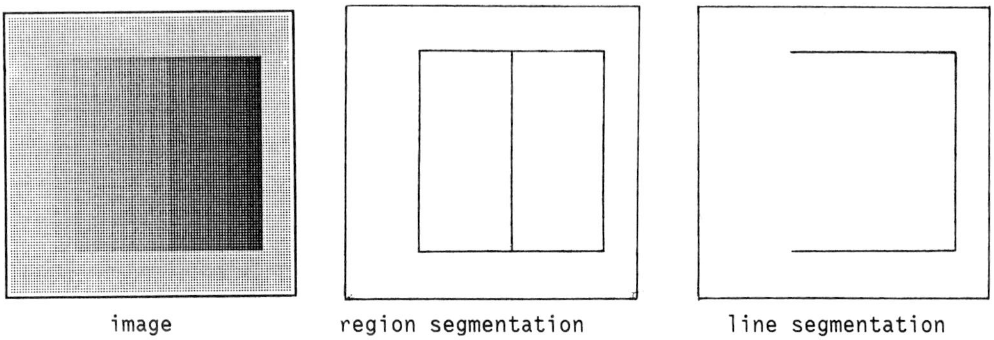

fig. 7 Line based segmentation versus region based segmentation

The disadvantages of the region based segmentations also hold for hybrid methods which use both lines of strong gradients and region based segmentation, since also here the region based segmentation can place lines at positions where there are only weak gradients.

2.2. Examples

The images of fig. 3 and 5 have also been segmented using a region growing algorithm. The start points for the regions were pixels with a small standard deviation of the grey values in their surrounding. A region was expanded by a four-connected pixel if the difference between the average of the grey values of the pixels in the region and the grey value of the selected pixel was below a certain threshold. The expansion stops if there are no more unlabeled pixels left which fulfil this demand. This is repeated until all start points have been processed. After that a morphological closing was performed to eliminate small gaps between the regions.

The resulting region boundaries are shown in fig. 8 and 9. Similar results were obtained by a histogram segmentation method (Ohlander et. al., 1978). The segmentation of the wall images succeeded quite well, because the contents of the bricks, the joints and the black dot are very homogeneous. In the cup images, however, many spurious lines are found in parts where there is a slow change in grey value due to illumination irregularities. Such lines can not be used in a matching process.

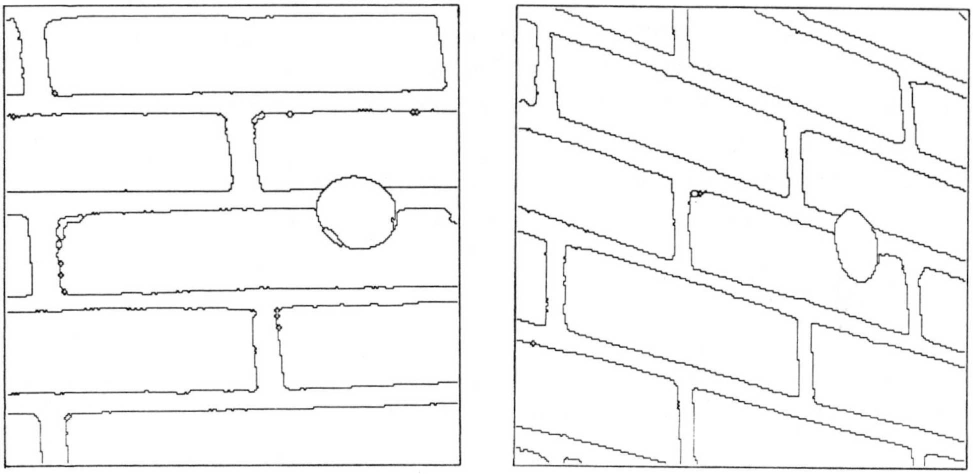

Fig. 8 Region segmentation of the wall images

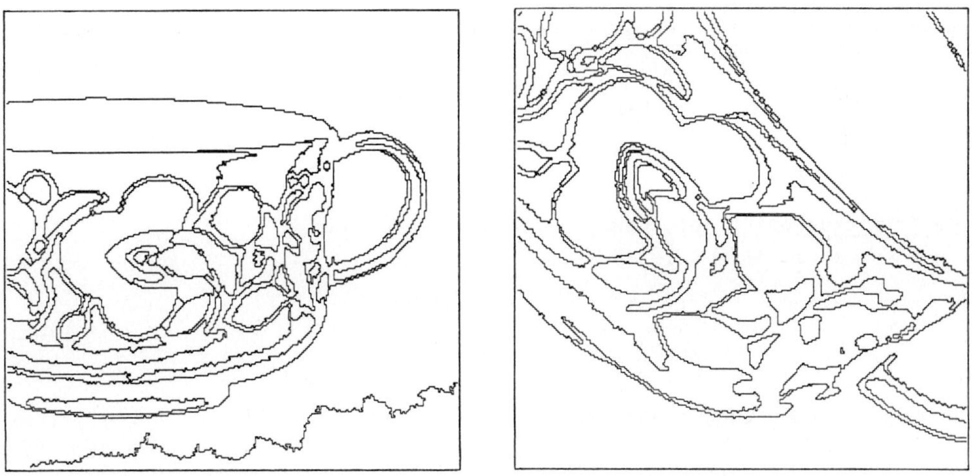

Fig. 9 Region segmentation of the cup images

3. Relations and representation

The main ideas in this section are due to a review paper by Molenaar (Molenaar, 1988). Some small adaptions have been made to suit our data types.

Each line is said to start and to end at a node. The geometry of the line is described by a coordinate list of the line pixels between the start and the end node. Each node has a node identifier and a coordinate pair. Such a node may be a point primitive (second line following termination criterion), but this is not necessary. Each line also has one region at its lefthand side and one region at its righthand side. Finally, an extra point-region link is needed to relate the regions to the points within it. These few relations, which already fully describe the topology, can be stored efficiently in a relational model.

In a relational model, the named tuples may represent entities and relations. The tuples are lists of attributes:

    name(attribute1, attribute2, ...)

The information on the topology and characteristics of the primitives can be represented in the following model:

```
region primitive(region id., char.)

line primitive(line id., node id., node id,. region id., region id., char.)
                            ↑         ↑           ↑             ↑
                          start      end         left          right
                          node       node        region        region

point primitive(point id., node id., region id., char.)
                                         ↑
                                     surrounding
                                       region

node(node id., row coordinate, column coordinate)
```

"char." is a pointer to a list of characteristics of the regions, resp., the lines or the points. The list of line characteristics also contains (a pointer to) a coordinate list of the line pixels. In a strict sense, this representation is not a relational model, since there is some redundancy in it. In its use, however, this representation will be more convenient.

To save the storage of the point coordinates in the list of point characteristics, it is usefull to add a "node" tuple for all points which are not connected to any line. Since almost all points are already nodes, the effort to do so is relatively small.

Having three types of primitives, there are six groups of relations. They can be easily extracted from the above described relational model:

- line-line relation
  * connected : two lines are connected if they have a node in common.
- point-line relation
  * connected : a point is connected to a line if it is a node of that line
- point-point relation
  * connected : this relation is defined using the point-line relation: two points are connected if they are connected to the same line
- line-region relations
  * border : a line is a part of a region border if the region is situated at one side of the line.
  * contained : a region contains a line, if the region is situated at both sides of the line.
- region-region relations
  * connected : this relation uses the line-region relation: two regions are connected if they have a border line in common.
  * island : a region is an island in another region if it is only connected to that region.
- point-region relations
  * border : a point is situated on a region border if it is connected to a line which is a part of the border of that region.
  * contained : if a point is not connected to a border line, it is contained in a region. In this case, the region identifier in the point primitive tuple is used to extract this relation. If the point is connected to a border line, the region identifier will be a zero-identifier.

4. Discussion

It has been shown that, for matching purposes, region based segmentations can not be used to extract symbolic image descriptions. The results of those segmentation methods are often very unstable and can not be matched reliably. Safer image descriptions can be obtained by a gradient based segmentation which also extracts points, lines and region, but allows a more general description, e.g. points do not have to be on a line and lines do not have to be region borders.

The six types of relations among the primitives can be easily extracted from a relational model which only contains a few elementary relations.

The characteristics of the primitives should be invariant to changes in the orientation of the image. A possible point characteristic is a type classification: corner or blob. Line characteristics can be: orientation invariant shape descriptions and a classification as line or edge, where a line is a pair of anti-parallel edges. Region characteristics can be: the Euler number, description of the grey value distribution and orientation invariant shape descriptions.

In further research we will use the presented image descriptions in relational matching. The relations and the characteristics of the primitives are used to limit the search space of all possible primitive to primitive mappings. Since we are primarily interested in the relative orientation of the images, the reliability of the determination of the orientation parameters should be used to select the best mapping among those which were not rejected by severe inconsistency of characteristics and relations.

5. Literature

Ballard, D.H. and C.M. Brown (1982); Computer Vision, Prentice-Hall, Englewood Cliffs, New Jersey, 1982.

Boyer, K.L. and A.C. Kak (1988); Structural Stereopsis for 3-D Vision, IEEE Transactions on Pattern Analysis and Machine Intelligence, vol. 10, no. 2, pp. 144-166, 1988.

Dreschler, L. (1981); Ermittlung markanter Punkte auf den Bildern bewegter Objekte und Berechnung einer 3D-Beschreibung auf dieser Grundlage, Dissertation Fachbereich Informatik, University of Hamburg, 1981.

Förstner, W. (1986); A Feature Based Correspondence Algorithm for Image Matching, ISPRS Comm. III Symposium, Rovaniemi, 1986.

Förstner, W. and E. Gülch (1987); A Fast Operator for Detection and Precise Location of Distinct Points, Corners and Centres of Circular Features, ISPRS Intercommission Workshop, Interlaken, June 1987.

Haralick, R.M. and L.G. Shapiro (1985); Image Segmentation Techniques, Computer Vision, Graphics and Image Processing, Vol. 29, pp. 100-132, 1985.

Haralick, R.M. and L.G. Shapiro (1987); Relational Matching, Applied Optics, vol. 26, pp. 1845-1851, May 15, 1987.

Molenaar, M. (1988); Single valued polygon maps. Paper to be presented at the XVIth ISPRS Congres, Kyoto, 1988.

Moravec, H.P. (1977); Towards Automatic Visual Obstacle Avoidance, Intern. Joint Conference on Artificial Intelligence, p. 584 , 1977.

Ohlander, R., K. Price and D.R. Reddy (1987); Picture segmentation using a recursive region splitting method, Computer Graphics and Image Processing, vol. 8, pp. 313-333, 1987.

Paderes, F.C., E.M. Mikhail and W. Förstner (1984); Rectification of Single and Multiple Frames of Satellite Scanner Imagery using Points and Edges as Control, NASA Symp. on Mathematical Pattern Recognition and Image Analysis, Houston, June 1984.

Pertl, A. (1984); Digital Image Correlation with the Analytical Plotter Planicomp C100, International Archives of Photogrammetry, Vol. 25-A3b, Rio de Janeiro, 1984.

E. CLOSE RANGE MENSURATION

Motorized Electronic Theodolites - High Precision Measurement Robots for Surveying

Gerhard Bayer
Inst. für Anwendungen der Geodäsie im Bauwesen, Universität Stuttgart

Summary

The guidance of a continuously moving construction machine requires informations about the desired trajectory. Nowadays the surveyor provides these informations by the determination of numerous fix points along the trajectory. The construction machine is guided along the fix points manually or automatically by following a wire or a laser beam connecting the discrete points.

The aim of the project "Tracking Systems" of the Sonderforschungsbereich 228 "Hochgenaue Navigation - Integration navigatorischer und geodätischer Methoden" described herein is to guide a slowly moving construction machine along a predetermined trajectory automatically without the necessity of a considerable a priori surveying.

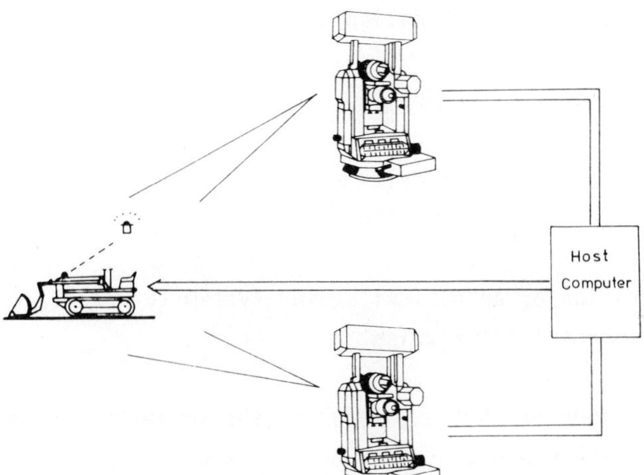

Fig. 1: Position control of moving targets

The realisation of such a target guidance system requires the use of an automatic, accurate and fast 3D-measurement system with a sufficient field of view. The presented tracker being in development is designed as a flexible, modular system, which can be adapted on different applications inside and outside the lab. The basis of the system is a motorized electronic theodolite (Fig. 1). The cost of the additional modules depends on the intended application.

1. Introduction

A theodolite is a very precise instrument for determining horizontal and vertical angles and is traditionally used in surveying. The classic use of a theodolite is the surveying of non-moving targets at distances of up to 100 km and with an accuracy better than 1 arcsec. The communication between user and theodolite is realized by a manual actuation and a visual reception of the angle information.

High precision measurements with theodolites base on repeated measurements of stationary targets. This time consuming measurement and evaluation process doesn't allow the surveying of moving targets. To overcome this system limitation the introduction of realtime techniques used for instance in robotics and the following steps for automation are necessary:

1. interfacing the theodolite to an intelligent machine;

2. provision of a motor drive system to enable the theodolite to be controlled by machine;

3. integration of an optical sensor system to detect environmental information.

By adapting these modules to the theodolite the surveying of moving targets becomes possible. The use of specific optical sensor types and evaluation techniques allow an automatic theodolite operation without loss of accuracy in case of surveying non-moving targets. Increasing target velocity generally results in a decrease of accuracy.

In the project "Tracking Systems" an electronic camera with an optical position sensor for a close range motion detection, e. g. robot motion, was developed in the last few years (Bayer, Krzystek 1986).

The following tracker is the combination of this camera and a motorized electronic theodolite (Wild T2000 MOT) with communication, movement and optical information reception functions.

Due to the substitution of human operation during measurement the automatic tracking system can be caracterized as surveying robot. Additionally, many problems and solutions are comparable with those in robotics.

2. Presentation of the theodolite based tracker

2.1 Hardware structure

The interfaces of the tracker are

- 2 serial outputs for the horizontal and vertical relative angle operating at 9600 baud,
- 1 serial output for the horizontal and vertical absolute angle operating at 2400 baud,
- 1 analog output for the optical position sensor signals,
- 2 analog inputs for the motor actuation and
- 2 serial inputs for the motor actuation (Fig. 2)

It is possible to select either an analog or a digital motor actuation. By using the digital actuation a microprocessor system is added to handle the communication between theodolite and computer and to control the required angular velocity.

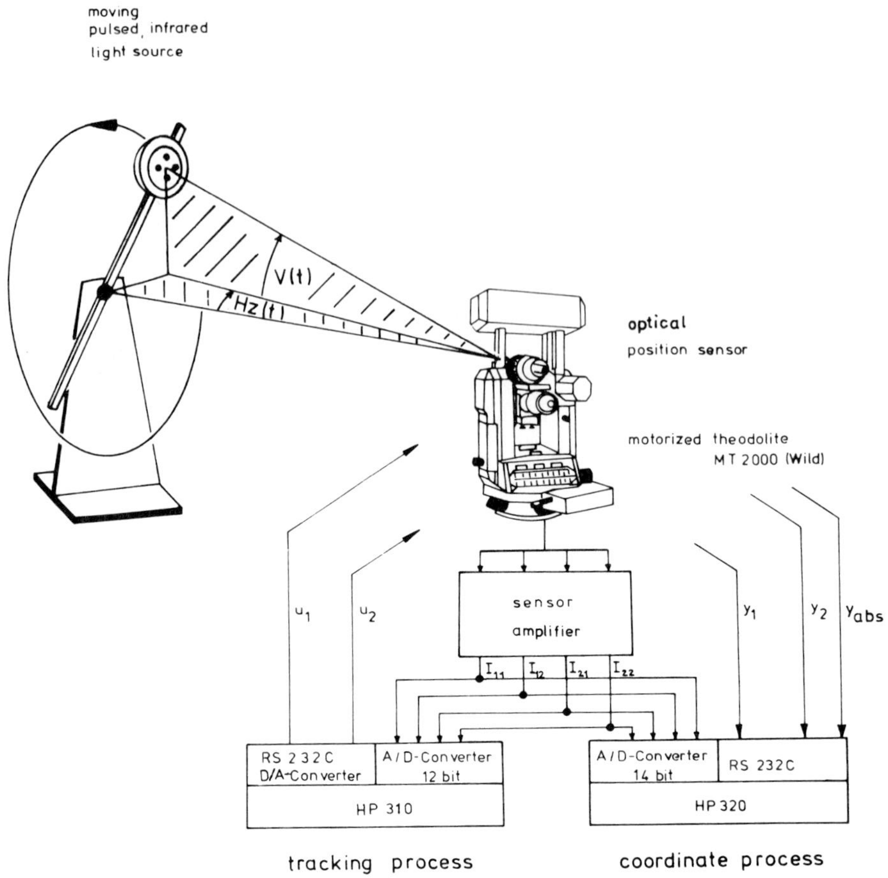

Fig. 2: Target tracking system

The drives consist of servo-motors with a gear ratio of 576 horizontal and 216 vertical. In comparison with industrial robots the main importance of the drive lay out is the positioning accuracy. The dynamic properties of the tracker don't satisfy completely the requirements of an industrial robot (see Fig. 3).

	industrial robot	theodolite based tracker
moment	10 - 400 Nm	0.02 Nm
angle	90 - 350°	horizontal: 360° vertical : 230°
velocity	30 - 230 °/sec	horizontal: 37 °/sec vertical : 42 °/sec
acceleration	10^3 - 10^7 °/sec^2	horizontal: 80 °/sec^2 vertical : 220 °/sec^2
positioning accuracy	± 15 - ± 1200 arcsec	horizontal: 0.6 arcsec vertical : 1.5 arcsec

Fig. 3: Comparison of the tracker and an industrial robot drive properties (compare Desoyer, Kopacek, Troch 1985)

The following measurement systems are utilised in the tracker:

- the optical position sensor,
- 2 incremental encoders,
- angle measurement system Wild T2000.

The optical position sensor is a lateral effect photodiode (tetra lateral). A 1000 cps pulsed infrared target mark produces currents at 4 electrodes, located on opposing sides of the quadratic sensor surface. An amplifier converts the currents to voltages, amplifies the voltages and determines the peak amplitudes. With aid of the sensor model the image coordinates are calculated from the electrode currents. The image coordinates can be transformed to angles. Thus the optical position sensor can provide the deviation between trakker and target orientation (Bayer, Krzystek, Möhlenbrink 1988).

The incremental encoders, each with a 4000 steps resolution are used for measuring the tracker orientation. The second angle measurement system Wild T2000 with an accuracy of 0.5 arcsec serves as an absolute reference measurement system.

2.2 Functional structure

The tracker operation is divided into two parallel working processes, both with differing requirements - the tracking process with high dynamic requirements, and the coordinate process with high accuracy requirements (Fig. 4).

The control of the horizontal and vertical axes is also realized as two separate processes. The two controllers must eliminate the influences of the mechanical coupling during movement.

In the following text the horizontal axis is caracterized by the index "1", the vertical axis by the index "2". Without index the variable is valid for both axes.

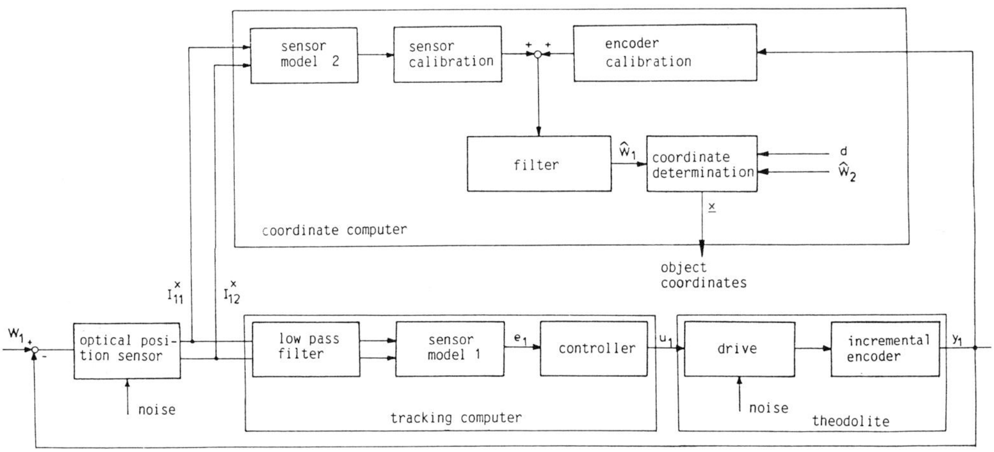

w_1 ... reference variable (target angle)

e_1 ... deviation (sensor angle)

u_1 ... manipulated variable (motor voltage)

y_1 ... controlled variable (theodolite angle)

$I_{11}^x\ I_{12}^x$ ... sensor voltage

$\hat{w}_{1,2}$ ... estimated reference variables

d ... distance between image and target movement plane

Fig. 4: Block diagram of the horizontal tracker axis

Seen from a control engineering perspective the goal of the tracker is to reconstruct unknown reference variables by measuring deviations and controlled variables in a closed control loop.

The tracking process consists of 3 modules - the optical position sensor, the tracking computer and the motorized theodolite. The position sensor provides electrode currents. The tracking computer filters digitally the high frequency noise of these signals and determines an image coordinate which is directly proportional to the deviation from the target angle.

The controller has to manipulate the motor voltage u depending on the deviation e. The stability and guidance qualities of the tracking process are influenced by the design of the controller.

The determination of the coordinate x is based on the accurate measurement of the theodolite angle y and the deviation e. High accuracy is attained by improving the sensor model, calibrating the non-linearities of both measurement systems, and using optimum filtering methods. The tracking and the coordinate process are performed for the horizontal and vertical axes of the motorized theodolite. Together with distance measurement it is possible to compute the object coordinates of the target.

3. Tracking Process

The goal of the tracking process design is to minimize the deviation between target angle and theodolite angle. The main aspects to reach this goal are the design of the controller, that is the control algorithm, and the design of the hardware to minimize the computing and the communication time.

The determination of the optimum control algorithm requires knowledge about the properties of the control system, that is the theodolite, derived from analytical and/or experimental methods.

Each tracker axis consists of several parts: the motor, the gearing system and the incremental encoder (Fig. 5; Desoyer, Kopacek, Troch 1985).

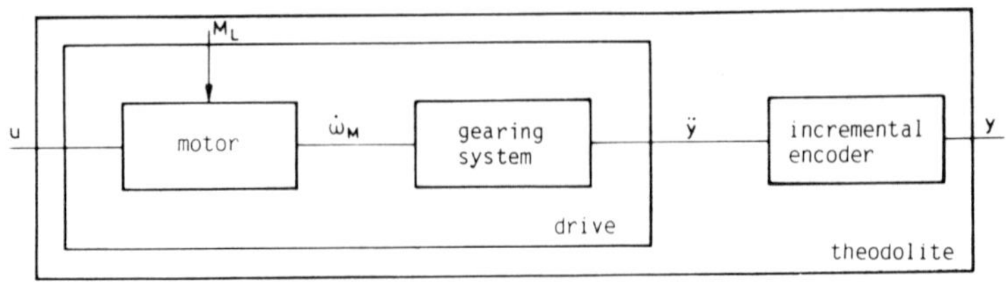

u	motor voltage	y	theodolite angle
ω_M	angular velocity of the motor shaft	M_L	load moment

Fig. 5: Tracker axis

The diagram of the constant current servo motor is presented in Fig. 6.

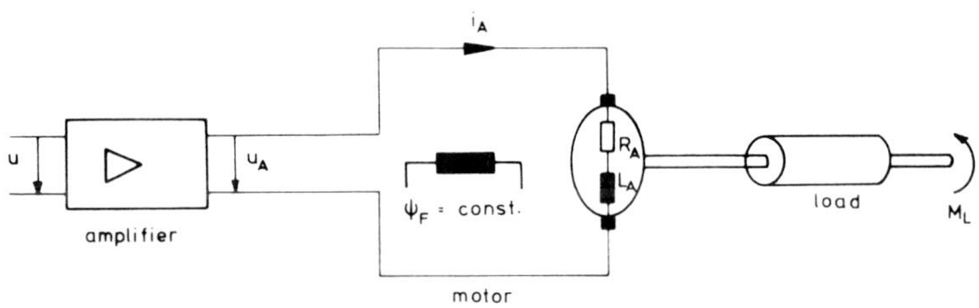

Fig. 6: Constant current servo motor

The signal transmission of the motor is described by following differential equation:

$$L_A I_M \ddot{\omega}_M + R_A I_M \dot{\omega}_M + c^2 \Psi_F^2 \omega_M = c \Psi_F^2 K u - R_A M_L - L_A \dot{M}_L$$

L_A	armature inductance	c	constant
I_M	moment of inertia	ψ_F	magnetic flux
R_A	armature resistance	K	amplifier gain

In order to simplify the mathematical model of the motor it is allowed to neglect the armature inductance L_A and the load moment M_L. The gravity of the masses should be in the rotating axes at any time.

$$R_A I_M \dot{\omega}_M + c^2 \Psi_F^2 \omega_M = c \Psi_F K u \qquad (1)$$

The differential equation (1) shows that the motor has a DT_1-transfer function

$$G_M(s) = \frac{\dot{\omega}_M}{u} = \frac{K_A s}{T_A s + 1}$$

The gearing system has a proportional transfer function with the amplification ν and the incremental encoder is a double integrating system. Thus the drive system has a PT_1-transfer function:

$$G_D(s) = \frac{K_A \nu}{s(T_A s + 1)}$$

The model simplifications make the theoretical mathematical analysis easier but the model inaccuracies increase. The variables which are neglected or not taken into account are considered as disturbances, which must be reduced by an optimum controller design.

The structure of the drive system can be verified by the experimental determination of the step response.

The time constant T_A = 30 ms, the horizontal gain $K_{A_1} \cdot v_1$ = 7 [°/sec · Volt] and the vertical again $K_{A_2} \cdot v_2$ = 12.4 [°/sec · Volt] (Fig. 7).

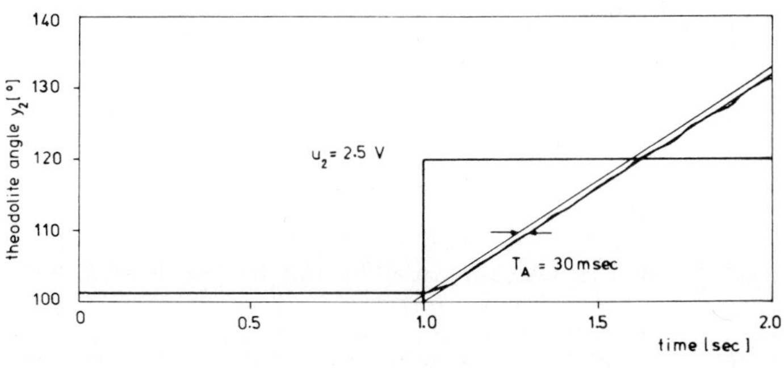

Fig. 7: Vertical step response

The most important criteria for the selection of the control algorithm is the computing time required. For applications requiring not much computing time a PID-algorithm is recommentable. Due to the time constant and the non-linearities of the control system, e. g. the motor starting voltage and the angle-dependend load moment of the vertically eccentric attached sensor box, it is necessary to implement a second controller with a PID-structure to improve the adjustment of the required angular velocity. Such a cascade control concept guarantees small deviations in spite of nonlinearities and disturbing influences on the control loop.

The controller design is the first step towards an efficient tracking system. The second step, which is equally important, is the increase of the scanning frequency. This increase is directly related to the target velocity. By the use of a HP 320-computer and an analog actuation we attain a scanning frequency of f = 50 cps, which is high enough to provide good control over the whole velocity range of the tracker. Fig. 8 shows the control quality of the tracker depending on the maximum angular velocity of the tracker axes by tracking a rotating target in a distance of 3 m (Fig. 2). The tracking computer is a HP 310 (scanning frequency f = 16 cps).

maximum angular velocity [°/sec]	average horizontal deviation [°]	average horizontal motor actuation [°]
5.5	0.008	0.14
11	0.027	0.16
16.5	0.060	0.18
22	0.112	0.46

Fig. 8: Control quality of the tracker in relation to the target velocity

Under these experimental conditions we achieve a maximum angular velocity of 30 °/sec. If the linearity of the optical sensor for measuring the deviations is guaranteed the target velocity does not affect the accuracy of the target positioning. However with increasing velocity the distance between the measured positions also increases and influences the quality of the coordinate process.

4. Coordinate Process

The accuracy of 0.6 arcsec by focussing on non-moving targets automatically is achieved

- by the use of the very precise angle measurement system Wild T2000,

- by measuring in the sensor center to avoid non-linearities and

- by reapeated measurements to reduce the influence of noise.

This measurement techniques however cannot be utilised for the tracking of moving objects. The angle measurement system requires a stationary theodolite and 1 sec of time for measurement. The use of a dynamic angle measurement system, consisting e. g. of incremental or absolute encoders or resolvers, can necessitate calibration to achieve a sufficient accuracy.

During the tracking process, deviations from the sensor center will always occur. The consequence is that non-linearities of the sensor must also be calibrated.

type of errors	standard deviation (arcsec) horizontal	vertical
1) sensor		
- non-linearity	30	30
- noise	100	100
2) incremental encoders		
- non-linearity	80	20
- free motion	2	20

Fig. 9: Types and present values of errors when determining target angles

Repeated measurements are only feasible if the measurement frequency is high enough. Other filtering methods e. g. Kalman filter must be examined. Fig. 9 shows the type and the present quantities of errors. These errors must be minimized successively to achieve high precision measurement systems.

The next step is to analyse the accuracy of the whole coordinate process. For that purpose a motion simulator with a predetermined trajectory has been designed. 5 LEDs are attached to a motor-driven, rotating arm (Fig. 2). The arm movement describes a circle with a radius of 1 m.

Due to the orientation of the tracker with respect to the rotating arm, the target describes an ellipse in relation to the tracker. We thus obtain a measure of accuracy by fitting an ellipse to the calculated object coordinates and determining the standard deviation.

With a target frequency of 0.02 cps, that is a maximum angular velocity of the tracker of 2.6 °/sec, we obtain an accuracy of +/- 80 arcsec and +/- 1.5 mm relative to the adjusted ellipse coordinates. Fig. 10 shows the measured horizontal target angle of the motion simulator arm.

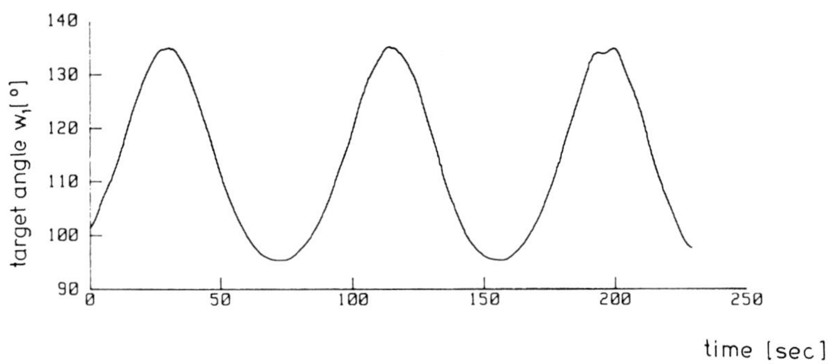

Fig. 10: Horizontal target angle of the rotating arm

5. Further research areas

High precision control of moving targets requires high precision and high speed coordinate processing by the tracking system. Further study will improve as well the accuracy of the sensor and the incremental encoders by reducing the various error influences as the coordinate calculation time by use of a suitable processing concept. The momentary measuring frequency is 10 cps, which is too low to guide a target with a maximum velocity of 1 m/sec and with an accuracy of better than 1 cm.

Operating under environmental conditions in the open air it is important to optimize the optical signal transmission with the aim of avoiding the influence of disturbing light, e. g. the sunlight, and of achieving a transmission distance of some hundreds of meter. Investigations are necessary with respect to

- the power of the emitting light source,

- the modulation of the light signal,

- the use of atmospheric windows and

- the use of image processing with CCD.

The intended application is the guidance of a target along a predetermined trajectory. Therefore the 3D-target position must be measured in a closed control loop.

To realize such a target guidance system additional processes have to be integrated:

- a second target angle measurement system (tracker, fixed camera system) or

- distance measurement and

- further controllers for minimum deviations between target and predetermined trajectory.

The additional processes cause further time and accuracy problems which have to be solved for achieving the desired target properties during movement.

6. References

Bayer, G.; Krzystek, P. (1986): Arbeiten und Ergebnisse der Teilprojekte D1/D3. Arbeits- und Ergebnisbericht 1984 - 1986, Sonderforschungsbereich 228 "Hochgenaue Navigation", Stuttgart.

Bayer, G.; Krzystek, P.; Möhlenbrink, W. (1988): Realtime Positioning of Moving Objects By Dynamic Target Tracking. 16th International Congress of Photogrammetry and Remote Sensing, Com. V, Kyoto, July 1988.

Desoyer, K.; Kopacek, P.; Troch, I. (1985): Industrieroboter und Handhabungsgeräte. R. Oldenbourg Verlag München Wien.

High Speed Infrared Tracking

Yves Castanet
Société Anonyme de Télécommunications Paris, France

Summary

The MINILIR System is a high speed infrared tracking system developed by SAT since 1975. Its new version Mark 2 has been improved by the addition of software control based on the use of a microprocessor. This system is built around a small turret giving high accuracy and high dynamic performances. IR tracking is using three different wavelengths that allows a highly accurate tracking in day or night operation, and trajectory results are delivered in real time. This system, already used in various countries, offers a good compromise between cost and operational capabilities with simple operating procedures and maintainance.

1. Foreword

Societe Anonyme de Telecommunication (SAT) is a well known company specialized in infrared-detectors manufacturing and the devolopment of systems using these detectors. We are going to present you the MINILIR System which is an infrared tracking system used for real-time automatic trajectory determination. This system has several applications for both, civil and military activities.

1.1 - Tracking requirements

The tracking system must have an high accuracy and deliver results of trajectory computation in real time. It must be able to track highly accelerated targets such as missiles. It also must be able to track both, cooperative targets using an IR source such as a projector and non-cooperative targets having only an IR signature sufficient for acquisition and trakking. The range for tracking will be as large as possible and operations must be possible during day or night.

1.2 - Tracking solutions

Passive infrared sensors are well suited for such requirements:

- accurate tracking,
- tracking of cooperative or non-cooperative targets,
- large range capabilities,
- day and night operations.

In order to achieve high speed tracking, the system will be built arround a small turret which allows highly dynamic performances and an easy on-site installation.

1.3 - Infrared solution advantages

Dynamic tracking performances are better than cinetheodolites. In fact the accuracy is comparable but the trajectory results are available in real time. Several sensors are available in the infrared bands. Range capability is larger than that of TV in good visibility conditions due to better IR contrast and far larger than that in poor visibility condition or during night.

System-mechanical characteristics allows easy transportation and installation. Compared to theodolits of identical accuracy, the cost factor is also one of the advantages.

2 - SAT's experience in Infrafed tracking

SAT has developed various IR tracker systems such as :

- K400, an IR tracker for CONTRAVES cinetheodolites;
  this system is in operation since 1969 in the Kourou space center,

- Some examples of range versus visibility are given :

Visibility	Range
800 m	2000 m
5000 m	15000 m
10000 m	100000 m

- T17, an IR tracker for CONTRAVES EOTS-F Cinetheodolites in use at weapon test ranges,
- PIRANA, a bispectral IR tracker for French anti-aircraft Corvettes,
- SEID, a differential IR tracker for naval anti-missile crotale system in operation in the French Navy,
- MINILIR, an IR high speed tracker which is in operation since the mid 70's in :
  . France
  . USA
  . UK
  . Netherland
  . Indonesia
  . Kenya.

3 - MINILIR System

3.1 - System definition

The MINILIR system has been developed in 1975 and first manufactured as Mark 1; the Mark 2 version now being manufactured incorporates a microprocessor and 360°-rotation capability. It is mainly composed of a turret supporting the Infrared head mounted on a tripod associated with two electronic units.

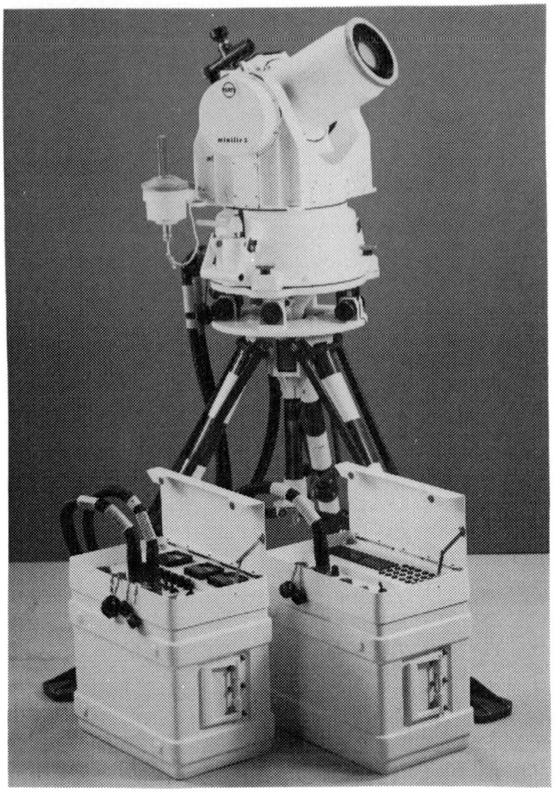

3.2 - System description

The MINILIR Mark 2 system is mainly composed of the following equipments:

- a turret including a base with access connectors, a yoke moving in azimuth, an elevation ring-mounted on the yoke holds the IR telescope; the telescope which is the IR sensor, is able to move in azimuth and elevation during the tracking of the target (Movement is performed by use of torque motors and angular position is given by angular resolvers),
- an electronic unit called IR unit used for IR signal processing and servo loop electronic circuits,
- an electronic unit called Encoding Unit used for angular measurement encoding, for system processing and system interface (This unit includes a microprocessor for the system control and process),
- a tripod for the mounting of the turret,
- a cooling system when the detector needs to be cooled.

3.3 - System operation

Pre-aquisitation of the target is performed in the manual mode with the use of a joystick using a TV camera or a telescope for aiming the IR head, or in the automatic mode by target designation from the encoding unit or from a remote computer. Target acquisition is automatic as soon as the IR level exceeds a present threshold. Target tracking is automatic after acquisition. An aiming correction can be added during tracking by use of the joystick. A speed memory capability is available by a push button when the target is temporarily hidden.

The output message is a digital type which contains :
- elevation and azimuth turret angular position,
- angular offset measurments,
- IR level measurements,
- system status.

Special features

Evolution of the software allows now special features to be implemented. Target acquisition can be commanded on the encoding unit by a remote computer after computation of an acquisition threshold. The servo loop can be digitalized by the use of the encoding unit working in designation mode. In this case digital filters can be initialized at encoding unit level or by remote computer. The servo loop characteristics are under software control; this feature allows acquisition at high angular velocities, and the adaptation of the servo loop response to the target characteristics.

3.4 - System characteristics

3.4.1 - IR characteristics

Three IR wavelengths are available by IR head replacement, the following table shows the characteristics of these heads.

Type of Target	Wavelength	Type of Cell	Sensitivity	Cooling
Cooperative	1.8-3 μm	PbS	2.10^{-12} W/cm²	No
Non-Cooperative	3.2-5 μm	InSb	10^{-12} W/cm²	Yes
Non-Cooperative	8-12 μm	HgCdTe	5.10^{-12} W/cm²	Yes

The telescope used has the following characteristics :

- Entrance pupil diameter : 127 mm
- Focal length : 500 mm
- Field of view : ±8 mrad (circular)

The cooling system uses a Joule-Thomson mini-cooler with high-pressure pure nitrogen. The three wavelengths correspond to regions where the radiation transmittance of the atmosphere presents a high value. The tracking range depends on the type of IR source, its intensity, weather conditions, visibility, elevation angle and the type of detector used.

3.4.2 - Tracking characteristics

The system and mainly the turret have been designed in order to reach high dynamic performances which are the following :

- angular velocity : 100°/s
  for azimuth or elevation in all operating modes
- maximum angular acceleration in tracking mode :
  . azimuth : 450°/s²
  . elevation : 1100°/s²
- static aiming error : less than 1 second of arc
- velocity coefficient : KV = 10^5
- acceleration coefficient : K_γ : 250
- angular limitation :
  . azimuth : none
  . elevation : -10° + 85° (Electrical)
  -10° + 180° (Mechanical)

3.4.3 - Mechanical characteristics

The mechanical characteristics are given by the following table.

	Turret	IR Unit	Encoding Unit	Tripod
Weight (Lb/kg)	73/33	40/18	26/12	44/20
Lenght (inch/mm)		14/356	14/356	
Width (inch/mm)		9.5/241	9.5/241	
Height (inch/mm)	23.5/597	14.5/368	14.5/368	43.5/1105
Diameter (inch/mm)	14/356			11.75/298

All the units are water resistant.
Operating temperature range : -20 to +50 °C.
Easily transportable in truck or aircraft.

3.4.4 - Interfaces

The system is supplied by the main line 115 V or 220 V 50/60 Hz, the power consumption is 300 VA.

The message output is composed of the following data :

* Angular position measurements :
  - resolution : 2.5 seconds of arc
  - accuracy : 10 seconds of arc (50 μrad);

  This accuracy corresponds at 2 inches at 3000 ft. (5 cm at 914 m)

* The format of the message output is the following :
  - angular position azimuth : 19 bits
  - angular position elevation : 19 bits
  - IR level measurement : 10 bits
  - X and Y angular offset : 16 bits
  - turret status : 3 bits

* The incoming message to the system has following format :
  - turret command : 4 bits
  - azimuth designation data : 16 bits
  - elevation designation data : 16 bits

Internal or external synchronization of the system is available. Various digital standard interfaces are available such as RS 232, RS 449, IEE.488, FSK modulation plus specific interfaces for test ranges.

4 - System advantages

This system presents a lot of advantages compared to heavy other tracking systems.

- The accuracy is comparable to other trajectory determination systems. The unit is able to deliver the trajectory data in realtime from launch.

- The tracking range is better than TV systems in poor or medium visibility conditions, during day or night, under low angles and close to the sun (10°C).
- The sizes of the turret allows the tracking of highly accelerated targets.
- System operation has been proven in the field under adverse weather condition (over 25 systems are in operation).
- The system modularity allows to change the wavelength within 15 minutes. Set-up time for the entire system is less than 1 hour.
- The turret is able to carry a TV camera or a laser head or both without any special feature.
- The system maintainance is low, due to the 10 years life time. We obtain a MTBF of 20 000 hours.

5 - System applications

5.1 - Actual applications

Various applications can be divided in three main groups: air, sea and ground systems.

- Air applications :
  navigation aids calibration (ILS - VOR - MLS),
  weapon test range trajectory.

- Sea applications :
  ship on board system reset (COSMAR),
  dragging,
  navigation bassin survey (Sogreah),
  constructive works (operation DELTA Netherland).

- Ground applications :
  tank weapon guidance evaluation,
  Eiffel tower top motion record.

5.2 - Future applications

For the same groups various applications are foreseen :

- Air applications :
  other landing aids calibration such as space shuttle, helicopters, aircraft recovery on carriers,
  trajectory improvements such as trajectory prediction or the acquisition of highly accelerated targets,
  development of IR thermal camera allowing thermal imaging, multi-target processing, differential tracking, weapon guidance.

- Sea applications :
  inertial navigation system calibration,
  weapon guidance,
  sea floor mapping (war mine).

- Ground applications :
  radar turret aid in case of high accelerated targets during launching phase,
  weapon guidance,
  building survey.

6 - MINILIR system sales

The MINILIR system has been sold to various countries for civil and military applications:

- 1 US NAVY (NATC - Pax River)
- 1 NASA (Space Shuttle MSBLS flight inspection)
- 1 Hazeltine Corp NY (MLS flight inspection)
- 1 Litton Canada (ILS flight inspection)
- 2 British CAA (Civil Aviation Authority Flight Inspection)
- 18 France mainly to test ranges (14)
- 6 Netherlands (Construction works)
- 1 Indonesia (ILS Flight inspection)
- 1 Kenya (ILS Flight inspection).

High Precision Surveying of Moving Objects by Electronic Cameras

Peter Krzystek
Inst. für Anwendungen der Geodäsie im Bauwesen, Universität Stuttgart

Summary

The development and accuracy aspects of an optoelectronic position measurement system with fixed electronic cameras is described. The electronic cameras are equipped with lateral-effect photodiodes providing excellent dynamic characteristics. The system is applied to a dynamic robot survey.

1. Introduction

For high precision navigation of moving objects a modular dynamic surveying system was developed in the Special Research Center on "High Precision Navigation". The key elements of that system are electronic cameras, which are equipped with lateral-effect photodiodes (PSD-sensors). The electronic cameras are either located in a fixed arrangement as an optoelectronic position measurement system (OPM) or mounted on a motorized theodolite as a target tracking system (Bayer, 1988). The general aim of the research work is guidance and control of slowly moving construction machines on construction sites in the close range with an accuracy of 1 cm/km (Bayer, Krzystek,1986).

This paper describes the development and some accuracy aspects of the optoelectronic position measurement system (OPM) with fixed electronic cameras. The OPM can be applied in a mono- and stereophotogrammetric mode. Möhlenbrink (1984) used a prototype of the OPM for the survey of a pipejacking with an accuracy of 2cm/230m. Thus, the OPM is a consequent development of that application of a PSD-sensor-based position measurement system. Special emphasis is made on the elimination of reflections producing large image distortions. Furthermore, the influence of the accuracy loss of the electronic cameras due to increasing camera-to-subject distance is demonstrated in a simple two-camera-arrangement. The system is applied to the dynamic survey of an industrial robot with an accuracy of ± 0.2 mm.

Compared to the well-known CCD-sensor cameras the PSD-sensor-based electronic cameras need only a light emitting target (e.g. LED) and have excellent dynamic characteristics due to the simple evaluation of the sensor coordinates. Thus, no large image processing systems and time consuming techniques of object identification and extraction are necessary.

PSD-sensor-based 3D-measurement systems were first developed and applied in biomedical engineering for human motion study (Woltring,1976) and in dynamic analysis of lightweight structures (Kirschstein,1980). These works demonstrate, that realtime 3D-measurement is possible within 0.1 - 1 kcps.

2. Description of PSD-sensor-based electronic cameras

The electronic cameras are characterized by a lateral effect photodiode (PSD-Sensor), which is located in the image plane and replace the conventional analog film. An active light source is projected on to the image plane via a common lens. The principle of the PSD-Sensors can be described as follows (Figure 1).

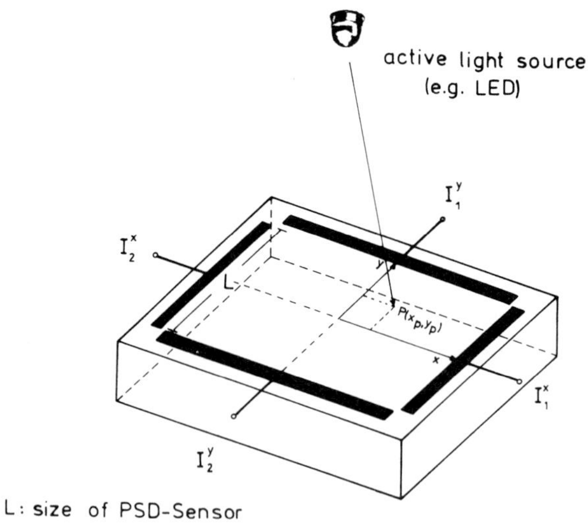

Figure 1: Tetralateral PSD-Sensor

The difference of currents produced by the radiation of an active light source is a measure of the position of the projected light source on the sensor surface. In practice it is common to introduce a value for the sensor coordinates with

$$x_p = \frac{L}{2} \frac{I_1^x - I_2^x}{I_1^x + I_2^x}$$

$$y_p = \frac{L}{2} \frac{I_1^y - I_2^y}{I_1^y + I_2^y} \qquad (1)$$

All commercially available sensor types have a high sensorspecific distortion due to the nonlinearity of the currents $I_j^{x(y)}$. The sensor coordinates in (1) deviate from the true position of the projected light spot between 0.1 - 10 % depending on the sensor type, operating mode and light spot position. Additionally the currents of the sensor correspond to the centroid of the light spot (Janocha, Marquardt, 1979), so that irregularities and disturbances of the light intensity cause imaging errors in the whole imaging system. By pulsing the light source and using specially designed optical filters, steady background ilumination will not affect the measurement precision. The influence of reflections of the light source on reflecting surfaces, however, still remain because reflections are not eliminated by electronic and optical filter methods.

The advantages of the PSD-sensors are the high dynamic characteristics of about 10 µs in rise time and high position resolution of less than 1 µm. The electronic cameras require only manipulation of a very small amount of data. If a 32 bit computer is used, 4*32 bit per camera are necessary to compute the coordinates of a projected target point in (1). Due to the high dynamics of the PSD-sensors, multi-point processing also becomes feasible by time-multiplexing of several light sources. Thus, this sensor-technique is a very simple image processing compared to the complicated and time consuming algorithms used in conjunction with CCD-cameras requiring 16000 times more data. Furthermore, CCD-camera-systems still have problems with optimal illumination and target size. Thus, CCD-array systems have limited realtime capabilities when applied in a 3D-measurement system to surveying and controlling of objects moving with high speed.

3. Problems of reflections

The coordinates of the light spot are shifted if a reflected beam is projected on to the sensor. The coordinate shift depends on the reflectivity function of the reflecting surface describing the intensity of the reflection. on the sensor (Figure 2)

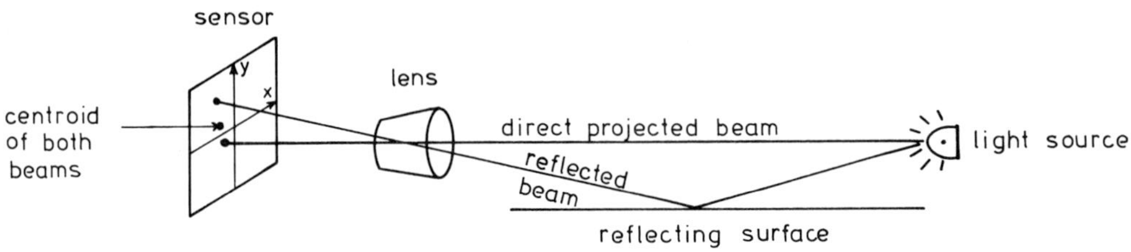

Figure 2: Reflection of a light source on the surface

The amount of sensor coordinate distortion due to reflections can be computed by the intensity ratio p and the distance l between the direct projected and reflected light spot. With the simple formula

$$\Delta l = \frac{p}{1+p}\, l \tag{2}$$

follows, that already a reflection of 1% and a distance of l=1 mm cause a coordinate shift of 10 µm. Figure 3 illustrates the shifting of the light spot position due to the intensity and the position of the secondary maximum of the reflection. Experiments have actually produced coordinate shifts of some millimeters in full accordance with the theory.

For compensation of reflections there exits several proposals and techniques. Some of them are the use of absorbing surfaces in the spectrum of the light source and the use of polarized light sources. These methods however cannot be applied to a flexible position measurement system or are not efficient enough for a total and precise compensation.

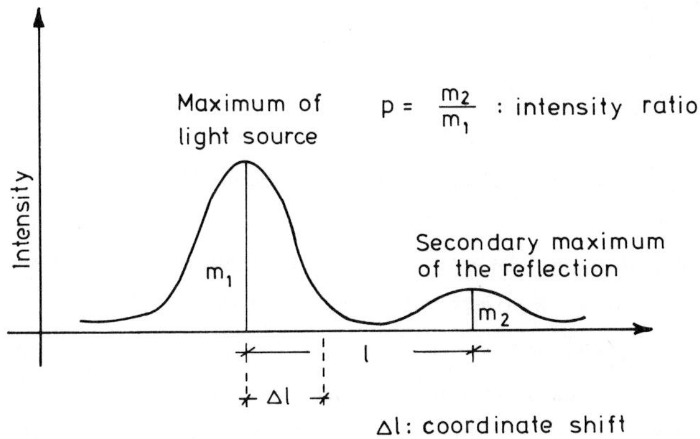

Figure 3: Coordinate shift due to secondary maximum of a reflection

A better approach is to shield the reflections in front of the sensor (Platz,1977).This principle idea has to be modified by an operational technique. Thus, for reflection elimination a special rotating diaphragm disc was installed in the optical imaging system. Figure 4 shows the principle of this specific technical part of the electronic cameras.

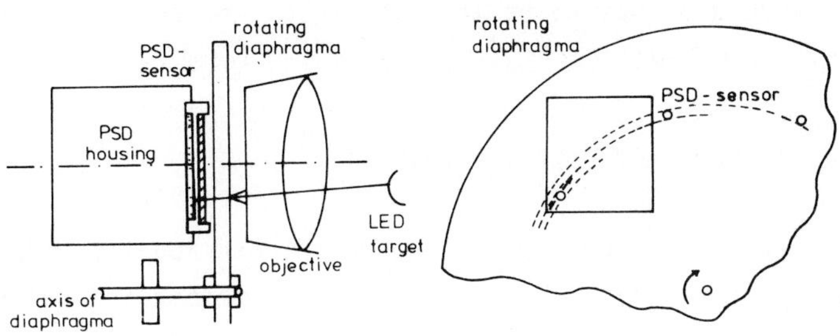

Figure 4: Rotating diaphragm disc for reflection compensation

Experiments showed that reflections could be eliminated with an accuracy of 1 - 2 micrometers in the image plane. That was a considerable improvement, because positioning by recently known PSD-cameras systems were heavily influenced by undetected reflections. This technique of reflection elimination seems to be an operational method. Therefore, the characteristic parts of the OPM-system are the PSD-sensor-based cameras, which are insensitive to reflections.

4. System-configuration

4.1 Hardware

In the stereo photogrammetric mode the OPM-system consists of at least two PSD-cameras with large area tetralateral photodiodes. As active light sources we use high powered LEDs. In order to avoid long cables, the signals of the electronic cameras are transmitted to a desktop computer of the HP 200/300 series via a 14 bit telemetry system. The maximum sample rate is 800 cps per camera with no time lags between the cameras. In this special hardware configuration the OPM-system computes 3D-positions with 25 cps in realtime. This is a remarkable feature if one considers, that the OPM works with a very small computer facility. Initial bench mark tests with a µVAX even showed that an improvement by a factor of 10 is to be attained. Thus, it should be pointed out, that the realtime characteristics of the OPM can be increased by the capabilities of the data aquisation system, the computer system and the process handling. But there are no time consuming algorithm problems to compute the corresponding sensor coordinates in the electronic cameras. For high speed applications the mentioned technique for reflection compensation by a rotating diaphragm disc has to be modified (e.g. controlled diaphragm)

4.2 Mathematical model and accuracy aspects

The modell of central perspective is chosen to describe the projection of a light source into the electronic cameras. The corresponding colinearity equations are modified with special correction functions for compensation of systematic errors.

$$x_i - x_o = -c \, \frac{A_{11}(X_i - X_o) + A_{12}(Y_i - Y_o) + A_{13}(Z_i - Z_o)}{A_{31}(X_i - X_o) + A_{32}(Y_i - Y_o) + A_{33}(Z_i - Z_o)} + dx(\underset{\sim}{\alpha}, \bar{x}, \bar{y})$$

$$y_i - y_o = -c \, \frac{A_{21}(X_i - X_o) + A_{22}(Y_i - Y_o) + A_{23}(Z_i - Z_o)}{A_{31}(X_i - X_o) + A_{32}(Y_i - Y_o) + A_{33}(Z_i - Z_o)} + dy(\underset{\sim}{\beta}, \bar{x}, \bar{y}) \qquad (3)$$

with

$$\bar{x} := x_i - x_o$$

$$\bar{y} : \quad y_i - y_o$$

where $\underset{\sim}{\alpha}$, $\underset{\sim}{\beta}$ are the self-calibrating parameters of the electronic cameras.

The correction functions in (3) are characterized by polynominals, which compensate for non-perpendicularity, scale differences, lens distortions and sensor-specific errors. The least-squares estimation of the orientation and calibration as well as the object coordinates are devided into space resection and intersection. A known calibration frame represents the ground control for orientation and calibration. Calibration with (3) and a bilinear interpolation lead to a sensor accuracy of 2 - 3 µm (Bayer et al,1988). With a bi-cubic spline functions approach the optical imaging system was calibrated with an accuracy of less than 1 µm. This calibration result renders possible the application of the OPM-system for high precision surveying.

Besides the systematic errors due to sensor distortion, lens distortion and non-linearities of the electronic equipment, the electronic cameras of the OPM have stochastic errors caused by quantization and signal noise. Möhlenbrink (1988) shows, that for a fixed sensor coordinate and quantization level the standard deviation increases with the square of the camera-to-subject distance s.

$$\sigma_{xp} \sim s^2 \tag{4}$$

This can be easily derived by error propagation with respect to (1) and the law of inverse squares. Thus, compared to a traditional theodolite-based surveying system, the electronic cameras of the OPM have a non-uniform angular accuracy if the camera-to-subject distance varies. Figure 5 demonstrates the influence of that special accuracy aspect of the OPM-cameras in a simple 2-camera-arrangement. For comparison the isolines of Helmert's error of position are plotted both for the OPM-system and a standard point intersection in the stereo-coverage of the fixed cameras. The intersection point of the optical axis of the fixed cameras defines the reference point with a normalized unit position error.

Figure 5 shows, that the OPM system has different isolines of position error than a well-known point intersection system with an uniform angular accuracy. The accuracy loss could be compensated by intensity control of the light source.

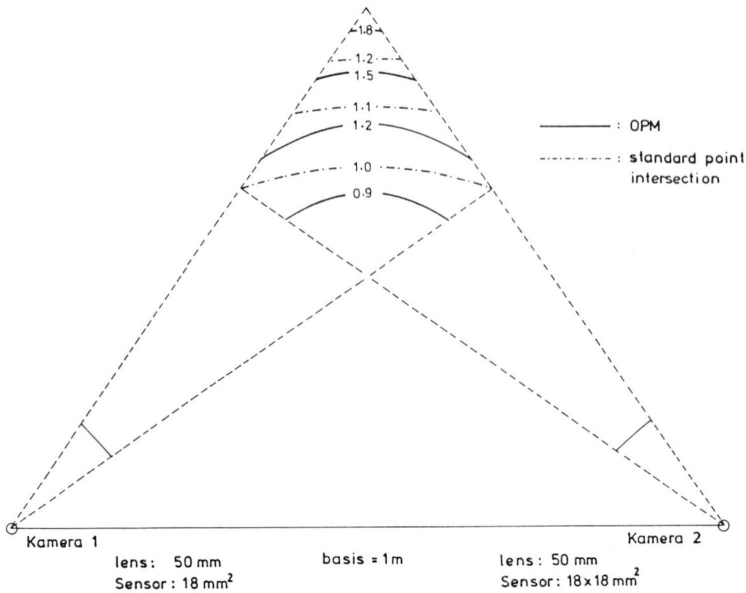

Figure 5: Isolines of Helmert's position error of the OPM-system with respect to the reference point

5. Surveying of robots

5.1 Existing techniques

Robotics becomes more and more interesting as an application field for highly precise measurement systems. Both manufactures and users need them to acquire geometric, kinematic, static and dynamic quantities about robots. From a general point of view the requirements for a measurement system usable for robot surveying and calibrating are high precision (0.1 - 0.15 mm) and resolution (0.01 mm), non-contacting and transportability (Scheffer,1982). The dynamic requirements are high and result from cycle times of the robot control loops of at least 20 msec (= 50 cps) .The existing non-contacting optical measurement systems can be devided mainly into stereophotogrammetric methods, theodolite-based methods and laser measuring methods, which are described in detail by (Schüßler,1987). Special methods are, for example, ultrasonic methods. Laser measuring methods have excellent accuracy and resolution features. An interesting system attains an accuracy of several micrometers in object space (Hof,Pfeiffer,1985). The traditional theodolite-based measurement systems with manual guidance of the theodolites have also high accuracy and resolution, but rather poor processing cycles. The automation of that type of measurement system with CCD-cameras (Gottwald,1987) led to a significant improvement of the time consuming point location process. Indeed, due to the slow realtime capabilities of the image processing subsystems, the application of such a theodolite-based system is presently limited to the slow motion surveying and calibration of robots. Bayer (1988) shows, that a theodolite-based tracking system with a PSD-camera is capable of surveying a moving object with 10 cps by a very small computer facility. The consequent development of that tracking concept with two theodolites might lead to a high precision measurement system with excellent realtime capabilities. The photogrammetric systems with PSD-cameras have the advantage of excellent on-line and realtime characteristics together with high resolution. The accuracy depends on the sensortype used, the calibration quality and the influence of reflections. Compared to laser measuring systems and theodolite systems this technique has a less accuracy due to the non-linearities of the optical imaging system . In addition the working space is limited by the stereo-coverage of the fixed cameras. It is well suited for surveying threedimensional trajectories of robots moving with high speed. If CCD-array-cameras are used the realtime capability of the stereophotogrammetric method is limited again by the time consuming image processing.

5.2 Robot survey with the OPM

The OPM was applied in the dynamic survey of an industrial robot. Figure 6 and 7 show the endeffector of the robot with the calibration frame for ground control as well as the camera locations.

With the OPM-system the accuracy of the calibration frame of ± 0.2 mm was attained. Experiments under laboratory conditions proved that the accuracy of the OPM is mainly limited by the accuracy of the calibration frame. Thus, a higher accuracy of the calibration frame might have led to higher accuracy of this robot survey. For this application the accuracy of the OPM was sufficient with respect to the position error of the robot. One of the intended tasks of the robot survey was to check the rectilinear movement in dependence of the robot velocity. Thus, several straight lines were programmed in the robot coordinate system with a mean moving direction towards the camera equipped with the telephoto lens. The OPM measured the robot trajectories for these defined straight lines within a velocity range of 50 mm/sec to 500 mm/-sec. The adjustment indicated significantly deviations from the straight line as a function of the velocity. The deviations vary for this type of robot between 0.2 mm and 5 mm. Figure 8 shows the typical robot trajectories of a programmed rectilinear movement for several velocities.

Figure 6: Industrial robot and the calibration frame

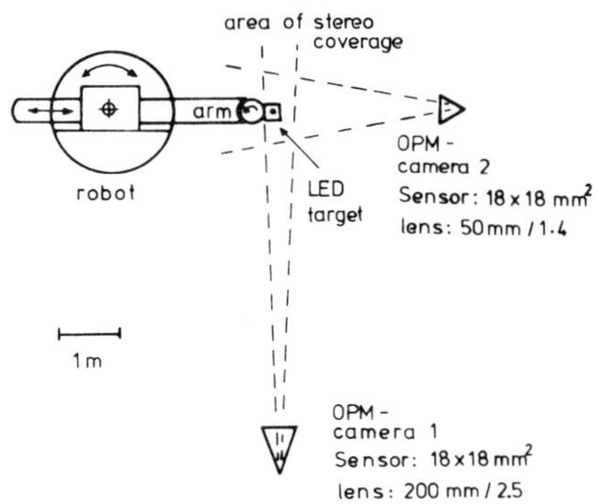

Figure 7: Camera locations for robot survey

This practical application verified the forecasted high precision capability of the OPM with respect to the theory and rather good results under optimal laboratory conditions. Especially in alignment mode the OPM is able to take advantage of the PSD-sensors and measure deviations from a rectilinear movement with an accuracy of several micrometers in object space.

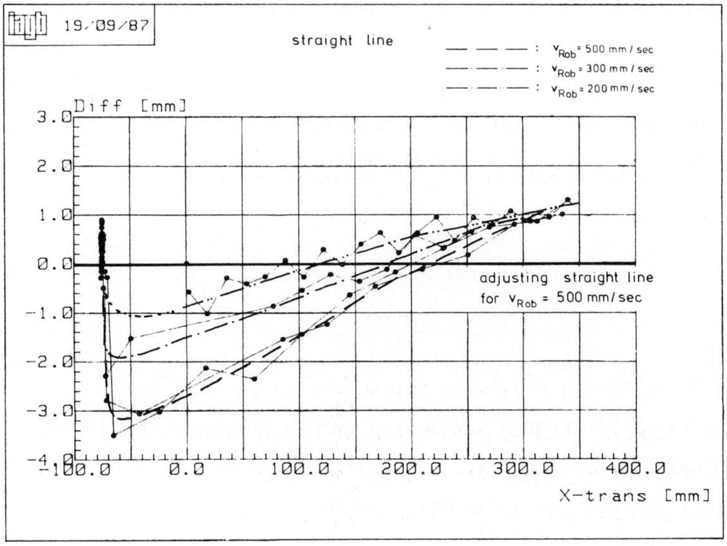

Figure 8: Deviations from rectilinear robot movement

6. Conclusion

PSD-sensor-based electronic cameras are still a cheap alternative to the forecoming CCD-cameras, because the latter are incorporated with larger computer systems and time consuming algorithms. For some application fields the use of an active light source as a target might be more interesting than an expensive image processing system and computer facility. The capabilities of the desribed OPM photogrammetric system are a sensor calibration accuracy of 1 - 3 µm and a realtime characteristics of 25 cps. The electronic cameras are insensitive to reflections due to the installation of a special rotating diaphragm disc. The aim for further work is to apply a larger computer facility and parallel processing to achieve a realtime characteristic of 1 kcps. This would be interesting with respect to high speed applications in robot surveying and controlling.

Acknowledgement

I would like to thank Prof. Dr.-Ing. G. Pritschow (Institut für Steuerungstechniken der Werkzeugmaschinen und Fertigungseinrichtungen) and Dipl.Ing. U. Rentschler for their support and the disposal of the industrial robot.

7. References

Bayer, G.,Krzystek, P. (1986): Arbeiten und Ergebnisse der Teilprojekte D1/D3. Arbeits- und Ergebnisbericht 1984 -1986, Sonderforschungsbereich 228 "Hochgenaue Navigation",Stuttgart

Bayer, G.,Krzystek, P.,Möhlenbrink, W. (1988): Realtime Positioning of Moving Objects by Dynamic Target Tracking.16th International Congress of Photogrammetry and Remote Sensing,Com. V,Kyoto,July

Bayer, G. (1988) : Motorized Electronic Theodolite - High Precision Measurement Robots for Surveying,Proc. of the International Workshop "High Precision Navigation",Stuttgart,May 17th-20th

Gottwald, R. (1987) : Kern E2-SE - Ein neues Instrument nicht nur für die Industrievermessung ?.AVN 4,S. 147-154

Hof, A.,Pfeiffer, T. (1985): Räumliches Wegmeßsystem,3D-Interferometer GMR - Bericht Nr. 6,S. 151 - 165,März

Janocha, H.,Marquardt, R. (1979): Universell einsetzbares Wegmeßsystem mit analog anzeigenden, positionsempfindlichen Photodioden. Teil 1: Eigenschaften der verwendeten zweiachsigen Photodioden. Technisches Messen tm, Heft 10,S. 369 - 373

Kirschstein, M. (1980): Verfahren zur berührungslosen dreidimensionalen Form- und Verformungsmessung,Dissertaion,Universität Stuttgart

Möhlenbrink, W. (1984): Abschlußbericht zum Einsatz von Positionssensoren zur Vermessung von Rohrvorpressungen, Forschungsgesellschaft für geodätische Meß- und Rechentechnik, GmbH, Stuttgart,unveröffentlicht

Möhlenbrink, W. (1988): Zielverfolgende Theodolite - ein Hilfsmittel für die Steuerung von Baumaschinen, DGK, Reihe B, in Vorbereitung

Platz, U. (1977): Untersuchung positionsempfindlicher analog anzeigender Fotodioden im Hinblick auf den Anwendungsfall räumlicher Koordinatenmessungen,Dissertation,Technische Universität Hannover

Scheffer, B. (1982): Geometric control and calibration method of an industrial robot.Proc. of 12th International Symposium on Industrial Robots,June 9th - 11th,Paris

Schüßler, H.-H. (1987): Meßverfahren zur Prüfung der Genauigkeit von Industrierobotern. Vorträge zum Ausssprachetag "Indutrieroboter messen und prüfen",GMA-Bericht Nr. 14,Stuttgart,November

Woltring, H.J. (1976): Calibration and measurement in 3D-monitoring of human motion by opto-electronin means.II Experimental results and discussion. Biotelemetry,3,pp.65-97

Pulsed Laser Radar for Close Range Operation in Space

Sigmund Manhart, Peter Autenzeller, Walter Braumandl
Messerschmitt-Bölkow-Blohm GmbH, München

Summary

A high resolution laser diode rangefinder was developed which is part of an autonomous target acquisition and tracking system for rendezvous and docking manoeuvres in space. The laser radar system is capable of measuring distances to retroreflector targets from 2 m up to 20 km. At a single point measuring speed of 0.5 s the ranging accuracy is 2...3 cm and the short term stability and reproducibility 2...3 mm. The ranging unit is based on the time-of-flight measurement of a short laser pulse. To achieve the cm-accuracy and the mm-resolution some sophisticated signal evaluation techniques have to be used:

* A constant fraction discriminator reduces the influence of signal amplitude variations on the electronic trigger of the time evaluation circuit
* A dual-slope time stretching method has been introduced to support precise time interval measurements
* An internal optical fiber is used for the calibration of the ranging system
* The target return signals are adjusted to convenient levels by a piezoelectric attenuation unit

1. Introduction

Laser radar systems are expected to play an important role in future space projects. Applications, in the near future, are seen in rendezvous and docking (RVD) manoeuvres, alignment of large structures, and formation keeping of satellite clusters.

Therefore ESA has initiated the development of a laser diode rangefinder for medium and short range applications. The German BMFT has supported this work with the intention to have a specific RVD sensor with automatic target acquisition and tracking capability.

The development of a new technology, however, is too costly if directed to one task only. Therefore the development goal of the laser radar presented in this paper was aimed towards a versatile and adaptable instrument which can be used for a variety of mission requirements. The baseline specifications for such an instruments are:

- total ranging distance : 2 m to 20 km
  (to retroreflector targets)
- ranging accuracy: 2 - 3 cm *)
- ranging stability and reproducibility: 2 - 3 mm *)
  *) in the close range from 2 m to 1 km
- capability to discriminate individual targets
- capability to acquire and track moving targets

Although each of the above specifications has been achieved in laboratory or even by commercially available instruments, it is not easy to obtain them altogether in a single, autonomously operating instrument of moderate weight, size and power consumption.

2. System description

A pulsed laser radar utilizing direct time-of-flight measurement was chosen instead of a cw phase measurement radar, mainly for two reasons:

* The better SNR characteristics allows for higher target distances at short target dwell times.

* An internal calibration can be performed simply by time gating between target returns and optical fiber reference pulses

To achieve the envisaged ranging accuracy and measuring speed three conditions had to be fulfilled:

a) The electronic trigger of the time-of-flight measurement has to be related to a well-defined position on the laser pulse. It should be independent of signal amplitude variations.
b) The time interval measurement must have picosecond resolution.
c) Internal calibration has to be foreseen to eliminate electronic drift and aging effects.

2.1 Precise triggering

Precise triggering of the time-of-flight measurement is accomplished by a high-speed comparator (Plessey SP 9685) as part of a constant fraction discriminator (CFD). The circuit diagram is shown in Fig.1. The electrical delay 2 and the resistor network at the Y input cause the comparator to switch at half-maximum of the receiver pulse, independent of its amplitude. Two additional comparators (COMP A and COMP B) are used to control a piezoelectric driven optomechanical signal attenuator (Fig.8).

The laser pulse shape is shown in Fig.2. Since the leading edge is used for triggering, emphasis was given to a fast pulse rise time. But the pulse duration must not be too short. Otherwise the pulse energy is too low for far distance ranging. The drive circuit was optimized to yield a pulse rise time of less than 500 ps and a pulse width of approximately 10 ns. It contains three avalanche transistors which provide the laser diode forward bias current by rapidly discharging high-voltage capacitors. Laser diode and drive circuit are arranged in a rf-shielded box (Fig.3).

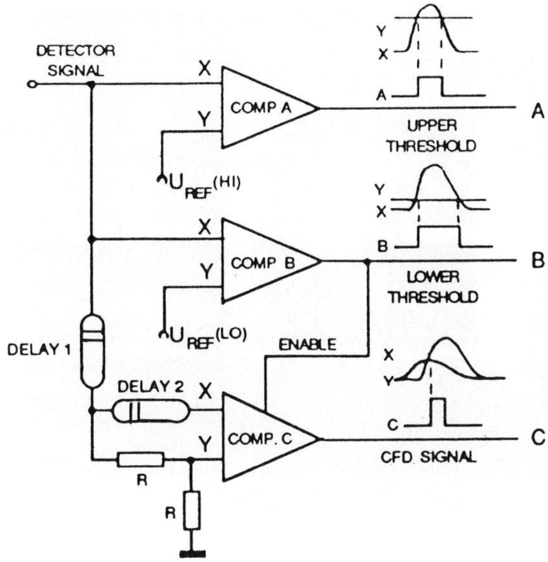

Fig.1
Circuit diagram
of the constant
fraction discriminator

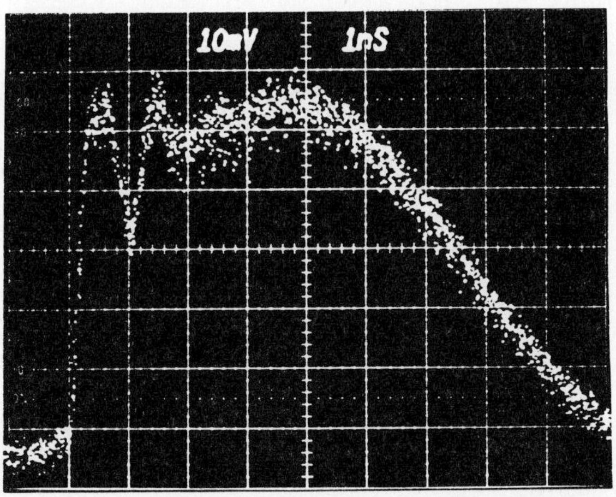

Fig.2 Laser output signal

wavelength	905 nm
peak power	10 W
rise time	200 ps
pulse width	8 ns
rep. frequ.	5 kHz

Tab.1 Laser Parameters

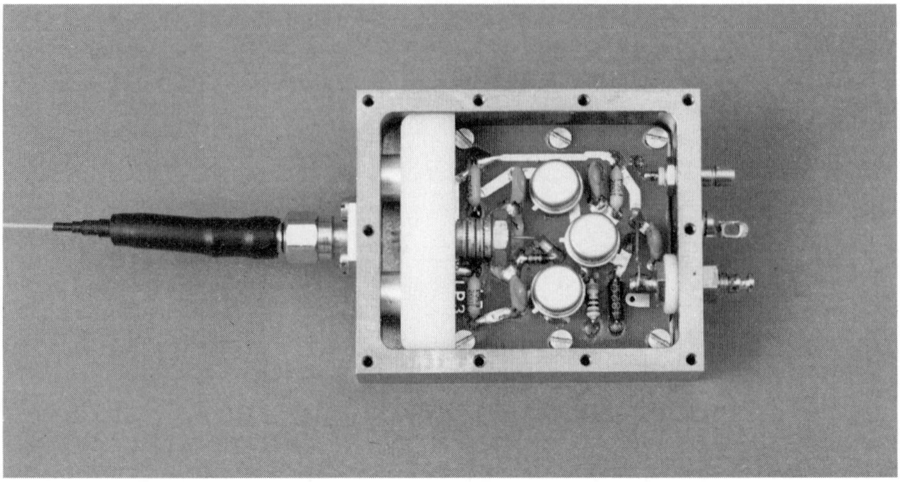

Fig.3 Laser transmitter

2.2 Time interval measurement

The time interval measurement is initiated by an ECL trigger signal SYNC which is synchronized to the outgoing laser signal. The "stop" signal is generated either by a reference signal (from the internal optical fiber) or by the target return signal. It depends on the position of an electronic time gate if the reference or the target signal is evaluated. The digitizing of the time interval is accomplished by two binary counters, a "coarse" interval counter and a "fine" interval counter. Both counters are controlled by a 50 MHz quartz clock. The coarse interval is started by the SYNC signal, the fine interval is started by the CFD trigger output. Both counters are stopped by the second quartz clock following the CFD trigger output. Since the SYNC signal is correlated to the quartz clock it is sufficient to have the integer numbers of clock pulses from the coarse counter and to measure precisely the residual time interval between CFD trigger and the next clock pulse. The difference between coarse interval and fine interval corresponds to the range value. This is illustrated in the timing diagram of Fig.4.

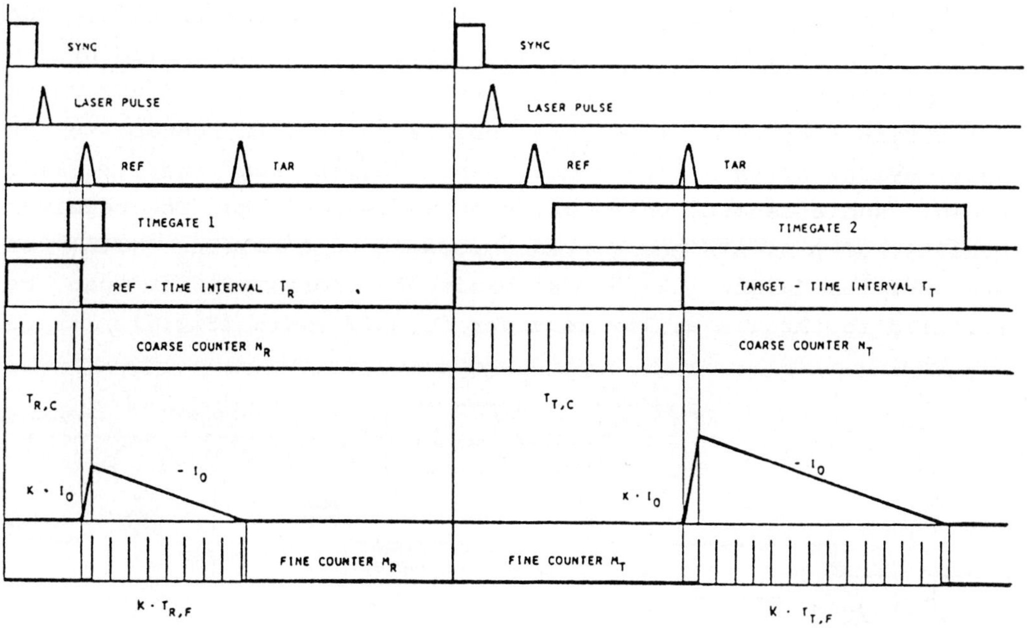

Fig.4 Timing diagram, coarse counter interval and fine counter interval

Precise measurement of the fine counter interval is made possible by stretching it electronically before it is digitized. During the time of interest a capacitor is charged by a constant current I_1. Afterwards it is discharged by a constant current I_2. I_2 is much smaller than I_1. Thus, the time interval is stretched by a factor of I_1/I_2. The conversion factor was chosen to be 256 in our model. The coarse counter is operated as an up-counter, the fine counter as a down-counter. The difference between coarse interval and fine interval is obtained by adding the binary results as shown in Fig.5.

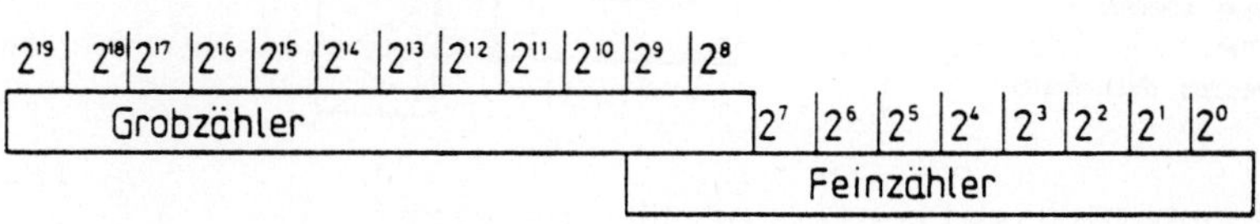

Fig.5 Addition of coarse counter and fine counter results

Fig.6 shows the block diagram of the entire electronics. The transmitter trigger and the range digitizer are controlled by the 50 MHz quartz clock. The transmitter is operated at 5 kHz pulse repetition frequency. This a reasonable value with respect to the duty cycle of the laser diode, the acquisition and tracking speed, and a ranging unambiguity of at least 20 km. The receiver consits of a Si-APD and a transimpedance preamplifier (TIA) of 500 MHz bandwidth. A 50 MHz bandwidth postamplifier can be switched to the TIA at low receiver signal levels (Fig.7).

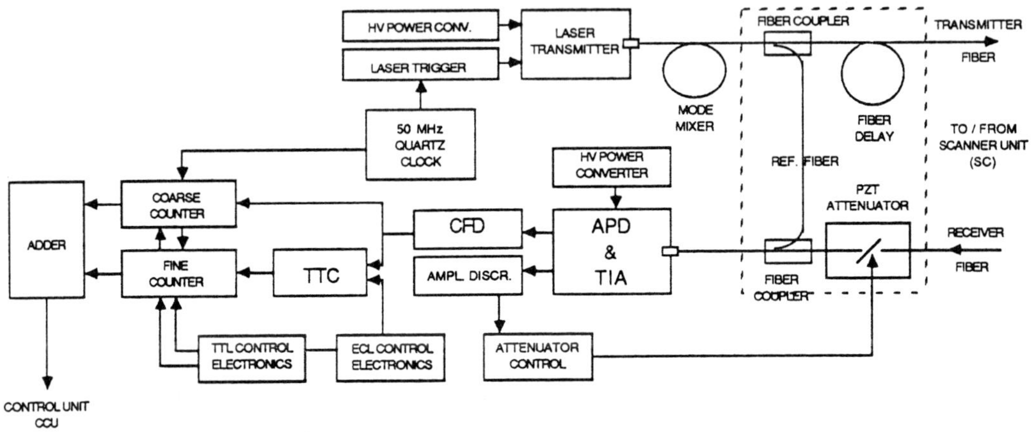

<u>Fig.6</u> Block diagram of the entire rangefinder circuit

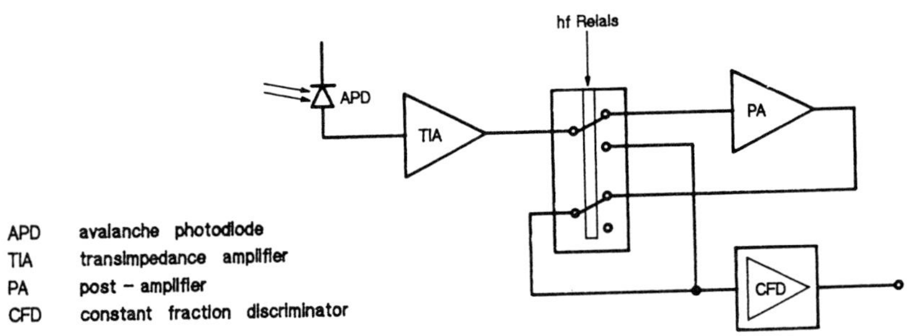

<u>Fig.7</u> Gain-bandwidth switching of the receiver

2.3 Internal calibration

Fig.8 shows the optical concept of the laser rangefinder. The laser diode signal is focused into a step index optical fiber. The "mode mixer" consists of an optical fiber coil of 1.5 m fiber length. It eliminates the far field and near field inhomogeneities of the multimode laser. A small fraction of the laser signal is fed directly to the receiver via an auxiliary fiber to provide a reference signal for the range evaluation. The major part of the transmitter signal goes on to the transmitter optics and the scanner unit. The fiber delay assures that the target signal always arrives later at the detector than the reference signal. A piezoelectric controlled signal attenuator in the receiver path automatically adjusts the receiver signals to convenient levels. It is controlled by the threshold comparators of the CFD circuit (Fig.1).

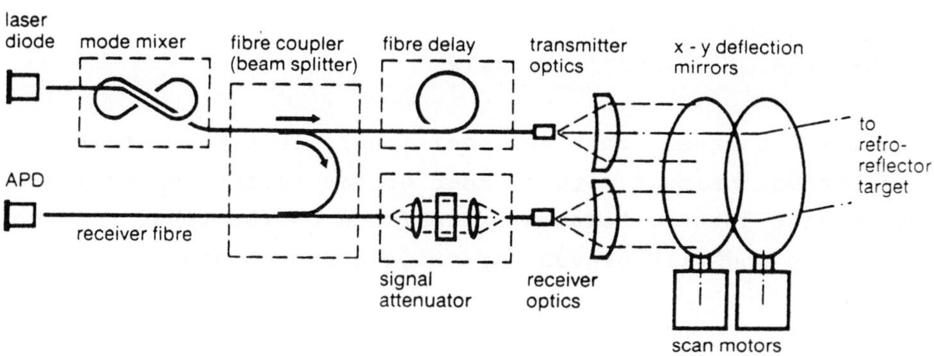

Fig.8 Fiber optic concept of the laser rangefinder

3. Error analysis

Since target return signals and fiber reference signals are going to the same detector and are processed by the same electronics, all electronic drifts can be compensated for by calibration measurements. Systematic errors caused by drift and aging effects can thus be eliminated. Statistic errors can be reduced by averaging a large number of individual single measurements.

The remaining error sources are:

a) the time walk of the CFD, and
b) the nonlinearity of the TTC

The time walk of the CFD, i.e the amplitude dependence of the comparator propagation delay, was measured to be approximately 140 ps for a detector signal variation of 40 dB. The nonlinearity of the TTC is approximately 0.3 % within the maximum time span of 20 ns to be stretched. This corresponds to 60 ps input error.

Both error sources, in principle, could be further reduced by:

* a higher clock frequency
* calibration of the CFD time walk

It was felt, however, that the additional electronic effort is not justified, because second order errors (quartz clock instability, trigger jitter, temperature effects on the fiber reference) will come up and will prevent dramatic improvements.

4. Test results

The laser rangefinder was tested with respect to ranging accuracy, linearity, short term stability and reproducibility. Fig.9 shows the statistical spread of individual range data at 2 uW and 200 uW detector input power. Averaging yields 2 mm range value stability and reproducibility. The total ranging accuracy was verified to be within 3 cm from 2 m to 100 m. Fig.10 shows the linearity of the range measurements over 4 m measured in steps of 1 cm and 5 cm.

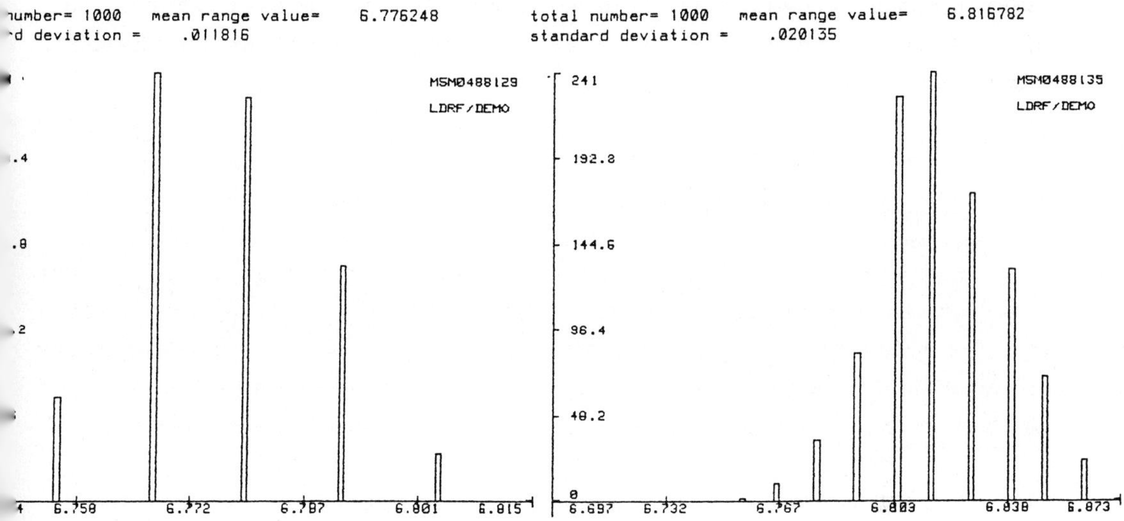

Fig.9 Statistical distributions of single range data at
a) 100 uW detector power b) 1 uW detector power

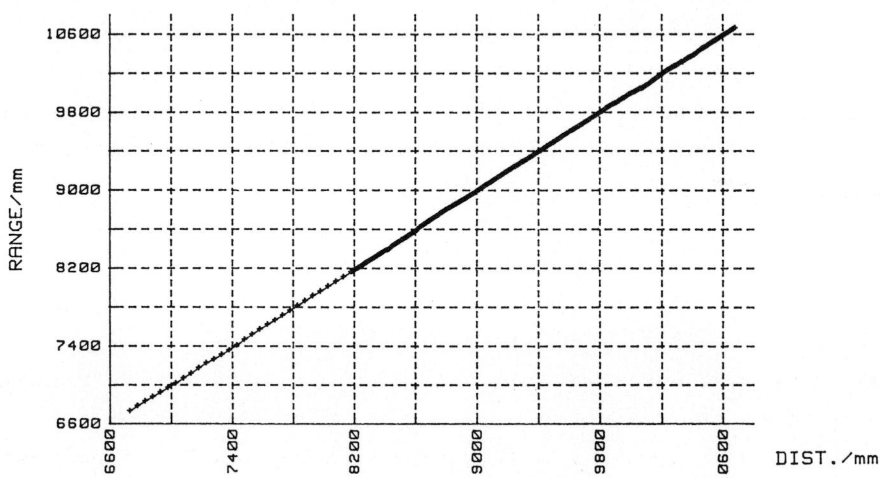

Fig.10 Linearity of the range measurements

Acknowledgement

The pulsed laser diode radar system was developed under ESA and BMFT contracts. We would like to thank Mr. P. Roussel (ESTEC) and Mr. Seher (DFVLR) for their stimulating discussions and helpful contributions.

Realtime Surveying in Close Range Area with Inertial Navigation Systems and Optical Target Tracking Techniques

Wolfgang Möhlenbrink
Inst. für Anwendungen der Geodäsie im Bauwesen, Universität Stuttgart

Summary

Realtime dynamic surveying methods on construction sites are treated in this paper. With the development of an optoelectronic measurement system the positioning of construction machines becomes possible. The measurement system is realized by the combination of realtime cameras which are equipped with position sensitive devices and classical photogrammetric approaches. A target tracking theodolite with automatic angle measurement capability is used for dynamic positioning of slowly moving objects. For tunneling and pipe jacking techniques a measurement methodology is proposed to determine position and orientation with an inertial measurement system using a special step technique. This technique provides a refractionless measurement tool for bad visible conditions.

1. Introduction

Automatic measuring techniques are becoming more and more important in geodesy, surveying, photogrammetry and engineering surveying. Most of these techniques relate to positioning systems of sea and land navigation. Realtime and long range properties are the main advantages of the global positioning systems TRANSIT and NAVSTAR. Initial results with sufficient accuracy for land surveying purposes were obtained approx. in 1975 by geodetic inertial platforms in the USA and Canada.

In the last five years excellent results for geodetic networks were obtained in the km-range from the GPS NAVSTAR and the use of inertial navigation systems decreased. One important point from these previous results is that since 1981 (BANFF-Symposium) the accuracy threshold of one dm has not been broken.

So the GPS-system NAVSTAR seems to be the realtime sureying tool of the future for global and local surveying networks.

Most of the daily surveying business in Germany however, takes place in urban areas and on construction sites within ranges less than 300 - 1000 m. The condition of a 5 m nonreflecting area around a GPS receiver antenna and free field of view above 15° to 20° elevation is rarely attainable.

Target tracking theodolites and modular optical positioning systems should work well in this close range, connecting directly to the surveyed control points located close to the area of interest.

Optical positioning techniques reach their limits in areas with bad visibility and obstacles in the field of view. For tunneling and pipe jacking specifically the non-refraction inertial measurement technique may provide operational positioning techniques in close range areas less than 1 km in the future.

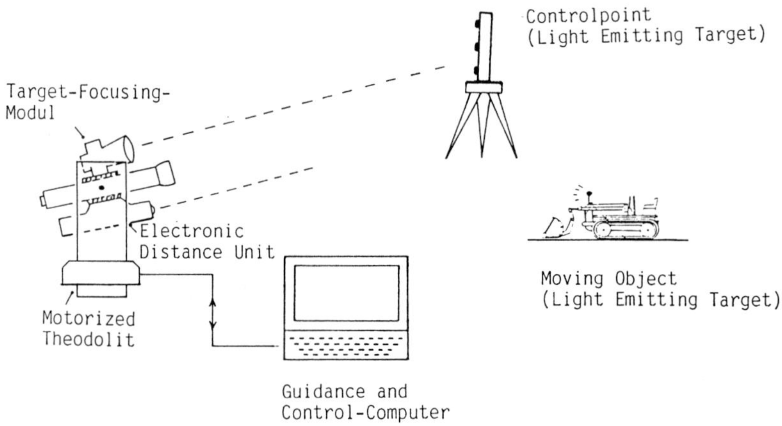

Figure 1: Modular Optical Positioning System (MOPS)

2. Positioning and guidance problems on construction sites

In the Sonderforschungsbereich 228 (SFB 228, Special Research Area "High Precision Navigation"), the Institute for Applied Geodesy is concerned with the development of automatic surveying and positioning systems for construction sites. The Modular Optical Positioning System (MOPS) consists of an

Optical Position Measurement system (OPM), a motorized theodolite and a guidance and control processor (Fig.2).

In many civil engineering situations the continuous and precise positioning and guidance of machines, e.g. graders, will lead to economic machine use, and a guaranteed level of quality (e.g. paving machines). In many cases, realtime surveying techniques could be integrated into the building process. (Fig.3)

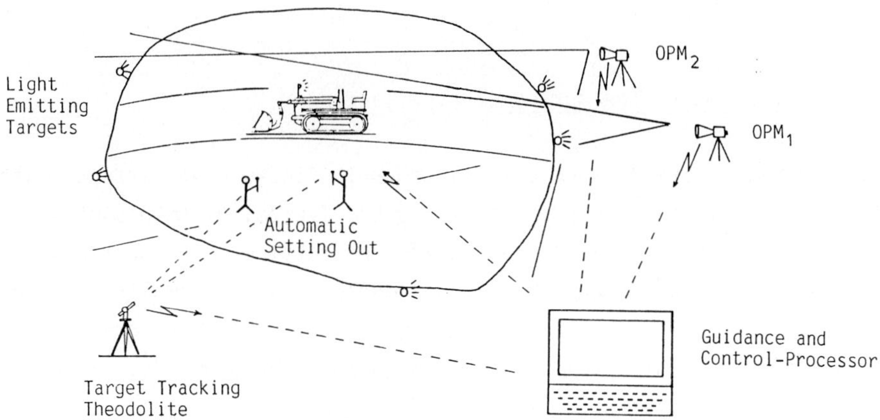

Figure 2: Realtime positioning on construction sites

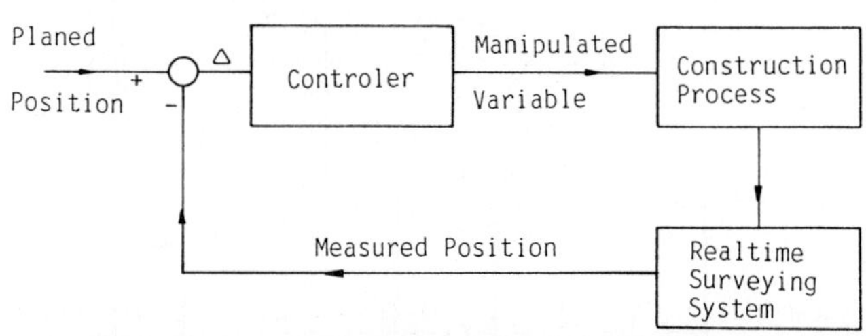

Figure 3: Realtime surveying system as component of the guidance and control circle

The automatic techniques range from simple distance sensing to complex spatial coordinate and direction determination. Examples are presented below for the problem of pipe jacking.

Pipe jacking is a highly automated technique for building, for example, waste water and storm water pipelines. The automatic surveying of the cutting shoe of the whole pipe system is demonstrated with an electronic OPM-camera and computer system. The main disadvantage of such a system is the refraction problem in the pipes and obstacles present in the relatively narrow pipe cross sections. These restrictions could be overcome with the use of inertial techniques. The basic idea of using inertial navigation systems in pipe jacking are presented below.

3. Accuracy problems in the survey of pipe jacking

The situation at a pipe jacking site is shown in figure 4. A vertical shaft is installed at the starting point of the pipe. Typical diameters of the shaft are between 5 and 10 m. In this shaft the cutting shoe and the machine tube are installed and pushed into the surrounding soil. Normally the shaft is so deep that the pipe is below the ground water level therefore a pressure lock must be installed to avoid water flowing into the pipe.

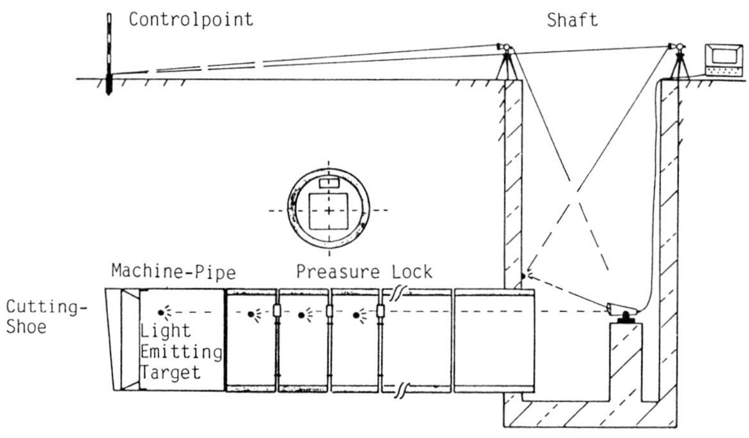

Figure 4: Pipe jacking

High pressure equipment is installed in the shaft and preconstructed pipes of 3 - 4 m of length are pushed behind the cutting shoe and machine pipe. The whole pipe is jacked through the ground to the target shaft. The necessary accuracy in this construction method is approximately 5 cm/km. The guidance, especially the heading of the pipe, is a rather difficult surveying problem. Depending on the consistency of the soil, the cutting shoe must be controlled for heading and position in realtime during the jacking process. This surveying problem contains three main components:

- the surveying network of the surface:
  the azimuth between start and end point of the pipeline could be obtained from the local surveying network, but is that sufficient? Pipe jacking is very suitable in urban areas. Distant control points can rarely be used. The case of a range of 100 m an accuracy of ± 1 cm in the coordinates the standard deviation of the azimuth will be in the order of 30 arcsec. That is not sufficient for the deviation error from the traverse of 5 cm/km. Therefore a special engineering surveying network has to be installed between the start and the end point of the pipeline.

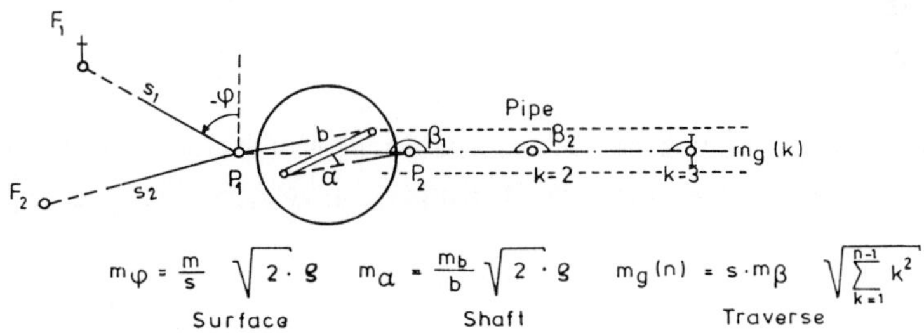

Figure 5: Horizontal network

- Azimuth transfer into the shaft: The second survey problem is the transfer of the azimuth from the earth surface to the level of the pipe. The coordinate standarddeviation at the pipe level should be between 0.2 and 0.5 mm within a basis of 5 - 8 m. If no gyroscopic theodolites and special plumb instruments are available the survey is performed with theodolites. Therefore the main error of the baseline points results from the inclination of the vertical axis of the theodolite with approx. 0.1 mm/10 m of depth. The azimuth error from the baseline is somewhere between 15 arcsec and 30 arcsec, this being the main error source in pipe jacking.

- Deviation error of pipe traverse:
  For the determination of the heading of the machine tube, the azimuth is transferred from the base line B to the machine base line B by traversing. Hence, angles within the tube are measured at well defined distances. The error propagation with "n" points and distance "S" with angular error m_β and baseline error $p \cdot m_\beta$ is approximately.

$$m(n) = s \cdot m_\beta \sqrt{p^2 + \frac{1}{3} n^3 + \ldots}$$

  With an accuracy of 3 - 6 arcsec for the traverse angles, deviation errors of 5 cm/km can be obtained.

  This error estimation indicates the influence of the azimuth accuracy on the system accuracy.

An optical position measurement system based upon a MOPS camera can produce very quick azimuth information at the cutting shoe if pipes are bent. With connection to the pipe jacking controller, this realtime positioning system may provide an excellent tool for automatic pipe jacking, guidance and control. Deformation of the start shaft particularly decreases accuracy during the jacking process.

4. Pipe jacking survey by MOPS

The measurement of the pipe-traverse by standard surveying techniques is a time consuming process. With a daily jacking rate of max. 25 m the orientation of the machine tube should be controlled up to 4 times a day. In long pipelines of several km the whole survey requires a construction break of 1 to 2 hours. Therefore, the reduction of this time interval could increase the economy of this construction technique. The survey in tube diameters less than 80 cm is particularly critical so the use of a pipecamera as used for the MOPS with the advantage of real-time traverse measurement could be of some benefit.

The basic idea is to observe the moving traverse in the pipe by a referenced tunnel camera with direct process control.

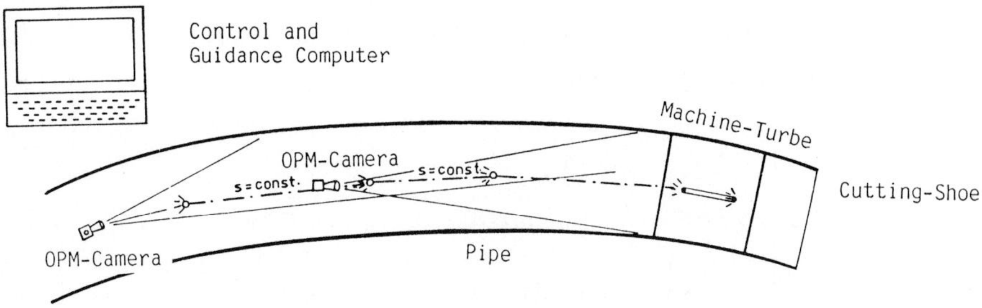

Figure 6: Pipe survey with MOPS-camera

First results with this techniques in 1984 showed an accuracy of 2 cm/230 m. Three main problems appeared:

1) Continuous control of the reference point was necessary (shaft deformation)

2) Refraction problems. The line of sight was installed on top of the pipe. Warm air from heavy machines in the front of the pipe disturbed the line of sight. Many obstacles were not avoidable in the narrow tube.

3) The optimum transfer through the pressure lock (special glass windows in the lockwalls) lead to azimuth errors and was time consuming.

The bending of the pipe can be overcome by a combination of a series of cameras. This method is naturally most efficient in straight pipes. Points 2 and 3 lead to the idea of using inertial techniques for the guidance of the system. The installation of gyro-compasses is already documented. The problem is that the drift needs to be updated by classic traverse measurements. The development of inertial measurement techniques, their limits and possibilities are treated in the project "Inertial techniques in close range area".

5. A concept for inertial positioning in pipejacking

Normally the inertial accelerations in the pipejacking process are indetectably small if the INS is fixed on the cutting machine. This disadvantage would be negligible by using a step technique to measure the position of special force centering points in the pipe.

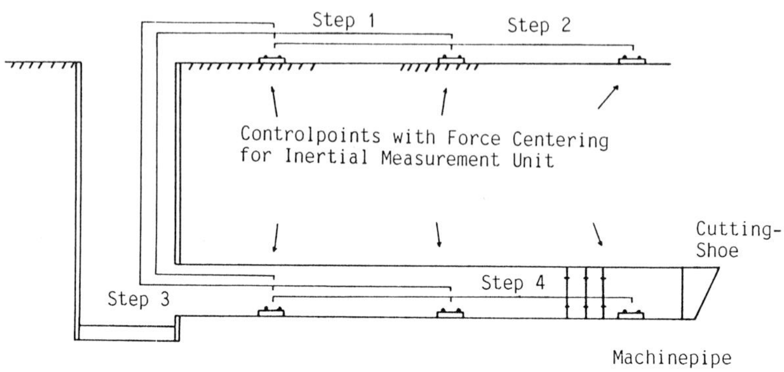

Figure 7: Step technique for pipe jacking

This step technique provides excellent coordinate, heading, bias and scale factor update conditions.

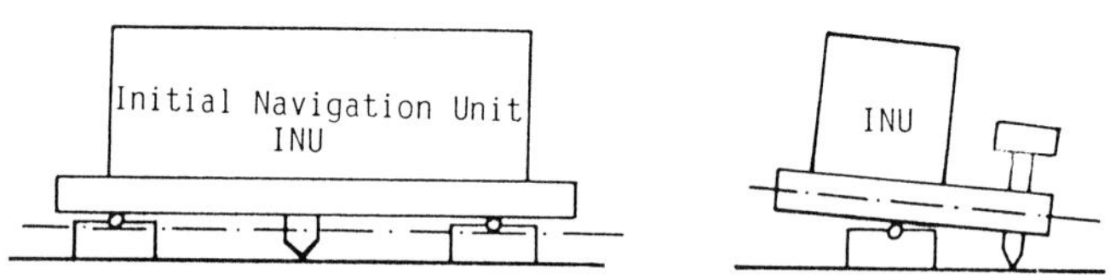

Figure 8: Force centering of INU

The force centering equipment consists of a plate with 2 or 4 spheres using the principle of sphere-cone force centering.

A position accuracy and repeatibility of 1 micrometer and an orientation stability of 2 arcsec is easily obtained. With 2 or 4 reverse order measurements combined with the described step-techniques (Fig.7), the following update conditions are at least able to be performed.

Bias:
Accelerometer: bias observation on every point
Gyro: bias between every pair of points

Scale factor:
Accelerometer: scale factor stability from point to point
Gyro: scale factor stability between points.

Coordinate update: position drift between unknown points
position update at known points.

The formulation of post mission strap-down algorithms containing constraints of this type provide new accuracy dimensions for these geodetic measurement techniques.

6. Models for strapdown systems

The dynamics of inertial sensors, such as one and two degrees of freedom Gyros and Servo-pendulum accelerometers, are well known and described for inertial navigation systems in aircraft and ship navigation.

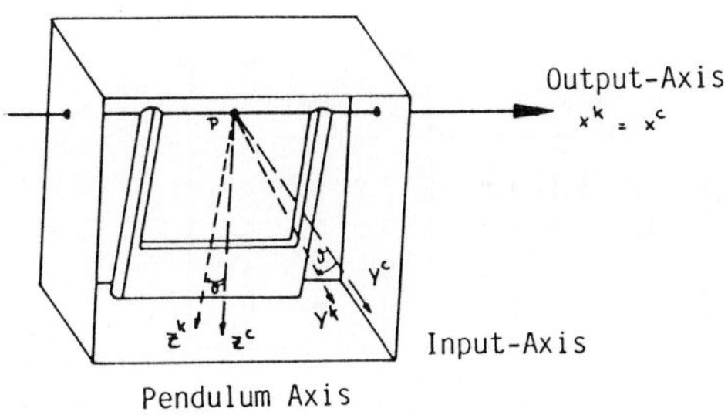

Figure 9: Pendulum accelerometer

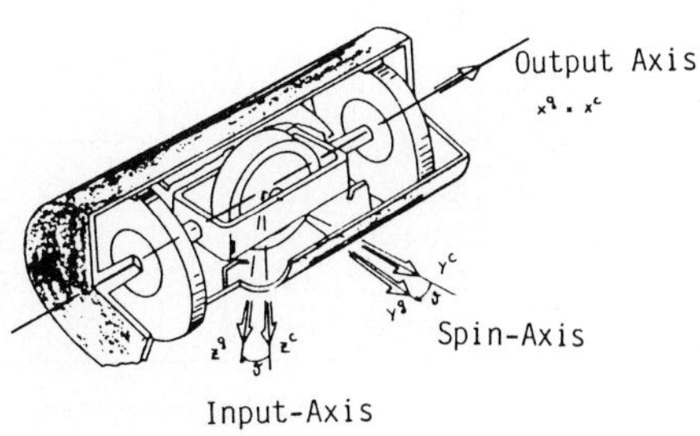

Figure 10: Gyro with a single-degree-of-freedom input axis

The dynamic sensor models used for navigation are based upon linearized equations of motion of gyros and accelerometers. R. Czommer developed a program system for nonlinear sensor-dynamics of a mechanical strapdown platform. It is possible to simulate the sensor signals of an arbitrary inertial track with different noise generators.

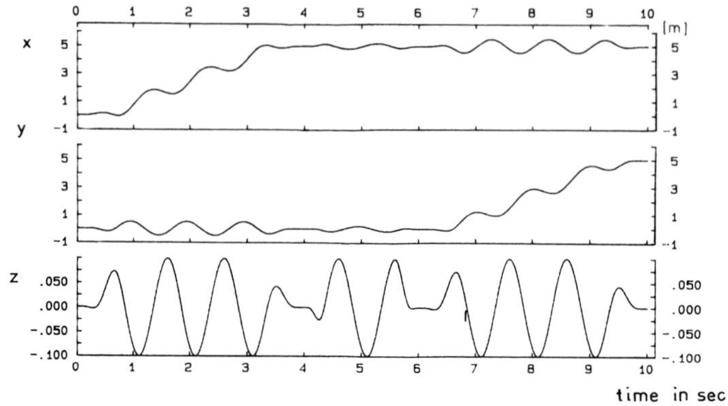

Figure 11: Straight line inertial track simulation

An important result was that with different misalignements of the sensors the nonlinear cross-coupling of mechanical gyros combined with an accelerometer model proposed by Savage resulted in smallest systematic position errors.

Misalignement / Model Gyro/Accel.	0'			5'			10'		
	dx	dy (cm)	dz	dx	dy (cm)	dz	dx	dy (cm)	dz
Joos / linear	4,2	4,8	4,6	4,0	5,1	4,6	3,9	5,6	4,6
Stieler / linear	4,2	4,8	4,6	4,0	5,1	4,6	3,9	5,6	4,6
Stieler / Savage	0,08	0,01	0,0	0,3	0,6	0,02	0,5	0,9	0,04
Czommer / Savage	0	0	0	0	0	0	0	0	0

These results of simulation underline the necessity of expanded dynamic models for inertial geodesy. The adaption of the software to laser-gyro-strapdown systems is the next step in this research work.

A common dynamic model for gyros and accelerometers with modular nonlinearity terms is developed covering most of the published sensor dynamics.

$$J\ddot{\vartheta} + D\dot{\vartheta} + K\vartheta = b_1\dot{\varphi}_x + b_2\dot{\varphi}_y + b_3\dot{\varphi}_z + b_4\ddot{x} + b_5\ddot{y} + b_6\ddot{z} +$$

$$+ b_7\ddot{\varphi}_x + b_8\ddot{\varphi}_y + b_9\ddot{\varphi}_z + b_{10}\dot{\varphi}_x^2 + b_{11}\dot{\varphi}_y^2 + b_{12}\dot{\varphi}_z^2 +$$

$$+ b_{13}\dot{\varphi}_x\dot{\varphi}_y + b_{14}\dot{\varphi}_x\dot{\varphi}_z + b_{15}\dot{\varphi}_y\dot{\varphi}_z,$$

$$b_1 = L_R(a_{31}\cos\vartheta + a_{21}\sin\vartheta)$$

$$b_2 = L_R(a_{32}\cos\vartheta + a_{22}\sin\vartheta)$$

$$b_3 = L_R(a_{33}\cos\vartheta + a_{23}\sin\vartheta)$$

$$b_4 = -m_S(a_{21}\cos\vartheta + a_{31}\sin\vartheta)$$

$$b_5 = -m_S(a_{22}\cos\vartheta + a_{32}\sin\vartheta)$$

$$b_6 = -m_S(a_{23}\cos\vartheta + a_{33}\sin\vartheta)$$

$$b_7 = J\cdot a_{11}$$

$$b_8 = J\cdot a_{12}$$

$$b_9 = J\cdot a_{13}$$

$$b_{10} = \Delta J\cdot[\sin\vartheta\cos\vartheta\,(a_{31}^2 - a_{21}^2) + (\cos^2\vartheta - \sin^2\vartheta)\,a_{21}a_{31}]$$

$$b_{11} = \Delta J\cdot[\sin\vartheta\cos\vartheta\,(a_{32}^2 - a_{22}^2) + (\cos^2\vartheta - \sin^2\vartheta)\,a_{22}a_{32}]$$

$$b_{12} = \Delta J\cdot[\sin\vartheta\cos\vartheta\,(a_{33}^2 - a_{23}^2) + (\cos^2\vartheta - \sin^2\vartheta)\,a_{23}a_{33}]$$

$$b_{13} = \Delta J\cdot[2\sin\vartheta\cos\vartheta\,(a_{31}a_{32} - a_{21}a_{22}) + (\cos^2\vartheta - \sin^2\vartheta)(a_{21}a_{32} + a_{22}a_{31})]$$

$$b_{14} = \Delta J\cdot[2\sin\vartheta\cos\vartheta\,(a_{31}a_{33} - a_{21}a_{23}) + (\cos^2\vartheta - \sin^2\vartheta)(a_{21}a_{33} + a_{23}a_{32})]$$

$$b_{15} = \Delta J\cdot[2\sin\vartheta\cos\vartheta\,(a_{32}a_{33} - a_{22}a_{23}) + (\cos^2\vartheta - \sin^2\vartheta)(a_{22}a_{33} + a_{23}a_{32})]\,.$$

7. Automatic surveying systems

The use of the described positioning systems in local areas is interesting in two aspects:

a) Modular optical positioning system (MOPS)
In many cases the realtime positioning leads to an increased operating speed of machines. The most attractive realization is the positioning by a tracker. With fixed installed theodolites and MOPS-cameras realtime positioning on the construction site is no longer a technique of the future.

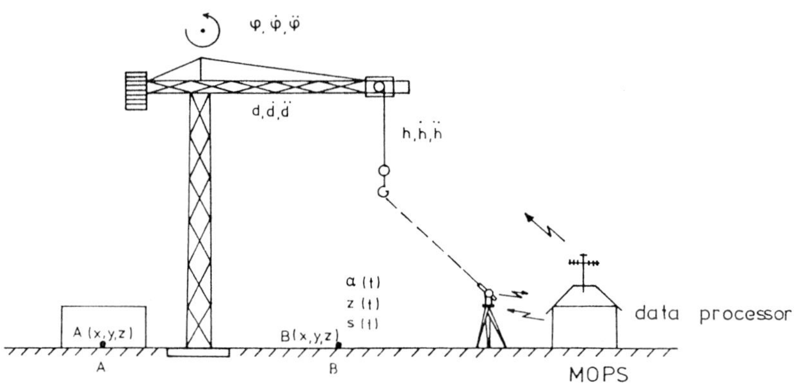

Figure 12: Guidance and control of crane position

b) Inertial Navigation System (INS)
The use of INS systems in combination with spatial force centering points as explained before, could become a useful tool for surveying. Therefore, it is necessary to construct small inertial measurement units no larger than a theodolite. Carrying these INU's from point to point and collecting the data could provide a better measurement technique. The post mission adjustment can be done off-line but also quite rapidly on the construction site. To develop this method is one important aim in the SFB 228.

8. References

Baumann, H.: Fehleranalyse und Simulation eines Trägheitsnavigationssystem mit fahrzeugfesten Sensoren. Dissertation, Technische Universität Braunschweig, 1976.

Bayer, G.: Motorized Electronic Theodolite - High Precision Measurement Robots for Surveying. Proc. of the International Workshop "High Precision Navigation", Stuttgart, May 17th-20th, 1988.

Britting, K.: Inertial Navigation Systems Analysis, New York, John Wiley & Sons, 1971.

Czommer, R.: Arbeits- und Ergebnisbericht 1984 - 1986 des Sonderforschungsbereichs 228 "Hochgenaue Navigation", Teilprojekt E3 "Inertialtechnik im Nahbereich".

Hiller, M.: Mechanische Systeme, Springer Verlag, Berlin 1983

Joos, D. K.: Comparison of Typical Gyro Errors for Strapdown Applications, Symposium über Kreiseltechnik, Stuttgart, 1977.

Krzystek, P.: High Precision Surveying of Moving Objects by Electronic Cameras. Proc. of the International Workshop "High Precision Navigation", Stuttgart, May 17th-20th, 1988.

Magnus, K.: Grundlagen der technischen Mechanik, Stuttgart, 1974

Magnus, K.: Kreisel: Theorie und Anwendungen, Berlin, Springer, 1971.

Möhlenbrink, W.: Abschlußbericht zum Einsatz von Positionssensoren zur Vermessung von Rohrvorpressungen, Forschungsgesellschaft für geodätische Meß- und Rechentechnik, GmbH, Stuttgart 1984 (unveröffentlicht).

Möhlenbrink, W.: Nonlinearities in the Dynamic Model of Inertial Sensors, Third International Symposium on Inertial Technology for Surveying and Geodesy, Banff, Canada, September 16-20, 1985.

Möhlenbrink, W.: Drift Efects in Inertial Measurement Systems (Resulting from Nonlinear Terms in the Equations of Motions), Proceedings: Inertial, Doppler and GPS Measurements for National and Engineering Surveys, München, 1985.

Oppelt, W.: Kleines Handbuch technischer Regelvorgänge, Weinheim Verlag Chemie GmbH, 1972.

Sauer, R., Szabo, I.: Mathematische Hilfsmittel des Ingenieurs, Teil I, Berlin, Springer, 1967.

Savage, P. G.: Strapdown Sensors, AGARD Lecture Series No. 95, "Strapdown Inertial Systems", 1978.

Savage, P. G.: Strapdown System Algorithmus, AGARD Lecture Series No. 133, "Advances in Strapdown Inertial Systems, 1984.

Schöne, A.: Simulation technischer Systeme, Band I, Carl Hanser Verlag, München, 1974.

Stieler, B., Winter, H.: Gyroscopic Instruments and their Application to Flight Testing, Agardograph, No. 160, Vol. 15, 1983.

Welsch, D.: Aktions-Reaktions-Analyse zur objektorientierten Verarbeitung von Meßdaten statischer und dynamischer Vorgänge an Bauwerken. Dissertation Universität Stuttgart, eingereicht 1988.

Wetzig, V.: Ermittlung der Richtungsreferenz im Strapdown-System, Vortrag CCG, Braunschweig, 1985.

Wittenburg, J.: Dynamics of Systems of Rigid Bodies, Teubner Verlag, Stuttgart, 1987.

Wiegner, S.: Fehleranalyse und Simulation eines Proportional-Wendekreisels für die Strapdown-Anwendung in einem Inertialnavigationssystem. Selbständige Geodät. Arbeit am IAGB, Stuttgart, 1985.

Multiple Target Tracking for Automatic Inland Navigation

Thomas Plocher, Klaus Mezger, Reinhard Neul, Ernst-Dieter Gilles
Institut für Systemdynamik und Regelungstechnik, Universität Stuttgart

Summary

The algorithm introduced is capable of tracking multiple targets accounting for clutter and missing reports. It is based on sequences of radar images and implemented in a real time environment. The tracking algorithm supplements an auto-pilot, using estimated target position and velocity to facilitate trajectory planning for inland ships. Each radar image supplies a number of measurements; hypotheses considered are association of measurements with previously known targets, initiation of new targets, classification of measurements as noise, and classification of previously known targets as not detected or as false alarms. Target states are estimated from each such data-association hypothesis using a Kalman filter. In order to keep the number of hypotheses reasonable and to allow real time application tracking gates are introduced, dividing the entire set of targets and measurements into independent clusters. Furthermore, the algorithm automatically eliminates less important hypotheses and combines hypotheses having a similar measurement 'history'.

1. Introduction

A radar based autopilot for inland ships

In order to increase the security of transport on inland waterways, it is recommended to introduce an automatic ship steering system.

This is especially desirable with respect to the possibility of environmental pollution. The autopilot developed at the Department of System Dynamic and Control is based on analysis and evaluation of radar images by a microcomputer.

Inland ships are usually equipped with the radar system required in order to make night and bad weather operation possible. The image processing results in a trajectory, taking into account the current traffic and environmental situation. A state controller ensures that the ship follows the generated trajectory.

The goal of the project is to support the steersman allowing him to concentrate on the supervision of traffic activities. The ship's radar and a rate gyro or directional gyroscope are the only sensor devices required in the first phase of the development. Specially designed interfaces are used to connect the sensor devices to the computer, which on the other hand is acting on the steering servo.

The implementation of the autopilot is relatively cheap, since all sensors are present on modern inland ships. Such a system can inform the steersman about other ships and issue an alarm in dangerous situations.

Presently the second phase of the autopilot is developed. A map of the river stored in the computer is taken into consideration. The continuous matching of the map with radar images is used to determine the current position of the own ship.

Knowing the position allows the usage of *a priori* information in the process of generating the ship's trajectory. Introducing additional sensors like Doppler-Log and/or GPS-receiver simplifies the matching process to a great extent.

Furthermore in this second phase of the project a multi-target tracking algorithm identifies ships radar buoys and other objects occuring in the radar image. Their estimated position and velocity facilitates trajectory planning for the own ship.

The multi-target tracking algorithm

Multiple target tracking is one of the topics of interest in our research, concentrating on a self-sufficient guidance system for inland ships. It is based on recursive analysis of radar image sequences. Every two seconds the radar equipment scans its 360°-range supplying an image, i.e. a grid of polar coordinates. These two seconds yield the upper bound on available computation time.

Target tracking and estimation of the target type — i.e. one of the discrete states *ship*, *river bank* or *radar buoy* — facilitates optimal trajectory computation, taking into account collision risk and rudder activity. This trajectory provides the reference input for a state-feedback control loop, determining the rudder angle.

The radar image shown in figure 1 originates from the river Rhine. The position of our own ship is marked with an accurately scaled ship symbol (label **W**). The occurring

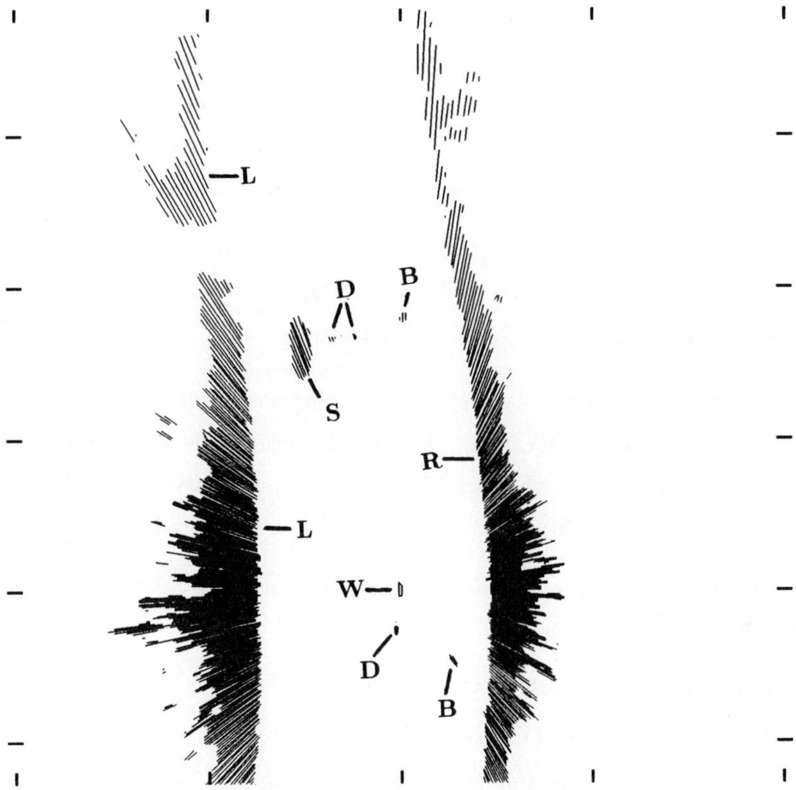

Figure 1: Sample radar image

objects are labeled **L** and **R** denoting left and right *river banks* and **S**, **B** and **D** for *ships*, *buoys* and *disturbances* respectively.

2. Problem statement

It is assumed that each *target* (n) is represented by a vector $\mathbf{x}^{(n)}$ of three state variables. Trajectory and measurements are described by a *linear process* (cf. Appendix)

$$\mathbf{x}_{k+1}^{(n)} = \Phi_k^{(n)}\,\mathbf{x}_k^{(n)} + \mathbf{w}_k^{(n)}, \qquad \mathbf{w}_k^{(n)} \simeq \mathrm{N}(\mathbf{0}, \mathbf{Q}_k^{(n)}),$$
$$\mathbf{y}_k^{(n)} = H\,\mathbf{x}_k^{(n)} + \mathbf{v}_k^{(n)}, \qquad \mathbf{v}_k^{(n)} \simeq \mathrm{N}(\mathbf{0}, \mathbf{R}_k^{(n)}).$$

Let m_k denote the number of measurements occuring with radar image k and n_{k-1} the number of targets assumed after image $(k-1)$. The evaluation of *a posteriori* densities and hypothesis probabilities, after associating measurements with targets, will be of dominating interest. To facilitate notation, the following (ordered) sets are introduced.

- The current *set of measurements* resulting from image k:

  $Y_k = \{\mathbf{y}_k^{(1)}, \mathbf{y}_k^{(2)}, \cdots, \mathbf{y}_k^{(m_k)}\}$

- The *cumulative data-set* up to the image k:

  $\mathbf{Y}_k = \{(Y_1), (Y_2), \cdots, (Y_k)\}$

- The *set of targets* resulting from image $(k-1)$:

  $\Theta_{k-1} = \{\theta_{k-1}^{(n)} \mid n = 1, 2, \ldots, n_{k-1}\}$

  (Where $\theta_{k-1}^{(n)}$ symbolizes all data characterizing target n of image $(k-1)$.)

- The i-th *association hypothesis* interpreting the current set of measurements, based on some prior hypothesis $\Omega_{k-1}^{(p_i)}$:

  $\omega_k^{(i)} = \omega_k^{(i,1)} \cup \omega_k^{(i,2)}$

  $\omega_k^{(i,1)} = \{n(1), n(2), \cdots, n(m_k)\}$

  ($n(m) = 0$ denotes that the m-th measurement is noise *or* a new target.

  $1 \le n(m) \le n_{k-1}$ associates target $n(m)$ with measurement m.)

  $\omega_k^{(i,2)} = \{n \mid 1 \le n \le n_{k-1},\ n \notin \omega_k^{(i,1)}\}$

  (Set of targets assumed not detected or false alarms.)

- *Cumulative association hypothesis* interpreting the whole data-set:

  $\Omega_k^{(i)} = \Omega_{k-1}^{(p_i)} \cup (\omega_k^{(i)}) = \{(\omega_1^{(i)}), (\omega_2^{(i)}), \cdots, (\omega_{k-1}^{(i)}), (\omega_k^{(i)})\}$

(The second part of the equation defines $\omega_1^{(i)}, \ldots, \omega_{k-1}^{(i)}$.)

- The *set of all cumulative hypotheses* implied:

$$\Omega_k = \{[\Omega_k^{(1)}], [\Omega_k^{(2)}], \cdots, [\Omega_k^{(i_k)}]\}$$

Example

$m_1 = 2$: $Y_1 = \{\mathbf{y}_1^{(1)}, \mathbf{y}_1^{(2)}\}$, $\mathbf{Y}_1 = \{(\mathbf{y}_1^{(1)}, \mathbf{y}_1^{(2)})\}$,
$\omega_1^{(1)} = \{0,0\}$, $\Omega_1^{(1)} = \{(0,0)\}$, $\Omega_1 = \{[(0,0)]\}$, $i_1 = 1$,
$\Theta_1 = \{\theta_1^{(1)}, \theta_1^{(2)}\}$, $n_1 = 2$.
$m_2 = 1$: $Y_2 = \{\mathbf{y}_2^{(1)}\}$, $\mathbf{Y}_2 = \{(\mathbf{y}_1^{(1)}, \mathbf{y}_1^{(2)}), (\mathbf{y}_2^{(1)})\}$,
$\omega_2^{(1)} = \{1,2\}$, $\omega_2^{(2)} = \{2,1\}$, $\omega_2^{(3)} = \{0,1,2\}$,
$\Omega_2^{(1)} = \{(0,0),(1,2)\}$, $\Omega_2^{(2)} = \{(0,0),(2,1)\}$, $\Omega_2^{(3)} = \{(0,0),(0,1,2)\}$,
$\Omega_2 = \{[(0,0),(1,2)], [(0,0),(2,1)], [(0,0),(0,1,2)]\}$, $i_2 = 3$,
$\Theta_2 = \{\theta_2^{(1)}, \theta_2^{(2)}, \cdots, \theta_2^{(7)}\}$, $n_2 = 7$.
E.g. $\omega_2^{(3)} = \{0\} \cup \{1,2\}$ denotes $\mathbf{y}_2^{(1)}$ as noise or as a new target and assumes targets $\theta_1^{(1)}$ and $\theta_1^{(2)}$ are not detected.

3. State estimation and probabilities of hypotheses

The *a posteriori* probability distribution of the target state variables, given by the Kalman filter (Equations (4)-(8)), is a multivariate normal distribution. If target and measurement are associated, the mean and covariance of this distribution are given by:

$$\hat{\mathbf{x}}_k^{(n,i)+} = \mathrm{E}[\mathbf{x}_k^{(n)} \mid \Omega_k^{(i)}, \mathbf{Y}_k]$$

$$P_k^{(n,i)+} = \mathrm{E}[(\mathbf{x}_k^{(n)} - \hat{\mathbf{x}}_k^{(n,i)+})(\mathbf{x}_k^{(n)} - \hat{\mathbf{x}}_k^{(n,i)+})^T \mid \Omega_k^{(i)}, \mathbf{Y}_k]$$

The *a priori* density is determined by:

$$\hat{\mathbf{x}}_{k+1}^{(n,i)-} = \mathrm{E}[\mathbf{x}_{k+1}^{(n)} \mid \Omega_k^{(i)}, \mathbf{Y}_k]$$

$$P_{k+1}^{(n,i)-} = \mathrm{E}[(\mathbf{x}_{k+1}^{(n)} - \hat{\mathbf{x}}_{k+1}^{(n,i)-})(\mathbf{x}_{k+1}^{(n)} - \hat{\mathbf{x}}_{k+1}^{(n,i)-})^T \mid \Omega_k^{(i)}, \mathbf{Y}_k]$$

Given all measurements up to time k we are interested in the probability of hypothesis $\Omega_k^{(i)}$. By use of Bayes' equation we obtain the following recursive relationship:

$$\Pr(\Omega_k^{(i)} \mid \mathbf{Y}_k) = \Pr(\Omega_{k-1}^{(p_i)}, \omega_k^{(i)} \mid \mathbf{Y}_{k-1}, Y_k) = \tag{1}$$

$$\frac{p(Y_k \mid \Omega_k^{(i)}, \mathbf{Y}_{k-1}) \Pr(\Omega_k^{(i)} \mid \mathbf{Y}_{k-1})}{p(Y_k \mid \mathbf{Y}_{k-1})} =$$

$$\frac{p(Y_k \mid \Omega_k^{(i)}, \mathbf{Y}_{k-1}) \Pr(\omega_k^{(i)} \mid \Omega_{k-1}^{(p_i)}, \mathbf{Y}_{k-1}) \Pr(\Omega_{k-1}^{(p_i)} \mid \mathbf{Y}_{k-1})}{p(Y_k \mid \mathbf{Y}_{k-1})} = \frac{\langle 1 \rangle \cdot \langle 2 \rangle \cdot \langle 3 \rangle}{\langle 4 \rangle}$$

⟨1⟩ The first term is the *likelihood* of the measurements Y_k, given the association hypothesis $\Omega_k^{(i)}$ and all past data $\mathbf{Y}_{k-1}$.

$$p(Y_k \mid \Omega_k^{(i)}, \mathbf{Y}_{k-1}) = \prod_{m=1}^{m_k} p(\mathbf{y}_k^{(m)} \mid \Omega_k^{(i)}, \mathbf{Y}_{k-1})$$

$$p(\mathbf{y}_k^{(m)} \mid \Omega_k^{(i)}, \mathbf{Y}_{k-1}) = \begin{cases} N(\mathbf{y}_k^{(m)};\ H\hat{\mathbf{x}}_k^{(n(m),i)-},\ (HP_k^{(n(m),i)-}H^T + R_k^{(n(m),i)})); & n(m) \neq 0 \\ p_{mn}(\mathbf{y}_k^{(m)}) + p_{nt}(\mathbf{y}_k^{(m)}); & n(m) = 0 \end{cases}$$

In the first case target $n(m)$ and measurement m have been assigned, the latter takes care of $n(m) = 0$, i.e. the m-th measurement is noise or a new target. The densities p_{mn} and p_{nt} describe the *a priori likelihood* of measurement noise and new target occurance and are determined empirically.
($\int p_{mn}\, d\mathbf{y} = \int p_{nt}\, d\mathbf{y} = 1$)

⟨2⟩ This is the probability of the current association hypothesis $\omega_k^{(i)}$, given the prior hypothesis $\Omega_{k-1}^{(p_i)}$ and $\mathbf{Y}_{k-1}$.

$$\Pr(\omega_k^{(i)} \mid \Omega_{k-1}^{(p_i)}, \mathbf{Y}_{k-1}) = \frac{e^{-(\lambda_{mn}+\lambda_{nt})}(\lambda_{mn} + \lambda_{nt})^{n_{mn}+n_{nt}}}{m_k!} \prod_{n\in\omega_k^{(i,1)}} P_D^{(n)} \prod_{n\in\omega_k^{(i,2)}} (1 - P_D^{(n)})$$

Here n_{mn} and n_{nt} are introduced, denoting the number of measurements $\omega_k^{(i)}$ consideres as measurement noise and as new targets respectively. This yields $n_{dt} = m_k - n_{mn} - n_{nt}$; the number of detected targets implied by hypothesis $\omega_k^{(i)}$. It is assumed that n_{dt} is given by a binomial distribution and n_{mn} as well as n_{nt} follow a Poisson distribution with parameter λ_{mn} and λ_{nt} respectively. Finally, $P_D^{(n)}$ is the probability of detection for target n. ($P_D^{(0)} = 1$ is defined.)

⟨3⟩ The probability of hypothesis $\Omega_{k-1}^{(p_i)}$ is the result of the previous time-step, i.e. image $(k-1)$.

⟨4⟩ $p(Y_k \mid \mathbf{Y}_{k-1})$ can be considered as *normalizing factor*, found by summing the numerator over the values of $i = 1, 2, \ldots, i_k$.

4. Practical algorithm

The theoretical concept considered so far is not suitable for a *real time problem* application. First of all, the data structure implied is too complex and secondly the number of hypotheses will grow immensly if all possible target measurement associations are taken into account. A practical algorithm must limit the number of hypotheses, pruning away unlikely associations and binding together 'almost' identical hypotheses. Furthermore it is not possible to store the process history up to an arbitrary depth, so suboptimal strategies have to be found.

Two approximation steps are introduced. The first step will be called *structural approximation*, defining a compact data structure. Exponential growth of memory demand will be eliminated in step two, combining several existing techniques.

Structural approximation

Assigning measurements with targets, hypotheses of image $(k-1)$ are continued with possible outcomes in image k. The evaluation of probabilities follows according to equation (1). Rather than keeping track of all hypotheses and their probabilities and of implied target state estimates and their covariance matrices, we will directly assign probabilities to target measurement associations — losing the generating hypotheses. These probabilities are obtained, summing up all hypothesis probabilities implying the data-association. Additionally the data-association history shall be kept up to a certain depth.

Second approximation step

The following criteria are used to limit the number of target entries emerging from the data-association:

- Introducing a *validation region* or *confidence ellipsoid* for each previous target prevents consideration of unlikely hypotheses. These '*innovation gates*' are determined by the mean and covariance of the target state estimate.

  $$\{\mathbf{y} \mid (\mathbf{y} - H\,\hat{\mathbf{x}}_k^-)(HP_k^-H^T + R_k)^{-1}(\mathbf{y} - H\,\hat{\mathbf{x}}_k^-)^T \le \rho^2 = -2\ln(1 - P_{det})\}$$

  (P_{det} is the probability that the *correct* assignment lies within the validation region.)

- Two hypotheses are combined if they have the same history during k images. (*k-scan-back algorithm*)

- Hypotheses are eliminated by dropping those implying a not detected target for several images or having probabilities below a given threshold.
- If the number of hypotheses supported is still above a given maximum, those least significant are dropped. The main factors determining the importance of hypotheses are the distance and the velocity of the implied target, in relation to the own ship.

5. Experimental results

The multi-target tracking algorithm is presently integrated within a radar auto-pilot for automatic ship steering. It can be tested off-line using recorded images, saved from previous experiments. On-line testing follows to investigate the closed-loop behavior.

Figure 2 presents first results, showing images originating from the river Rhine. A rectangle approximates the confidence ellipse belonging to the predicted measurement of the next image. The straight line originating from the center of each rectangle symbolizes the target velocity predicted.

6. Conclusions

This paper outlined a multi-target tracking algorithm incorporated in a ship steering auto-pilot. The aim is to estimate position and velocity of ships, buoys and other objects present using a Kalman filter for each track. The algorithm accounts for track initiation, false alarms, missing measurements and measurement noise; data association is based on a recursive Bayesian approach. Main approximations for real time application are the formation of independent clusters defining tracking gates, and the elimination or combination of hypotheses fulfilling certain criteria.

Appendix

Filter equations:

Target state variables (cf. figure 3):

- Position (x,y)
- Velocity v

Figure 2: Radar image sequence

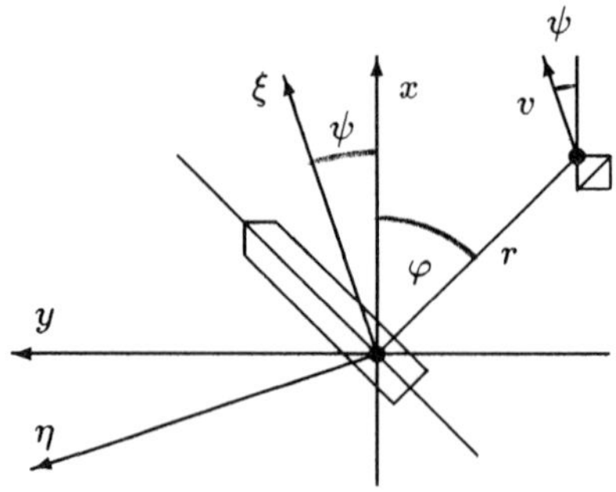

Figure 3: Target coordinate system

$v_x = v\cos\psi \quad$ and $\quad v_y = v\sin\psi; \qquad \psi \in]-\frac{\pi}{2}, \frac{\pi}{2}]$.
(ψ: angle of approximated river axis at target position)

- $$\begin{bmatrix} x(t) \\ y(t) \\ v(t) \end{bmatrix} = \begin{bmatrix} \cos\psi(t) & -\sin\psi(t) & 0 \\ \sin\psi(t) & \cos\psi(t) & 0 \\ 0 & 0 & 1 \end{bmatrix} \begin{bmatrix} \xi(t) \\ \eta(t) \\ \dot{\xi}(t) \end{bmatrix}$$

- $\dot{\eta} = w_\eta, \qquad \ddot{\xi} = w_v,$
  $E[w_\eta(t_1)\, w_\eta(t_2)] = \sigma_\eta^2\ \delta(t_1 - t_2),$
  $E[w_v(t_1)\, w_v(t_2)] = \sigma_v^2\ \delta(t_1 - t_2)$ and
  $E[w_\eta(t_1)\, w_v(t_2)] = 0 \quad \forall \quad t_1, t_2 \in \mathrm{R}$.

Continuous state equation:

$$\dot{\mathbf{x}}(t) = \frac{d}{dt}\begin{bmatrix} x(t) & y(t) & v(t) \end{bmatrix}^T = F(t)\,\mathbf{x}(t) + G(t)\,\mathbf{w}(t),$$

$$F(t) = \begin{bmatrix} 0 & 0 & \cos\psi(t) \\ 0 & 0 & \sin\psi(t) \\ 0 & 0 & 0 \end{bmatrix}, \quad G(t) = \begin{bmatrix} -\sin\psi(t) & 0 \\ \cos\psi(t) & 0 \\ 0 & 1 \end{bmatrix}, \quad \mathbf{w} = \begin{bmatrix} w_\eta(t) \\ w_v(t) \end{bmatrix}.$$

Discrete state equation:

$$\mathbf{x}_{k+1} = \Phi_k\, \mathbf{x}_k + \mathbf{w}_k, \qquad \mathbf{w}_k \simeq \mathrm{N}(\mathbf{0}, \mathbf{Q}_k). \tag{2}$$

I.e. $\mathbf{w}_k$ is a white noise sequence of normal random variables with zero mean and covariance Q_k. Φ_k is the state transition matrix:

$$\Phi_k = \Phi(t_{k+1}, t_k) \approx I + F(t_k)\Delta t = \begin{bmatrix} 1 & 0 & \Delta t \cos\psi_k \\ 0 & 1 & \Delta t \sin\psi_k \\ 0 & 0 & 1 \end{bmatrix}, \qquad \Delta t = t_{k+1} - t_k,$$

$$\mathbf{w}_k = \int_{t_k}^{t_{k+1}} \Phi(t_{k+1}, t)\ G(t)\ \mathbf{w}(t)\ dt \approx \begin{bmatrix} w_v \frac{\Delta t^2}{2} \cos\psi_k - w_\eta \Delta t \sin\psi_k \\ w_v \frac{\Delta t^2}{2} \sin\psi_k + w_\eta \Delta t \cos\psi_k \\ w_v \Delta t \end{bmatrix},$$

$$Q_k = E[\mathbf{w}_k \mathbf{w}_k^T] = \int_{t_k}^{t_{k+1}}\int_{t_k}^{t_{k+1}} \Phi(t_{k+1}, t)\ G(t)\ E[\mathbf{w}(t)\,\mathbf{w}^T(s)]\ G(s)^T\ \Phi(t_{k+1}, s)\ ds\, dt$$

$$\approx \begin{bmatrix} \sigma_v^2 \frac{\Delta t^3}{3} \cos^2\psi_k + \sigma_\eta^2 \Delta t \sin^2\psi_k & (\sigma_v^2 \frac{\Delta t^3}{3} - \sigma_\eta^2 \Delta t) \cos\psi_k \sin\psi_k & \sigma_v^2 \frac{\Delta t^2}{2} \cos\psi_k \\ (\sigma_v^2 \frac{\Delta t^3}{3} - \sigma_\eta^2 \Delta t) \cos\psi_k \sin\psi_k & \sigma_v^2 \frac{\Delta t^3}{3} \sin^2\psi_k + \sigma_\eta^2 \Delta t \cos^2\psi_k & \sigma_v^2 \frac{\Delta t^2}{2} \sin\psi_k \\ \sigma_v^2 \frac{\Delta t^2}{2} \cos\psi_k & \sigma_v^2 \frac{\Delta t^2}{2} \sin\psi_k & \sigma_v^2 \Delta t \end{bmatrix}.$$

Measurements:

Polar coordinates (r_k, φ_k) of true target position (r_k^*, φ_k^*), with noise assumption

$\Delta r_k = r_k^* - r_k \simeq \mathrm{N}(0, \sigma_r^2)$,

$\Delta \varphi_k = \varphi_k^* - \varphi_k \simeq \mathrm{N}(0, \sigma_\varphi^2)$.

The measurement equation in karthesian coordinates follows:

$$\mathbf{y}_k = \begin{bmatrix} r_k \cos\varphi_k & r_k \sin\varphi_k \end{bmatrix}^T = H\,\mathbf{x}_k + \mathbf{v}_k, \tag{3}$$

$$H = \begin{bmatrix} 1 & 0 & 0 \\ 0 & 1 & 0 \end{bmatrix}, \qquad \mathbf{v}_k \simeq \mathrm{N}(\mathbf{0}, \mathbf{R}_k),$$

$$R_k = E[\mathbf{v}_k \mathbf{v}_k^T] \approx \begin{bmatrix} \sigma_r^2 \cos^2\varphi_k + r_k^2 \sigma_\varphi^2 \sin^2\varphi_k & (\sigma_r^2 - r_k^2\sigma_\varphi^2) \sin\varphi_k \cos\varphi_k \\ (\sigma_r^2 - r_k^2\sigma_\varphi^2) \sin\varphi_k \cos\varphi_k & \sigma_r^2 \sin^2\varphi_k + r_k^2 \sigma_\varphi^2 \cos^2\varphi_k \end{bmatrix}.$$

Kalman filter equation:

- Extrapolation:

$$\hat{\mathbf{x}}_k^- = \Phi_{k-1}\, \hat{\mathbf{x}}_{k-1}^+ \tag{4}$$

$$P_k^- = \Phi_{k-1} P_{k-1}^+ \Phi_{k-1}^T + Q_{k-1} \tag{5}$$

- Update:

$$K_k = P_k^- H^T \left[H P_k^- H^T + R_k \right]^{-1} \tag{6}$$

$$P_k^+ = [I - K_k H] P_k^- \tag{7}$$

$$\hat{\mathbf{x}}_k^+ = \hat{\mathbf{x}}_k^- + K_k \left[\mathbf{y}_k - H \hat{\mathbf{x}}_k^- \right] \tag{8}$$

References

[1] Gilles, E.D., Mezger, K., Raisch, J., und Wehlan, H., *Rekursive Klassifikation und Verfolgung von Objekten in Radarbildfolgen*, Automatisierungstechnik 35, pp. 224-234, 1987.

[2] Reid, D.B., *An Algorithm for Tracking Multiple Targets*, IEEE Trans. Automat. Contr., vol. AC-24, pp. 843-854, Dec. 1979.

[3] Bar-Shalom, Y., *Tracking Methods in a Multitarget Environment*, IEEE Trans. Automat. Contr., vol. AC-23, pp. 618-626, Aug. 1978.

[4] Smith, P., and Buechler, G., *A Branching Algorithm for Discriminating and Tracking Multiple Objects*, IEEE Trans. Automat. Contr., vol. AC-20, pp. 101-104, Feb. 1975.

[5] Singer, R.A., Sea, R.G., and Housewright, K.B., *Derivation and Evaluation of Improved Tracking Filters for Use in Dense Multitarget Environments*, IEEE Trans. Inform. Theory, vol. IT-20, pp. 423-432, July 1974.

[6] Alspach, D.L., *A Gaussian Sum Approach to the Multi-Target Identification-Tracking Problem*, Automatica, vol. 11, pp. 285-296, May 1975.

[7] Sorenson, H.W., and Alspach, D.L., *Recursive Bayesian Estimation Using Gaussian Sums*, Automatica, vol. 7, pp. 465-479, May 1971.

3D-Mapping by a Semiconductor Laser Scanner, Description of an Experimental Setup

Aloysius Wehr
Institut für Navigation, Universität Stuttgart

SUMMARY

An experimental setup of a semiconductor laser scanner for 3D-mapping is described. First measurement results are presented. The experimental results are discussed and estimations for a future remote sensing system are given. Finally the main possibilities for applications of a semiconductor laser mapper are identified.

1. INTRODUCTION

In the field of remote sensing the optical imaging technique applies in most cases multispectral systems with either opto-mechanical and opto-electronical devices (SPOT-Scanner, TM-Scanner), or it uses photogrammetric systems, as well known. These instruments use "passive sensors", which collect and process electromagnetic energy originating from the sun and reflected by the targets to be observed. In the thermal infrared spectrum the emitted radiation of the target itself is the signal source.

In addition to the receiving device active sensors have also a transmitter. The imaging laser sensor which is described in the following, belongs to the family of active sensors and offers the following advantages

- The transmitted signal is monochromatic and can be highly collimated. Therefore, extremely high directivity can be achieved.

- The laser's signal can be modulated allowing to measure the two-way distance transmitter - target - receiver.

- The measurement geometry is well defined and the received echo is determined by the backscatter values of the targets illuminated by the laser.

- Laser ranging can be performed independent of daytime, because the laser is the relevant light source.

- The laser setup allows to measure 3-dimensional geometric features of the target in a strightforward manner and to perform the evaluation with relatively simple devices. In addition one can perform multi-spectral mapping in well defined spectral lines, if either a tuneable laser or serveral narrow band lasers are used.

In the following, some results of an experimental laser measurement setup are presented which was developed to obtain design parameters for a precise scanning ranging system. This test design is presently used in near range measurement tasks on earth. It allows various practical applications which range from CAD to scientific data collection, robotics and use in daily life.

2. PRINCIPLE MEASUREMENT SETUP

In order to map a scene in three dimensions we have to scan a scene with a special device and to sample the x- and y-coordinate as well as the slant range. The grey levels of a scene are normally imaged by a tv-camera. In this case we obtain the x,y coordinates of the scene and the associated intensity of the reflected light for each measured spot, but it is impossible to determine the slant range without extra measurement effort.

We developed a device which maps a scene in three dimensions with high geometric and range resolution. As we talk about an active sensor we need a unit which illuminates the target and measures synchronously the range. This can be carried out by laser range finders used for geodetic applications. They measure distances to diffuse reflecting targets of about 100 m, dependent on atmospheric conditions and on the reflectivity of the target. As the light of these laser range finders is well collimated they fulfill the requirement that a well defined area is measured. Scanning the transmitted and received beam we realize a laser mapper.

Therefore, a laser mapper consists of

1) a range finder

and 2) a scanning device which scans the received and transmitted light beam

As shown in fig. 2-1, the range finder measures the R-component and the opto-mechanical scanner scans the beam in x- and y-direction. Such a laser mapper was developed by using the DIOR 3002 from Wild and two General Scanning Scanners and an IBM-AT (s. fig. 2-2). The computer controlled the laser range finder and the scanners, calculated the Cartesian coordinates from the rotation angles of the scanner and the measured slant range, stored the data and displayed graphically the results. For example we measured the contour of a METEOSAT antenna (s. fig. 2-3).

The experiments carried out by this system demonstrated that

- higher geometric resolution

and - much shorter measuring time

are required to measure the contour of continuums. The measuring time of the DIOR was about 0.7 sec for one range measurement with an accuracy of 1 cm. In our experiments we sampled 32 x 32 points. This means a total mapping time of about 12 minutes which is very long.

The following chapters will deal with an improved laser mapper which features much shorter measuring time although achieving higher geometrical and range resolution.

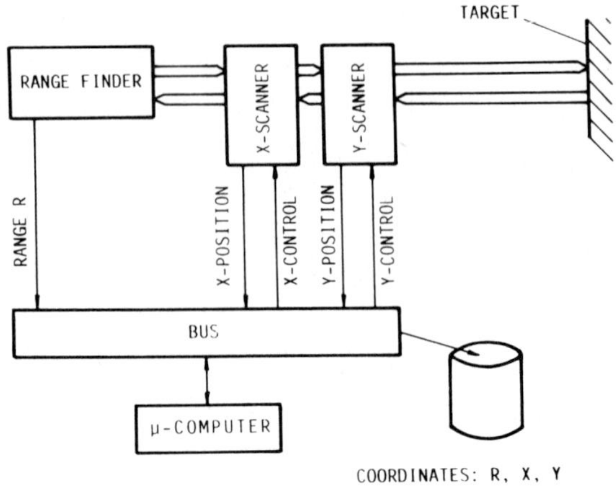

Fig. 2-1: Principle Blockchart of a Laser Mapper

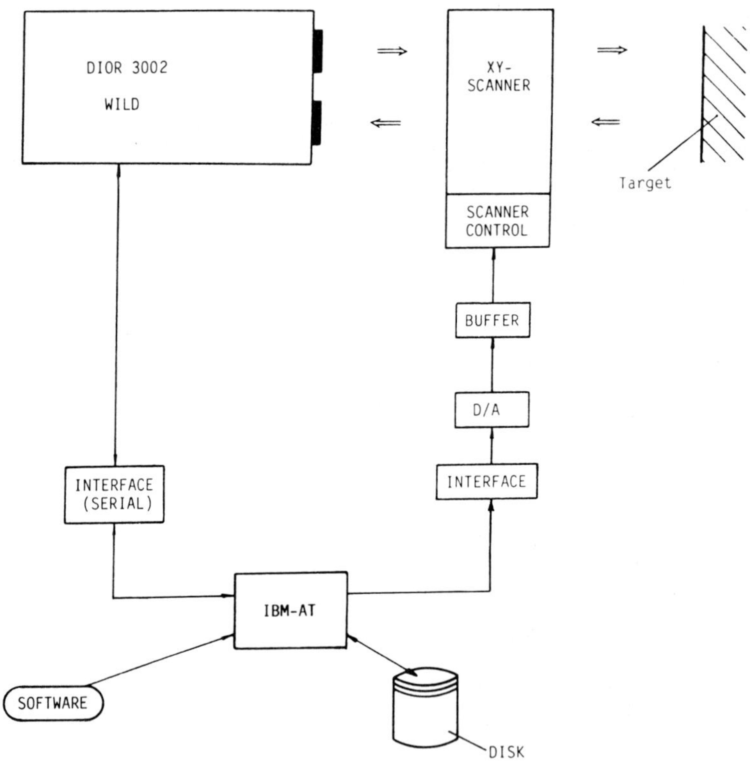

Fig. 2-2: Laser Mapper Setup with DIOR 3002 of Wild

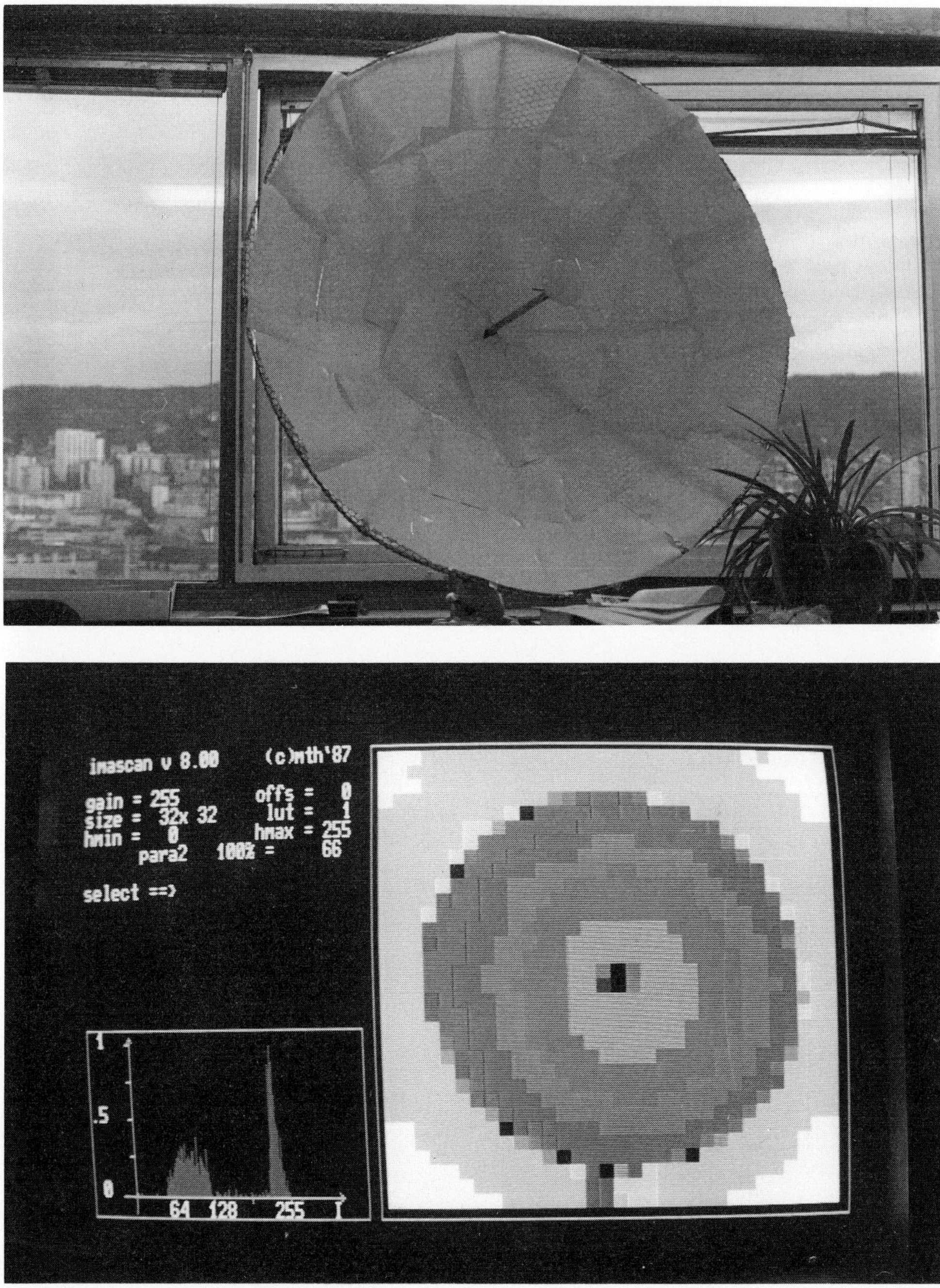

<u>Fig. 2-3:</u> Contour of a METEOSAT Antenna

3. RANGE MEASUREMENT

For the laser mapper we used a semiconductor laser as a transmitter which transmits at 830nm and an avalanche photodiode as a receiver. In order to perform very fast range measurements with millimeter accuracy, we made the decision to use a continuous wave measurement principle. The described measurements were carried out by side tone ranging.

In our experimental setup we modulated the optical intensity of the cw-laser by modulating the drive current with 403 MHz. At the avalanche photodiode the received optical power was detected, and the 403 MHz tone was demodulated. However, the received signal was delayed by

$$t = 2*R/c \qquad (3\text{-}1)$$

where R is the range to the target and c is the speed of light ($3*10^8$ m/s). The time delay corresponds to a phaseshift of

$$\theta = 2\pi f \cdot t = 4\pi f \; R \; / \; c \qquad (3\text{-}2)$$

where f is the frequency of the tone signal.

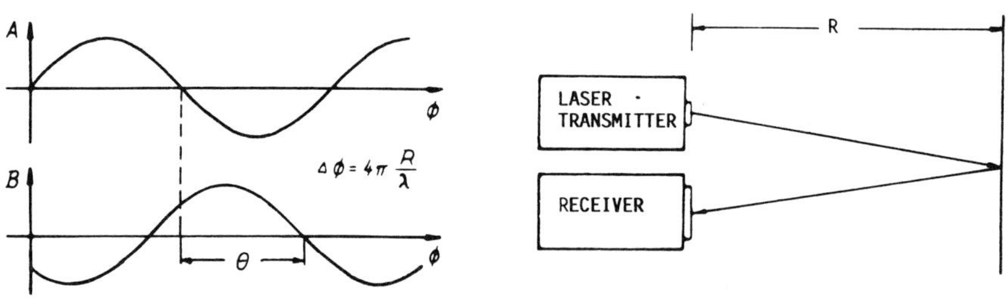

Fig. 3-1: Ranging Method

These formulas show that we obtain the range R by measuring the phase difference θ between the received and transmitted signal. The ranging principle is cleared in fig. 3-1. Formula 3-2 shows that the ranging resolution is theoretically given by

$$\Delta R = \Delta\theta \cdot c \;/\; (4\pi f) \qquad (3\text{-}3)$$

Using a tone of 403 MHz and assuming a phase resolution $\Delta\theta$ of 1° the range resolution is valid 1 mm.

These formulas make obvious that much better accuracies can be achieved with the sidetone ranging principle than by measuring the travelling time of a light pulse. Our experimental setup demonstrates that this high resolution is obtained in a much shorter time than that of range finders which work with a pulse laser system.

4. EXPERIMENTAL SETUP

The blockdiagram of our experimental setup is shown in fig. 4-1. This setup consisted of two galvanometer scanners which deflected the transmit laser beam and the receiving field of view in x- and y- direction. The receiving field of view corresponds to the transmitting field of view and was about 0.1^{o}. With the galvanometer scanners we were able to scan over a total field of view of 30°x30^{o}. This total field of view could be decreased independently in each direction optionally.

The 3D laser scanner measured the range by the side tone ranging principle. The intensity of the transmitted light was modulated by modulating the drive current of the laser with a side tone of 403 MHz. We used an avalanche photodiode as a photodetector and the receiving clear aperture was about 10 mm in diameter. After preamplification of the received signal, the side tone signal was regenerated by a phase locked loop circuit (PLL). The phase shift θ between the transmitted and the received signal was measured by the phase-detector (s. fig. 4-1). The intensity of the reflected laser light was determined by an intensity detector.

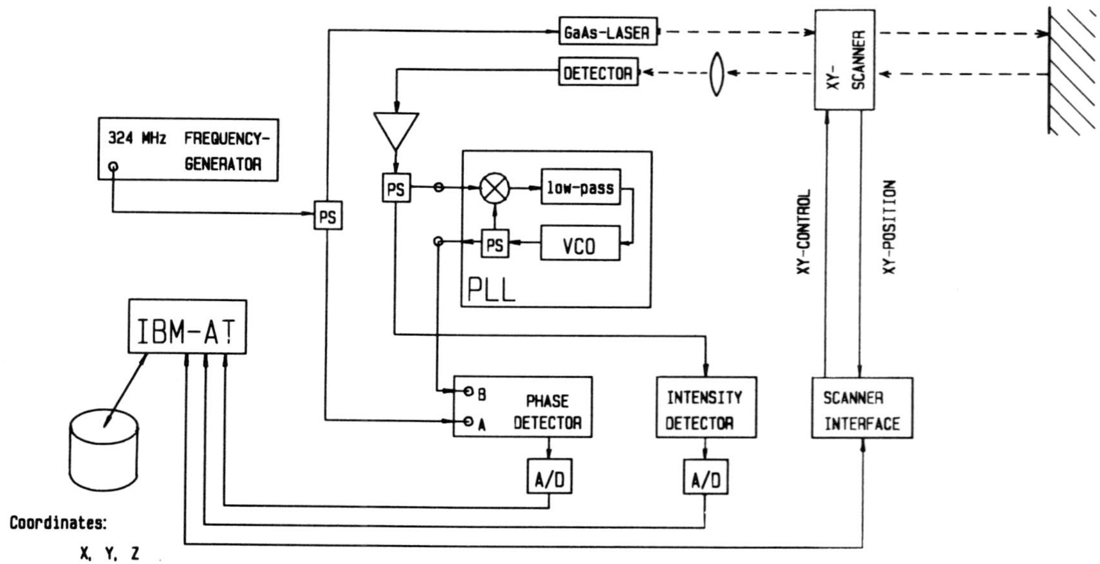

Fig. 4-1: Blockchart of Experimental Setup

Scanning a scene the synchronously measured range and intensity values were stored in and processed by an IBM-AT (s. fig. 4-1). The software for the laser scanner was implemented on the IBM-AT. Scanning a scene of 200x200 pixels took about 1.5 minutes. This means each range measurement lasts 2 msec. We used different types of objects for the experimental measurements for example boxes,cups, plates and heads. The different targets were positioned in a distance of about two meters.

5. MEASUREMENTS

With our experimental setup we carried out various measurements which are presented in the following. These chosen experiments will demonstrate the performance of such a tool.

First we scanned a simple plate which was positioned oblique-angled to the scanner. Fig. 5-1 shows a picture of the scene. Fig. 5-2 displays a laser intensity image of the scanned object. The image is obliterated, because the light spot of the laserbeam was about 5 mm in diameter. The range image of the same scene is given in fig. 5-3. This image displays 256 different grey levels. Each grey level represents a different distance. A more impressive image (s. fig. 5-4) was obtained by defining different grey levels for various ranges. The following range intervals were defined:

Fig. 5-1: Plate

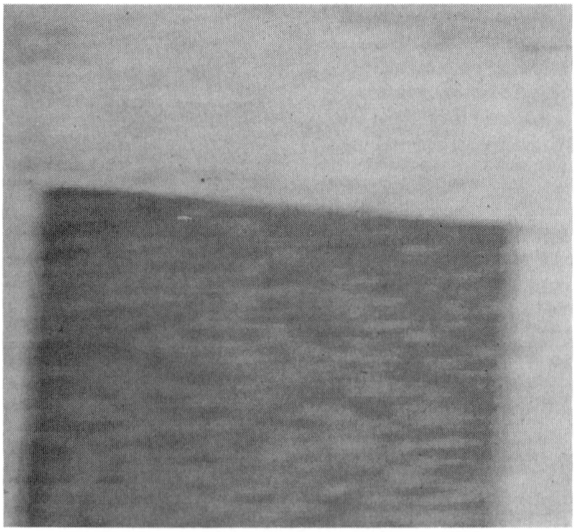

Fig. 5-2: Intensity Image of the Plate

Fig. 5-3: Range Image of the Plate

Fig. 5-4: Range Image of the Plate with Discrete Grey

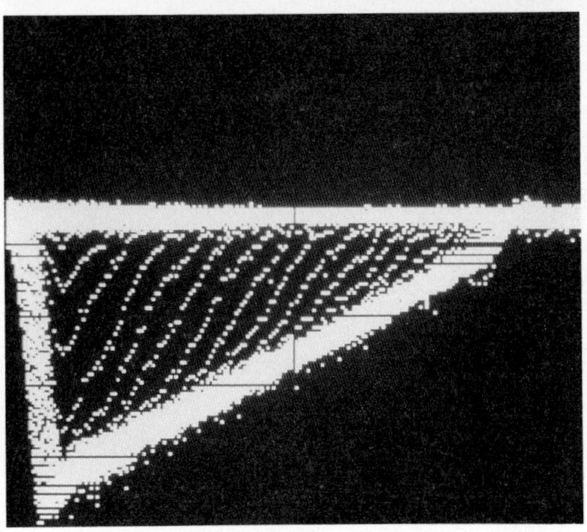

Fig. 5-5: 3D-View of the Range Image of the Plate

To demonstrate that we really obtained three dimensional data we processed our measured results with a special 3D-software on our Aries Image Processing System. In fig. 5-5 we look from above to the plate. This means we rotated the scene by 90 degrees about the x-axis. This image shows very clearly the slant plane and the accuracy of this laser mapper. In additional experiments we measured more complex structures.

Very impressive images were obtained by mapping an artificial head made of yellow glass (s. fig. 5-6). Fig. 5-7 displays the range image coded in discrete grey levels. In fig. 5-8 the scene was processed by 3D-software and rotated by 30 degrees, 45 dedrees and 0 degrees. Fig. 5-9 and -10 show that deformations can be detected and measured very precisely with our laser scanner.

The 3D-image of a cup (s. fig. 5-12) shows that the curvature of the cup can be well measured. We see clearly that at the upper end of the handle the laser beam interacts with the wall and the handle at the same time due to its finite dimensions.

With regard to the field of application of robotics we mapped a hand-drilling machine (s. fig. 5-13 and -14) in 2 m distance. We observed that the machine's contour is displayed in fine details. This fact can be exploited for the recognition of objects in robotics.

Next, we mapped a piece of brownstone which was used in historic buildings and churches (s. fig. 5-15). We see impressively that even in 2 m distance the fine structure could be resolved very well. In fig. 5-16 the upper right half of the stone was measured in zoomed mode. This means, we adjusted the scanning field of view so that only this small part was sampled. As the distance to the object had not been changed we really performed a zoomed measurement. This experiment demonstrated the feasibility of the laser mapper for measuring and mapping historic buildings, statues and churches.

Fig. 5-6: Yellow Glass Head

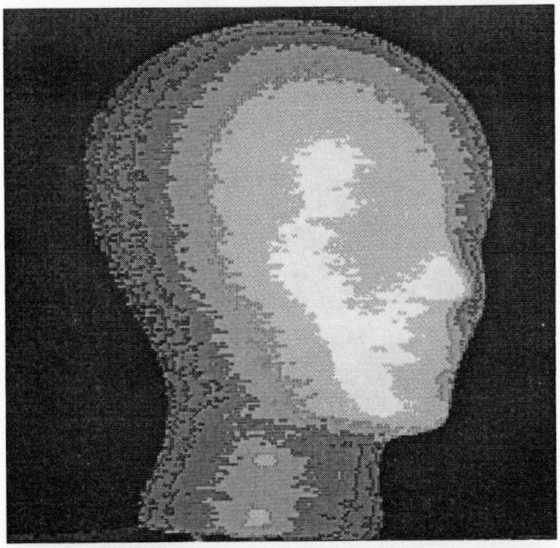

Fig. 5-7: Range Image of the Glass Head Coded in Grey Levels

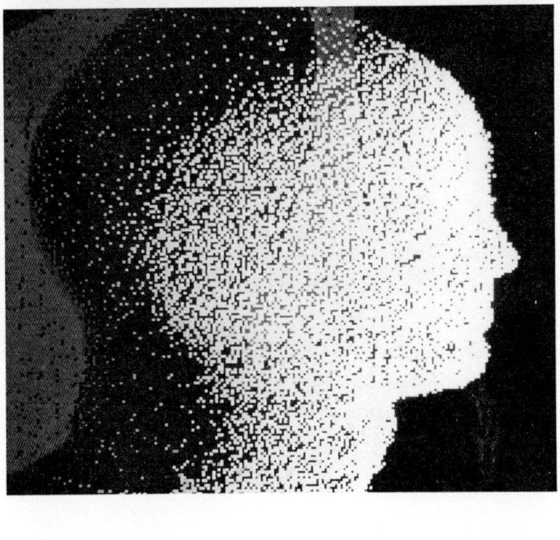

Fig. 5-8: 3D-Processed Glass Head

Fig. 5-9: Deformed Bottle

Fig. 5-10: Range Image of Deformed Bottle

Fig. 5-11: Cup

Fig. 5-12: 3D-View of the Cup

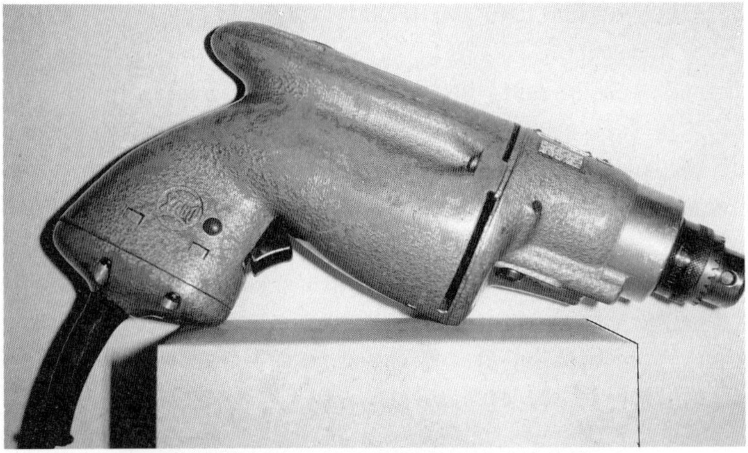

Fig. 5-13: Hand-Drilling Machine

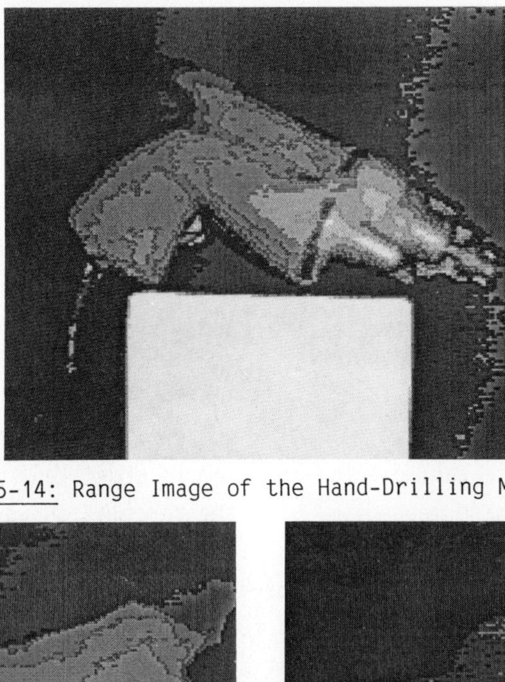

Fig. 5-14: Range Image of the Hand-Drilling Machine

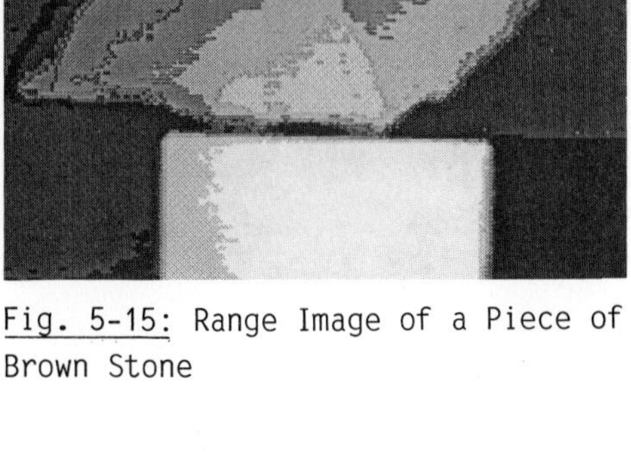

Fig. 5-15: Range Image of a Piece of Brown Stone

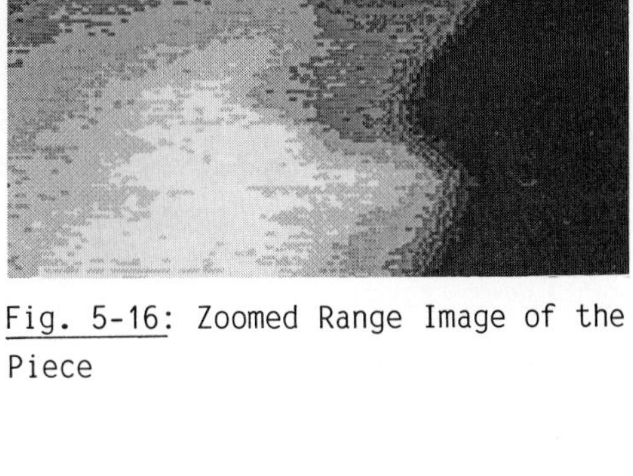

Fig. 5-16: Zoomed Range Image of the Piece

6. REMOTE SENSING APPLICATIONS AND TERRAIN PROFILING

The experimental measurements demonstrated the potential of this new measurement tool. In this paragraph we carried out calculations for a first estimation of the applicability of a 3D-mapper in remote sensing and terrain profiling. We distinguished three different cases:

1) diffuse reflecting target e.g. earth's surface
2) retroreflector as a target
3) one-way transmission (laser-transmitter e.g. on earth and receiver in the airplane)

We assumed a laser mapper system with the following specifications:

cw laser power at 834 nm	25	mW
transmitter beam divergence	0.1	deg
receiving aperture	78.0	mm
instantaneous field of view IFOV	0.1	deg
optical bandwidth	10.0	nm
responsivity of avalanche photodiode with gain M=100	37.0	A/W
electrical bandwidth	10	kHz
total transmission of the optical system	0.27	

The receiving aperture was selected to 78 mm in order to limit the technical effort for the scanning mechanism. For the calculation we assumed that we look from an airplane down to nadir at clear weather conditions. The typical reflectivity of the earth's surface was about 20%. The total one-way transmission through the atmosphere was set to 0.7. The received background radiation was calculated for a first estimate to P_b= -53.58 dBm regarding the transmission of the atmosphere of 0.7 and assuming a reflectivity of 0.2. The background radiation as a function of the target's reflectivity is shown in fig. 6-1. In this figure the transmission of the atmosphere was not taken into account. Figures 6-2 and 6-3 show the signal-to-noise ratio as a function of distance for the assumed system parameters. Fig. 6-2 shows that this system can only measure over a distance of about 50 m, if the target has a reflectivity of 20% and a signal-to-noise ratio of 10 dB is required. However, using a corner cube with an effective radar cross section of 0.55 km^2 we are able to measure distances of 10 km. Carrying out one-way measurements much greater distances can be observed as plotted in fig. 6-3, if the IFOV of the receiver is 0.1^o. However, in this case an additional scanner at the receiver's site is required.

A better comparable calculation between case 2) and 3) is carried out in the second graph, where an receiver's IFOV of 30° was assumed. This is the same angular range a corner cube can be used. Looking at fig. 6-3, we see that up to a distance of 7 km the corner-cube offers a better performance even taking not into account the extra effort of a second scanner for the IFOV of 0.1^o.

These calculations demonstrated the possibility of carrying out remote sensing tasks with such a system. However, measuring diffuse reflecting continuums larger receiving apertures and higher laser power are required. The influence of the background radiation can be minimized either by reducing the IFOV or by implementing narrow-band filters in the receiving optical path.

The calculations proofed that buildings can be sensed remotely with the 78 mm aperture. Scanning buildings deformations can be detected. By analyzing these measurement results stress-conditions can be detected. Our experiments showed that we need 90s for 40000 samples. The total measuring time of 90s is to long for detecting time dependent variations. Assuming that buildings behave as a continuum which mechanical behaviour can be described by e.g. finite methods, it is not necessary to measure the whole continuum. E.g. assuming a frequency of 2 Hz, we have to sample by at least 4 Hz. Assuming a measurement time of 4 ms for each sample we can measure 63 points. The selection of the 63 critical points of the structure must be carried out by pre-

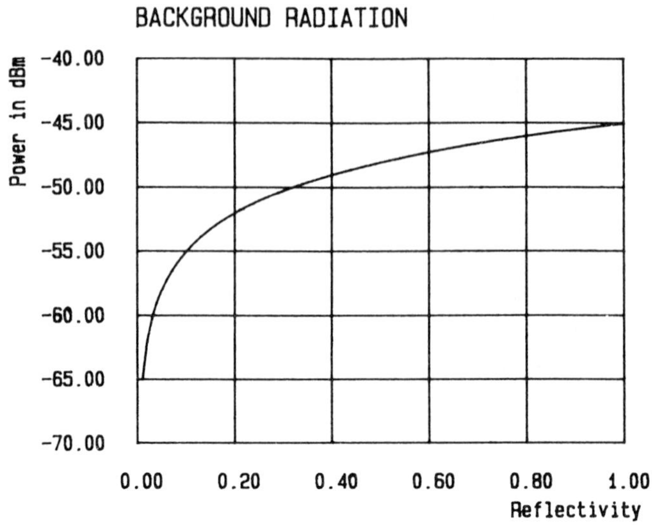

Fig. 6-1: Background Radiation Caused by Sunlight Reflected from a Diffuse Reflecting Target

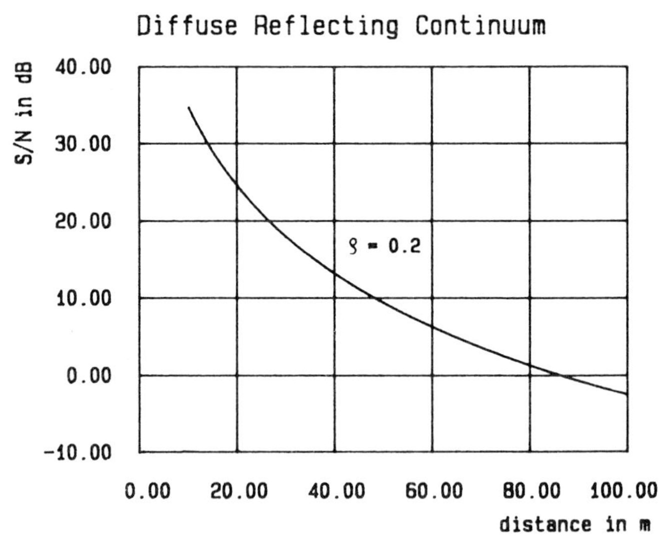

Fig. 6-2: Signal-to-Noise Ratio for a Diffuse Reflecting Continuum

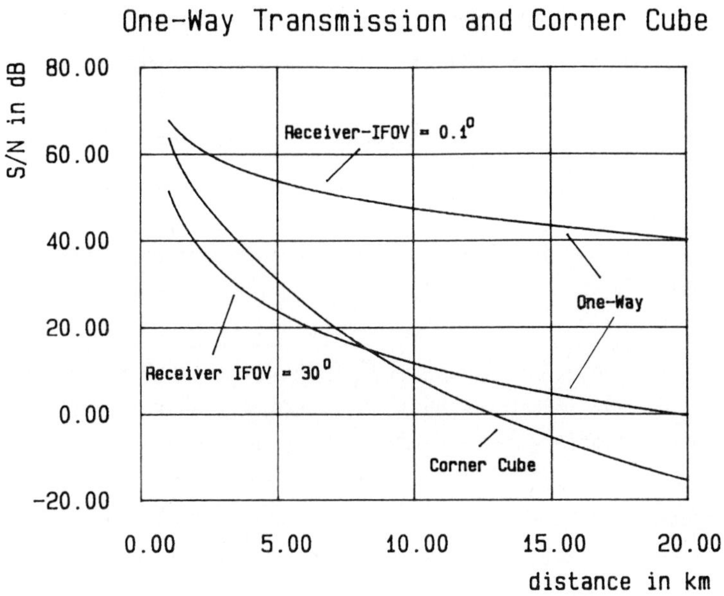

Fig. 6-3: Comparison between One-Way Transmission and Corner Cube

calculations. These points are scanned one after another over a certain periode of time. Experiments with galvanometer scanners demonstrated, that the repeatability was sufficient for such measurements. Therefore, even dynamics of extended structures could be measured by the 3D-laser mapper.

7. CONCLUSIONS

The experiments showed the high measurement performance of the developed laser mapper. It was demonstrated that it was possible to map extended continuums in a reasonable time with millimeter accuracy.

Many applicabilities can be imagined for the laser mapper. It can be used for precise contour measurements of extented objects e.g. large antennas. Especially in future space missions large antennas must be deployed in space. As the antennas will be designed for very short wavelengths and should collimate the beam sharply the surface of such antennas must be controlled very precisely. This will require very accurate contour measuring devices. The laser mapper is such an item. It is small in size, has low weight and has low power consumption.

However, the described laser mapper can also solve a lot of measurement problems on earth which occur in short distances, e.g. a few decimeters, as well as in longer distances, e.g. more than 100 m.

Work pieces can be measured during manufacturing inspection in industry. Very often it is required to sample manufacturing data for work pieces where only a master piece exists and no drawings are available. Here, the 3D-laser mapper can sample the data for computer integrated manufacturing (CIM).

In car industry it takes a long time to obtain the coordinates from a model car for further processing in a CAD-system. Today, photogrammetric methods are used which require complicated and time consuming measurement procedures and very long computation times which can be circumvented by the described laser system.

An application in anatomy is the measurement of the head in order to detect variations.

With powerful lasers and sophisticated detector systems it will be possible to measure extended objects like buildings and even remote sensing tasks could be fulfilled.

REFERENCES:

1. Bachman, C.G., 1979. Laser Radar Systems and Techniques. Artech House, Inc., Massachusetts.

2. Wolfe, W.L., Zissis, G.J., 1978. The Infrared Handbook. Environmental Research Institute of Michigan.

Acknowledgements: I would like to thank Ms. R. Schlothan who carried out the calculations on our ARIES Image Processing System.

F. INERTIAL TECHNIQUES

A Model for Highly Precise Inertial-Survey Adjustments (Summary)

Wolfgang Hausch
Institut für Physikalische Geodäsie, Technische Hochschule Darmstadt

Post-processing of unfiltered inertial data is usually done in two separate steps; the error velocity correction and, subsequently, the coordinate adjustment (Schwarz and Gonthier, 1982; Huddle, 1986). These techniques are in some cases not sufficient for inertial traverses which extend over time periods of two hours. To overcome this limitation one should consider the dynamics of the platform to get a more realistic model for the error behaviour of the inertial survey system.

The post-misson adjustment model proposed in this paper is based on the linearized inhomogeneous system of differential equations for an earth-oriented slaved geodetic inertial measurement system (e.g. Schröder et al., 1987), which is derived for a rotational-ellipsoidal representation of the coordinates of the earth's surface. In state-space notation the dynamical system is given by

$$\dot{z} = F(t)\ z + G(t)\ u(t)\ ,$$

with the dynamic matrix F(t), the state vector z and the coefficient matrix G(t) of the forcing vector u(t). The solution of the system of differential equations

$$z(t) = \phi(t,t_0)\ z(t_0) + \int_{t_0}^{t} \phi(t,\tau)\ G(\tau)\ u(\tau)\ d\tau$$

provides the error propagation of the inertial measurement unit. The state vector contains the misalignments, position errors, error velocities and constant parameters of the sensor errors. The forcing vector includes the stochastic sensor errors and the disturbance of the gravitation field. The transition matrix ϕ can be determined by a matrix exponential function. This approach is approximated by using a fifth-order Taylor series, where the dynamic matrix F is considered as constant over short time intervals of about ten seconds. The solution of the system of differential equations for the position errors is combined with the coordinates of surveyed stations in order to obtain observation equations for the unfiltered inertial coordinates, which are calculated by numerical integration in the inertial measurement unit. Together with the solution for error velocities, a set of observation equations is derived for the ZUPT measurements at the vehicle stops and the inertial coordinates at surveyed stations. The initial state vector $z(t_0)$ and the forcing vector u(t) are solved by least-squares collocation, using all inertial observations and the geodetic coordinates of the control stations.

The model is applied to a stationary operating Litton LTN-72 aircraft platform, where a uniform solution is obtained independently of the considered observation time interval or the different types of observations (error velocities and/or position errors). Moreover, the model is used for the adjustment of observations carried out by the Ferranti Inertial Land Surveyor platform in the Edinburgh inertial test network on an L-shaped traverse. The results are compared with results of a conventional empirical adjustment model (Boedecker, 1987), and a model refined for the position errors caused by gyro drift rates and the disturbance of the gravitation field (Hausch, 1988). The accuracies of the adjusted station coordinates are for all three models in the same order of magnitude. Systematic errors in the adjusted coordinates may be caused by nonlinear terms, which are usually neglected in the preprocessing of the inertial measurement unit (Möhlenbrink, 1985). The proposed one step model gives the opportunity to reduce or replace the zero velocity updates by GPS coordinate updates using differential GPS positioning techniques, for instance in a moving vehicle. If the sensor noise is sufficiently low, it may be possible to separate sensor errors and the disturbance of gravitation. A detailed discussion of the model and results will be given in a future publication.

Literature

Boedecker, G., Zur Ausgleichung von Inertialnetzen. Zeitschrift für Vermessungswesen, 112. Jg., Heft 1, S. 21, 1987.

Hausch, W., A refined adjustment model for inertial surveying. Manuscripta geodaetica 13, 13-28, 1988.

Huddle, J.R., Historical perspective on estimation techniques for position and gravity survey with inertial systems. J. Guidance 9, No. 3, 257-267, 1986.

Möhlenbrink, W., Drift effects in inertial measurement systems. Schriftenreihe der Universität der Bundeswehr München Studiengang Vermessungswesen, Heft 20-1, S. 51, 1985.

Schröder, D., S. Wiegner, E. Grafarend, B. Schaffrin, Simulation eines geodätischen Inertialsystems. Schriftenreihe der Universität der Bundeswehr München Studiengang Vermessungswesen, Heft 22, S. 197, 1987.

Schwarz, K-P. and M. Gonthier, Adjustment problems in inertial positioning. Deutsche Geodätische Kommission, Reihe B, Heft 258/IV, S. 97, 1982.

Advances in Strapdown Systems for Geodetic Applications

James R. Huddle
Litton Guidance and Control Systems Division, Woodland Hills, California, USA

Summary

This paper discusses recent developments in strapdown inertial navigation systems that have been pursued to obtain high accuracy. In particular a strapdown system with a single gimbal axis to obtain controlled rotation of the inertial sensor assembly is discussed. An approximate theoretical error model of this hybrid, strapdown inertial system is derived which permits its error characteristics to be compared with those of a conventional strapdown system in the vehicle coordinate frame. The high accuracy potential of the hybrid mechanization relative to a conventional strapdown inertial system is evident from a comparison of the respective error models. Testing of single and multiple gimbal hybrid strapdown inertial systems has confirmed the high accuracy potential of such systems in certain applications.

Introduction

In the last 10 years, pure or fully strapped down inertial systems have appeared in the marketplace displacing the older multiple gimbal, isolated inertial sensor or platform systems. These strapdown systems have for the most part been realized using two different technologies, the mechanical tuned-rotor gyro[1,2] and the optical dithered ring laser gyro.[3] The first application of these strapdown technologies has been in commercial aircraft. The evolution to strapdown systems has been driven by the lower weight, size, cost and higher reliability potential of this technology relative to gimbal stabilized platform systems.

However, fully strapped down inertial systems have accuracy disadvantages relative to platform systems. These disadvantages stem directly from the fact that the inertial sensors are fixed in orientation relative to the vehicle for which the navigation function is being provided. One of these disadvantages is the lack of ability to calibrate the system gyros and accelerometers prior to use. This can be achieved by use of the gimbal axes of a platform system during initial alignment, to obtain a change in orientation of the sensor assembly with respect to the local gravity and earth rate vectors. This action allows the observation and removal of turn-on bias, scale factor error and angular misalignments associated with the inertial instruments.

A second disadvantage of a full strapdown system is the increased sensitivity of error in the navigation variables to vehicle motion or trajectory characteristics. In that the sensor assembly is not isolated from spurious vehicle angular changes as with a platform system, two detrimental effects occur. The first of these effects is an often rapid de-correlation of error sources after initial alignment as for example occurs between initial tilt and accelerometer bias during leveling. A subsequent change in the heading of the vehicle for a strapdown system results in rotation of the accelerometer bias in space destroying the equilibrium condition with tilt which was realized during alignment. This results in the creation of transient navigation error due to both the accelerometer bias and initial tilt. For a platform isolated sensor assembly which is maintained in fixed orientation with respect to the earth, no such error effects occur.

Another detrimental error effect of strapdown systems occurs through exposure of the inertial instruments to the full extent of vehicle angular motion. This results in stringent requirements on instrument scale factor error and relative mechanical mounting alignment to suppress transient errors resulting directly from such motion. In fact certain vehicle motion profiles exist which cause additive effects for gyro and accelerometer errors in strapdown systems, resulting in rapid build-up of error in the computed orientation of the instrument reference frame and velocity error. However, most vehicle motion profiles cause random transient attitude and velocity error accumulation due to strapdown system generalized linear scale factor errors. This results in attitude and velocity error increase proportional to $(\text{time})^{1/2}$ due to the scale factor error effects.

As a result of such accuracy disadvantages of a pure strapdown inertial system, hybrid strapdown systems with rotation[1,4,5] axes are being investigated for high accuracy applications as exemplified by position and gravity vector survey. In such systems, the inertial sensor assembly is rotated in an a priori determined manner with the objective of reducing error propagation. Indeed it can be shown theoretically, that for a sufficient number of rotation axes the generalized gyro drift rate can be reduced to an essentially white noise process in inertial space. Such a mechanization obtains the optimal lower limit for the rate of accumulation of attitude error, in proportion to $(\text{time})^{1/2}$, due to all gyro drift rate error sources.

For these hybrid systems, the computation is still performed in a strapdown manner thereby reducing the necessity for high performance isolation characteristics obtained with the gimbal control technology on older platform systems. This obtains the desired error reduction attribute with minimal increase in system complexity (i.e. cost, size, etc.). Further, although multiple gimbal arrangements have been investigated, a single axis gimbal "strapdown" system appears to have great practicality as it is the simplest to implement of the hybrid class while realizing a number of accuracy enhancement properties.

This paper derives the approximate theoretical error model for a single-axis, periodic reversing rotation, strapdown inertial system. This error model is examined to illustrate the beneficial effects of such a system relative to a pure strapdown system for a high accuracy application such as position and gravity vector survey.

Inertial System Error Model

The general equations describing error propagation in an inertial navigation system are well-known and summarized in detail in Appendix I for a local-level system mechanization.[6] In summary, the vectorial form for these equations is as follows:

- Error in the system computed angular orientation of the navigation frame with respect to the earth ($\delta\theta$):

$$\delta\dot{\theta} = \delta\rho + \delta\theta \times \rho \tag{1}$$

  where ρ is the vector angular rate of the navigation frame with respect to the earth.

- Error in system computed velocity of the navigation frame (vehicle) with respect to the earth (δV):

$$\delta\dot{V} = -[\phi \times f] + \delta C + \delta\gamma + {}_N T_S\,[\delta a] \qquad (2)$$

where:

f	is the accelerometer measurement vector (specific force) expressed in the navigation frame
δC	is the error in the system computed Coriolis acceleration vector
$\delta\gamma$	is the error in the system computed gravity vector model
${}_N T_S$	is the transformation from the orthogonal inertial sensor reference frame to the navigation frame
δa	is the vector of errors in the specific force measured by the accelerometer triad

- Error in the knowledge of the angular orientation of the orthogonal inertial sensor reference frame with respect to the actual navigation frame (ϕ):

$$\dot{\phi} = \delta\omega + \phi \times \omega + {}_N T_S\,[\delta d] \qquad (3)$$

where:

ω	is the vector spatial angular rate of the navigation frame with respect to inertial space
$\delta\omega$	is the error in the system computed spatial angular rate vector
δd	is the vector of errors in the angular rate measured by the gyro triad

Characteristics of the Transformation, ${}_N T_S$

For a pure strapdown system, the coordinate transformation between the orthogonal reference frame for the inertial sensor assembly and the orthogonal navigation reference frame ${}_N T_S$, changes with the typically rapid angular motion of the vehicle in pitch, roll, and yaw. For a platform system mechanized to maintain the inertial sensor assembly aligned with the selected navigation frame through a high performance platform gimbal isolation system, ${}_{N}T_{S}$ is constant and typically the identity matrix. For a hybrid single gimbal axis strapdown system, ${}_{N}T_{S}$ varies in an a priori prescribed manner about this gimbal axis while reflecting the effects of vehicle angular motion in the plane orthogonal to the gimbal axis.

Inertial Sensor Error Models

The generic equations for the accelerometer measurement error model in the orthogonal inertial sensor reference frame are:

$$\delta a = \begin{bmatrix} \delta a_x \\ \delta a_y \\ \delta a_z \end{bmatrix}_S = \nabla + [\delta S]\, f_S + \delta F + \eta \tag{4}$$

where:

$\nabla = \begin{bmatrix} \nabla_x \\ \nabla_y \\ \nabla_z \end{bmatrix}_S$ is a vector of 3 constant accelerometer bias errors along the inertial sensor reference axes $[x, y, z]_S$

$[\delta S] = \begin{bmatrix} \delta S_{xx} & \delta S_{xy} & \delta S_{xz} \\ \delta S_{yx} & \delta S_{yy} & \delta S_{yz} \\ \delta S_{zx} & \delta S_{zy} & \delta S_{zz} \end{bmatrix}$ is the generalized linear scale factor error matrix due to individual accelerometer scale factors errors and mounting misalignments with respect to the inertial sensor reference frame.

$\delta F = \begin{bmatrix} \delta F_x[a_x, a_y, a_z] \\ \delta F_y[a_x, a_y, a_z] \\ \delta F_z[a_x, a_y, a_z] \end{bmatrix}$ is a vector of accelerometer measurement errors due to scale factor non-linearities and sensitivities to products of force (e.g. compliances) along the inertial sensor reference axes.

$f_S = \begin{bmatrix} f_x \\ f_y \\ f_z \end{bmatrix}_S$ is the force measured by the accelerometer triad in sensor axes

$\eta = \begin{bmatrix} \eta_x \\ \eta_y \\ \eta_z \end{bmatrix}_s$ is a vector representing the sum of accelerometer white and correlated noise components induced by various random environmental effects (temperature change and gradients, random vibration, magnetic fields, etc.) along the inertial sensor reference axes.

Similarly, the generic equations for the gyro measurement error model in the inertial sensor reference axes are:

$$\delta d = \begin{bmatrix} \delta d_x \\ \delta d_y \\ \delta d_z \end{bmatrix}_s = b + [\delta D]\ \omega_s + \delta B + \xi \qquad (5)$$

where:

$b = \begin{bmatrix} b_x \\ b_y \\ b_z \end{bmatrix}_s$ is a vector of 3 constant gyro bias errors along the inertial sensor references axes $[x, y, z]_s$

$\delta D = \begin{bmatrix} \delta D_{xx} & \delta D_{xy} & \delta D_{xz} \\ \delta D_{yx} & \delta D_{yy} & \delta D_{yz} \\ \delta D_{zx} & \delta D_{zy} & \delta D_{zz} \end{bmatrix}$ is the generalized linear scale factor error matrix due to individual gyro scale factor errors and mounting misalignments with respect to the inertial sensor reference frame.

$\omega_s = \begin{bmatrix} \omega_x \\ \omega_y \\ \omega_z \end{bmatrix}_s$ is the angular rate of the sensor frame with respect to inertial space expressed in the sensor axes.

$\delta B = \begin{bmatrix} \delta B_x(\omega_{x,y,z}, a_{x,y,z}) \\ \delta B_y(\omega_{x,y,z}, a_{x,y,z}) \\ \delta B_z(\omega_{x,y,z}, a_{x,y,z}) \end{bmatrix}$ is a vector of gyro measurement errors due to scale factor non-linearities and sensitivities to products of force (e.g. compliance) and angular rate in the inertial sensor reference axes.

$$\xi = \begin{bmatrix} \xi_x \\ \xi_y \\ \xi_z \end{bmatrix}_s$$ is a vector representing the sum of gyro white and correlated drift rate noise components induced by random environmental effects along the inertial sensor reference axes.

Pure Strapdown Inertial System Error Characteristics

As discussed above, pure strapdown inertial systems exhibit performance disadvantages relative to gimbaled stabilized platform systems due to the following effects:

- De-correlation or disturbance of the equilibrium between errors obtained during initial alignment due to subsequent vehicle angular motion
- Limited ability to calibrate the inertial instrument errors during initial alignment
- Direct exposure of the inertial instruments to the full dynamic range of vehicle angular motion

During initial alignment with a pure strapdown system, the time derivative of velocity errors and the angular rates associated with the level navigation axes are nulled. Thus from (2,3):

$$\dot{\phi}_x = \dot{\phi}_y = 0$$

$$\delta\dot{V}_x = \delta\dot{V}_y = 0$$

this yields tilts due to accelerometer measurement error:

$$\phi_E(o) = < {}_{G_E}T_s, \delta a > . \gamma^{-1}$$

$$\phi_N(o) = - < {}_{G_N}T_s, \delta a > . \gamma^{-1} \qquad \gamma \text{ is the gravity intensity} \qquad (6)$$

where $< {}_{G_{E,N}}T_s$ projects the accelerometer measurement error δa, along the level east (E) and north (N) navigation axes. Further, azimuth error is caused by the gyro measurement δd:

$$\phi_z(o) = < {}_{G_E}T_s, \delta d > . \Omega_N^{-1} \qquad \Omega_N \text{ is north earth rate} \qquad (7)$$

where $<_{G_E}T_S$ projects the gyro measurement error δd, along the local east axis.

Clearly a subsequent change in vehicle heading results in changes in the projection vectors $<_{G_{E,N}}T_S$ disturbing the equilibrium conditions implied by (6,7) and causing spurious error propagation due to the initial alignment errors $\phi(o)_{E,N,Z}$ and the current projection of the instrument measurement errors.

If a hybrid strapdown system with a single gimbal nominally along the local vertical were implemented, rotation about this axis during initial alignment could be obtained. This action provides variation in the transformation $_GT_S$ in (2,3) during the alignment process, allowing observation and removal of some gyro and accelerometer measurement errors and elimination of most of the initial alignment errors.

Properties of a Hybrid Strapdown System with Single Axis for Rotation

In the previous section the disadvantages of a pure strapdown system for high accuracy applications have been discussed. The error characteristics of a hybrid strapdown system which has a single gimbal axis to obtain controlled, periodic-reversing rotation of the inertial sensor assembly with respect to inertial space are derived next. The result of this derivation is an approximate drift rate and accelerometer measurement error model for the hybrid strapdown inertial system. Comparison of this approximate error model with that for the pure strapdown system illustrates the accuracy benefits of the hybrid strapdown system.

Gyro Drift Rate for a Hybrid Single Gimbal Axis Strapdown System

The accumulative effects of gyro drift rate for an inertial system are most clearly evident when examined in an inertial frame. For generalized gyro drift δd, expressed in the inertial sensor reference frame in (5) above, the incremental inertial drift angle over a time increment $2\Delta\tau$ is written:

$$\Delta\delta\theta_I = \int_{\tau}^{\tau+2\Delta\tau} {}_IT_S \; \delta d \; dt \tag{8}$$

where:

$_I T_s(t)$ is the transformation from the inertial sensor assembly reference frame s, to an inertially-fixed frame, I.

Imposition of a prescribed rotation rate ω_R, about a single gimbal axis of a hybrid strapdown system yields

$$_I T_s = {_I T_B}\ {_B T_s} \tag{9}$$

where

$_I T_B(t)$ is the transformation from the rotating vehicle body axes to the fixed inertial frame

$$_B T_s(t) = \begin{bmatrix} \cos\zeta & -\sin\zeta & 0 \\ \sin\zeta & \cos\zeta & 0 \\ 0 & 0 & 1 \end{bmatrix}$$ is the transformation from the inertial sensor assembly reference axes to the vehicle body axes

$$\zeta = \int [\omega_R - \omega_{z_B}]\ dt$$ is the angle between the inertial sensor assembly axes and the vehicle body axes about the gimbal axis z_B, in this discussion.

Using these definitions an expanded expression for the incremental inertial drift angle (8) is:

$$\Delta\delta\theta_I = \int_{\tau}^{\tau+2\Delta\tau} {_I T_B}\ {_B T_s} \left\{ \varepsilon + [\delta D]\ [{_B T_s^T}\ \omega_B + \vec{\omega}_R] \right\} dt \tag{10}$$

where:

$$\varepsilon = [b + \xi + \delta B]$$

Inertial Drift Rate Effect of ε Under Controlled Rotation

For a rapid rotation rate ω_R, about the gimbal axis of the hybrid strap-down inertial system relative to the vehicle body rotation rate components:

$$\left|\omega_R\right| >> \left|\omega_{x,y_B}\right| \tag{11}$$

then the integrated effect of the gyro drift rate bias (b), noise (ξ) and drift rate due to non-linearities (δB) assumed relatively constant in the inertial sensor reference frame, are minimized in the plane orthogonal to the z_B axis.

The effect of the rotation rate ω_R with respect to inertial space, is to translate the frequency spectrum associated with the drift rate errors ε, projected onto the $[x,y]_S$ plane, typically concentrated in a base-band near zero frequency, to a higher frequency noise when viewed from inertial space. This converts what would be constant gyro drifts in inertial space, whose integrals are unbounded, to sinusoids which have bounded integrals. Further, the low frequency baseband drift rate noise is converted to high frequency noise centered about the rotation rate ω_R, which then only integrates at a minimum rate in inertial space; in the limit this is proportional to the square-root of time.

In Appendix II an approximation to the integral of the transformation ${}_IT_S$ in (8) is evaluated over a time increment $2\Delta\tau$. Here a positive rotation of 2π with respect to inertial space[1] about z_B is made using the control rate ω_R in time $\Delta\tau$, followed by a negative rotation of 2π in the subsequent $\Delta\tau$ interval. The time average of the result of this evaluation over $2\Delta\tau$ is:

$$\overline{I}_1 = \left[\frac{I_1}{2\Delta\tau}\right] = \frac{1}{2\Delta\tau}\int_{\tau}^{\tau+2\Delta\tau} {}_IT_S\,(\tau+t)\,dt = {}_IT_B(\tau)\begin{bmatrix} 0 & 0 & \omega_{y_B}\cdot\Delta\tau \\ 0 & 0 & -\omega_{x_B}\cdot\Delta\tau \\ 0 & 0 & 1 \end{bmatrix} \tag{12}$$

(1) Note these rotations are <u>not</u> with respect to the rotating body axes

when it can be assumed ω_{x,y_B} have relatively small change in $2\Delta\tau$.

Application of $\overline{I}_1$ to a constant drift rate vector in the inertial sensor assembly reference frame:

$$b = \begin{bmatrix} b_x \\ b_y \\ b_z \end{bmatrix}_s \tag{13}$$

results in the approximate (first-order) average drift rate with respect to inertial space over $[\tau, \tau + 2\Delta\tau]$ of:

$$\delta\omega_{I_b} = {}_I T_B \begin{bmatrix} \omega_{y_B} \cdot \Delta\tau \\ -\omega_{x_B} \cdot \Delta\tau \\ 1 \end{bmatrix} b_{z_s} \approx \begin{bmatrix} 0 \\ 0 \\ b_{z_s} \end{bmatrix} \tag{14}$$

using the assumption (11).

The result (14) indicates that to first-order:

1. The drift rates b_{x,y_s} have no average effect due to rotation about z_B.

   Note however that if the rotation direction is not reversed for the second cycle these drift rate components cause the following <u>additional</u> second-order average drift rate in inertial space along z_B over $[o, 2\Delta\tau]$:

$$\delta\omega_{I_b} = \left[\frac{\Delta\tau}{2\pi}\right] {}_I T_B(\tau) \begin{bmatrix} 0 \\ 0 \\ -\omega_{x_B} \cdot b_{x_s} - \omega_{y_B} \cdot b_{y_s} \end{bmatrix} \tag{15}$$

This result is evident from the results (18,19) of Appendix II.

2. The drift rate b_{z_s} appears with full effect along z_B.

3. The drift rate b_{z_s} causes second-order drifts along $[x,y]_B$ due to rotation of the z_B^s axis by ω_{x,y_B} over the time increment $2\Delta\tau$.

Inertial Drift Rate Effect of [δD] Under Controlled Rotation

From (10), the inertial drift angle due to generalized linear scale factor error [δD]:

$$\Delta\delta\Theta_{I_{\delta D}} = \int_{\tau}^{\tau+2\Delta\tau} {}_I T_B \; {}_B T_s \left\{ [\delta D][{}_B T_s^T \omega_B + \vec{\omega}_R] \; dt \right\} \tag{16}$$

occurs thru two mechanisms.

1. The simplest of these mechanisms yields the inertial drift rate:

$$ {}_I T_B \; {}_B T_s \; [\delta D] \; \vec{\omega}_R \tag{17}$$

Here $\vec{\omega}_R$ thru [δD] causes constant drifts along the inertial sensor assembly axes which are then transformed to the body axes ${}_B T_s$ which itself is a function of ω_R. The drifts are subsequently transformed to inertial space thru ${}_I T_B$ which is a function of ω_{x,y,z_B}. Due to $\pm$ 2π inertial rotation of the sensor assembly about z_B by ω_R, the drift rate effects in inertial space over $2\Delta\tau$ will be null except for the effects of a small rotation of the z_B axis due to ω_{x,y_B}. Details are presented below.

2. The more complex contribution of drift rate due to [δD] is expressed:

$$ {}_I T_B \; {}_B T_s \; [\delta D] \; {}_B T_s^T \; \omega_B \tag{18}$$

Here the ω_{x,y_B} body rates are first transformed to the sensor assembly frame, where thru [δD], drifts are caused which are then transformed back to the body axes thru ${}_B T_S$ and subsequently to inertial space thru ${}_I T_B$. Since the product:

$$ {}_B T_S \, [\delta D] \, {}_B T_S^T \tag{19} $$

involves squares of sines and cosines of the rotation angle, rectification of certain generalized linear scale factor errors occurs. Details are presented below.

Using the results of Appendix II, in particular expression (8), the inertial drift angle (16) can be expressed:

$$ \Delta\delta\Theta_{I_{\delta D}} = {}_I T_B\,(\tau) \int_{\tau}^{\tau + 2\Delta\tau} dt \begin{bmatrix} \cos(\omega_R t) & -\sin(\omega_R t) & \omega_{y_B} \cdot t \\ \sin(\omega_R t) & \cos(\omega_R t) & -\omega_{x_B} \cdot t \\ [\omega_{x_B} \cdot \sin(\omega_R t) - \omega_{y_B} \cdot \cos(\omega_R t)] \cdot t & [\omega_{y_B} \cdot \sin(\omega_R t) + \omega_{x_B} \cos(\omega_R t)] \cdot t & 1 \end{bmatrix} $$

$$ \begin{bmatrix} \delta D_{xx} & \delta D_{xy} & \delta D_{xz} \\ \delta D_{yx} & \delta D_{yy} & \delta D_{yz} \\ \delta D_{zx} & \delta D_{zy} & \delta D_{zz} \end{bmatrix} \cdot \begin{bmatrix} [\omega_{x_B} \cos(\zeta) + \omega_{y_B} \sin(\zeta)] \\ [-\,\omega_{x_B} \sin(\zeta) + \omega_{y_B} \cos(\zeta)] \\ \omega_R \end{bmatrix} \tag{20} $$

The third component of the angular rate vector in (20) is shown as ω_R. The body rate ω_{z_B} which exists about the z_B axis, is not impressed on the inertial sensor assembly. This occurs as isolation is implemented with the gimbal axis control policy which realizes the controlled angular rate ω_R, about this axis.

Evaluation of (20) can be simplified by considering it in two parts. The first part considers only the effect of the third component of the angular rate ω_R, yielding:

$$ \Delta\delta\Theta_{I_{\delta D_{\omega_R}}} = {}_I T_B\,(\tau) \int {}_{B_\tau} T_S\,(t)\,[\delta D_3>]\,\omega_R\,dt \tag{21} $$

where:

$_{B_\tau}T_S\ (t)$ is the transformation from the sensor frame at time t, to the body frame at time, τ .

$$\delta D_3 > = \begin{bmatrix} \delta D_{xz} \\ \delta D_{yz} \\ \delta D_{zz} \end{bmatrix} \tag{22}$$

Since ω_R is positive during [o, $\Delta\tau$] and negative during [$\Delta\tau$, $2\Delta\tau$], then: (21) can be expressed as:

$$\left|\omega_R\right|\ {}_IT_B\ (\tau) \left[\int_0^{\Delta\tau} {}_{B_\tau}T_S(t)\ dt - \int_{\Delta\tau}^{2\Delta\tau} {}_{B_\tau}T_S(t)\ dt \right] \delta D_3> \tag{23}$$

The two integrals involved in (23) are evaluated in Appendix II yielding:

$$\Delta\delta\Theta_{I_{\delta D_{\omega_R}}} = \left|\omega_R\right|\ {}_IT_B\ (\tau)\ [I_3 - I_4]\ \delta D_3> \tag{24}$$

which since:

$$I_3 - I_4 = \begin{bmatrix} 0 & 0 & -\omega_{y_B}(\Delta\tau)^2 \\ 0 & 0 & \omega_{x_B}(\Delta\tau)^2 \\ -\omega_{x_B}\dfrac{(\Delta\tau)^2}{\pi} & -\omega_{y_B}\dfrac{(\Delta\tau)^2}{\pi} & 0 \end{bmatrix} \tag{25}$$

yields:

$$\Delta\delta\Theta_{I_{\delta D_{\omega_R}}} = (2\Delta\tau)\ {}_I T_B(\tau) \begin{bmatrix} -\pi . \omega_{y_B} . \delta D_{zz} \\ \pi . \omega_{x_B} . \delta D_{zz} \\ -\omega_{x_B} . \delta D_{xz} - \omega_{y_B} . \delta D_{yz} \end{bmatrix} \tag{26}$$

or an average inertial drift rate over [τ, τ + 2Δτ] of:

$$\delta\omega_{I_{\delta D_{\omega_R}}} = {}_I T_B(\tau) \begin{bmatrix} -\pi . \omega_{y_B} . \delta D_{zz} \\ \pi . \omega_{x_B} . \delta D_{zz} \\ -\omega_{x_B} . \delta D_{xz} - \omega_{y_B} . \delta D_{yz} \end{bmatrix} \tag{27}$$

Note the third component in (27) is first-order but similar in form to that of (15). This result occurs as the reversal of rotation direction via ω_R during the second 2π cycle causes the large drift rates induced by $[\delta D_{xz}, \delta D_{yz}]$ to reverse sign. This is equivalent to constant drift rate biases $[b_{x,y_s}]$ rotating in the same direction during the second cycle (no reversal of ω_R). The reason the third component here is first-order is that the induced drift rate is proportional to ω_R during the cyclic rotation (see 21) and hence is much larger than the constant drift rates of (13,15).

The second part of (20) which excludes the direct effect of the third component of the angular rate vector ω_R, is evaluated as (9) in Appendix III. The result for the inertial drift angle due to the direct effect of ω_{x,y_B} in (20) is to first-order:

$$\Delta\delta\Theta_{I_{\delta D_{\omega_{x,y_B}}}} = [2\Delta\tau]\ {}_I T_B(\tau) \begin{bmatrix} \overline{\delta D} . \omega_{x_B} + \gamma_z . \omega_{y_B} \\ \overline{\delta D} . \omega_{y_B} - \gamma_z . \omega_{x_B} \\ 0 \end{bmatrix} \tag{28}$$

where it is assumed ω_{x,y_B} have small changes over $2\Delta\tau$ and:

$$\overline{\delta D} \triangleq .5\,[\delta D_{xx} + \delta D_{yy}] \qquad \gamma_z \triangleq .5\,[\delta D_{xy} - \delta D_{yx}] \tag{29}$$

The result (28) yields an average inertial drift rate over $(\tau,\ \tau + 2\Delta\tau]$ of:

$$\delta\omega_{I_{\delta D_{\omega_{x,y_B}}}} = {}_I T_B(\tau) \begin{bmatrix} \overline{\delta D}\cdot\omega_{x_B} + \gamma_z\cdot\omega_{y_B} \\ \overline{\delta D}\cdot\omega_{y_B} - \gamma_z\cdot\omega_{x_B} \\ 0 \end{bmatrix} \tag{30}$$

Combining the results (27, 30), the first-order expression for drift rate about body axes due to generalized linear scale factor error is:

$$\delta\omega_{I_{\delta D}} = {}_I T_B(\tau) \begin{bmatrix} \overline{\delta D} & \Gamma_z & 0 \\ -\Gamma_z & \overline{\delta D} & 0 \\ -\delta D_{xz} & -\delta D_{yz} & 0 \end{bmatrix} \begin{bmatrix} \omega_{x_B} \\ \omega_{y_B} \\ \omega_{z_B} \end{bmatrix} \tag{31}$$

where:

$$\Gamma_z \triangleq \left[\gamma_z - \pi\ \delta D_{zz}\right] \tag{32}$$

The result (31) for the approximate gyro drift rate in body axes due to linear scale factor error is interesting in two respects:

1. The equivalent scale factor error along x_B and y_B is the same and equal to the average of the scale factor errors along x_s and y_s.

2. There exists a simple rotation misalignment Γ_z, about the z_B axis. Since this is not a non-orthogonality misalignment, there is no drift rate induced by this error.

Approximate Gyro Drift Rate Error Model for a Hybrid Single Gimbal Axis Strapdown System

Combining the results of (14) and (31), the approximate drift rate model in body axes for a single gimbal axis strapdown system with controlled rotation about z_B, is:

$$\delta d_B = \begin{bmatrix} \overline{\delta D} & \Gamma_z & 0 \\ -\Gamma_z & \overline{\delta D} & 0 \\ 0 & 0 & 0 \end{bmatrix} \begin{bmatrix} \omega_x \\ \omega_y \\ \omega_z \end{bmatrix}_B + \begin{bmatrix} 0 \\ 0 \\ \overline{b}_z \end{bmatrix} + \varepsilon_{g_B} \tag{33}$$

where also using (5):

$$\overline{b}_z = b_{z_s} - [\delta D_{xz} \cdot \omega_{x_B} + \delta D_{yz} \cdot \omega_{y_B}] + \delta B_z \tag{34}$$

is a composite low-frequency drift along the z_B axis and

$$\varepsilon_{g_B} = \begin{bmatrix} \varepsilon_{g_x} \\ \varepsilon_{g_y} \\ \xi_z \end{bmatrix}_B \quad \text{are high frequency drift rate components} \tag{35}$$

The components $\varepsilon_{g_{x,y_B}}$ result from the shift of low frequency drift rate components to high frequency due to the rotation rate, $\pm |\omega_R|$, over successive $\pm\, 2\pi$ rotations with respect to inertial space.

The corresponding drift rate model for a pure strapdown system (5) is seen to have a significant number of more deleterious terms than that for the hybrid system expressed in (33).

Approximate Force Measurement Error Model for a Hybrid Single Gimbal Axis Strapdown System

The accelerometer measurement error for the system described above is expressed in an inertial frame using (4,8,10) as:

$$\delta a_I = \int_{\tau}^{\tau+2\Delta\tau} {}^{I}T_B \; {}^{B}T_S \left\{ \varepsilon_A + [\delta S] \; {}_{B}T_S^{\;T} \; f_s \right\} dt \tag{36}$$

where:

$$\varepsilon_A = \nabla + \delta F + \eta$$

It is immediately evident by analogy to the results derived above for corresponding terms between (10) and (36) that the approximate force measurement error model in body axes for the single gimbal axis system is:

$$\delta a_B = \begin{bmatrix} \overline{\delta S} & \mu_z & 0 \\ -\mu_z & \overline{\delta S} & 0 \\ 0 & 0 & \delta S_{zz} \end{bmatrix} \begin{bmatrix} f_x \\ f_y \\ f_z \end{bmatrix}_B + \begin{bmatrix} 0 \\ 0 \\ \overline{\nabla}_z \end{bmatrix} + \xi_{a_B} \tag{37}$$

when it can be assumed f_{x,y_B} have relatively small change in $2\Delta\tau$ and

where:

$$\overline{\delta S} = .5 \, [\delta S_{xx} + \delta S_{yy}] \tag{38}$$

$$\mu_z = .5 \, [\delta S_{xy} - \delta S_{yx}]$$

$\overline{\nabla}_z = \nabla_z + \delta F_z$ is the composite low-frequency accelerometer error component along z_B

$$\varepsilon_{a_B} = \begin{bmatrix} \varepsilon_{a_x} \\ \varepsilon_{a_y} \\ \eta_z \end{bmatrix}_B$$

are the high frequency accelerometer noise components along $[x,y]_B$ due to ω_R.

Note the orthogonal rotation misalignment μ_Z differs in form from Γ_Z as the drift rate expression (10) has an additional effect due to $\vec{\omega}_R$.

Again, the corresponding accelerometer measurement error model for a pure strapdown system (4), is seen to have a significant number of more deleterious terms than that for the hybrid system expressed in (37).

Summary of Error Characteristics of Conventional and Hybrid Single Gimbal Axis Strapdown Inertial Systems

The error characteristics of the single gimbal axis and conventional strapdown inertial systems are summarized in Table 1. For the hybrid system, the controlled inertial rotation rate ω_R is applied about the nominally vertical z_B axis and is reversed after a full cycle of rotation with respect to inertial space. One advantage of the hybrid system is that relatively small alignment errors in tilt and azimuth are realized after initial alignment relative to the conventional strapdown system due to both instrument bias calibration and reduction through averaging by the rotation. Further, reduced error propagation due to low frequency noise errors in the plane orthogonal to the gimbal axis is realized after initial alignment as these errors become higher frequency due to the rotation rate, ω_R. Scale factor error effects are present in both systems but are different in form in body axes. The effects of non-orthogonal misalignments of the conventional strapdown system are eliminated or converted to an orthogonal misalignment in body axes by the presence of the rotation rate ω_R in the hybrid system.

Litton Hybrid Strapdown Inertial System

Single and multiple gimbal axis hybrid strapdown inertial systems have been under development for the last several years. Prototype hardware testing

Table 1. Characteristics of Conventional Strapdown and Hybrid Single Rotational Axis Strapdown Systems

Error Source	Hybrid Strapdown Gimbal Axis Along z_B	Pure Strapdown
• Accelerometer Biases	• Biases in plane orthogonal to gimbal axis are calibratible during alignment • Small initial alignment tilts	• Not calibratible and cause tilts in initial alignment
• Gyro Biases	• Biases in plane orthogonal to gimbal axis are calibratible • Small initial azimuth misalignment[(1)]	• North directed bias can be calibrated during alignment • Uncalibrated east component causes azimuth error in initial alignment
• Low Frequency Instrument Noise	• Translated to high frequency about rotation rate ω_R, for those components in plane orthogonal to the gimbal axis • Small effect on initial alignment • Components along z_B have full effect on error propagation	• Full error propagation effect for all 3 axes
• High Frequency Instrument Noise	• Full error effect	• Full error effect
• Linear scale factor errors along the $[x,y]_S$	• Equal scale factor errors $[\overline{\delta D}, \overline{\delta S}]$ along the $[x,y]_B$ axes $\overline{\delta D} = .5\ [\delta D_{xx}+\delta D_{yy}]$ $\overline{\delta S} = .5\ [\delta S_{xx}+\delta S_{yy}]$	• Non-equal scale factor errors along the $[x,y]_B$ axes • $[\delta D_{xx}, \delta D_{yy}]$ for the gyros • $[\delta S_{xx}, \delta S_{yy}]$ for the accelerometers
• Linear scale factor errors (misalignments) along cross axes $[y,x]_S$	• Simple orthogonal rotation misalignment about z_B: $\Gamma_z = .5\ [\delta D_{xy}-\delta D_{yx}]\ -\pi\delta D_{zz}$ for the gyros $\mu_z = .5\ [\delta S_{xy}-\delta S_{yz}]$ for the accelerometers	• Nonorthogonal rotations about z_B $[\delta D_{xy}, \delta D_{yx}]$ for the gyros $[\delta S_{xy}, \delta S_{yx}]$ for the accelerometers

(1) For z_B axis vertical

has validated the potential of these configurations as high accuracy positioning systems. Figure 1 depicts the inertial sensor assembly of such a system in operation with the rotation axis in the vertical direction. This particular sensor assembly contains Litton 28cm ring laser gyros and an A-4 accelerometer triad. Figure 2 depicts in two views the full system configuration which includes required high speed computer processors, power supplies, control and input-output electronics. Flight tests of this system indicate performance is limited by the effects of the uncompensated gravity disturbance vector.

Conclusion

This paper has derived an approximate theoretical error model for a hybrid, single gimbal axis strapdown inertial system which permits its error characteristics to be compared with those of a conventional strapdown system in a common coordinate frame, the vehicle body axes. The high accuracy potential of the hybrid mechanization relative to a conventional strapdown inertial system is evident from a comparison of the respective error models. Testing

Figure 1. Inertial Sensor Assembly

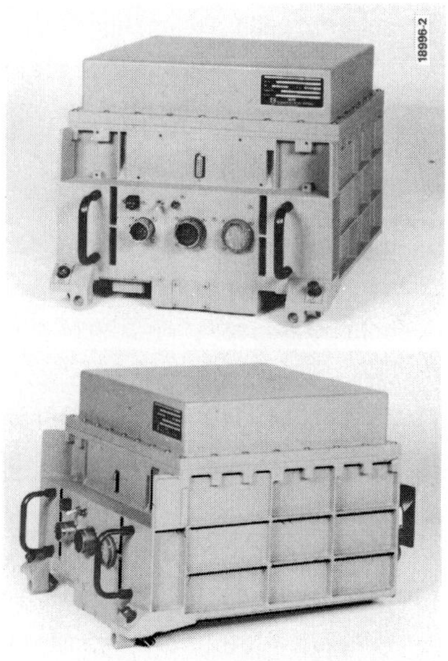

Figure 2. Laser Gyro Inertial Navigation System - LN-94R

of single and multiple gimbal hybrid strapdown inertial systems has confirmed the high accuracy potential of such systems in certain applications.

References

1. Lottman, D., et.al., "Kalman Filter Implementation in the Litton LR-80 AHRS", IEEE 1984 National Aerospace and Electronics Conference Proceedings, Dayton, Ohio, May, 1984.

2. Sturza, et.al., "GPS/AHRS: A Synergistic Mix", IEEE 1984 National Aerospace and Electronics Conference Proceedings, Dayton, Ohio, May, 1984.

3. Mark, J., et.al., "Design of RLG Inertial Systems for High Vibration", IEEE Position Location and Navigation Systems Conference, Atlantic City, New Jersey, December 1982.

4. Matthews, A., et.al., "Low Cost Testing of High Accuracy INS Using GPS", Institute of Navigation National Technical Meeting, Long Beach, California, January, 1986.

5. Levinson, E., "Accuracy Enhancement Techniques Applied to the Marine Ring Laser Gyro Inertial Navigator (MARLIN)", Institute of Navigation National Technical Meeting Proceedings, Anaheim, California, January, 1987.

6. Huddle, J.R., "Inertial Navigation System Error Model Considerations in Kalman Filtering Applications", Chapter 6 in Volume 20 of Control and Dynamic Systems, C.T. Leondes, Editor, Pp. 293–339, Academic Press, 1983.

APPENDIX I GENERAL EQUATIONS DESCRIBING ERROR PROPAGATION IN AN INERTIAL SYSTEM SPECIALIZED TO THE LOCAL-LEVEL MECHANIZATION CASE

The general equations describing error propagation in an inertial navigation system are well known and are derived as for example in [3]. These equations are summarized here in component form as a reference for the above discussion.

- Propagation of angular error in the system computed navigation frame orientation

$$\begin{aligned} \dot{\delta\theta}_x &= \delta\rho_x + \rho_z\,\delta\theta_y - \rho_y\,\delta\theta_z \\ \dot{\delta\theta}_y &= \delta\rho_y + \rho_x\,\delta\theta_z - \rho_z\,\delta\theta_x \\ \dot{\delta\theta}_z &= \delta\rho_z + \rho_y\,\delta\theta_x - \rho_x\,\delta\theta_y \end{aligned} \tag{1}$$

where

$\delta\theta_{x,y,z}$ represent angular error about the x,y,z axes of the navigation frame

$\rho_{x,y,z}$ represent angular rate of the navigation coordinates with respect to the earth where for a "local-level" system mechanization:

$$\rho_x = -V_y/R_y \qquad \delta\rho_x = -\delta V_y/R_y$$
$$\rho_y = V_x/R_x \qquad \delta\rho_y = \delta V_x/R_x$$

$V_{x,y,z}$ are the vehicle velocity components with respect to the earth

$R_{x,y}$ are the radii of curvature of the reference gravity field model of the earth at the vehicle position.

$\delta V_{x,y,z}$ represent the errors in the computed vehicle velocity components

- Elevation error propagation

$$\delta\dot{h} = \delta V_z$$

- Propagation of error in system computed velocity:

$$\delta\dot{V}_x = \phi_z a_y - \phi_y (a_z + \gamma) + \delta C_x + \delta\gamma_x + \langle {}_N T_{s_x}, \delta a \rangle$$

$$\delta\dot{V}_y = -\phi_z a_x + \phi_x (a_z + \gamma) + \delta C_y + \delta\gamma_y + \langle {}_N T_{s_y}, \delta a \rangle \quad (2)$$

$$\delta\dot{V}_z = \phi_y a_x - \phi_x a_y + \delta C_z + \delta\gamma_z + \langle {}_N T_{s_z}, \delta a \rangle$$

where

$\phi_{x,y,z}$ represent the error in the knowledge of the angular orientation of an orthogonal inertial sensor reference frame about the navigation frame axes

$a_{x,y,z}$ represent vehicle acceleration components along the navigation frame axes

γ represents the gravity intensity for a normal ellipsoidal reference model of the earth where the gravity vector is along the vertical navigation axis for a local-level mechanization of the navigation system.

$f_{x,y,z}$ represents the specific force components measured by the accelerometers expressed along the navigation coordinate axes. Note $f_z = a_z + \gamma$ for the local-level mechanization.

$\delta\gamma_{x,y,z}$ represent the error in the system computed gravity components.

$$\delta C_x = (\delta\rho + 2\delta\Omega)_z V_y - (\delta\rho + 2\delta\Omega)_y V_z + (\rho + 2\Omega)_z \delta V_y - (\rho + 2\Omega)_y \delta V_z$$

are the errors in the system computed Coriolis acceleration corrections which are required to be applied to acceleration measurements transformed

$$\delta C_y = (\delta\rho + 2\delta\Omega)_x V_z - (\delta\rho + 2\delta\Omega)_z V_x + (\rho + 2\Omega)_x \delta V_z - (\rho + 2\Omega)_z \delta V_x$$

$$\delta C_z = (\delta\rho + 2\delta\Omega)_y V_x - (\delta\rho + 2\delta\Omega)_x V_y + (\rho + 2\Omega)_y \delta V_x - (\rho + 2\Omega)_x \delta V_y$$

to the navigation frame to account for the rotation of the earth (Ω), and the navigation frame with respect to the earth (ρ). These corrections are used to obtain the time derivative of vehicle velocity with respect to the earth, which is taken with respect to the navigation frame.

$$\delta\Omega_x = \Omega_y\delta\theta_z - \Omega_z\delta\theta_y$$
$$\delta\Omega_y = \Omega_z\delta\theta_x - \Omega_x\delta\theta_z$$
$$\delta\Omega_z = \Omega_x\delta\theta_y - \Omega_y\delta\theta_x$$

are the errors in system computed components of earth rate about the navigation axes due to error in the system computed navigation frame orientation.

$$N^T S = \begin{bmatrix} < N^T S_x \\ < N^T S_y \\ < N^T S_z \end{bmatrix}$$

is the transformation between the orthogonal reference frame for the inertial sensors and the navigation frame

$$\delta a = \begin{bmatrix} \delta a_x \\ \delta a_y \\ \delta a_z \end{bmatrix}_S$$

is the generalized error in the specific force measured by the accelerometers expressed in the inertial sensor reference frame

- Propagation of error in knowledge of the orthogonal inertial sensor reference frame orientation with respect to the actual navigation frame

$$\dot{\phi}_x = \delta\omega_x + \omega_z \phi_y - \omega_y \phi_z + < {}_N T_{s_x}, \delta d >$$

$$\dot{\phi}_y = \delta\omega_y + \omega_x \phi_z - \omega_z \phi_x + < {}_N T_{s_y}, \delta d > \tag{3}$$

$$\dot{\phi}_z = \delta\omega_z + \omega_y \phi_x - \omega_x \phi_y + < {}_N T_{s_z}, \delta d >$$

where:

$\omega_{x,y,z} = [\rho_{x,y,z} + \Omega_{x,y,z}]$ are the spatial angular rates of the navigation frame with respect to inertial space which is the sum of the angular rate of the navigation frame with respect to the earth($\rho_{x,y,z}$) and the earth with respect to inertial space, ($\Omega_{x,y,z}$)

$\delta\omega_{x,y,z} = \delta\rho_{x,y,z} + \delta\Omega_{x,y,z}$ are the errors in the system computed spatial angular rates

$\delta d = \begin{bmatrix} \delta d_x \\ \delta d_y \\ \delta d_z \end{bmatrix}_s$ is the generalized error in angular rate measured by the gyros expressed in the inertial sensor reference frame

APPENDIX II APPROXIMATION OF THE INTEGRAL OF THE TRANSFORMATION ${}_I T_s$

The transformation between the orthogonal inertial sensor assembly reference frame and an inertially-fixed frame can be written at any point in time as:

$$ {}_I T_s = {}_I T_B \; {}_B T_s \tag{1}$$

where:

${}_I T_B$ is the time-varying transformation from the rotating vehicle body frame (B) to an inertially-fixed frame (I)

${}_B T_s$ is the time-varying transformation from the inertial sensor assembly reference frame(s) to the body frame.

The time rate of change of the transformation ${}_I T_B$ can be written:

$$ {}_I\dot{T}_B = {}_I T_B \, [\omega] \tag{2} $$

where:

$$ [\omega] = \begin{bmatrix} 0 & -\omega_z & \omega_y \\ \omega_z & 0 & -\omega_x \\ -\omega_y & \omega_x & 0 \end{bmatrix}_B \tag{3} $$

and ω_{x,y,z_B} are the angular rates of the vehicle with respect to inertial space expressed in the body axes, assuming the right hand rule.
For small angular change about the body axes over time, t:

$$ \alpha_{x,y,z} \triangleq \int_0^t \omega_{x,y,z_B} \, dt \quad << 1 \tag{4} $$

then ${}_I T_B$ at a later point in time may be written:

$$ {}_I T_B (\tau + t) \approx {}_I T_B(\tau) \, [I + [\alpha]] \tag{5} $$

where:

$$ I + [\alpha] \approx \begin{bmatrix} 1 & -\alpha_z & \alpha_y \\ \alpha_z & 1 & -\alpha_x \\ -\alpha_y & \alpha_x & 1 \end{bmatrix} $$

The transformation ${}_B T_S$ at a later point in time may be written:

$$ {}_{B}T_{S}(\tau+t) = {}_{B}T_{S}\,(\zeta) = \begin{bmatrix} \cos(\zeta) & -\sin(\zeta) & 0 \\ \sin(\zeta) & \cos(\zeta) & 0 \\ 0 & 0 & 1 \end{bmatrix} \tag{6} $$

where:

$$ \zeta = \int_{\tau}^{\tau+t} [\omega_R - \omega_{z_B}]\, dt \qquad \text{and} \qquad \zeta = 0 \text{ at } t = 0 $$

is the angle from the body axes to the inertial sensor assembly reference axes measured about the z_B axis. Note ω_R is an angular control rate with respect to inertial space which is applied about the z_B axis such that the inertial sensor assembly moves at an essentially constant inertial rate (plus and minus) through complete $\pm\, 2\pi$ rotations with respect to inertial space about the z_B axis. Hence the inertial sensor assembly is isolated from the body rate ω_{z_B} but does experience in addition to ω_R, the body angular rates ω_{x,y_B} about the $[x,y]_B$ axes respectively, – all these rates being with respect to inertial space. Combining the above, then, ${}_{I}T_{S}$ at a later point in time may be approximated as:

$$ {}_{I}T_{S}\,(\tau+t) \approx {}_{I}T_{B}\,(\tau)\, [I + [\alpha]]\; {}_{B}T_{S}(\zeta) \tag{7} $$

where:

$$ {}_{B_\tau}T_{S}\,(t) \triangleq [I + [\alpha]]\; {}_{B}T_{S}\,(\zeta) = \begin{bmatrix} 1 & -\alpha_z & \alpha_y \\ \alpha_z & 1 & -\alpha_y \\ -\alpha_y & \alpha_x & 1 \end{bmatrix} \begin{bmatrix} \cos(\zeta) & -\sin(\zeta) & 0 \\ \sin(\zeta) & \cos(\zeta) & 0 \\ 0 & 0 & 1 \end{bmatrix} \tag{8} $$

$$ \approx \begin{bmatrix} \cos(\nu) & -\sin(\nu) & \alpha_y \\ \sin(\nu) & \cos(\nu) & -\alpha_x \\ [\alpha_x \sin(\nu) - \alpha_y \cos(\nu)] & [\ \alpha_y \sin(\nu) + \alpha_x \cos(\nu)] & 1 \end{bmatrix} $$

and:

$$\nu = \int_{\tau}^{\tau+t} \omega_R \, dt \quad \text{and} \quad \nu=0 \quad \text{at} \quad t=0 \tag{9}$$

is the change in the orientation of the inertial sensor assembly with respect to inertial space about the axis z_B, which is rotating at ω_{x,y_B} respectively about x,y_B. Here, small angle approximations for α_z have been used to obtain:

$$\begin{aligned} \cos(\nu) &\triangleq \cos(\zeta+\alpha_z) \approx \cos(\zeta) - \alpha_z \sin(\zeta) \\ \sin(\nu) &\triangleq \sin(\zeta+\alpha_z) \approx \sin(\zeta) + \alpha_z \cos(\zeta) \end{aligned} \tag{10}$$

The integral of the transformation ${}_IT_S$ over a time period $2\Delta\tau$, wherein the inertial sensor assembly is first rotated thru a + 2π change in ν via ω_R with respect to inertial space about z_B in the subsequent time increment $\Delta\tau$, can now be written using (7,8) as:

$$\int_{\tau}^{\tau+2\Delta\tau} {}_IT_S\,(\tau+t)\, dt = {}_IT_B\,(\tau) \int_{0}^{2\Delta\tau} {}_BT_S^{\tau}\,(t)\, dt \tag{11}$$

where:

$$\Delta\tau \triangleq \frac{2\pi}{|\omega_R|} \tag{12}$$

ω_{x,y_B} is assumed constant over $[0,\ 2\Delta\tau]$

The full expression for the integral in (11) is:

$$I_1 = \int_0^{2\Delta\tau} dt \begin{bmatrix} \cos(\omega_R t) & -\sin(\omega_R t) & \omega_{y_B}.t \\ \sin(\omega_R t) & \cos(\omega_R t) & -\omega_{x_B}.t \\ [\omega_{x_B} \cdot \sin(\omega_R t) - \omega_{y_B} \cdot \cos(\omega_R t)].t & [\omega_{y_B} \cdot \sin(\omega_R t) + \omega_{x_B} \cdot \cos(\omega_R t)].t & 1 \end{bmatrix} \tag{13}$$

The result of the integration (13) is:

$$I_2 = \begin{bmatrix} \frac{\sin(\omega_R t)}{\omega_R} & \frac{\cos(\omega_R t)}{\omega_R} & .5\omega_{y_B}.t^2 \\ -\frac{\cos(\omega_R t)}{\omega_R} & \frac{\sin(\omega_R t)}{\omega_R} & -.5\omega_{x_B}.t^2 \\ [\omega_{x_B}.\ I_s - \omega_{y_B}\ .\ I_c] & [\omega_{y_B}\ .I_s + \omega_{x_B}\ .\ I_c] & t \end{bmatrix} \quad (14)$$

where:

$$I_s = \frac{\sin(\omega_R t) - (\omega_R t)\ .\ \cos(\omega_R t)}{\omega_R^2}$$

$$I_c = \frac{\cos(\omega_R t) + (\omega_R t)\ .\ \sin(\omega_R t)}{\omega_R^2} \quad (15)$$

Due to the reversal of the sign of ω_R at $\Delta\tau$, (14) is evaluated as:

$$I_1 = I_2 \Big[_{0,\ \omega_R>0}^{\Delta\tau} + I_2 \Big[_{\Delta\tau,\ \omega_R<0}^{2\Delta\tau} \triangleq I_3 + I_4 \quad (16)$$

Noting:

$$I_s \Big[_0^{\Delta\tau} = \frac{-2\pi}{\omega_R^2} = \frac{-\Delta\tau}{|\omega_R|} = \frac{-(\Delta\tau)^2}{2\pi} = -I_s \Big[_{\Delta\tau}^{2\Delta\tau}$$

$$I_c \Big[_0^{\Delta\tau} = I_c \Big[_{\Delta\tau}^{2\Delta\tau} = 0 \quad (17)$$

then:

$$I_3 = \begin{bmatrix} 0 & 0 & .5\omega_{y_B}(\Delta\tau)^2 \\ 0 & 0 & -.5\omega_{x_B}(\Delta\tau)^2 \\ \omega_{x_B}\left(\frac{-2\pi}{\omega_R^2}\right) & \omega_{y_B}\left(\frac{-2\pi}{\omega_R^2}\right) & \Delta\tau \end{bmatrix} \quad (18)$$

$$I_4 = \begin{bmatrix} 0 & 0 & 1.5\omega_{y_B}(\Delta\tau)^2 \\ 0 & 0 & -1.5\omega_{x_B}(\Delta\tau)^2 \\ \omega_{x_B}\left(\frac{2\pi}{\omega_R^2}\right) & \omega_{y_B}\left(\frac{2\pi}{\omega_R^2}\right) & \Delta\tau \end{bmatrix} \quad (19)$$

Consequently summing (18, 19), the result of the integration (11), of ${}_I T_S$ over sequential $\pm$ 2π rotations about z_B with respect to inertial space is:

$$I_1 = \int_{\tau}^{\tau+2\Delta\tau} {}_I T_S\,(\tau+t)\,dt = {}_I T_B\,(\tau) \begin{bmatrix} 0 & 0 & 2\omega_{y_B}(\Delta\tau)^2 \\ 0 & 0 & -2\omega_{x_B}(\Delta\tau)^2 \\ 0 & 0 & 2\Delta\tau \end{bmatrix} \quad (20)$$

The average of the integral (20) over the integration period is then:

$$\left[- \frac{I_1}{2\Delta\tau} \right] = {}_I T_B\,(\tau) \begin{bmatrix} 0 & 0 & \omega_{y_B}\cdot\Delta\tau \\ 0 & 0 & -\omega_{x_B}\cdot\Delta\tau \\ 0 & 0 & 1 \end{bmatrix} \quad (21)$$

APPENDIX III EVALUATION OF AN INTEGRAL

In this appendix the integral form (20) of the text (excluding ω_R) is evaluated for the effect of the body rate components, ω_{x,y_B}. The simplified form of the integrand of interest is:

$$
{}_5 = \begin{bmatrix} \cos(\omega_R t) & -\sin(\omega_R t) & \omega_{y_B}.t \\ \sin(\omega_R t) & \cos(\omega_R t) & -\omega_{x_B}.t \\ [\omega_{x_B}\sin(\omega_R t) - \omega_{y_B}\cos(\omega_R t)].t & [\omega_{y_B}\sin(\omega_R t) + \omega_{x_B}\cos(\omega_R t)].t & 1 \end{bmatrix} \begin{bmatrix} \delta D_{xx}[\omega_{x_B}\cos(\omega_R t) + \omega_{y_B}\sin(\omega_R t)] \\ + \delta D_{xy}[-\omega_{x_B}\sin(\omega_R t) + \omega_{y_B}\cos(\omega_R t)] \\ \delta D_{yx}[\omega_{x_B}\cos(\omega_R t) + \omega_{y_B}\sin(\omega_R t)] \\ + \delta D_{yy}[-\omega_{x_B}\sin(\omega_R t) + \omega_{y_B}\cos(\omega_R t)] \\ \delta D_{zx}[\omega_{x_B}\cos(\omega_R t) + \omega_{y_B}\sin(\omega_R t)] \\ + \delta D_{zy}[-\omega_{x_B}\sin(\omega_R t) + \omega_{y_B}\cos(\omega_R t)] \end{bmatrix} \qquad (1)
$$

Here the fact that products of $(\omega_{z_B}.t)$ and elements of $[\delta D]$ are second-order has been used to simplify the terms containing $\sin(\zeta)$, $\cos(\zeta)$. See (9,10) of Appendix II.

The following terms are evaluated:

$$\int_0^{\Delta\tau} \cos^2(\omega_R t)\, dt = \int_{\Delta\tau}^{2\Delta\tau} \cos^2(\omega_R t)\, dt = .5\Delta\tau$$

$$\int_0^{\Delta\tau} \sin^2(\omega_R t)\, dt = \int_{\Delta\tau}^{2\Delta\tau} \sin^2(\omega_R t)\, dt = .5\Delta\tau$$

$$\int_0^{\Delta\tau} \sin^2(\omega_R t)\cos(\omega_R t)\, dt = \int_{\Delta\tau}^{2\Delta\tau} \sin^2(\omega_R t)\cos(\omega_R t)\, dt = 0 \qquad (2)$$

$$\int_{\Delta\tau}^{2\Delta\tau} t.\cos^2(\omega_R t)\, dt = 3\int_0^{\Delta\tau} t.\cos^2(\omega_R t)\, dt = \frac{3(\Delta\tau)^2}{4}$$

$$\int_{\Delta\tau}^{2\,\Delta} t.\sin^2(\omega_R t)\, dt = 3\int_0^{\Delta\tau} t.\sin^2(\omega_R t)\, dt = \frac{3(\Delta\tau)^2}{4}$$

Here:

$$\Delta\tau = \frac{2\pi}{|\omega_R|} \tag{3}$$

$$\omega_R = \begin{cases} |\omega_R| \; , & t \in [0, \Delta\tau] \\ -|\omega_R| \; , & t \in [\Delta \; , \; 2\Delta\tau] \end{cases}$$

Note the results (2) have no dependence on the sign of ω_R.

Integral results for terms of the form:

$$\int_{0,\omega_R > 0}^{\Delta\tau} \sin(\omega_R t)\cos(\omega_R t).t \; dt = -\int_{\Delta\tau,\; \omega_R > 0}^{2\Delta\tau} \sin(\omega_R t)\cos(\omega_R t).t \; dt = -\frac{(\Delta\tau)^2}{8\pi} \tag{4}$$

have a dependence on the sign of ω_R.

Further from Appendix II, recall:

$$\begin{aligned} I_c\Big[_0^{\Delta\tau} &= I_c\Big[_{\Delta\tau}^{2\Delta\tau} = 0 \\ I_s\Big[_{0,\omega_R > 0}^{\Delta\tau} &= -I_s\Big[_{\Delta\tau,\omega_R < 0}^{2\Delta\tau} = -\frac{(\Delta\tau)^2}{2\pi} \end{aligned} \tag{5}$$

where for I_s there is a dependence on the sign of ω_R.

The integral of (1) is evaluated in two parts as:

$$I_8 = \int_0^{2\Delta\tau} I_5 \; dt = \int_0^{\Delta\tau} I_5 \; dt + \int_{\Delta\tau}^{2\Delta\tau} I_5 \; dt \triangleq I_6 + I_7 \tag{6}$$

Using the results (2,3,4) with inspection of (1) for non-zero products yields:

$$I_6 = \left[\begin{array}{l} (\omega_{x_B}\cdot\Delta\tau)\cdot\left[\dfrac{\delta D_{xx}+\delta D_{yy}}{2}\right]+(\omega_{y_B}\cdot\Delta\tau)\cdot\left[\dfrac{\delta D_{xy}-\delta D_{yx}}{2}\right] \\ -(\omega_{y_B}\cdot\Delta\tau)^2\cdot\left[\dfrac{\delta D_{zx}}{2}\right]+\omega_{x_B}\cdot\omega_{y_B}(\Delta\tau)^2\cdot\left[\dfrac{\delta D_{zy}}{2\pi}\right] \\ \hline (\omega_{y_B}\cdot\Delta\tau)\cdot\left[\dfrac{\delta D_{yy}+\delta D_{xx}}{2}\right]+(\omega_{x_B}\cdot\Delta\tau)\cdot\left[\dfrac{\delta D_{yx}-\delta D_{xy}}{2}\right] \\ -(\omega_{x_B}\cdot\Delta\tau)^2\cdot\left[\dfrac{\delta D_{zy}}{2\pi}\right]+\omega_{x_B}\cdot\omega_{y_B}\cdot(\Delta\tau)^2\cdot\left[\dfrac{\delta D_{zx}}{2\pi}\right] \\ \hline \dfrac{(\Delta\tau)^2}{4}\left[-\left[\dfrac{\omega_{x_B}^2-\omega_{y_B}^2}{2\pi}\right][\delta D_{xx}-\delta D_{yy}]+(\omega_{x_B}^2+\omega_{y_B}^2)[\delta D_{yx}-\delta D_{xy}]\right. \\ \left.-\left[\dfrac{\omega_{x_B}\cdot\omega_{y_B}}{\pi}\right][\delta D_{xy}+\delta D_{yx}]\right] \end{array}\right] \tag{7}$$

where it is assumed ω_{x,y_B} are constant over $2\Delta\tau$.

and:

$$I_7 = \left[\begin{array}{l} (\omega_{x_B} \cdot \Delta\tau) \cdot \left[\dfrac{\delta D_{xx} + \delta D_{yy}}{2} \right] + (\omega_{y_B} \cdot \Delta\tau) \cdot \left[\dfrac{\delta D_{xy} - \delta D_{yx}}{2} \right] \\ + (\omega_{y_B} \cdot \Delta\tau)^2 \cdot \left[\dfrac{\delta D_{zx}}{2\pi} \right] - \omega_{x_B} \cdot \omega_{y_B} (\Delta\tau)^2 \cdot \left[\dfrac{\delta D_{zy}}{2\pi} \right] \\ \hline (\omega_{y_B} \cdot \Delta\tau) \left[\dfrac{\delta D_{yy} + \delta D_{xx}}{2} \right] + (\omega_{x_B} \cdot \Delta\tau) \left[\dfrac{\delta D_{yx} - \delta D_{xy}}{2} \right] \\ + (\omega_{x_B} \cdot \Delta\tau)^2 \left[\dfrac{\delta D_{zy}}{2\pi} \right] - \omega_{x_B} \cdot \omega_{y_B} (\Delta\tau)^2 \left[\dfrac{\delta D_{zx}}{2\pi} \right] \\ \hline \dfrac{(\Delta\tau)^2}{4} \left[\dfrac{\omega_{x_B}^2 - \omega_{y_B}^2}{2\pi} \right] [\delta D_{xx} - \delta D_{yy}] + 3(\omega_{x_B}^2 + \omega_{y_B}^2) [\delta D_{yx} - \delta D_{xy}] \\ \quad + \left[\dfrac{\omega_{x_B} \cdot \omega_{y_B}}{\pi} \right] [\delta D_{xy} + \delta D_{yx}] \end{array} \right] \tag{8}$$

The sum of the results (7,8) is:

$$I_8 = [2\Delta\tau] \cdot \left[\begin{array}{l} \omega_{x_B} \cdot \left[\dfrac{\delta D_{xx} + \delta D_{yy}}{2} \right] + \omega_{y_B} \cdot \left[\dfrac{\delta D_{xy} - \delta D_{yx}}{2} \right] \\ \omega_{y_B} \cdot \left[\dfrac{\delta D_{xx} + \delta D_{yy}}{2} \right] - \omega_{x_B} \cdot \left[\dfrac{\delta D_{xy} - \delta D_{yx}}{2} \right] \\ \Delta\tau[\omega_{x_B}^2 + \omega_{y_B}^2] \cdot \left[\dfrac{\delta D_{yx} - \delta D_{xy}}{2} \right] \end{array} \right] \tag{9}$$

The third component is second-order for prior assumptions (11) of the text.

Some Experiences with the LTN-72 Aircraft Platform in Terrestrial Applications

Dieter Keller
Institut für Physikalische Geodäsie, Technische Hochschule Darmstadt

Abstract

The LTN-72 aircraft platform was tested on two courses of the Odenwald testarea, with the inertial system being mounted in a Mercedes Unimog van. The experiences gained with the INS (Inertial Navigation System) data-registration and -processing, rather than in the evaluation of the analog vertical acceleration signal of the platform, are presented. Furthermore the results of the post-processing of two missions on the same profile are compared.

1. Introduction

To test the application of a classical platform of medium accuracy requirements with the aid of modern computer possibilities, the LTN-72 platform was used in the Rheingraben/Odenwald testarea. The results presented in this paper are based on the N-S profile beginning at Darmstadt, leading through the Odenwald area and ending at Hirschhorn/Neckar (Fig. 1).

The INS missions on this profile typically lasted 4 ½ hours and employed 1.5 to 2 minutes ZUPT-registration. Consequently the intervals between the approximately evenly spaced ZUPTs range from 5 to 6 minutes.

2. Data Recording and Interface Problems

According to the contract of loan the only possibility of recording data from the LTN-72 platform is to use the ARINC standard interface (Fig. 2). The ARINC interface of the LTN-72, the ARINC-561 s, is of older type and has a decisive disadvantage in comparison to the new interface types (e.g. ARINC-429): the data of this interface are not updated with constant update rates.

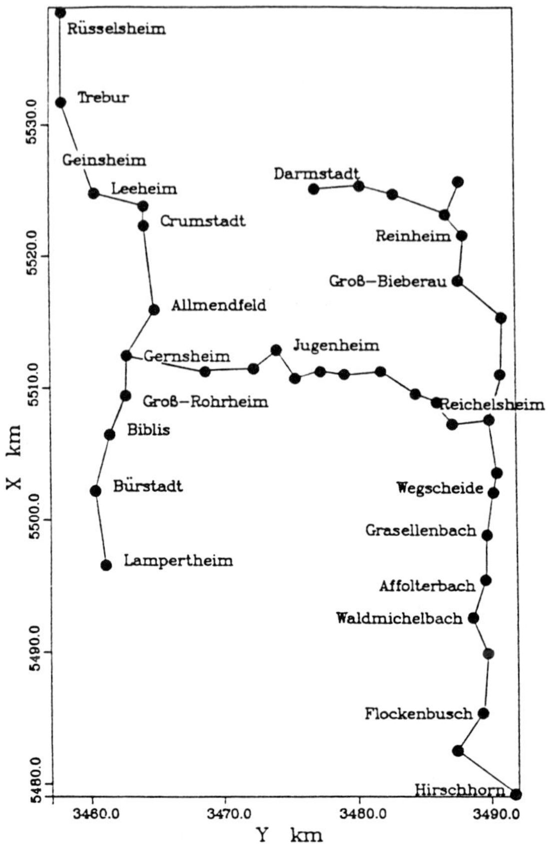

Fig. 1 Rheingraben/Odenwald testarea

Using the ARINC interface, the horizontal velocities v_N, v_E, the ellipsoidal positions φ, λ and the azimuth α were recorded by a HP 9320 computer system. In addition to this analog vertical acceleration data was made accessible by a Flucke 8840A A/D converter (Fig. 2).

At the very beginning of our LTN-72 testseries, a couple of problems arose which were related to the online recording of data on ¼ " magnetic tapes. In spite of relatively low data rates - about 60 values plus timecount per second (~500 bytes/s) - the ¼ " tapes proved to be unsuitable. Due to the long lasting campaigns (4 ½ hours), the asynchronous behaviour of the ARINC-

561 s bus and the computer system design constraints, the tapes had to registrate the data in start/stop mode. In this operation mode the tape was to slow, and this caused loss of data. Only for a comparison of data rates it should be mentioned here that the output data rate for geodetic relevant quantities (accelerations, angular rates, time, temperature etc.) of the Honeywell LINS is about ten times higher. The situation completely changed after the purchase of a 2×20 MB Bering 5840 Bernoulli removable cartridge disk.

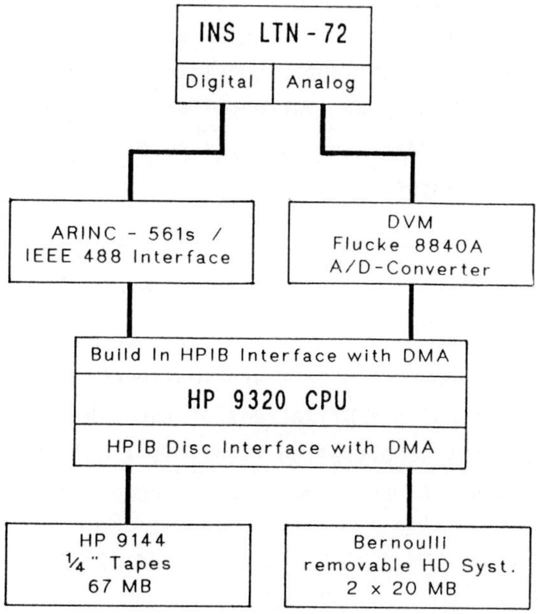

Fig. 2 Sketch of data recording equipement

Since the LTN-72 is originally designed as an aircraft platform, ZUPTs (zero velocity updates) have to be implemented in such a way that ZUPT data can be separated from the data in case the vehicle is moving. In the initial registration program the separation was implemented by using standard interrupt procedures provided by the computer operation system. ZUPTs were marked by striking different keys on the keyboard which caused a system interrupt, which wrote a mark on the data storage medium. The secondary effects arising with this simple procedure are wrong marking of ZUPTs caused by the operators, and disturbances in the output buffer to the hard disc due to the interrupt priority. These effects cost a lot of time in the subsequent data processing. The problem could be solved by writing own interrupt routines beside the operating system interrupts. In these routines you mainly have to manage the interrupt chaining and the interrupt priorities. The interrupt handling also plays an important role when GPS data would be combined online with INS data.

3. Analysis of the Analog Vertical Acceleration Signal

The analog vertical acceleration signal is originated from a branch just before the quantizers of the Litton platform. The signal was prefiltered by a passive RC-filter of first order. The computer recorded the digital data coming out of the A/D converter.

3.1 Manual Specifications

To understand better the accuracies that are mentioned in the following chapters, we look into the manual specifications of the LTN-72 system. The accuracy of the analog vertical acceleration signal is 1 gal, the resolution is 100 mgal. The threshold of the built in A2 accelerometers is 1 mgal (Litton Aero Products, 1976).

3.2 Hardware Restrictions and Filter Problems

According to the incoming vertical acceleration signal we had to switch the A/D converter's maximal signal range to 2V. With a maximal sampling rate of 16.67 Hz and an accuracy of 5 ½ digits the resolution is 10 μV. After applying a scale factor the resolution of 10 μV is adequate to 1.635 mgal which is a more convenient quantity to geodesists.

Generally in the A/D conversion we have to take into consideration the famous Nyquist/Shammon theorem. This theorem states that the original signal only can be recovered from digitized data when the signal frequency is smaller than half the sampling frequency (Best, 1982).

A practical consequence of the theorem is to prefilter the analog signal in order to avoid the superposition of noise effects with the effective signal. In our case the effective signal is found between 0 - 2.5 Hz. To reduce the noise effects of 14.17-16.67 Hz, we used a passive RC low pass filter of 1st order. The damping of such filter types amounts to ~14 dB at $5\omega_o$ (ω_o = cutoff angular frequency). With a cutoff frequency of 3 Hz and a constant signal amplitude of 0.06 gal the noise effects at 15 Hz reach 12 mgal. The efficient signal is superposed by these noise amplitudes.

In Fig. 3 the results of a LTN-72 gravity measurement are compared with the gravimetrically determined ZUPT points. The INS gravity was obtained by analog prefiltering the vertical acceleration signal with a passive RC low pass filter mentioned above. After A/D conversion, the acceleration signal was filtered digitally with a Butterworth low pass filter of 6th order and cutoff frequency of 2.5 Hz. Finally the signal was averaged over the whole ZUPT interval (forward and backward run of the traverse) and due to a fixed accelerometer bias an offset is added which minimizes the differences between the INS gravity measurements and the gravity values of the ZUPT points. It is readily seen in Fig. 3 that the more theoretical considerations of the preceding chapter are confirmed. Under static environmental conditions of the gravity measurements at ZUPTs the main error source is the inadequate analog filter design. In dynamic applications, we have to take into account additionally the vibratory environment. After applying gravity-, Coriolis- and accelerometer drift corrections, we obtain INS-heights by double integration of the corrected vertical acceleration. Fig. 4 demonstrates the effects of error accumulation in a vibratory environment. The INS-heights between two ZUPT stations are compared to those provided by our program system SCOP, in which linear prediction is used for height interpolation (for further details see Kling et al., 1987). Due to the selected analog filter and the vibratory environmental conditions, the height difference increases rapidly until finally reaching ~12 meters, when arriving at the next ZUPT.

To end up with better results as shown in Fig. 3 and Fig. 4, we should increase the resolution of the A/D converter up to 6 ½ digits even in the case of higher data rates. The passive RC-filter of first order should be replaced by an analog Butterworth low pass filter of 4th or 6th order.

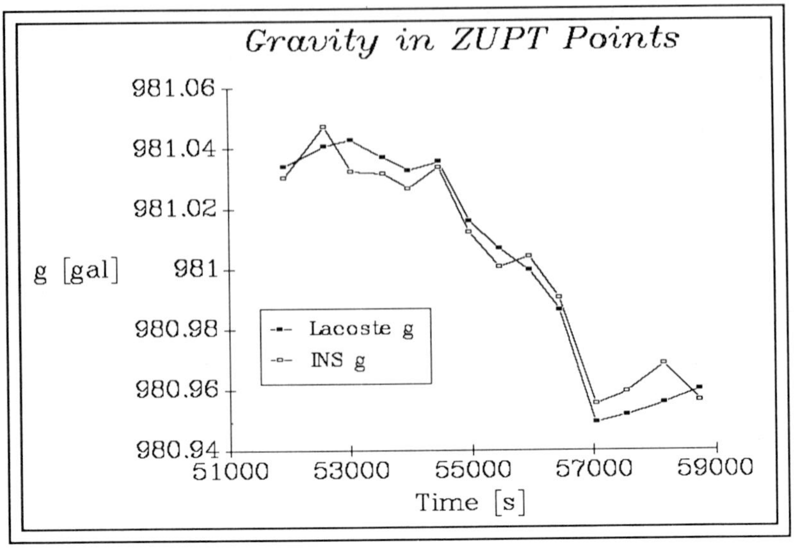

Fig. 3 INS gravity versus Lacoste-Romberg measured gravity

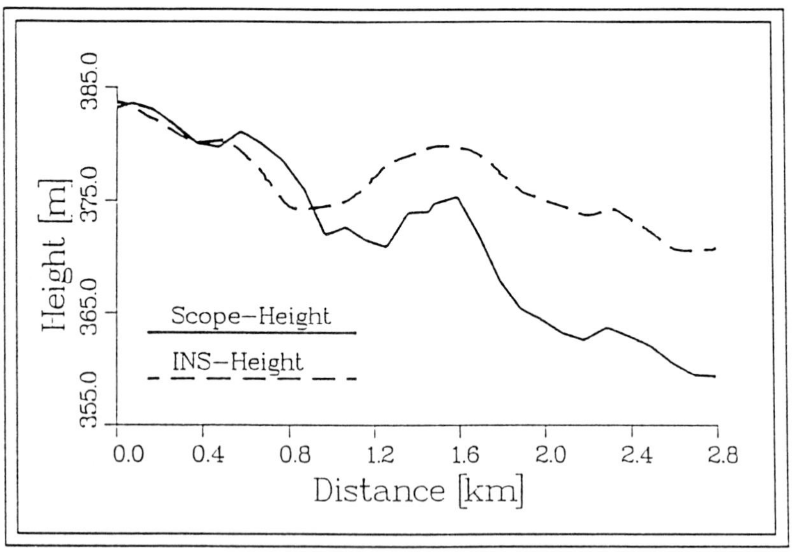

Fig. 4 Measured INS height versus SCOP interpolated height

4. Analysis of Horizontal ARINC 561 s Data

For the post processing we adopted the classical two step modell (Boedecker, 1987). Despite of small variations the first step is analogous to the Ferranti (FILS2/FILS3) internal post processing (Cross, 1985). In the Ferranti system a curve is fitted to sets of sixteen (thirty - two in FILS2) velocity measurement at ZUPTs for each of the three orthogonal accelerometers. For the LTN-72 system it has been found best to take a maximum of three velocity measurements, depending on the frequency of the steps of the digital velocity output. A cubic spline was used to fit the consecutive velocity registrations at ZUPT points. Analytical integration of the splines yields the position corrections which are added to the uncorrected online positions or the integrated online velocities of the LTN-72 ARINC output.

In the second step of the post-processing an adjustment procedure with a Gauss-Markov error modell, as first proposed by (Hannah, 1982) and extended by (Boedecker, 1987), was applied. The error equations of the INS coordinates are written in the following form

$$\varphi_i^i + v_{\varphi i} = \varphi_i + a_0 + a_1 (\varphi_i^0 - \varphi_1^0) + a_2 (\lambda_i^0 - \lambda_1^0) + a_3 \sum_{k=1}^{i-1} t_k (\lambda_{k+1}^0 - \lambda_k^0) + a_4 \sum_{k=1}^{i-1} (t_{k+1} - t_k)^2$$

$$\lambda_i^i + v_{\varphi i} = \lambda_i + b_0 + b_1 (\varphi_i^0 - \varphi_1^0) + b_2 (\lambda_i^0 - \lambda_1^0) + b_3 \sum_{k=1}^{i-1} t_k (\varphi_{k+1}^0 - \varphi_k^0) + b_4 \sum_{k=1}^{i-1} (t_{k+1} - t_k)^2$$

with

φ^0, λ^0 : approximated coordinates
$v_{\varphi i}, v_{\lambda i}$: residuals
φ^i, λ^i : inertial coordinates

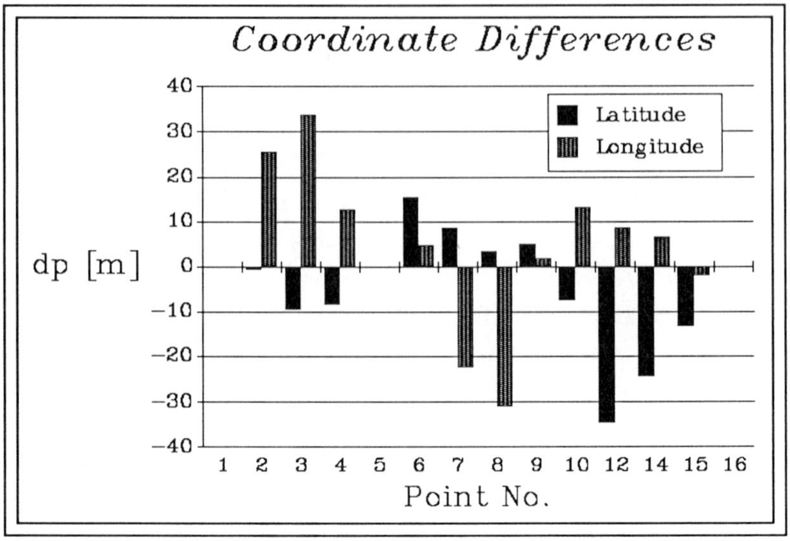

Fig. 5 Coordinate differences of INS- and geodetic-coordinates of the first mission based on ARINC velocity data

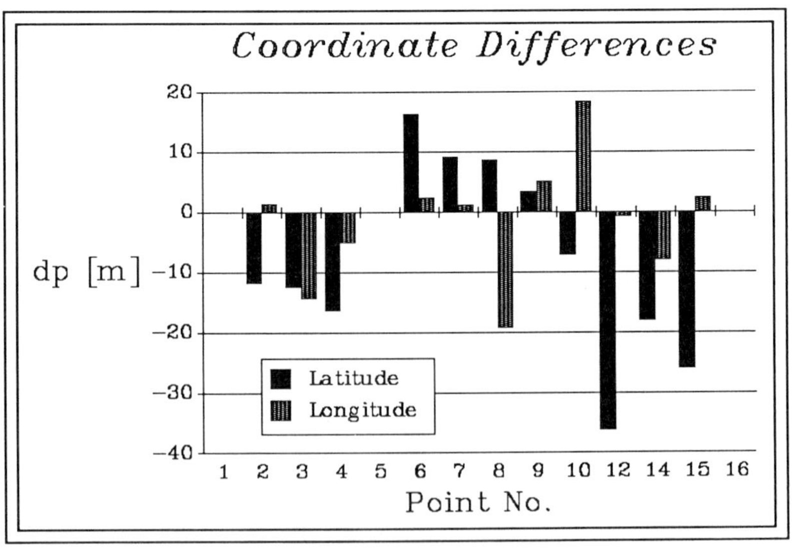

Fig. 6 Coordinate differences of INS- and geodetic-coordinates of the first mission based on ARINC position data

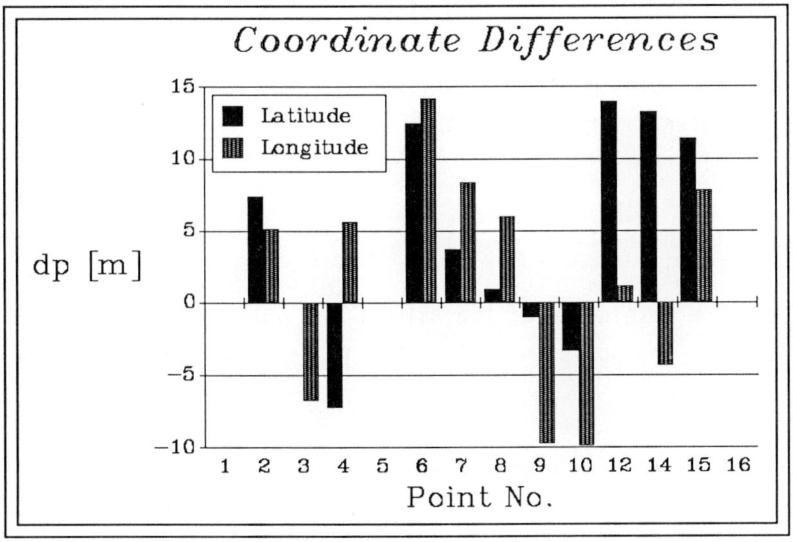

Fig. 7 Coordinate differences of INS- and geodetic-coordinates of the second mission based on ARINC velocity data

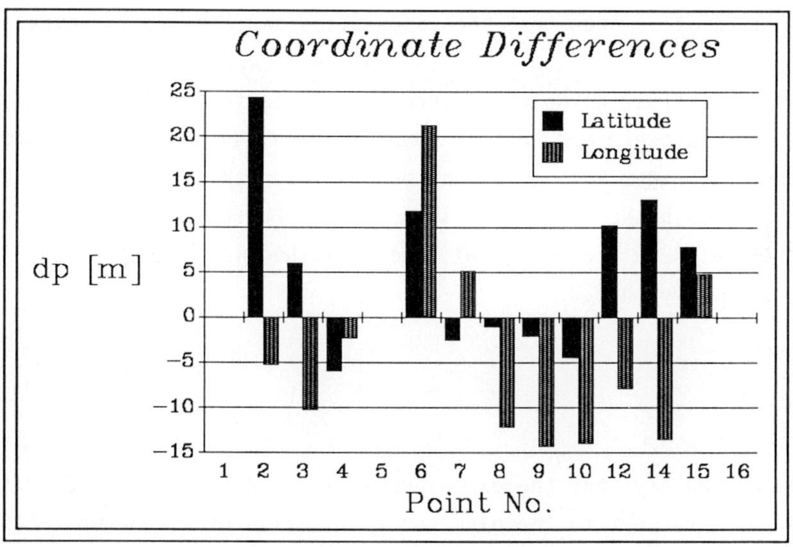

Fig. 8 Coordinate differences of INS- and geodetic-coordinates of the second mission based on ARINC position data

and the unknown parameters associated with

a_o, b_o : offset of trajectory starting point
a_1, b_2 : accelerometer scale factor
a_4, b_4 : accelerometer drift
a_2, b_1 : azimuth bias
a_3, b_3 : azimuth drift.

Because of the low resolution of the ARINC-561 s data - 5 cm/s for the velocities and 76 m for the positions - it seems to be absurd to use a more refined adjustment modell.

Fig. 5-8 shows the coordinate difference between the LTN-72 coordinates and the geodetic control coordinates. In Fig. 5 and Fig. 7 the ARINC velocity data are used for INS point determination. For the other two figures (Fig. 6, Fig. 8) the onboard position data is used for evaluation. In addition to the traverse end points, point no. 5 is kept fix in the adjustment. In principle, it can be thought as a position update. Introducing more position updates (e.g. from GPS) would of course improve the results and could reduce the coordinate differences below the 10 m level. Comparing Fig. 5 and Fig. 6 with Fig.7 and Fig. 8 it is seen that the second mission (Fig. 7, 8) reveals much better results. This is due to the fact that some improvements in the registration program (better keyboard interrupt routines to mark ZUPTs etc.) and measurement arrangement (e.g. better communication car-driver - operator) between the two campaigns were made. In the best case an accuracy of point determination within the 15 m level is achieved. Further results of the LTN-72 horizontal data evaluation will be published in the next report of the Sonderforschungsbereich 228.

Conclusion

Essentially we can say that the experiences gained with the LTN-72 system are very useful with respect to the expected application of a Honeywell LINS strapdown system in future. Of course, the results presented in this paper do not have geodetic accuracies but valuable hints are obtained especially concerning a registration software package that would satisfy geodetic demands. An urgent need to improve the performance of a medium accuracy system would be to get the system's raw data, in order to apply the various refined post processing methods.

References

Best, R.: Handbuch der analogen und digitalen Filterungstechnik, AT-Verlag, Aarau/Schweiz, 1982

Boedecker, G: Zur Ausgleichung von Inertialnetzen, ZfV, No. 1, 1987

Cross, P.A., P. Harrison: First Results from FILS3 Trials on the Edinburgh Inertial Test Network, Proc. of the Third Intern. Symp. on Inertial Technology for Surveying and Geodesy, Banff, 1985

Hannah, J.: The Development of Comprehensive Error Models and Network Adjustment Techniques for Inertial Surveys, Reports, Dep. Geod. Science, Ohio State Univ., No. 330, 1982

Kling, Th., M. Becker, H.-J. Euler, E. Groten: Studien zur detaillierten Geoidberechnung, DGK Reihe B, No. 285, München 1987

Kulhánek, O.: Introduction to Digital Filtering in Geophysics, Elsevier Scientific Publishing Company, New York, 1976

Lacroix, A: Digitale Filter, Oldenbourg, München, Wien, 1980

Litton Aero Products, Product Description Litton LTN-72: Inertial Navigation System for Transport Aircraft, Woodland Hills/California , 1976

Separation of Gravitational and Inertial Accelerations with a Combined Inertial Navigation and Gravity-Gradiometer System

Thomas Kling
Institut für Physikalische Geodäsie, Technische Hochschule Darmstadt

Abstract:
With the advent of high-precision gravity-gradiometers it seems to be necessary to reconsider whether it is advantageous to combine inertial navigation with gradiometer systems. For such a combined accelerometer-gradiometer system the equations of motion are presented. Especially the online integration of the gravitational-gradients has been investigated. For the Odenwald test-profile a numerical simulation study was carried out.

1. Introduction:

The application of Inertial Navigation Systems (INS) in geodesy is twofold: (1) if the gravitational vector is known along the profiled traverse, it is possible to determine the positions on this profile without assumptions and (2) with known positions the determination of the gravitational vector is possible. Usually an iterative algorithm is used to derive both, positions and the gravitational vectors. It was shown by Moritz (1967) that the second derivatives of the gravitational potential are not perturbed by inertial effects as long as the measuring system (gravity gradiometer) is inertially stabilized. This feature of the second derivatives can be exploited to determine independently the gravitation vector along the profile by combining the output of the Inertial Navigation System with the gradiometer measurements.

2. Basic formulae:

As described in (Britting, 1971) the specific force vector in the geographic frame (n) is of the form:

$$f^n = \dot{v}^n + (\Omega^n_{en} + 2\,\Omega^n_{ie})\, v^n + \Omega^n_{ie}\, \Omega^n_{ie}\, r^n + g^n \tag{1}$$

with: $f^n = [\ f_N\ f_E\ f_D\ ]^T$ specific force vector

g^n - vector of gravitational acceleration

Ω^n_{en} - skew symmetric matrix with the components of the angular rotation vector with respect to the earth (transport rate)

Ω^n_{ie} - skew symmetric matrix of the earth's inertially referenced angular velocity (earth rate)

v^n - velocity vector

r^n - position vector

V^n - tensor of gravitational gradients .

Similar to the derivation of this well known formula the gradiometer accelerometer equation was derived by Wang (1987) :

$$\dot{f}^n = \ddot{v}^n + 2\ \Omega^n_{in} \cdot \dot{v}^n + (\dot{\Omega}^n_{in} + \Omega^n_{in}\ \Omega^n_{in} - V^n)\ v^n + \Omega^n_{ne}\ f^n . \tag{2}$$

The transition to the first derivatives of the specific force vector is necessary in order to introduce the gravitational gradients V, because

$$\dot{g}^i(r(t)) = \frac{\delta g^i}{\delta r^i} \cdot \frac{\delta r^i}{\delta t} \tag{3}$$

and

$$V^i(r(t)) = \frac{\delta g^i}{\delta r^i} \quad . \tag{4}$$

By incorporating gravitational gradients in the navigation formula there is (theoretically) no need to know the gravitational acceleration along the traverse. This advantage is connected directly with the disadvantage that a more complex system of differential equations has to be solved. The gradiometer accelerometer formula (2) was derived for the non-inertially stabilized local level case. In case of inertially stabilized system these equations simplify considerably:

$$f^i = \ddot{r}^i - (g^i(t_0) + \int_{t_0}^{t} V^i(t)\ \dot{r}^i(t)\ dt) \quad . \tag{5}$$

This convenient formula was used for the error calculations.

3. The Gravity Gradiometer Survey System:

After several years of development the "Gravity Gradiometer Survey System" (GGSS) is now in the operational test phase. The system was originally designed for the determination of the intermediate wave-lengths (<< 500km) of the gravity field. The system is hosted in a van, by loading the van into an airplane it is possible to use the system for airborne applications. The gradiometer consists of a complete Inertial Measuring Unit (IMU) combined with a set of 12 accelerometers which form three Gravity Gradiometer Instruments (GGI). A GGI basically consists of a rotating disk. Four accelerometers are located mutually orthogonal at this disk (see Fig.1). The three GGI's are installed so that each of them senses the same part of the gravity vector (see Fig.2). This configuration is called the "umbrella" configuration. As mentioned in (Jekeli, 1988) the base of the gradiometer is the IMU (local level type). The IMU is formed by two two-degree-of-freedom Litton G1200 gyros and three orthogonal Bell VII accelerometers. With a system of three gimbals the platform including the three GGI's is kept to (north oriented) local level.

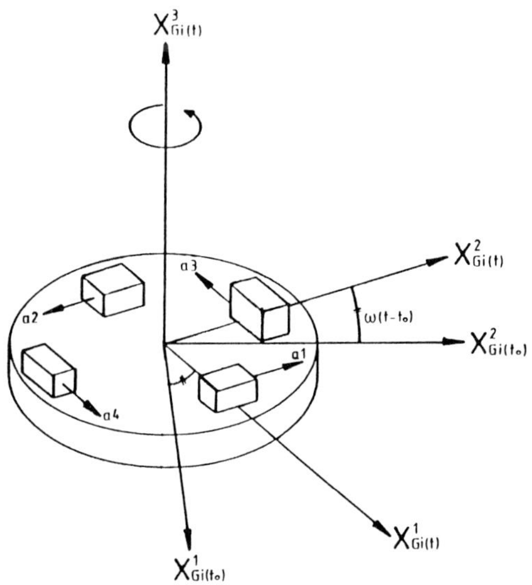

Fig.1: Gravity Gradiometer Instrument

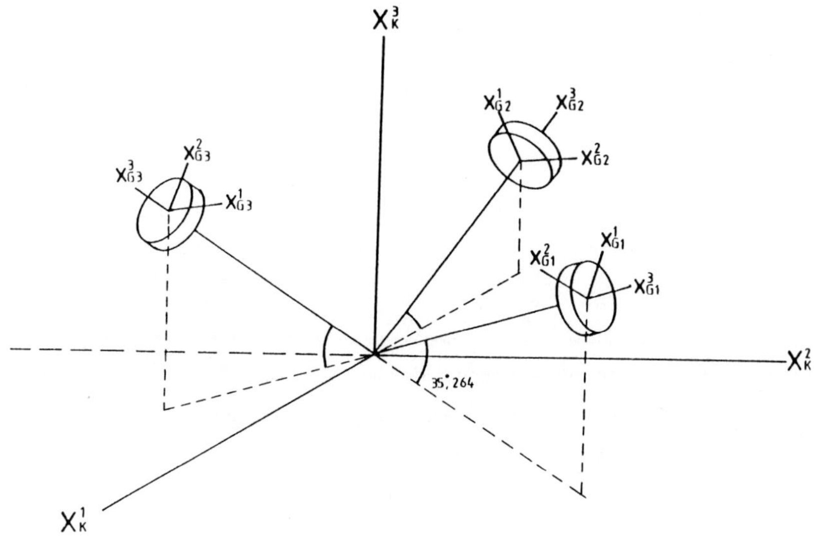

Fig.2: Three GGI's in umbrella configuration

4. Derivation of basic observation equations:

For the simplified model which was described in section 3 the equations of motion should be discussed. In this way a better understanding of the GGI's output signals ("inline" and "cross" components (see Metzger and Pfohl, 1987)) is aimed. In the first step the gravity vector is transformed to the sensor frames of the four rotating accelerometers for each GGI, respectively. Additional to the commonly used coordinate systems, (inertial, equatorial, horizontal and sensor frame see (Schröder et al., 1987)) two further systems have been introduced in the derivation of the observation equations. The first additional frame ($G_i(t_0)$) is non-rotating and located at the GGI platform with the origin at the centre of the rotating wheel of each GGI. The link between the GGI platform frame ($G_i(t_0)$) and the IMU platform system is given by the rotation matrix R_{GiK}, with:

$$\bar{R}_{GiK} = \bar{R}^3(\pi/2-(\alpha+\beta_i))\cdot\bar{R}^1(\pi/2-\gamma_i)\cdot\bar{R}^3(\kappa_i) \qquad (6)$$

- i - number of GGI ($i=1,2,3$)
- j - number of accelerometer at the GGI platform
- $\alpha_i, \beta_i, \gamma_i, \kappa_i$ - angles which determine the spatial orientation of the rotating wheels ($\beta_i = (i-1)\cdot 120°$, $\gamma_i = \cot^{-1}\sqrt{2}$).
- δ_j - orientation of the j'th accelerometer with respect to i'th GGI
- ω_i - rotating frequency of the i'th GGI (e.g. 0.25 Hz)

The rotation matrices $\bar{R}^3, \bar{R}^2, \bar{R}^1$ are applied as usual for right-handed coordinate systems. Furthermore, the connection between the non-rotating systems $G_i(t_0)$ and the rotating systems located at the j'th - accelerometer is described by:

$$\bar{R}_{AjGi} = \bar{R}^3(-\omega_i(t-t_0)) \cdot \bar{R}^3(-\delta_j) \cdot \bar{R}^3(-\pi/2) \tag{7}$$

Here the assumption that the x-components of the rotating systems are aligned to the corresponding accelerometers sensitiv axes was included. With these informations it is possible to derive a model for the GGI output signal. The rotational transformation between the accelerometer (i,j) and the terrestrial system is given by:

$$\bar{R}_{AijT} = \bar{R}_{KT} \cdot \bar{R}_{GiK} \cdot \bar{R}_{AjGi} , \tag{8}$$

where R_{KT} contains the misalignment angles between the platform and the earth fixed topocentric system . We see that the matrices R_{KT} and R_{AjGi} are time dependent in contrast to the matrix R_{GiK} which is installation dependent. The derivation of the observation equations is now performed in two steps. With the assumption that the gravity vector can be approximated linearly between the two corresponding accelerometer positions it can be written as:

$$\bar{g}_{ij} - \bar{g}_{i0} = \begin{vmatrix} W_{xx} & W_{xy} & W_{xz} \\ W_{yx} & W_{yy} & W_{yz} \\ W_{zx} & W_{zy} & W_{zz} \end{vmatrix} \cdot \begin{vmatrix} x_{ij}-x_{i0} \\ y_{ij}-y_{i0} \\ z_{ij}-z_{i0} \end{vmatrix} \tag{9}$$

with: $\bar{g}_{i0}$ - gravity vector at the centre of the i'th rotating wheel

x_{ij}, y_{ij}, z_{ij} - spatial coordinates of j'th accelerometer at i'th GGI

x_{i0}, y_{i0}, z_{i0} - coordinate of the centre at the i'th GGI .

Now we replace the coordinate differences $(x_{ij}-x_{i0},\ldots)$ by the coordinates of the centre $(\bar{x}'_{ij})$ with respect to the corresponding rotating accelerometer frame. Additionally the gravity vector is transformed to the accelerometer frame, therefore we obtain :

$$\bar{a}_{ij} = \bar{R}^T_{AjGi}\ \bar{R}^T_{GiK}\ \bar{R}^T_{KT}\ \bar{W}_{m,n}\ \ \bar{R}_{KT}\ \bar{R}_{GiK}\ \bar{R}_{AjGi}\ \bar{x}'_{ij} \ . \tag{10}$$

Assuming that the accelerometers are lying within a plane $\bar{x}'_{ij}$ we obtain $\bar{x}'_{ij} = [\ 0 \quad d_{ij} \quad 0\ ]^T$ (d_{ij} - distance centre of the wheel to the accelerometer (i,j)). In the following derivation we will disregard the misalignment between the platform system K and the terrestrial system ($\bar{R}_{KT}$-unit matrix).

As described in (Jekeli, 1988) the output of the opposite located accelerometers are summed up and the difference of these sums is formed for each GGI, respectively. This linear combination is the original measuring signal. We see that the signal can be modelled by:

$$\begin{aligned} a_i = 2d_{ij}\ [&(r_{11}^2 + r_{12}^2)\ W_{xx} + (r_{21}^2 + r_{22}^2)\ W_{yy} + (r_{31}^2 + r_{32}^2)\ W_{zz} \\ &- 2(-r_{11}\ r_{21} + r_{12}\ r_{22})\ W_{xy} - 2(-r_{11}\ r_{21} + r_{13}\ r_{23})\ W_{xz} \\ &- 2(-r_{12}\ r_{22} + r_{13}\ r_{23})\ W_{yz}]\ \sin(2\omega_i(t-t_0)) \\ -4d_{ij}\ [&r_{11}\ r_{12}\ W_{xx} + r_{21}\ r_{22}\ W_{yy} + r_{31}\ r_{32}\ W_{zz} + (r_{12}\ r_{21} - r_{11}\ r_{22})\ W_{xy} \\ &+(r_{13}\ r_{21} - r_{11}\ r_{23})\ W_{xz} + (r_{13}\ r_{22} - r_{12}\ r_{23})\ W_{yz}]\ \cos(2\omega_i(t-t_0)) \end{aligned} \tag{11}$$

We realize, that the signal is modulated at twice the rotating frequency (which is 0.25 Hz). By amplitude demodulation the "inline" component (amplitude of the sinus wave) and the "cross" component (amplitude of the cosinus wave) is obtained. These data are available with a rate of 16 Hz. The gravity gradients are obtained after calibration and filtering procedures with a data rate of 1 Hz. Due to the observation equation one can suppose that the errors of the estimated gravity gradients will be strongly correlated. In this derivation the following rotation matrices have been used:

$$R_{AjGi} = \begin{vmatrix} -\sin(\omega_i(t-t_0)+\delta_j) , & -\cos(\omega_i(t-t_0)+\delta_j) , & 0 \\ \cos(\omega_i(t-t_0)+\delta_j) , & -\sin(\omega_i(t-t_0)+\delta_j) , & 0 \\ 0 , & 0 , & 1 \end{vmatrix} \tag{12}$$

$$\begin{aligned} R_{GiK}(1,1) &= \cos\kappa_i\ \sin(\alpha+\beta_i) - \cos(\alpha+\beta_i)\ \sin\kappa_i\ \sin\gamma_i &= r_{11}^i \\ R_{GiK}(2,1) &= -\cos\kappa_i\ \cos(\alpha+\beta_i) - \sin(\alpha+\beta_i)\ \sin\kappa_i\ \sin\gamma_i &= r_{21}^i \\ R_{GiK}(3,1) &= \cos\gamma_i\ \sin\kappa_i &= r_{31}^i \\ R_{GiK}(1,2) &= \sin\kappa_i\ \sin(\alpha+\beta_i) + \cos(\alpha+\beta_i)\ \sin\gamma_i\ \cos\kappa_i &= r_{12}^i \\ R_{GiK}(2,2) &= -\sin\kappa_i\ \cos(\alpha+\beta_i) + \sin(\alpha+\beta_i)\ \sin\gamma_i\ \cos\kappa_i &= r_{22}^i \end{aligned}$$

$$
\begin{aligned}
R_{GiK}(3,2) &= -\cos\gamma_i \cos\kappa_i &= r^i_{32} \\
R_{GiK}(1,3) &= \cos(\alpha+\beta_i)\cos\gamma_i &= r^i_{13} \\
R_{GiK}(2,3) &= \sin(\alpha+\beta_i)\cos\gamma_i &= r^i_{23} \\
R_{GiK}(3,3) &= \sin\gamma_i &= r^i_{33}
\end{aligned}
\qquad (13)
$$

5. Numerical considerations:

For the "Odenwald" test profile (Keller,1988) a numerical simulation study has been carried out. In this investigation no attempt was made to solve the differential equations (1)+(2) in a fundamental theoretical manner but rather special effects have been calculated. The disturbing gravitational field was simulated by 3 layers of point masses, distributed on a grid around the test profile. The point masses were generated by a random number generator. The simulated values can be seen in Fig. 3 to Fig. 5. In order to estimate the error of the computed gravitational vector we assumed random as well as systematic gradient and velocity errors. So for instance the influence of the erroneous velocity measurements on the gravity vector was investigated. It is well known (Vassiliou,1984, Eissfeller,1986) that the systematic part of the velocity error of an INS can be modelled by a sum of trigonometric functions

$$
\begin{aligned}
\delta\dot{\varphi} &= a_1 \sin(\omega_e t) + a_2 \cos(\omega_e t) + a_3 \sin(\omega_s t) + a_4 \cos(\omega_s t) + \\
&\quad + a_5 t \sin(\omega_s t) + a_6 t \cos(\omega_s t) + a_7 \\
\delta\dot{\lambda} &= b_1 \sin(\omega_e t) + b_2 \cos(\omega_e t) + b_3 \sin(\omega_s t) + b_4 \cos(\omega_s t) + \\
&\quad + b_5 t \sin(\omega_s t) + b_6 t \cos(\omega_s t) + b_7 \\
\delta\dot{h} &= c_1 \sinh(zt) + c_2 \cosh(zt) + c_3 \cos(\omega_e t) + c_4 \sin(\omega_s t) + \\
&\quad + c_5 t \cos(\omega_s t) + c_6 t \sin(\omega_s t)
\end{aligned}
\qquad (14)
$$

with:

$z = \sqrt{2}\,\omega_s$

ω_s - Schuler frequency

ω_e - Earth frequency .

The effect of these velocity errors on the gravitational vector was calculated. It was seen, that the unbounded error increase of the height channel makes a useful estimation of the gravitational vector impossible. With a stabilized height velocity error the error growth of the estimated gravity vector will reduce considerably (see Fig. 6+7). Assuming that the random velocity error is of the order of ±5 mm/s for the horizontal and ±2 mm/s for the vertical channel the estimated gravity error can be seen in Fig. 8 with several assumptions of random gradient errors. The values are below 1.5 [mgal] for a time interval of 2500 seconds. This first error study was based on the assumptions that (1) no bias (neither in the velocities nor in the gradient measurements) is present and (2) white noise error behaviour exists. Both assumptions are not quite realistic. Especially a bias of the gradients substantially effects the accuracy. Therefore the influence of an offset and systematic gradient errors were also investigated. In Fig. 9 the resulting gravitation error due to gradient offsets are shown. It is seen, that even an offset of 1.0 [E] produces 1 [mgal] error of the gravity vector after 1000 seconds.

With data taken from (Metzger and Pfohl, 1987) the amplitude spectrum of the presently available gradient measurements was computed (see Fig. 10). Due to the limited information about details of these data the spectrum is only considered as a rough estimate of real spectra. Nevertheless to calculate the resulting error behaviour (see Fig. 10) it seems to be appropriate to use these data.

Besides the influence of erroneous velocity and gradient measurements two other interaction effects between INS and gradiometer measurements were investigated, namely (1) the effect of using gravity disturbed velocity within the integration of the gradients and (2) the effect of the platform misorientation on the measured gravity gradients.

The use of gravity disturbed velocities is an effect coming from the iterative solution of the equations of motion (1) and (2) as described by Moritz (1975). Assuming that no correction of the velocities by ZUPTs was applied this interaction effect amounts up to a few milligals (see Fig. 14). With ZUPT correction this term reduces below to 1 [mgal] (see Fig. 15).

As descibed in (Hausch, 1988) non-calibrated gyro drift rates are of the order of 50"/hour and could be larger. The influence on the gradients, for simulated gradient measurements, can be seen in Fig.12 and 13. An initial misalignment of 100" and an azimuth drift of 100"/hour was assumed. It is seen, that the influence on the gradients is well below 1 [E] which is the anticipated gradiometer accuracy.

6. Conclusions:

In this study the online integration of gravity gradients was investigated. It was shown, that by combining a Gravity Gradiometer with an Inertial Navigation System it is possible to determine the gravitational acceleration without assuming a particular (erroneous) gravity model. Furthermore it is realized that ZUPT's (Zero Velocity Updates) and CUPT's (Coordinate Updates) will still be essential in determining systematic velocity errors. Otherwise a decrease of system accuracy for the determination of the gravity vector up to several milligals can be presumed. A misorientation of the horizontal gyros which cannot be controlled by ZUPT's has an influence on the measured gradients of about 0.5 [E]. The most problematic error is certainly the long term drift and a possible bias of the gradiometer system. Further studies have to show whether and how these long term errors can be handled or not.

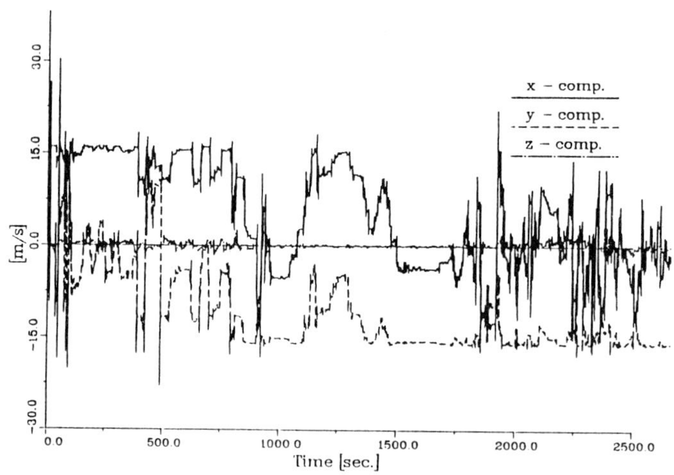

Fig.3: simulated velocities for the "Odenwald" profile

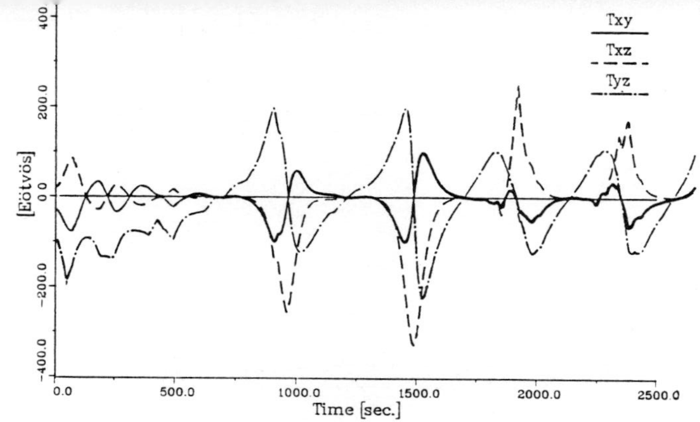

__Fig.4:__ simulated gravity gradients for the "Odenwald" profile

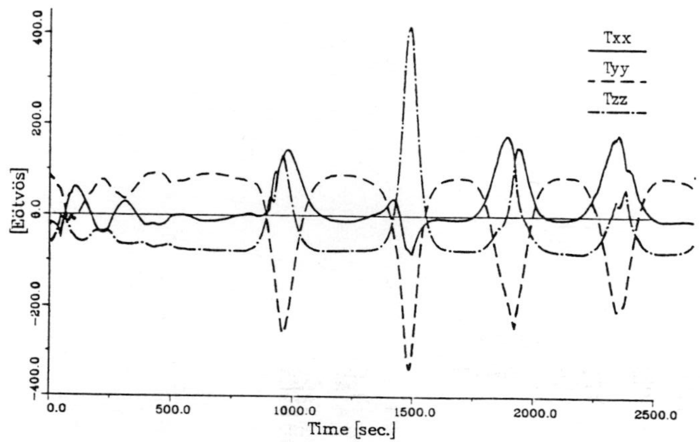

__Fig.5:__ simulated gravity gradients for the "Odenwald" profile

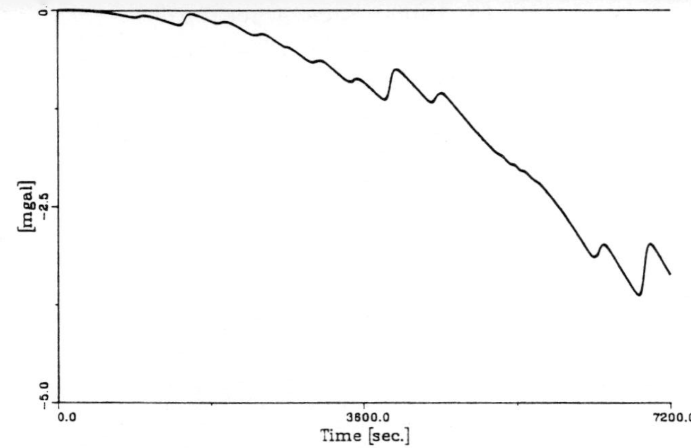

__Fig.6:__ Estimated gravity error due to systematic velocity error with height channel stabilization

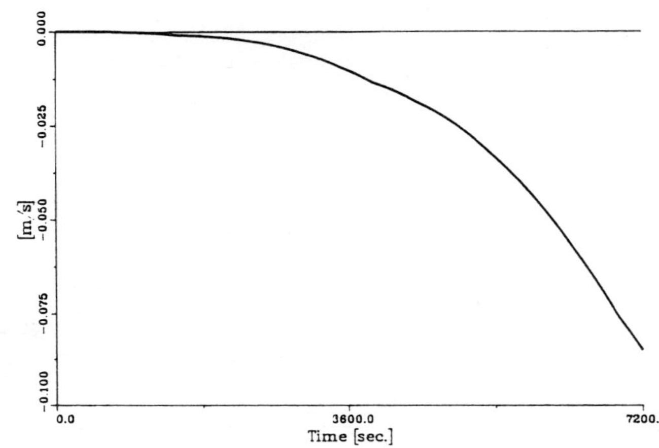

__Fig.7:__ velocity error due to error under Fig. 6

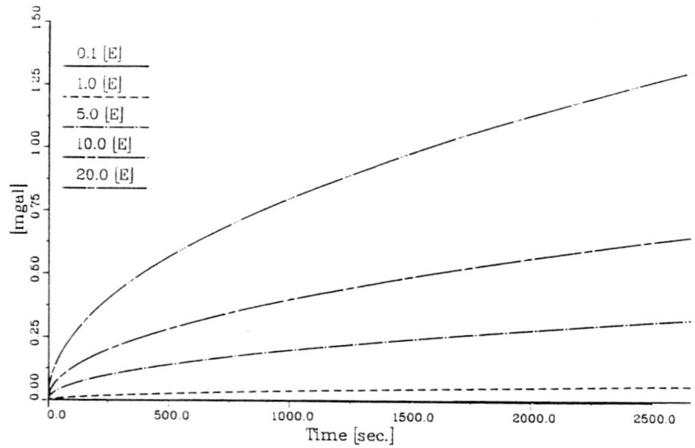

Fig.8: Estimated gravity error due to random velocity error

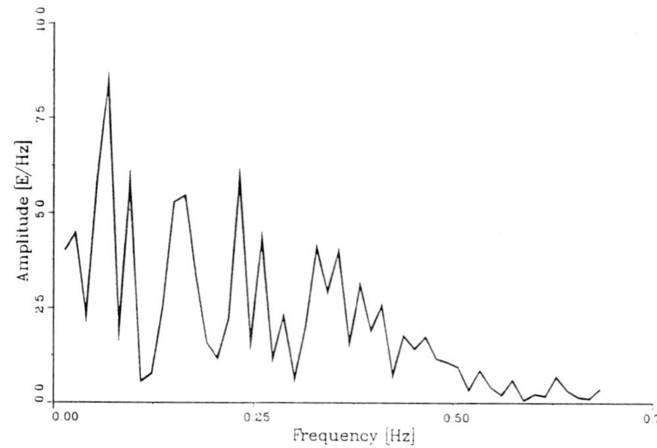

Fig.10: Amplitude spectrum of gradient measurement errors

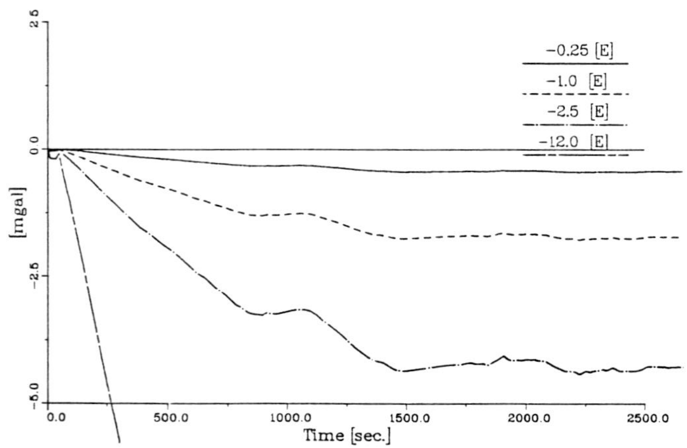

Fig.9: Estimated gravity error due to gradient offset

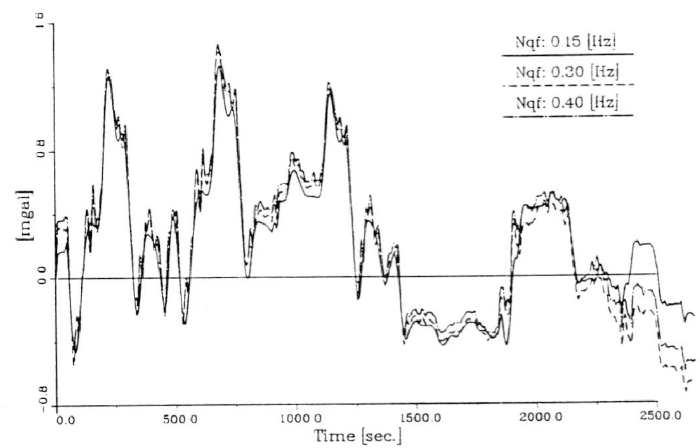

Fig.11: Estimated gravity error due to gradient errors (Fig. 10)

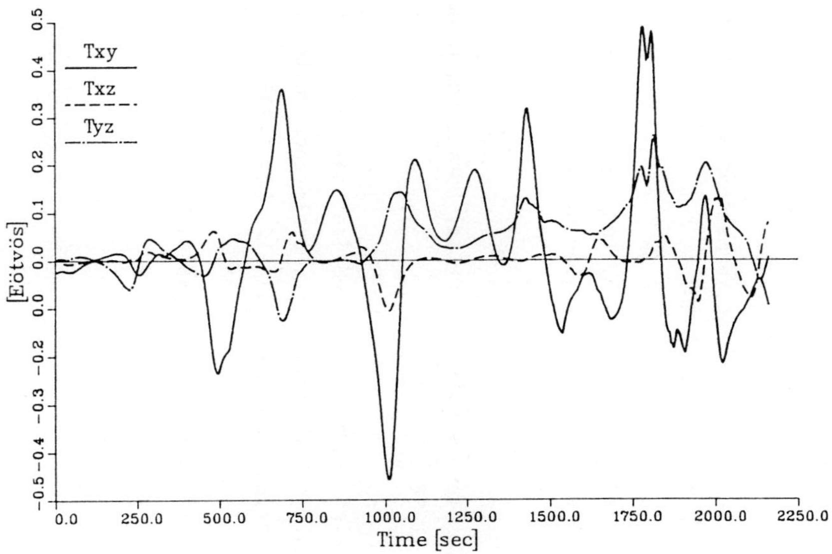

Fig.12: Gradient error due to azimuth drift

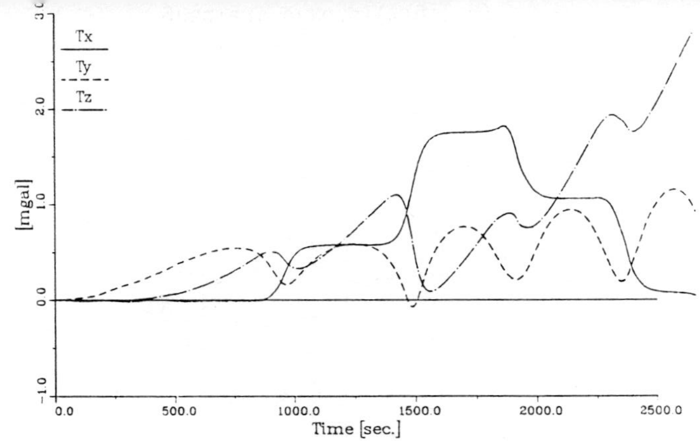

Fig.14: Interaction effect, no ZUPT correction

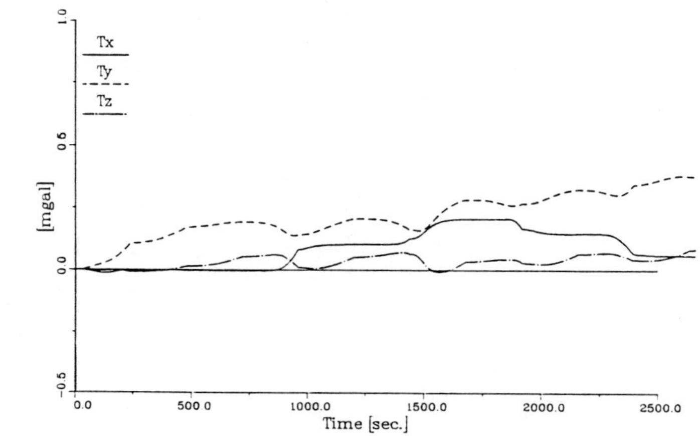

Fig.15: Interaction effect with ZUPT correction

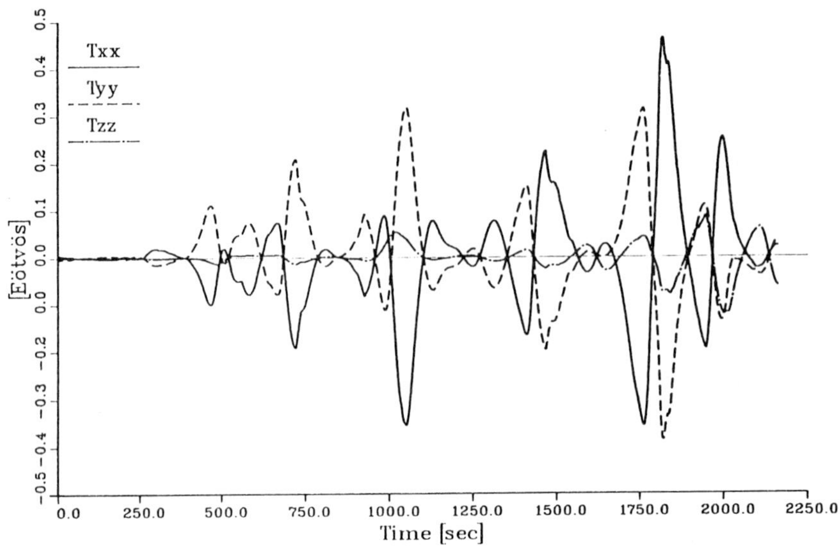

Fig.13: Gradient error due to azimuth drift

References:

Britting K.R., Inertial navigation systems analysis, Wiley-Interscience, New York, London, Sydney, Toronto, 1971

Eissfeller B., Kalman-Filter und andere Echzeitverfahren, in: Beiträge zur Inertialgeodäsie Schriftenreihe Studiengang Vermessungswesen Universität der Bundeswehr München, Heft 22 München, Mai 1987

Fundak T.J., The Gravity Gradiometer Survey System, in proceedings of the 14 th. annual gravity gradiometry conference United States Air Force Academy, Colorado Springs, Colorado, 11.-12. Feb. 1986

Hausch W., A refined adjustment model for Inertial Surveying, manuscripta geodaetica Vol.13, No 1, 1988

Heller W.G., Quick review of gradiometer-aided land navigation, in proceedings of the 14 th. annual gravity gradiometry conference United States Air Force Academy, Colorado Springs, Colorado, 11.-12. Feb. 1986.

Jekeli Ch., The Gravity Gradiometer Survey System (GGSS), in EOS Transactions, American Geophysical Union Vol. 69 No.8, 1988

Keller D., Some experiences with the LTN72 Aircraft-Platform in Terrestrial Applications, paper presented at the International Workshop High Precision Navigation May 17th - 20th, 1988, Stuttgart - Altensteig-Wart West Germany

Metzger E.H., L.L.Pfohl, BELL Aerospace gravity gradiometer survey system (GGSS) program review, in: proceedings of the 15 th. annual gravity gradiometry conference United States Air Force Academy, Colorado Springs, Colorado, 11.-13. Feb. 1987

Moritz H., Kinematical Geodesy, report presented at XIVth General Assembly International Union of Geodesy and Geophysics, Switzerland, September 25 - October 7, 1967

Moritz H., Combination of Aerial Gravimetry and Gradiometry, The Ohio State University Dept. of Geodetic Science, Report No. 223 July 1975

Schröder D., S. Wiegner, E. Grafarend, B. Schaffrin, Simulation eines geodätischen Inertialsystems, Beiträge zur Inertialgeodäsie, Geodätisches Seminar 25./26. September 1986, Schriftenreihe Studiengang Vermessungswesen, Universität der Bundeswehr München, Heft 22, Mai 1987

Schwarz K.P., A.A. Vassiliou, Requirements for the use of airborne gravity gradiometry in geophysical exploration, in proceedings of the 14 th. annual gravity gradiometry conference United States Air Force Academy, Colorado Springs, Colorado, 11.-12. Feb. 1986

Vassiliou A., Processing of unfiltered inertial data, The University of Calgary, Publ.20006, March 1984

Wang Y.M., The gradiometer-gravimeter equation in various coordinate systems. Bulletin Geodesique, 61, 2 1987

Zorn A.L., Observability of Laplace's equation using a torsion-type gravity gradiometer, in proceedings of the 15 th. annual gravity gradiometry conference United States Air Force Academy, Colorado Springs, Colorado, 11.-13. Feb. 1987

The Physics of NMR-Gyroscopes

Michael Mehring, Stephan Appelt, Bodo Menke, Peter Scheufler
2. Physikalisches Institut, Universität Stuttgart

Summary

We discuss the principles of NMR gyroscopes and their physical limitations. The Stuttgart NMR-gyro setup is described and novel detection modes are proposed.

1. Introduction

Rotations are a common feature in nature. They range from $\omega \simeq 10^{-15}$ rad s^{-1} for our galaxy and up to $\omega \simeq 10^{23}$ rad s^{-1} for a proton nucleus if its spin angular momentum $J=\frac{1}{2}\hbar$ is compared with its classical angular momentum $I\omega$, where I is the inertial moment of the proton. We may visualize the extremely small rate of rotation of our galaxy by the number of revolutions during the last 10 billion years, namely about 50. During the time mankind exists our galaxy has rotated by about 3°. It is evident, that no rotational sensor exists which is capable of detecting such weak rotational rates.

Let us look at something more realistic. The rotation of the earth corresponds to a 2π-rotation in about 24 hours or $\omega=2\pi\cdot 11\cdot 10^{-6}$ rad s^{-1}. This rate is small enough to be an appropriate reference for any rotational sensor. We will in the following discuss the possibility to detect this and smaller rotational rates with the help of nuclear spin precession, called Nuclear Magnetic Resonance (NMR)[1]. It should be noted that also electrons posses a spin $S=\frac{1}{2}\hbar$ and could be used as a rotational sensor as well, but

have some disadvantage in terms of resolution, as will become evident below. Their precession in a magnetic field is called Electron Spin Resonance (ESR).
The precession frequency of spins in a magnetic field is changed if the complete NMR setup is rotated around the magnetic field direction with respect to an inertial frame. This has been realized a long time ago and the exploitation of this fact as a rotational sensor was proposed already in the sixties [2-6] and some patents were issued somewhat later [7].

2. Basic Concepts

Consider a nuclear spin with angular momentum $J=I\hbar$. We will restrict ourselves in this discussion to $I=\frac{1}{2}$. Placed in a static magnetic field B_o this spin will start to precess as is scetched in Fig.2.1 due to its magnetic moment μ_I. The Larmor precession frequency is given by [1]

$$\omega_o = \gamma B_o \qquad (1)$$

where γ is the gyromagnetic ratio.

A nuclear spin, however, as a quantum object must be discussed in the framework of quantum mechanics. A spin 1/2 therefore has two quantum levels in a magnetic field B_o as is shown in Fig.2.2. Their level splitting turns out to be $\hbar\omega_o$, where ω_o is given by Eq.(1). Moreover the full quantum mechanical dynamics of a spin $\frac{1}{2}$ ensemble (not for $I > \frac{1}{2}$) can be exactly transformed into a classical vector picture as is scetched in Fig.2.1. The resulting equations are the so- called "Bloch equations", which will be discussed below [1].

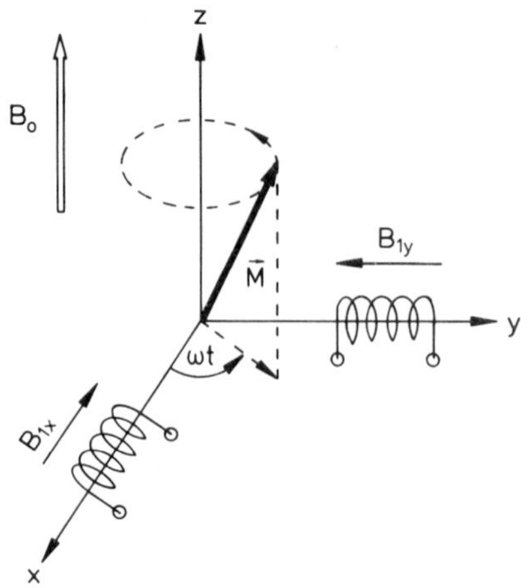

Fig.2.1

Principles of NMR. Classical vector precession picture, where the macroscopic magnetization M precesses around the magnetic field B_o with Larmor-frequency ω_o. Excitation and detection of the precession is achieved by rf-fields B_{1x} and/or B_{1y}.

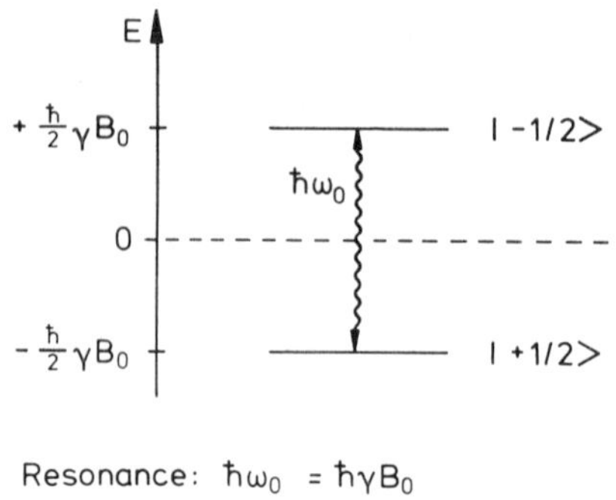

Fig. 2.2.

Quantum picture for a spin $\frac{1}{2}$ in a magnetic field B_o. Energy level spacing is equal to the Larmor frequency ω_o.

Let B_o be parallel to the z-direction of the vehicle frame. If the vehicle is rotating with respect to an inertial frame with $\mathbf{\Omega}=(\Omega_x, \Omega_y, \Omega_z)$, the observed rotational rate in the vehicle frame can be expressed as

$$\omega_3{}^2=\Omega_x{}^2+\Omega_y{}^2+(\gamma B_o+\Omega_z)^2 \tag{2}$$

For $\gamma B_o \gg |\mathbf{\Omega}|$ Eq.(2) can be approximated by

$$\omega_3 \simeq \gamma B_o+\Omega_z+(\Omega_x{}^2+\Omega_y{}^2)/(2\gamma B_o) \tag{3}$$

i.e. the effective rotational rate of the magnetization vector **M** is in first order determined by γB_o and Ω_z under these conditions. The full rotational vector $\boldsymbol{\Omega}$ can be determined by means of three mutually orthogonal gyroscopes as shown in Fig.2.3. The corresponding effective resonance frequencies can be expressed as

$$\omega_1{}^2 = (\gamma B_{o\,1}+\Omega_x)^2+\Omega_y+\Omega_z{}^2 \tag{4a}$$

$$\omega_2{}^2 = \Omega_x{}^2 + (\gamma B_{o\,2}+\Omega_y)^2+\Omega_z{}^2 \tag{4b}$$

$$\omega_3{}^2 = \Omega_x{}^2 + \Omega_y{}^2 + (\gamma B_{o\,3}+\Omega_z)^2 \tag{4c}$$

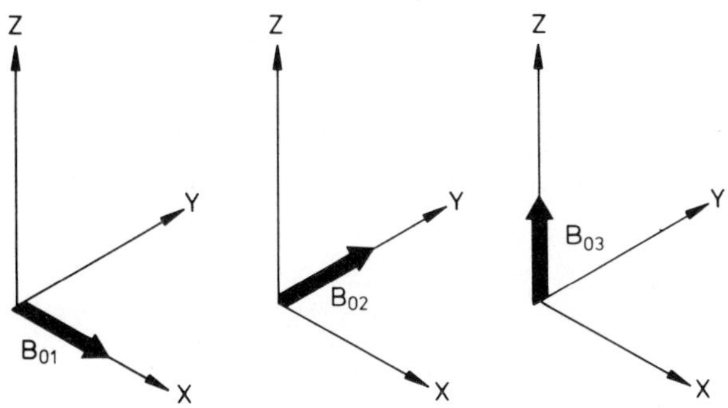

Fig.2.3 Three axis NMR gyro system comprised of three mutually orthogonal magnetic fields $B_{o\,1}$, $B_{o\,2}$, $B_{o\,3}$. This allows the full rotation vector (Ω_x, Ω_y, Ω_z) to be determined.

Let us again concentrate on a single NMR gyroscope where B_o is parallel to the z-direction. In order to observe the precession, the magnetization vector **M** must be tilted away from the z-direction. This can be achieved by an AC field B_1 in x-(or y-) direction which oscillates with frequency ω near ω_o. If B_{1x} is applied on-resonance ($\omega=\omega_o$) the magnetization vector **M** rotates about the x-direction with rate $\omega_{1x}=\gamma B_{1x}$ and of course at the same time about the z-direction (with rate $\omega_o=\gamma B_o$). The interaction of the spins with their surrounding is now introduced by two different relaxation times, namely

T_1: Relaxation of M_z towards equilibrium M_o

T_2: Relaxation of $M_{x,y}$ towards zero.

The magnetization is usually not observed in the vehicle frame, but rather in the "rotating frame", rotating at the frequency ω of the AC-field B_1 around the z-direction. The Bloch equations in this coordinate frame can be expressed as

$$\dot{\mathbf{M}} = \mathbf{M} \times \omega_{eff} - \frac{M_x + M_y}{T_2} - \frac{M_z - M_o}{T_1} \tag{5}$$

where

$$\omega_{eff} = \begin{bmatrix} o \\ o \\ \gamma B_o - \omega \end{bmatrix} + \begin{bmatrix} \gamma B_{1x} \\ \gamma B_{1y} \\ o \end{bmatrix} + \begin{bmatrix} o \\ o \\ \Omega_z \end{bmatrix} \tag{6}$$

From Eq.(6) follows, that the actually observed frequency in an NMR experiment is (under the condition of weak B_1-fields; no saturation)

$$\delta\omega = \gamma B_o - \omega + \Omega_z \tag{7}$$

This frequency $\delta\omega$ is directly observed as a frequency shift or can be related to the ratio of M_x and M_y in a continous wave (cw) or some pulsed NMR experiments as

$$\tan \phi = \frac{M_x}{M_y} = \delta\omega \cdot T_2 \tag{8}$$

if irradiation is performed only in x-direction ($B_{1x} \neq o$).

The reference frequency ω can be chosen to make $\delta\omega$ arbitrarily small or actually zero. If γB_o, ω and ϕ are considered independent random variables, the standard deviation for the measurement of the vehicle rotational rate Ω_z can be expressed as

$$\sigma_\Omega = \left(\sigma_\omega^{\,2} + \sigma_{\omega o}^{2} + \sigma_\phi^{\,2}/T_2^2 \right)^{1/2} \tag{9}$$

It is a technical demand to keep the fluctuation in the reference frequency and B_o as small as possible, since they contribute to the rotational error directly. Suppose this can be achieved, we are still left with the NMR-signal error

$$\sigma_{\Omega}^{NMR} = \left(\sigma_{\phi}{}^{2}/T_2^2 \right)^{1/2} \tag{10}$$

according to Eq.(9). Introducting the signal-to-noise ratio S/N of the NMR-signal, the signal power P_s and noise Power P_N, we obtain with σ_{ϕ}=N/S=$(P_N/P_S)^{1/2}$

$$\sigma_{\Omega}^{NMR} = \frac{1}{T_2} \cdot \frac{N}{S} = \frac{1}{T_2}\sqrt{\frac{P_N}{P_S}} \tag{11}$$

Introducting the detector bandwith b and the average noise power $\overline{P_N}(b)$ over the bandwith b, the S/N ratio observed during the "measuring" time t is given by

$$\frac{S}{N} = \left(\frac{P_S \cdot b \cdot t}{\overline{P}_N} \right)^{1/2} \tag{12}$$

Inserting this into Eq.(11) by using the spectral noise power density $R_{S/N}=(P_S b/\overline{P}_N)^{1/2}$ leads to

$$\sigma_{\Omega}^{NMR} = \frac{1}{T_2} \frac{1}{R_{S/N}} \cdot t^{-1/2} \tag{13}$$

In order to reduce the error σ_{Ω}^{NMR} a long T_2 time, a large signal-to-noise ratio and a long measuring (averaging) time t is required. It is evident, that the standard deviation of the rotational angle Φ of the vehicle, which is obtained by integrating $\Omega(t)$, is given by [4-6]

$$\sigma_{\Phi} = \frac{1}{T_2} \cdot \frac{1}{R_{S/N}} \cdot t^{1/2} \tag{14}$$

where $(T_2 R_{S/N})^{-1}$ is called the "random-walk" coefficient. In any case the random walk coefficient should be made very small.

Let us look at some numbers. We have achieved in our test setup (to be described below) $T_1 \simeq 1000$s for ^{129}Xe nuclei. A typical signal-to- noise ratio would be $R_{S/N} \simeq 500 \sqrt{Hz}$. This gives a random-walk coefficient of $2 \cdot 10^{-6}\ s^{-1/2}$. For a measuring time of t = 1s this corresponds to a rate error well below the earth rotational rate of $\omega_E = 2\pi\ 11\cdot10^{-6}$ rad s^{-1}. For a measuring time of 10 seconds this would already result in a relative error of less than 1%. The angular drift according to Eq.(14) would lead to $\sigma_\Phi \simeq 0.01°/\sqrt{h}$. This has not been realized in an actual gyro system yet, but should be feasable.

If we want to approach the ultimate theoretical limit of NMR gyroscopes we have to consider quantum noise which limits the maximum feasible signal-to-noise ratio to S/N 0 $\simeq 10^5$-10^6. Combined with a T_2- time of T_2 = 10h (possible with ^{3}He) leads to a random walk coefficient of $(3\text{-}30)10^{-11}s^{-1/2}$. The error in rotational rate after 1h measuring time would be according to Eq.(13)

$$\sigma_\Omega^{NMR} \cong (0.5\text{-}5)\ 10^{-12}\ \text{rad s}^{-1}$$

It will be very difficult to achieve this experimentally, however.

So far we have ignored σ_ω and $\sigma_{\omega o}$. Frequency sources can be made extremely stable, and magnetic fields can be stabilized to a high degree by using superconductors and superdonducting shields. Their fluctuations therefore can be reduced drastically with state of the art techniques. Nevertheless some fluctuations will always be present. However, if several different nuclei are used in the same NMR gyrocell, these errors can be eliminated by capitalizing on the fact (see Eq.(7)), that the vehicle rotational frequency Ω_z is the same for all nuclei, but ω_o - γB_o and ω are different for the different nuclei. Since long T_2 times are required for high precision operation, rare gases like, ^{3}He, ^{21}Ne, ^{83}Kr, ^{129}Xe are usually employed, which offer long relaxation times if $I=\frac{1}{2}$ because of their closed s-shell electronic states. A typical gas cell may contain 10^{16} atoms. In a 1 G field, corresponding to a Larmor frequency around 1 kHz, the signal-to-noise ratio for direct observation of the NMR signal would be exceedingly small and virtually impossible to detect at all. It is therefore required to first polarize the nuclei to a high degree. This is usually done by optical pumping of an open shell atom and spin exchange with the nucleus to be polarized [8-12].

In order to be more specific we consider the Rb/Xe system used also in the Stuttgart NMR gyro system (see section 3). Fig.2.4 shows the level scheme of a Rb atom. There are two different allowed optical transitions which connect the ground state with the excited P-states. If circularly polarized light is used, angular momentum is transmitted to the atom and after "pumping" for a while, a highly polarized groundstate is achieved.

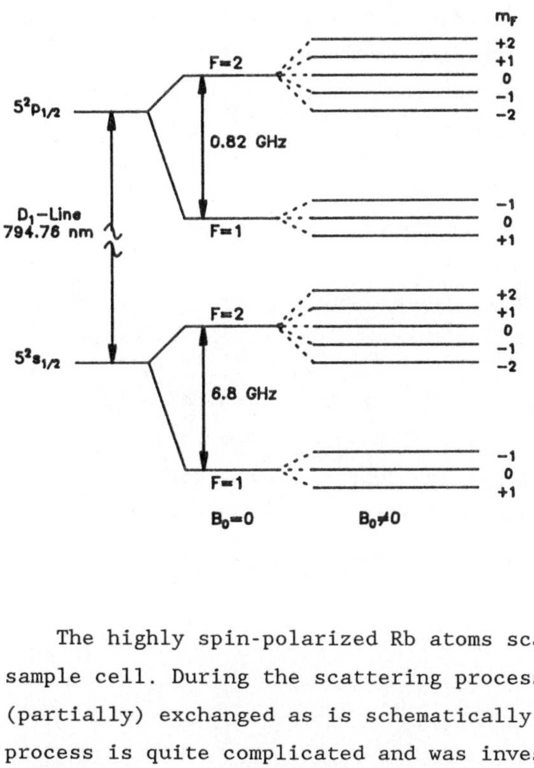

Fig.2.4 ^{87}Rb energy level scheme in a magnetic field B_o including hyperfine interactions. ^{87}Rb spin polarization in the F=2 groundstate is achieved by "pumping" with circularly polarized light on the D_1-line.

The highly spin-polarized Rb atoms scatter with the Xe atoms in the sample cell. During the scattering process the spin states can be (partially) exchanged as is schematically scetched in Fig.2.5. The actual process is quite complicated and was investigated in detail [11-13]. It involves the formations of a van der Waals complex and spin-exchange due to hyperfine interaction during the lifetime of this complex. In order to favour the formation of these van der Waals complexes a third scattering partner, e.q. N_2 is necessary.

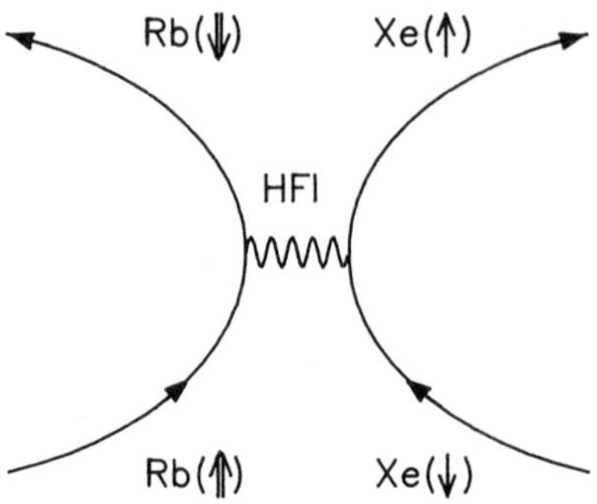

Fig.2.5 Pictorial representation of the spin exchange process between highly polarized Rb atoms and unpolarized Xe nuclei. Spin exchange due to hyperfine interaction during encounter.

The NMR signal of the poarized ^{129}Xe nuclei is usually not detected directly, but indirectly by observing changes in the optical transition of the Rb atoms. This "quantum transformation" leads to a drastic increase in signal-to-noise ratio.

3. The Stuttgart NMR Gyro Test Setup

It is quite clear from section 2 that long relaxation times T_2 and large signal-to-noise ratios are required for high precision NMR gyro operation. Both parameters depend in complicated ways on the details of the spin physics in the gyro sample cell. In order to investigate the complex spin exchange processes and the different relaxation mechanisms we have setup an NMR gyro test system, whose block diagram is shown in Fig.3.1.

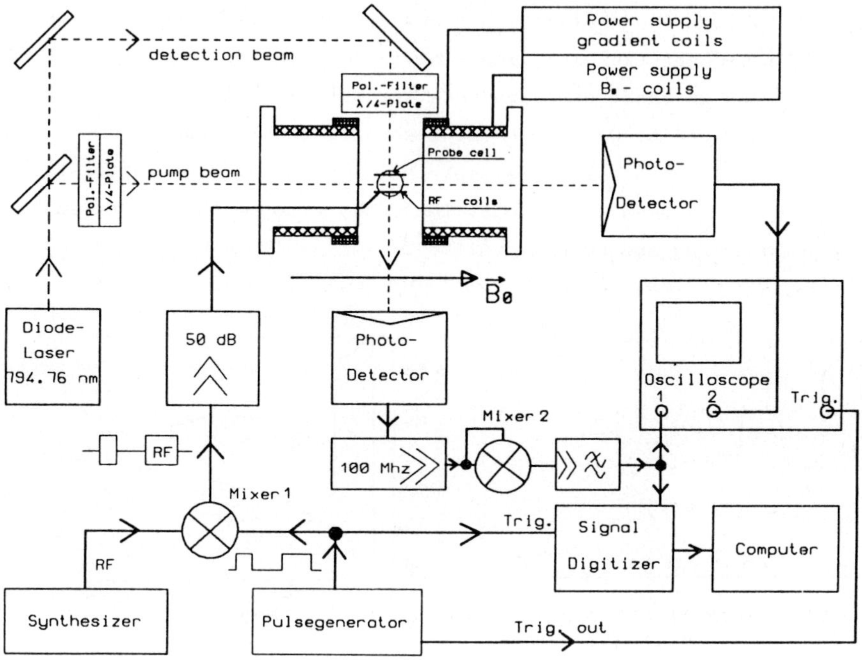

Fig.3.1 Block diagram of the Stuttgart NMR gyro test setup

The sample cell is placed in a specially designed solenoid with high homogeneity. The cell contains Rb-, ^{129}Xe and N_2- atoms with variable pressure. A particular Rb line is pumped selectively by a specially designed, highly stabilized, monomode infrared diode-laser. This unit is discussed in more detail in the lecture of Scheufler at al. in this book. Part of the light beam is split off for detecting any change in absorption and coherence effects to be discussed later. The different radio-frequency (rf) fields are supplied by a synthesizer and other frequency sources and are amplified and fed to different coils around the sample cell. A photodiode is used as a detector for changes in the Rb absorption signal as well as for the observation of coherent Raman beats (see next section).

Double μ-metal shielding is applied to avoid stray magnetic fields from the environment. With this setup we have observed ^{129}Xe NMR signals around 1 kHz in differently prepared sample cells. If special wall treatment is applied long relaxation times can be achieved as is shown in an adiabatic rapid passage experiment in Fig.3.2. The M_z magnetization of ^{129}Xe is observed in these experiments after optical polarization and using optical detection. Each time the resonance is crossed the magnetization is inverted. This is repeated a couple of times after the polarizing light has been turned off. The decay time of the magnetization is T_1 which in Xe-gas should be equal to T_2.

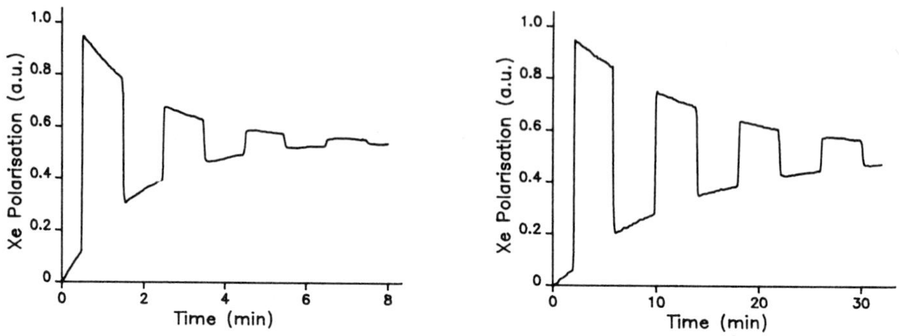

Fig.3.2 Decay of the nuclear spin polarization of ^{129}Xe
a) without wall coating and with high Rb pressure
b) with wall coating and low Rb pressure

If no special wall treatment is applied and with higher Rb concentration T_1 is much shorter as can be seen from Fig.3.2b. The understanding of the details of these relaxation mechanisms is of paramount importance for designing gyro cells with even longer T_1 and T_2 relaxation times.

Another line of investigations with this test setup deals with the improvement of the signal-to-noise ratio. Here different light sources including dye lasers are currently under investigation. Also more efficient pumping cycles and the details of the spin-exchange mechanisms are being investigated with our setup.

A significiant improvement in signal-to-noise ratio may be achieved with different detection schemes. In this context we have recently investigated the coherent Raman beat effect to be discussed briefly in the next section.

4. The Coherent Raman Beat Detection Scheme

The classical detection of optical-ESR or -NMR double resonance proceeds via the observation of changes in the absorption or fluorescence of light. Any variation, fluctuation or drift in these quantities leads to noise in the observation channel. This can be elegantly circumvented by observing the beating light due to the coherent interaction between the atomic sublevel coherence and the light absorption process [14]. Such a scheme can be understood on the basis of the four-level scheme shown in Fig.4.1.

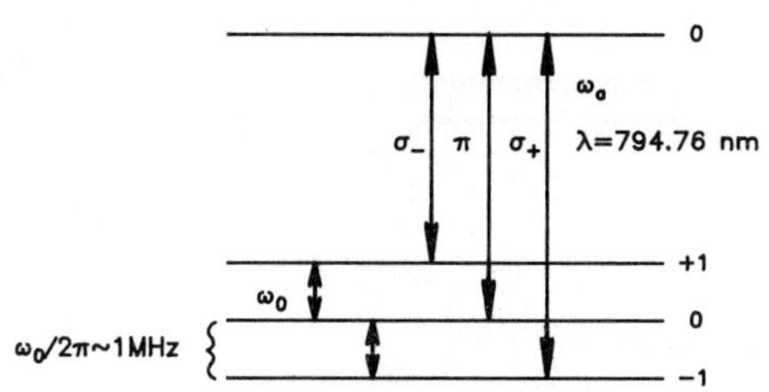

Fig.4.1 Priciple of Raman beat detection scheme. The Larmor frequency ω_o can be detected by beating $\sigma_{\pm}$-light with π-light, once a superposition of coherent states is created.

The experimental setup for Raman beat detection is scetched in Fig.4.2. The bandwidth of the photodetector and the following electronic components must cover the frequency range determined by the Larmor frequency.

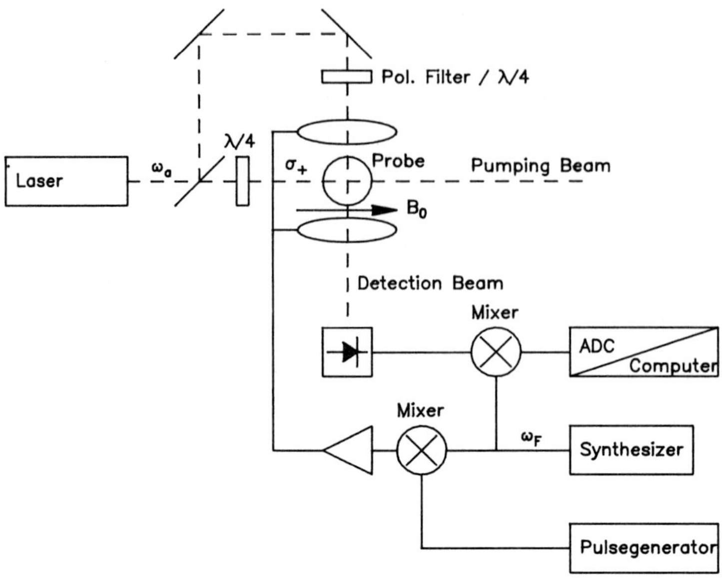

Fig.4.2 General setup for Raman beat detection

The three sublevels can be brought into coherent superposition by a radio-frequency field at frequency ω_F. Single- and double quantum beats can occur now between σ-and π-light and between σ_+- and σ_--light respectively. This effect shows up as an oscillating light intensity with the frequency ω_F and $2\omega_F$. We have observed such a behaviour in Rb atoms as is shown in Fig.4.3. The signal-to-noise ratio is very large and the precession of 10^{12} Rb atoms can be observed directly on the scope in a single shot experiment. Fig.4.3 shows the free induction decay (FID) of single and double quantum Raman beats excited by a 90°-pulse, along with the corresponding spectra after Fourier-transformation. Further details of these investigations will be published elsewhere.

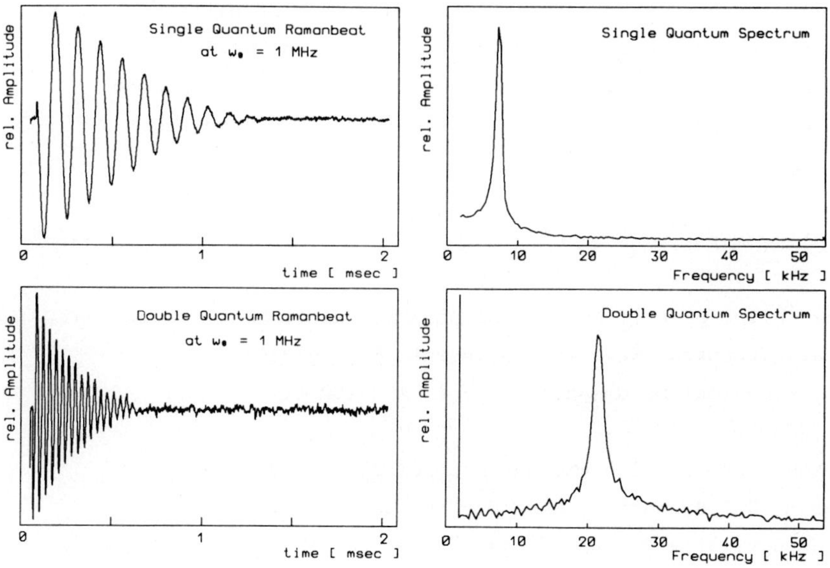

Fig.4.3 FID of single and double quantum Raman beats with corresponding spectra

5. Future Developments

In order to improve the precision of NMR gyroscopes it is required to increase the relaxation times and to improve the signal-to-noise ratio. We therefore plan to investigate the details of the spin-relaxation mechanisms in gases in more detail in particular the atom-wall interaction. The preparation of special cells with different wall coatings and the systematic investigation of relaxation effects on surfaces will be important.

The signal-to-noise ratio could be improved in two ways: (i) By obtaining higher nuclear spin polarization and (ii) by coherent detection schemes. The approach (i) requires stronger light sources and more important a better understanding of the spin-exchange process which might lead to alternative schemes with improved spin polarization.

Approach (ii) offers the possibility to avoid the 1/f noise of the photodetector and the preamplifier. Moreover a detailed knowledge about these subtle quantum coherence phenomena may lead to wastly improved signal-to-noise ratios. Research along these lines is in progress.

References

1. A.Abragam, Principles of Nuclear Magnetism, Oxford Univ. Press 1961.
2. W.H.Culver, Technical Note in ARS-Journal 23, 943 (1962).
3. R.Kramer, Technical Documentary Report, ASD-TDR-62-492, Vol.II.(1963).
4. J.H.Simpson, Astronautics and Aeronautics 2, 42 (1964).
5. J.H.Simpson, D.I.Shemoff, J.McKendry, Research and Development Program on Optically Pumped Nuclear Gyroscope, 1st and 2nd Quaterly Progress Report, Prep. for Bureau of Weapons, Contract No. NOw 65-0577-d Oct. 1965 and Jan. 1966.
6. J.H.Simpson, I.A.Greenwood, 6th Meeting, Guidance and Control Panel AGARD Nato, May 1968.
7. B.C.Grover, E.Kanegsberg, J.G.Mark, R.L.Meyer, Nuclear Magnetic Resonance Gyro, U.S. Patent 4, 157, 495, June 5, 1979.
8. F.A.Karwacki, Navigation: The Journal of The Institute of Navigation, 27, 72 (1980).
9. M.A.Bouchiat, T.R. Carver, and C.M.Varnum, Phys.Rev.Lett. 5, 373 (1960).
10. C.Cohen-Tannoudji, J.DuPont-Roc, S.Haroche, and F.Laloë, Phys.Rev.Lett. 22, 758 (1969).
11. C.H.Volk, T.M.Kwon, and J.G.Mark, Phys.Rev. A21, 1549 (1980).
12. W.Happer, Rev.Mod.Phys. 44, 169 (1972).
13. W.Happer, E.Miron, S.Schaefer, D.Schreiber, W.A.van Wijngaarden, and X.Zang, Phys.Rev. A29, 3092 (1984)
14. (a) J.Mlynek, N.C.Wong, R.G.DeVoe, E.S.Kintzer, and R.G.Brewer, Phys.Rev.Lett. 50, 993 (1983).
    (b) N.C.Wong, E.S.Kintzer, J.Mlynek, R.G.deVoe, and R.G.Brewer, Phys.Rev. B28, 4993 (1983).

Data Processing for GPS/INS Integration

Michael Napier
University of Nottingham, Dept. of Civil Engineering, Nottingham, U.K.

Summary

This paper gives an overview of the data processing for the integration of GPS and INS. The role of the Kalman filter as the appropriate mathematical algorithms is described, and the basic variations in Kalman filter architecture are discussed.

1. Introduction

The Global Positioning System offers an absolute positioning accuracy of 15m to 30m. Inertial navigation complements GPS in that it provides relative positioning and is totally self-contained. These two positioning sensors are ideally suited for system integration. Although there is no improvement in accuracy, the integration of GPS with inertial navigation does enable an increase in system performance.

This paper gives an overview of the data processing for the integration of GPS and INS. A brief overview of GPS and inertial navigation is given in §2 and §3. The reason behind integration is discussed in §4 and a definition of the integration problem outlined in §5.

The Kalman filter is such an essential element of GPS/INS integration that a brief description is given in §6. The use of the Kalman filter for both GPS and INS as separate systems is then discussed in §7 and §8.

The basic alternatives of filter architecture, their advantages and disadvantages are covered in §9, §10 and §11. Brief conclusions follow in §12.

2. Global Positioning System

The Global Positioning System (GPS) will offer a global, 24 hour, all weather, positioning capability for navigation and surveying, whether on land, at sea or in the air. A GPS receiver measures the range to three satellites to solve for three position unknowns. The ranges are determined by comparing the time of transmission of a coded signal with the time of arrival as measured by the receiver's clock. This measured time delay is multiplied by the velocity of light to obtain the so called 'pseudorange'. As the receiver clock is liable to be in error, a pseudorange to a fourth satellite is also observed in order to solve for the local clock error.

In addition to the pseudorange measurement, a navigation receiver also measures the rate at which the pseudorange is changing. This observable is called the 'range rate'. An alternative observable is the 'deltarange', the amount by which the range has changed between measurements. These observables and the pseudorange are obtained from the code tracking logic of the GPS receiver. Though not discussed in this paper, tracking of the carrier signal yields the 'carrier phase' and the 'carrier phase rate' observables.

The basic technique of pseudoranging gives an absolute positioning accuracy of 15m to 30m. The use of the carrier signal observables can yield relative accuracies of 10mm + 1ppm for static surveys.

3. Inertial Navigation Systems

Inertial navigation is a dead-reckoning method of navigation based on the integration of acceleration to determine velocity, and the integration of velocity to get position. Three orthogonally mounted accelerometers measure the vehicle acceleration and three orthogonally mounted gyroscopes measure the vehicle angular velocity. The latter provides the necessary information to stabilize the navigation reference frame, either physically in a gimbaled inertial system or analytically in a strapdown system. So long as the starting position is known, a continuous output of position, velocity and attitude is available independent of any outside agency, of the environmental conditions, or of the location.

Inertial navigation has two very significant advantages over all other navigation or positioning aids. Firstly, inertial navigation systems are totally self-contained, and, secondly, the full navigation solution (i.e. position, velocity, heading, pitch and roll) is provided. However, there are two major limitations: cost and an high intrinsic error (about 1800 m/hour for an unaided navigation quality IN). Nevertheless, inertial navigation complements an absolute positioning system such as GPS and is ideally suited for system integration.

4. Integration of GPS with INS

The GPS/INS integrated system maintains the advantages of both navigation systems without many of the limitations of either. The inertial navigation system benefits from the integration with GPS by the provision of fast in-flight alignment of the INS, and by the calibration of the inertial instruments (accelerometers and gyroscopes).

For GPS operation, the integration of GPS with inertial navigation enables an increase in system performance, but does not improve accuracy. The main advantages that arise are:

- position data for re-acquisition of satellites
- velocity aiding of code tracking loop
- navigational capabilities during periods of GPS outage.

These features are particularly advantageous for high dynamic navigation and for the battlefield scenario. For example, fighter aircraft frequently lose satellite signals during sharp turns and other manoeuvres and rapid re-acquisition of satellite signal is essential. Reduced susceptibility to jamming is possible by using the inertially derived velocity to aid the GPS tracking loops when there is a poor signal-to-noise ratio.

5. Definition of the Problem

The requirement for the data processing of GPS and inertial data is to mix all available sensor information to yield a best estimate of position, velocity and attitude (Figure 1). Here, 'best' means unbiased and satisfying some criterion of optimality such as the minimisation of the sum of the squares of the weighted residuals (i.e. 'least squares'). Inevitably, the sensor data will contain signal and noise. If the deterministic relationship between the signals is known and if there is *a priori* knowledge of the stochastic behaviour of the noise, only then can an optimal estimate be obtained.

The most commonly used set of algorithms for the processing of GPS or inertial data is the Kalman filter. For those familiar with 'survey least squares' the Kalman filter is in essence a sequential least squares approach to the estimation of parameters that change with time. For the benefit of those unfamiliar with Kalman filtering, a brief outline is given in §6.

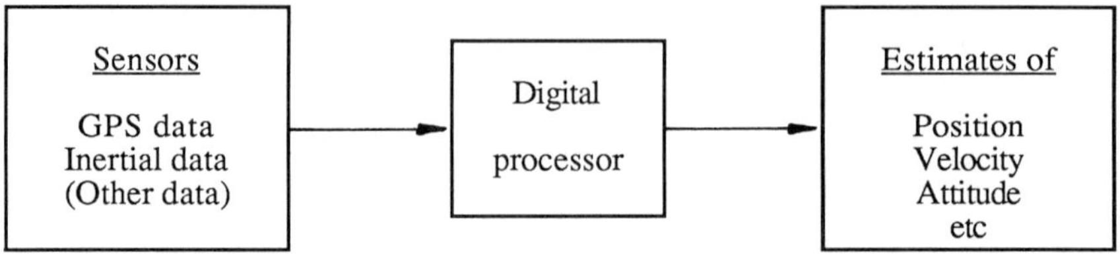

Figure 1: The Integration Problem

The Kalman filter is not the solution to the problem in itself. Consideration must also be given to a whole host of constraints and priorities, only some of which are listed here:

- speed of processing
- precision of the estimate
- robustness
- integration to other sensors.

There is a trade off to be gained between speed of processing and the precision of the results. For real-time navigation, it is the speed that is of utmost importance, whereas for precise surveying it is the precision. In the latter case it may be more appropriate to record all the observations and subsequently post-process.

Robustness of the integration software is of most concern to the real-time application. The data processing must be able to cater for sensor malfunction or failure. A fully integrated system may be susceptible to such problems, and alternative cascaded architectures become necessary.

Finally, some consideration must be given to the need to pass on the results of the integrated positioning system to other systems, such as weapon management systems, automatic landing systems, etc.

6. Kalman filter

Consider a vector of parameters (the state vector) changing with time, and sets of measurements observed at different epochs, t_j, that are related to the parameters by a linear model (measurement model, equation 1). If the way in which the parameters are changing with time can be modeled both for deterministic and stochastic effects (the system model, equation 2), then the Kalman filter provides a set of algorithms for the estimation of the state vector at _any_ point in time. The Kalman filter is an optimal estimator so the estimates of the state vector (i.e. the unknown parameters) are unbiased and have minimum variance.

There are three distinct estimation problems:

(1) FILTERING - the estimate of the state vector at time t_k using the measurements at all epochs up to and including time t_k

(2) PREDICTION - the estimate of the state vector at time t_j after the last set of measurements at epoch t_k ($t_j > t_k$)

(3) SMOOTHING - the estimate of the state vector at time t_i using _all_ the available sets of measurements from the first to last epochs at times t_1 and t_n respectively ($t_1 \leq t_i \leq t_n$).

All three estimation problems can be solved using the Kalman filter algorithms. However, only the filtering and prediction equations are given here (Table 1). The reader is referred to Chapter 5 of Gelb (1974) for the smoother algorithms.

Two equations are required to predict the state vector and its covariance matrix from one epoch to the next (Equations 6 and 7). This prediction might be made to some time at which the state is required to be known but at which time there are no measurements. In this case, the same equations (6 & 7) may be used again and again to compute the 'predicted' state at any number of epochs (with small or large time intervals between them) until another set of measurements is available.

When a set of measurements becomes available, the state vector must be extrapolated to that epoch using the prediction equations, in appropriate time intervals as in the above paragraph if so required. The three update equations (8, 9 and 10) are then used to compute the 'filtered' state (and its

KALMAN FILTER EQUATIONS

MODELS	measurement	$\underline{z}_k = \mathbf{H}_k \underline{x}_k + \underline{v}_k$	$\underline{v}_k \sim N(\underline{0},\mathbf{R}_k)$	(1)
	system	$\underline{x}_k = \Phi_{k-1} \underline{x}_{k-1} + \underline{w}_{k-1}$	$\underline{w}_{k-1} \sim N(\underline{0},\mathbf{Q}_{k-1})$	(2)
		$E[\underline{w}_k \underline{v}_j^T] = 0$ for all j, k		(3)
ALGORITHM	initialise	$\hat{\underline{x}}_k(+) = \underline{x}_o$		(4)
		$\mathbf{P}_k(+) = \mathbf{P}_o$		(5)
	prediction	$\hat{\underline{x}}_k(-) = \Phi_{k-1} \hat{\underline{x}}_{k-1}(+)$		(6)
		$\mathbf{P}_k(-) = \Phi_{k-1} \mathbf{P}_{k-1}(+) \Phi_{k-1}^T + \mathbf{Q}_{k-1}$		(7)
	update	$\mathbf{K}_k = \mathbf{P}_k(-) \mathbf{H}_k^T [\mathbf{H}_k \mathbf{P}_k(-) \mathbf{H}_k^T + \mathbf{R}_k]^{-1}$		(8)
		$\hat{\underline{x}}_k(+) = \hat{\underline{x}}_k(-) + \mathbf{K}_k [\underline{z}_k - \mathbf{H}_k \hat{\underline{x}}_k(-)]$		(9)
		$\mathbf{P}_k(+) = [\mathbf{I} - \mathbf{K}_k \mathbf{H}_k] \mathbf{P}_k(-)$		(10)

IMPLEMENTATION Easily programmed.

Application needs -		
	measurement model	$\mathbf{H}_k, \mathbf{R}_k$
	system model	$\Phi_{k-1}, \mathbf{Q}_{k-1}$
	initial values	$\underline{x}_o, \mathbf{P}_o$

NOTES
(1) Subscript k denotes at time t_k.
(2) The symbols (-) and (+) denote the instant before and after an update.

TABLE 1: Kalman filter equations

covariance matrix) which is effectively the weighted mean of the predicted state and that state that best fits the measurements at that epoch. If more than one measurement model is in use (one say for position measurements and another for velocity measurements) and observations pertaining to those models are available at the same time, then the update equations may be computed sequentially for each type of measurement. This clearly obviates the need for lots of measurement models to cater for any combination of observations made at once. For example, a measurement model for a single range measurement could be sequentially applied to update the state vector from any number of ranges available at one time.

Once all the measurements at an epoch have been processed, then the cycle starts again and the state is extrapolated to the next time at which a predicted estimate of the state is required.

The algorithms are very straight forward to program. The only matrix to be inverted is the term $[H_k P_k(-) H_k^T + R_k]$ that appears in the equation to compute the Kalman gain matrix K_k (equation 8). The order of this matrix is the number of observations being simultaneously updated. By adopting the sequential approach to updating with many observations, the order of this matrix can be kept low; order 3, 2 or even 1.

7. Kalman filter for GPS receiver

A GPS navigation receiver most commonly has a Kalman filter to process the raw pseudorange and range rate (or deltarange) measurement data obtained from the code tracker (Figure 2). An eight state filter is implemented to estimate position (3 components), velocity (3 components) and clock errors (clock bias and clock rate).

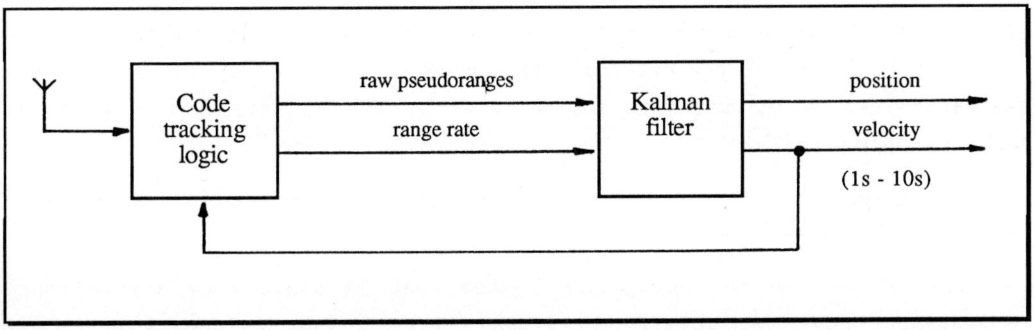

Figure 2: GPS Navigation Receiver

The filter will usually have three 'tuning' settings for high, medium and low dynamics, which may be factory set or externally switcheable, though more on this shortly. Filtered position and velocity values will be output at between

1 and 10 second intervals. The estimated velocity is fedback to aid the code tracking logic.

The system model for the Kalman filter is based on the vehicle kinematics. A constant velocity model disturbed by a constant acceleration between measurement updates is used. The constant acceleration is modeled as an unbiased, normally distributed random forcing function with variance of magnitude σ_a^2.

The significance in this model of the random forcing function is very important. The constant velocity model is unrealistic: no vehicle travels only in a straight line. The unbiased constant acceleration is telling the Kalman filter that for any period between measurement updates the expected acceleration is zero with variance σ_a^2. This ensures that the velocity elements of the state vector are changed during manoeuvres.

The magnitude of the acceleration variance determines how quickly the Kalman filter will respond to manoeuvres, and this variance is usually switcheable within the receiver to account for low, medium of high dynamic applications of the receiver. Altering the value of this variance (in comparison to R_k is sometimes called tuning the filter. If the variance is set low, then the system model will be weighted in favour of the measurements (slow response to manoeuvres). Whereas, if the variance is set very high, the system model is almost totally ignored and the Kalman filter will weight the solution almost entirely in favour of the external measurements. The ability to tune the GPS Kalman filter by the user has implications in GPS/INS integration design as will be seen later.

8. Kalman filter for INS

A stand alone inertial navigation system that is unaided by any external measurement does not utilize a Kalman filter. The navigation equations operate directly on measured accelerations (and angular velocities) to produce a near-continuous output of position, velocity and attitude. However, the intrinsic error behaviour of an inertial system leads to large velocity error in the order of 1-2m/s which results in a position error of about 1800m per hour. To overcome these large navigational errors, aiding from an additional external sensor is necessary.

A schematic of the Kalman filter implementation for an aided IN is shown in Figure 3. The Kalman filter operates on the inertial system errors and may well be external to the inertial navigation system itself. The state vector will consist of the nine inertial system errors in position, velocity and attitude and any sensor errors (such as gyro drift, accelerometer biases, etc) that are to be modeled. Details of the system model are beyond the scope of this paper.

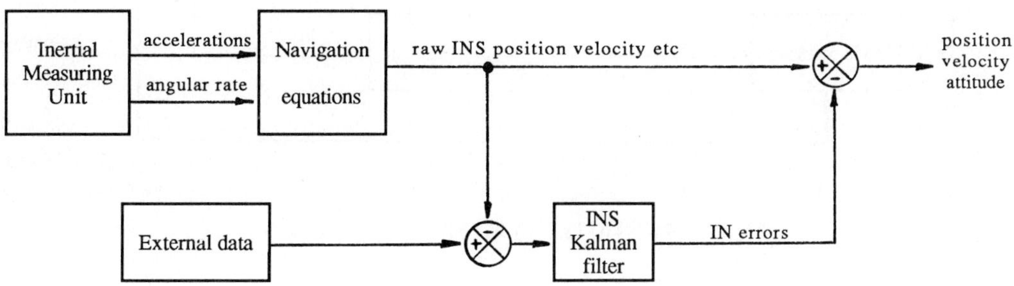

Figure 3: Aided Inertial Navigation System

9. How to integrate GPS and INS

For the optimal results, the best architecture is a fully integrated GPS and INS. This is merely an aided IN, as the schematic in Figure 4 shows, but using the pseudoranges and range rates directly from the GPS receiver. Note that there is no Kalman filter dedicated to processing the GPS data and that the GPS code tracking loop is now aided by the inertial velocity, not the GPS velocity.

In addition to offering the optimal solution, the fully integrated system has several other advantages as listed below.

- There is only one centralized Kalman filter to implement.
- The Kalman filter includes the effects of satellite geometry, and can model the GPS clock errors and INS errors.
- The Kalman filter can accept pseudoranges from less than four satellites.
- Bad GPS pseudoranges can be detected and rejected.

- The INS can provide full navigational capability during GPS outage.
- There is no correlation problem (see §11).

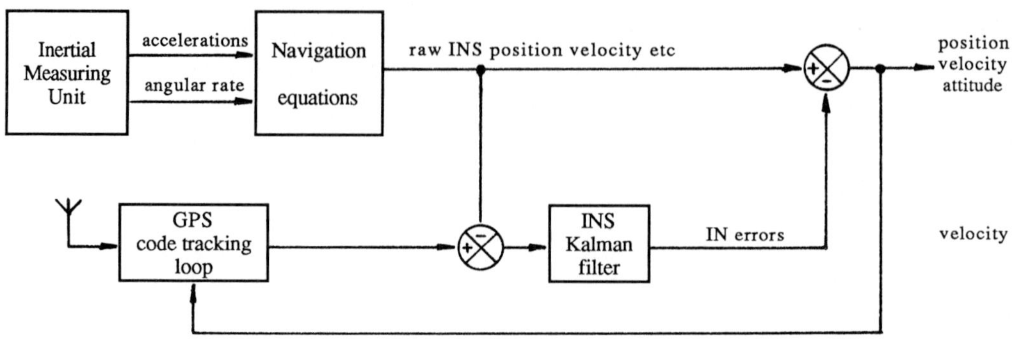

Figure 4: Fully integrated GPS and INS

However, there are some distinct disadvantages of the fully integrated system which are listed below. Comparison should be made with the 'cascaded' alternative discussed in §11.

- There is no separate Kalman filter to process the GPS observations. For this reason the IN Kalman filter must have access to the satellite ephemerides from the GPS receiver.
- The measurement equations are more complex.
- Due to the tight integration a special purpose GPS receiver would have to be used, not any off-the-shelf receiver.
- There is lack of robustness in that if the inertial system fails, either by malfunction of one or more inertial components, or by total failure, navigation capability is totally lost.

10. Open or closed loop architecture

Figure 4 shows the integration for an open loop configuration. Apart from the IN velocity aiding of the GPS code tracking loop, there is no feedback of the estimated position, velocity and attitude. In the closed loop configuration, the computed position, velocity and attitude are fedback to the inertial system. In the case of gimbaled systems, the platform is 'torqued' to remove the attitude errors, and for strapdown systems, the body-to-navigation rotation matrix is corrected. For both systems, the navigation equations are corrected for the position and velocity errors.

The closed loop configuration is very advantageous for strapdown inertial systems. The inertial system errors are kept small, therefore a linear error model is a good approximation. Moreover, the error dynamics are easier to handle. However, the system is unstable and very sensitive to bad measurements.

The open loop configuration is best for gimbaled inertial systems. In this case bad measurements do not affect the INS operation: only the Kalman filter is affected and normality will return on resumption of good measurements. In addition, the Kalman filter can be run external to the inertial system itself. However, the full error dynamics of the inertial system must be modeled and the state vector must be augmented with IN error driving functions such as gyro drift and accelerometer bias.

11. Cascaded Kalman filter approach to GPS/INS integration

An alternative to the fully integrated system is the cascaded Kalman filter approach to GPS/INS integration. A schematic of an open loop implementation of a cascaded filter is shown in Figure 5. Although it is not shown, velocity aiding of the GPS code tracking loop can be achieved by feedback from the GPS or the INS velocity.

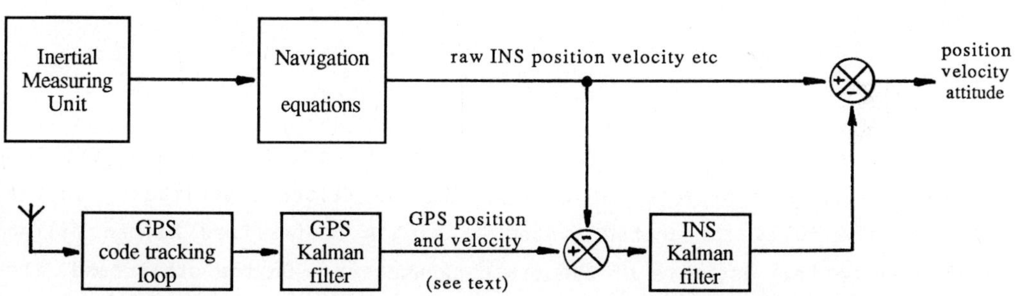

Figure 5: Cascaded Kalman filter for GPS/INS integration

The key feature of the cascaded Kalman filters is the fact that the output from one Kalman filter is used as the measurement input to the next. With this kind of integration architecture, great care has to be taken. To meet the statistical conditions required by the Kalman filter algorithms, the measurements must be uncorrelated between observing epochs (i.e. $E[\underline{v}_j \underline{v}_k^T] = 0$ for all $j \neq k$). However, this is not the case for a sequence of filtered positions output by the GPS receiver. Due to the nature of the GPS Kalman filter, the filtered velocity is effectively uncorrelated.

To overcome the problem of correlated measurements into the second Kalman filter, one of three options are possible:

- Take the raw pseudoranges as measurements for the INS Kalman filter. This is the same as having a fully integrated system.
- Detune the GPS Kalman filter. This will remove any correlation between position measurements.
- If neither of the previous two alternatives are possible, then the filtered position values output by the GPS receiver should only be used as measurements at infrequent epochs, such as every 100 seconds instead of the normal 1Hz rate.

In addition to the problem of position measurement correlation, the cascaded approach has the additional disadvantage that the GPS Kalman filter needs pseudoranges to four or more satellites to be able to accept a measurement update. Nevertheless, the cascaded filter is the simplest implementation. Not only may any GPS receiver and any INS be integrated, but also the GPS and INS Kalman filters are smaller than the filter required for a fully integrated system.

12. Conclusions

There are clearly two approaches to the data processing for GPS/INS integration. Both approaches utilize the Kalman filter algorithms. In the one case, the fully integrated system, a single centralized Kalman filter provides an optimal estimate of the state parameters. On the other hand, the cascaded approach offers a simpler implementation at the expense of suboptimal results.

For both approaches there is the option of an open or closed loop mechanization. The closed loop is ideally suited for strapdown inertial systems, and the open loop for gimbaled inertial systems.

Within the broad categories discussed above there are many detailed variations to overcome stability problems and to match the system to the constraints and priorities of the designer. Space has not allowed any further discussion in this interesting and important area.

13. Bibliography

BROWN A. (1987) *GPS/Inertial Integration.* Lecture notes for 'Tutorials on The Global Positioning System', Series 400: GPS/INS Integration, workshop. Presented at The Antlers Hotel, Colorado Springs, Colorado, September 1987 by Navigation Technology Seminars Inc.

CROSS P.A. (1983) *Advanced Least Squares Applied to Position-Fixing.* Working Paper No. 6, Department of Surveying, North East London Polytechnic, April 1983.

Geier G.J. (1987) *GPS.Integration in Practice.* Lecture notes for 'Tutorials on The Global Positioning System', Series 400: GPS/INS Integration, workshop. Presented at The Antlers Hotel, Colorado Springs, Colorado, September 1987 by Navigation Technology Seminars Inc.

GELB A. (1974) *Applied Optimal Estimation.* M.I.T. Press, 1974.

NAPIER M.E. (1987) *Kalman Filter Algorithms.* Lecture notes for a seminar on 'Kalman Filtering', Department of Civil Engineering, Nottingham University, February 1987.

NAPIER M.E. (1988) *Integration of GPS to Other Navigation Sensors.* Lecture notes for a 'Seminar on the Global Positioning System', Department of Civil Engineering, Nottingham University, April 1988.

Integration of INS, GPS and Doppler for Helicopter Navigation

Bruno M. Scherzinger, D. Blake Reid
Honeywell Limited, Toronto, Canada

Abstract

The Canadian Helicopter Integrated Navigation System (HINS), designed for an ASW helicopter, is described. The HINS Advanced Development Model is configured with an F-cubed RLG inertial navigator, GPS receiver and Doppler velocity sensor, and integrated using multiple Kalman filters and redundancy management algorithms. The system offers high levels of performance, fault tolerance and reliability. Performance characteristics are described, as are possible future developments and applications of the technology.

1. Introduction

The roles of military maritime helicopters include search and rescue (SAR), anti-submarine warfare(ASW) and anti-ship missile defence (ASMD). Many of the missions must be carried out at very low altitudes under all weather and visibility conditions. The increased range, speed and accuracy of modern weapons systems impose stringent accuracy and reliability requirements on the aircraft navigation system. The traditional manual and Doppler dead reckoning navigation methods no longer provide the required navigation accuracy, and also distract the crew unnecessarily from performing mission-critical functions.

Functional requirements for a navigation system which meets modern day maritime helicopter requirements include:

- accurate navigation over water, sometimes in "silent mode" (no electro-magnetic radiation),
- fast reaction capability,
- high degree of fault tolerance,
- tactical navigation relative to a floating sonobouy field or SAR target driven by wind and current,
- simple, yet fully functional, operator interface

The Helicopter Integrated Navigation System (HINS), currently being developed by Honeywell for the Canadian Department of National Defence (DND), is a high performance, redundant navigation system designed to satisfy these demanding requirements. While HINS is specifically targeted for DND's new ASW helicopter, the EH101, it also has application to other helicopter (and fixed wing) platforms, including the European NH90.

HINS integrates multiple navigation sensors, including a ring laser gyro Inertial Navigation System (INS), a 4-beam Doppler Velocity Sensor (DVS) and a high dynamics P-code Global Positioning System (GPS) receiver, to achieve a high degree of accuracy, reliability and fault tolerance. Integration is performed within the navigation computer, which implements (a) multiple Kalman filters for alignment, error estimation/control and sensor calibration, and (b) failure, detection, isolation and reconfiguration algorithms/logic for redundancy management.

This paper reviews the HINS Program, describing in some detail the Honeywell/DND Advanced Development Model (ADM), and concludes with a discussion of future developments of relevance to other NATO helicopter programs - in particular, the NH90.

2. HINS Feasibility Study

The HINS Program was initiated by DND in 1983 with the award to Honeywell Canada of a study contract for configuration design of the integrated nav system for Canada's ASW helicopter. The scope of this study, which was completed in 1985, included a tradeoff analysis of several candidate configurations, design of integration algorithms, development of simulation tools (covariance analysis and Monte-Carlo), and the evaluation of the performance of selected candidates by simulation. Figure 1 shows a "road map" of HINS Phase I activities.

Examples of Phase I covariance analysis results are given in Figure 3 for a typical sonobuoy mission of 3.3hr duration (Figure 2). Shown are radial position error (95%) time-histories comparing performances of two candidate HINS configurations:

a) Form-Fit-Function (F^3) RLG INS with 0.008 deg/hr gyro bias stability, or

b) Dry Tuned Gyro (DTG) Attitude and Heading Reference System (AHRS), with 0.2 deg/hr bias stability,

each with GPS and Doppler. Figure 3(a) shows results for a GPS in-air align with GPS outage at 1800s and Figure 3(b) plots performance for an in-air Doppler align with no GPS. A 1 knot sea bias was assumed in both cases. Performance of HINS with the F^3 INS is clearly superior providing a radial position error rate of about 0.75 nm/hr (95%). AHRS performance is particularly poor during the first hour of the Doppler in-air align scenerio. Here, HINS is unable to calibrate the AHRS gyro biases until the helicopter executes its first large heading change on completion of the outward transit leg. On execution of this maneuver, HINS is able to reduce the system's heading error by continued gyrocompassing, and pull-in the position error produced by the large heading error during the transit leg.

3. HINS Advanced Development Model

The Phase I sudy was followed by the award to Honeywell in 1987 of a contract for the implementation of the HINS Advanced Development Model. Phase II is a three year project in which the ADM will undergo extensive lab, van and flight trials following completion of hardware and software integration. The project will culminate in early 1990 with DND's evaluation flight tests at the U.S. Air Force Central Inertial Guidance Test Facility (CIGTF).

The following paragraphs describe the implementation of the HINS ADM in terms of system architecture, integration methodology and performance requirements.

3.1 System Architecture

The HINS architecture shown in Figure 4 uses two 1553B buses. The navigation bus interfaces the navigation computer system (NCS) with the primary sensors, these being the INS, GPS receiver and the DVS.The secondary sensors are the vertical gyro (VG), strapdown magnetometer and the air data system (ADS), and these interface directly with the NCS via ARINC 429 in the

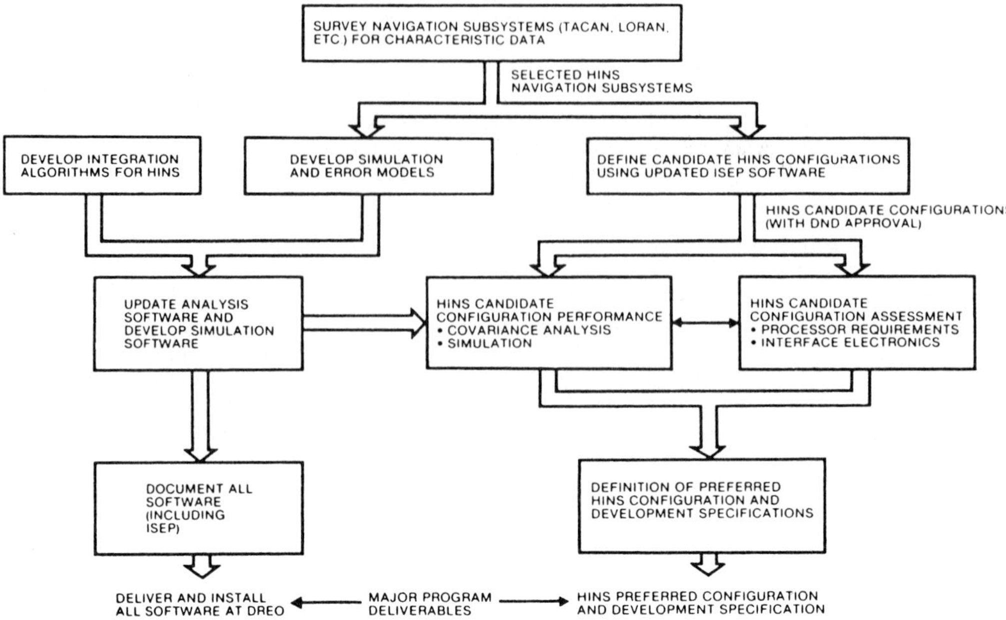

Figure 1: HINS Feasibility Study Tasks and Deliverable Items

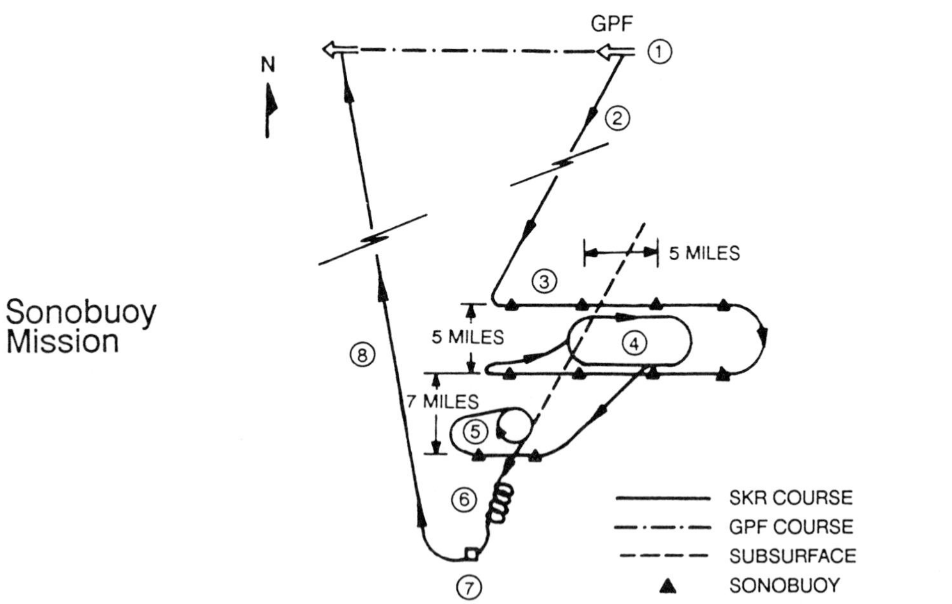

Figure 2: Mission Profile

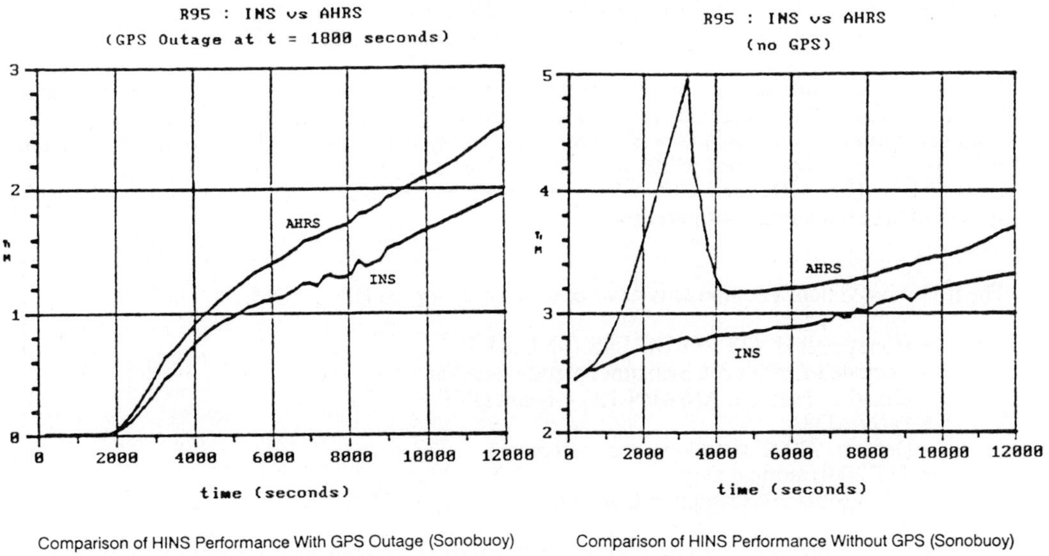

Figure 3: HINS Phase I Study Results

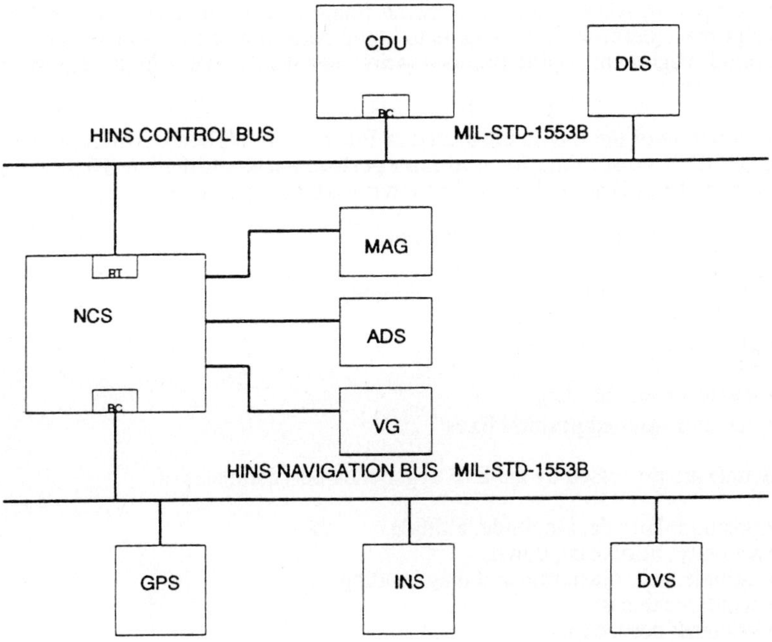

Figure 4: HINS ADM Architecture

case of the VG and ADS, and analog in the case of the mag. The CCS bus connects the NCS with the control computer system (CCS) and the control and display unit (CDU).

This division of hardware is along functional lines. The NCS performs the integrated navigation function (INF) in that it receives primary and secondary sensor inputs, computing the optimally blended navigation solution and performing redundancy management. The NCS also implements area navigation computations while the CCS provides a smart interface between the CDU and the Control Bus. (The CDU and CCS will be replaced by the helicopter's information and display system in the production configuration.)

The following vendor's equipments have been selected for the HINS ADM:

- Honeywell H-423 F3 RLG INS (SNU 84-1)
- Rockwell-Collins 3A 5-channel p-code receiver.
- Canadian Marconi AN/APN 135 4-beam DVS.
- Pacer ADS
- Delveco 9200C strapdown magnetometer.
- JET 204L vertical gyro.
- DY-4 M68020 Navigation Computer

It is anticipated that the VG and magnetometer will be relaced by a strapdown AHRS in the production configuration for back-up roll, pitch and heading.

3.2 Integrated Navigation Function

The HINS is a highly redundant system. To capitalize on this redundancy, the HINS implements a software function in ADA, referred to as the Integrated Navigation Function (INF), within the NCS. The primary purpose of the INF is to blend data from the HINS navigation sensor suite to achieve rapid alignment, optimum navigation accuracy, and a high degree of tolerance to subsystem failure.

The basic structure of the INF is illustrated in Figure 5. It implements multiple Kalman filters for blending of HINS sensor data for optimum performance, and Failure Detection, Isolation and Reconfiguration (FDIR) algorithms/logic for redundancy management.

The following sensor data are input to the INF:

- INS position, heading, velocity, attitude,
- GPS position and velocity
- DVS and ADS velocities
- baro altitude
- magnetometer heading
- operator inserted position fixes

The input data are processed by the INF to generate best estimates of:

- position: latitude, longitude, altitude,
- velocity: north,east, down,
- attitude: roll, pitch, true and mag heading.
- wind: north,east
- sea bias: north,east

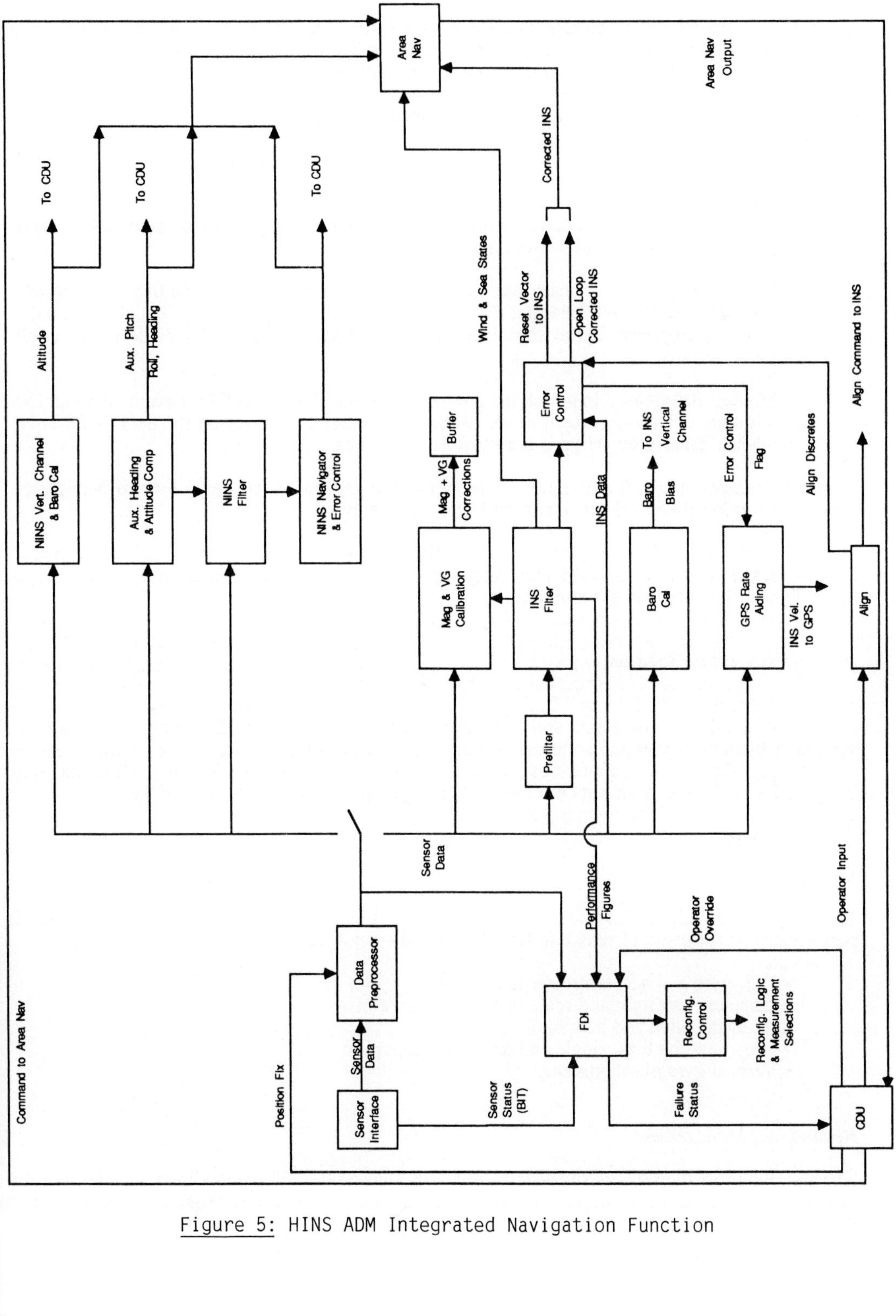

Figure 5: HINS ADM Integrated Navigation Function

which are output for use by functions internal and external to HINS. The INF also generates a reset vector for INS alignment, a velocity vector for GPS rate aiding, and figures of merit providing estimates of the accuracies of the computed navigation variables.

Specific functions performed by the INF are described briefly below:

Alignment

The INF directs the INS through one of three principle alignment sequences, based on the status of the aircraft and the availability of sensor data. These are:

- *Ground Align*, for stationary-base leveling and gyrocompassing of the INS. The operator may select the Stored Heading (SH) or Best Available True Heading (BATH) submodes for rapid response. In the absence of initial heading, the INF will execute a wide angle gyrocompass.

- *Moving-Base Align*, in which the INF employs Doppler and/or GPS data to effect an INS alignment. This alignment mode is normally entered in-air, or on the deck (GPS only) when a Transferfer Align is not possible or practical.

- *Transfer Align*, for on-deck alignment of the INS in which Ships Inertial Navigation System (SINS) velocities and positions are employed.

The INF implements a Kalman filter for each of these alignment modes, with the exception Ground Align which is internal to the INS. The alignment errors estimated by the filter are applied as resets to the INS for leveling and gyrocompassing.

Blended Navigation Solution

In Nav Mode, the INF Kalman filter computes estimates of INS position, velocity and attitude errors from available sensor data and applies these as corrections to the INS outputs to obtain the optimally blended navigation solution. (Resets are not applied to the INS in Nav mode to ensure the integrity of the INS computations.) If the INS fails, HINS enters the No INS (NINS) mode in which GPS, DVS and ADS data are blended, using magnetometer heading and VG roll/pitch, to obtain the best available navigation solution.

Calibration

The INF continuously calibrates HINS using available sensor data for enhanced reversionary mode performance in the event of sensor failure. Calibrated error sources include:

- INS, gyro and accelerometer biases
- Doppler, boresight and scale factor errors, sea biases
- Air Data, wind and baro bias
- Mag, heading bias, single and double cycle errors
- Vertical gyro misalignments

Redundancy Management

The INF monitors the sensors/subsystems of HINS, determines a fault condition, isolates a faulty sensor/subsystem, and reconfigures the remaining sensors into a new configuration for continued operation.

3.3 Kalman Filter Mechanization

The INF implements four Kalman filters:

- INS Filter, employed for blending GPS and Doppler measurements with INS outputs in Nav mode, Transfer Align mode and Moving-Base Align mode.
- *NINS (No INS) Filter*, employed for blending GPS, Doppler and air speed measurements if the INS fails, using (calibrated) heading from the magnetometer and roll/pitch from the VG. In this case, a simple dead reckoning computation (NINS Navigator) is implemented together with the NINS Filter to derive continuous best estimates of position and velocity.
- *Mag Cal Filter*, employed for automatic calibration of the magnetometer using INS roll, pitch and heading as reference to account for offset, hard iron (single-cycle) and soft-iron (double-cycle) errors (which are manifested as bias (3) and scale factor (9) errors in the strapdown magnetometer's outputs).
- *Baro Cal Filter*, employed for calibration of baro altitude using GPS altitude measurements

These filters are mechanized using the robust U-D factorized form of the Kalman algorithm, which guarantees the numerical stability of the computations and provides greater numerical precision than the more standard Kalman formulations.

The INS filter estimates 18 HINS error states in Nav and Moving-Base Align modes. These states are:

- x,y position errors (wander azimuth)
- x,y velocity errors (wander azimuth)
- x,y, z platform misalignments (wander azimuth)
- x,y accelerometer biases (body)
- x,y,z gyro biases (body)
- DVS x scale factor error
- DVS z boresight misalignment
- north, east sea biases
- north, east winds

The NINS filter is a 12-state mechanization similar to the INS filter. The Mag Cal filter estimates 12 magnetometer states: 3 field biases and 9 along-axis and across-axis field scale factor errors. The Baro Cal Filter is a simple 1-state mechanization which estimates baro bias on the basis of GPS altitude measurements. The INS is cycled at an interval of 10 seconds (nominal). The NINS and Baro Cal filters are cycled at an interval of 1 second. The Mag Cal filter is activated on power down of HINS or on entry in to NINS mode, when it processes mag and inertial data which have been acquired during the mission prior to one of these events.

A prefilter is implemented to construct measurements for processing by the INS filter. These measurements are derived by differencing GPS, DVS and ADS data with the positions and velocities computed by the INS. Measurements are computed at intervals of 1 second and averaged over 10 samples prior to processing by the filter. The effects of averaging on error dynamics and measurement geometry are accounted for in the filter formulation. In addition, DVS/INS velocity differences are integrated over the 1 second prefiltering interval to further attenuate Doppler fluctuation noise.

3.4 Redundancy Management

The redundancy management function of HINS is performed by the Software element called Failure Detection, Isolation and Reconfiguration (FDIR). This is shown in Figure 2-6. The HINS FDIR performs subsystem failure detection within the navigation computer to detect both hard and soft failures. When a subsystem failure is detected HINS will automatically reconfigure. The failed system data is isolated from the integrated solution to prevent corruption of the output data. The operator is able to override this automatic reconfiguration and either accept or exclude any of the sensor data.

Fault detection and isolation (FDI) is performed sequentially at 5 levels.

1. built in test,
2. direct reasonableness test,
3. comparison of subsystem outputs,
4. Kalman filter measurement processing,
5. indirect reasonableness test.

FDI can find and isolate hard failures, generally detected via BIT, and soft failures which are characterized by a slow deterioration of the navigation sensor data quality. The order of processing is such that the tests for hard failures are done fINSt, and once these are passed, the tests for soft failures are performed.

If any of the five tests detects and successfully isolates a subsystem failure, system reconfiguration is undertaken automatically.

Built in Test

Each primary subsystem used in HINS provides status flags indicating a failure/no-failure condition of components of the subsystem. Should a failure flag be raised, the corresponding sensor/subsystem will be removed from HINS and the overall system reconfigured. However, the operator can at any time reintroduce or remove a subsystem. Thus, if BIT indicates failure of some subsystem capability which the operator deems not to be essential for his mission, such as Doppler memory mode, he could override automatic FDIR. This capability also protects against spurious BIT signals causing permanent removal of a good sensor.

Direct Reasonableness Test

Direct tests of sensor reasonableness are performed by comparing a sensor quantity with a predefined bound. These quantities will be averaged over appropriate time intervals to remove undesired effects. Once a sensor exceeds the predefined bound, it is isolated.

Comparison of Subsystem Outputs

Comparison of subsystem outputs involves forming differences of navigation parameters and comparing these against a bound. This test, as with the tests of reasonableness, operates on raw quantities from navigation subsystems averaged over a short time. The following are examples of tests performed in this failure detection stage:

- HINS and GPS Ground Speed
- HINS and Doppler speed
- GPS and Doppler speed
- HINS and Magnetometer heading

Once a detection based on the comparison of subsystem outputs is made, the FDIR attempts to determine which subsystem failed and the subsystem is isolated and removed.

Kalman Filter Measurement Processing

HINS uses a measurement rejection mechanism in its Kalman filters. Like quantities output by the navigation subsystems (HINS, GPS, DVS, ADS, MAG) are compared against each other. This comparison is accomplished by monitoring filter measurements in the implemented Kalman filter. Once a measurement is rejected, it is not used in the state and covariance update at the time of the rejection. This reconfiguration is not permanent since a subsequent measurement may not be rejected.

A subsystem failure is declared after a measurement has been rejected, in sequence, a specified number of times. When this occurs, the subsystem is isolated/removed.

Indirect Reasonableness Test

The INS filter's estimates of accelerometer biases and gyro drift rates provide indications of HINS health. In addition, filter estimates of wind states, sea current, Doppler boresight and scale factor errors, and baro bias can serve as indirect checks on the Doppler and Air Data Systems.

System Reconfiguration

After isolation of a failed subsystem, this subsystem is removed from the integrated solution. The operator retains ultimate control over system reconfiguration and can override automatic reconfiguration at any time. The main reconfiguration modes are outlined below:

a) Reconfiguration Following INS Failure

If the INS fails, then HINS reverts to NINS mode in which GPS, DVS and ADS are integrated using the magnetometer as the heading reference and the VG for roll/pitch. Prior to this reconfiguration the operator may choose to impose a re-alignment of the INS if he deems the problem to poor alighnment rather than failure.

b) Reconfiguration Following Failure other than INS

The measurements using the failed sensor data in filter processing are not constructed for the remainder of the mission (unless otherwise commanded by the operator). The exceptions here are the GPS and the DVS. After isolation of a GPS failure and removal of the GPS from the system, the GPS status will be continuously monitored. Should GPS data of good quality once more become available, the GPS measurements will again be used. The same applies to the DVS.

Operator Override

The operator has the option of overriding automatic system reconfiguration either by reintroducing a subsystem into the HINS configuration which FDIR would eliminate, or by removing a subsystem. An override takes precedence over all FDIR reconfiguration actions, including those induced by BIT failures. Once done, no further automatic reconfiguration, other than that triggered by a BIT, shall be done by the system.

3.4 Performance Requirements

The performance of HINS will vary with the particular mode or sensor configuration. These modes may change either by operator choice or automatically under FDIR control. System performance with and without GPS is specified as follows:

With GPS

radial position error (95%)	0.016 nm
radial velocity error (95%)	0.47 knots
true heading error (rms)	0.1 deg at 60 degrees latitude

Without GPS (In-Air Align, 1Knot Sea Bias)

radial position error rate (95%)	1.0 nm/hr
radial velocity error (95%)	2.4 knots
true heading error (rms)	0.1 deg at 60 degrees latitude

4. Future Trends

This HINS design points to future trends in integrated navigation systems for military helicopters. Key design issues will include, in addition to the performance and functional versatility objectives of HINS, size, weight, life-cycle cost, reliability and expansion capability. The HINS experience shows that the RLG-based strapdown INS contributes to significant improvements in all of these categories.

An example of this is the Autonomous Navigation System (ANS) proposed by Honeywell for the European NH90 ASW helicopter. The architecture for the proposed system is shown in Figure 7. The ANS employs dual Honeywell H764 Inertial Reference Systems (IRSs). The H764 is a small medium accuracy IRS which currently uses GG1328 RLGs (0.02 deg/hr bias stability) and Sundstrand QA2000 accelerometers (50 micro-g bias stability). Honeywell plans to replace the GG1328 in the future with the new smaller, higher performance GG1320 RLG (0.008 deg/hr bias stability or better), which is now nearing completion of its development cycle. Salient features of the ANS architecture are the embedded GPS receiver and the embedded navigation computer which implements a modified version of the HINS Integrated Navigation Function. These features result in smaller size and lower weight, with only moderate decrease in performance when compared with the F-cubed HINS ADM.

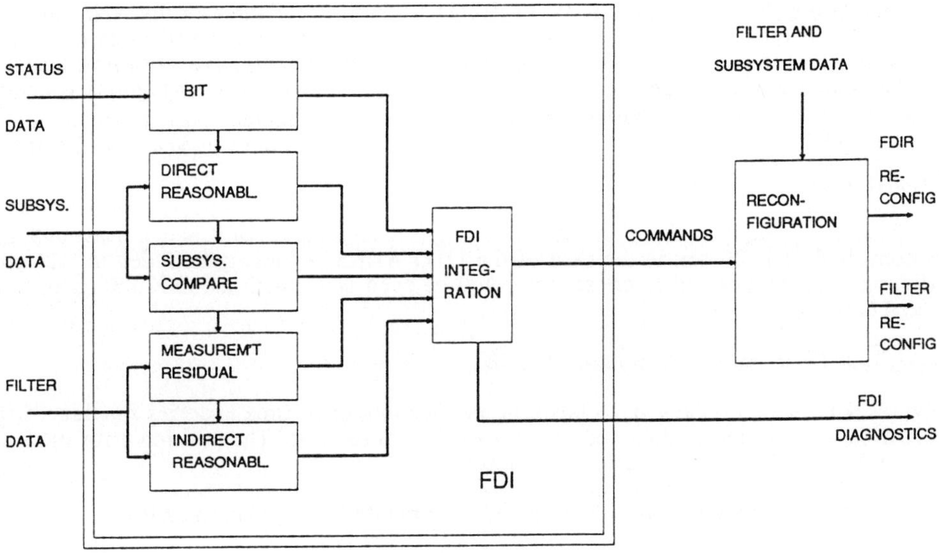

Figure 6 (2-6): HINS ADM Redundancy Management

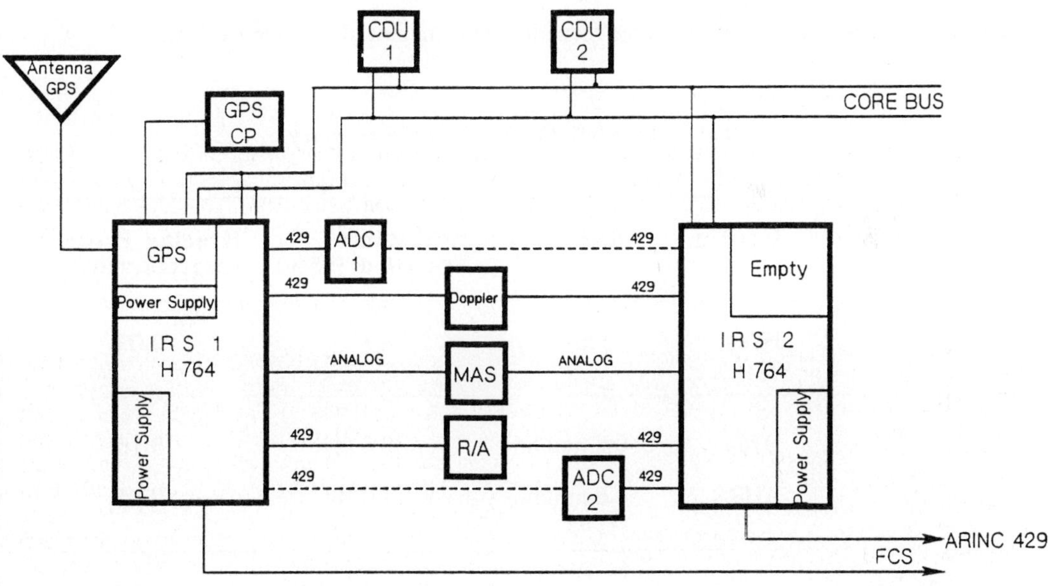

Figure 7: NH90 Autonomous Navigation System

The following graphs illustrate some projected performance figures of the ANS obtained by covariance analysis, illustrating for comparison the difference between the H764-based system and a system which implements a DTG AHRS (0.2 deg/hr bias stability). All error time histories are for the sonobouy mission profile shown in Figure 2, with the exception of Figure 8 which shows heading error for a stationary ground align.

Ground Align

As illustrated in Figure 8, the ANS with H764 aligns to an accuracy of 0.34 deg (rms) in 5 minutes as compared to 1. 26 deg in the case of the DTG AHRS, achieving 0.25 deg in 9.2 minutes. Alignment accuracies and response times will be even better with the GG1320 gyro, or using stored heading.

Navigation Performance - Sonobuoy Mission

Figures 9 through 11 show ANS position and heading error time histories for three different scenarios, all with in-air align and assuming a 1 knot sea bias. The position error plots are 95 percentile radial (R95); the heading error plots are rms.

1. Full-Up System: H764 integrated with GPS, DVS, Mag and ADS
2. GPS Failed: same as Scenario 1 except no GPS
3. GPS Align and Outage, DVS Failed: same as Scenario 1 except no Doppler and GPS present for only first half hour

Also shown in Figures 10 - 11 are corresponding results for a DTG AHRS replacing the H764 in Scenarios 2 and 3.

The results corresponding to these three scenarios are summarized in Table 1 below.

Table 1: ANS Performance Summary

Scenario	Config.	R95	% Distance Travelled (95%)	Heading Error (degrees rms)
1	H764	13 m	-	0.03
2	H764	0.8 nmi/hr	0.8 %	0.10
	AHRS	1.0 nmi/hr	1.0 %	0.45
3	H764	1.3 nmi/hr	1.4 %	0.08
	AHRS	4.4 nmi/hr	4.6 %	0.36

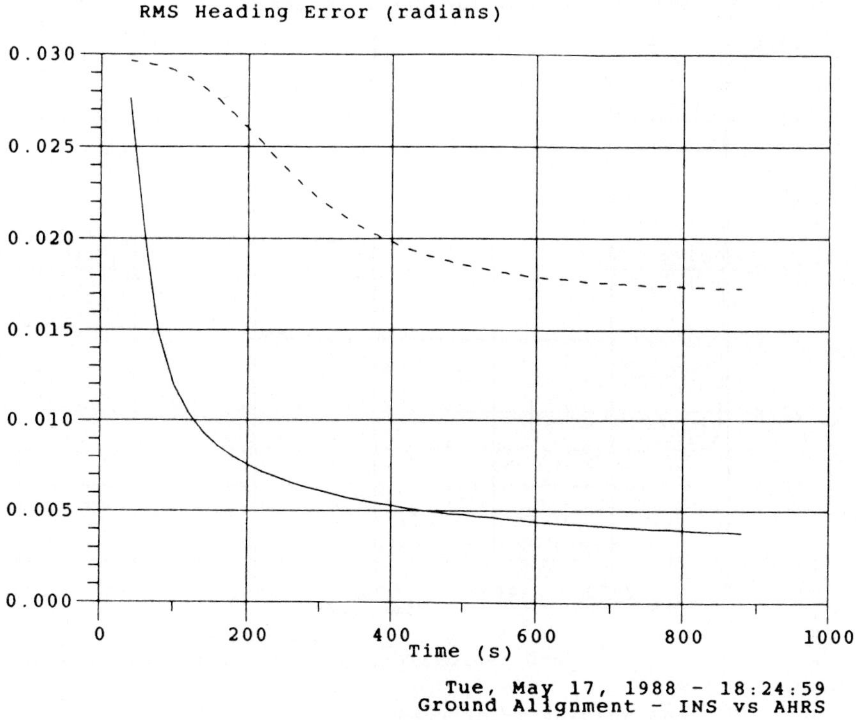

Figure 8: Heading Error Comparison: INS vs AHRS after a 10 minute ground alignment

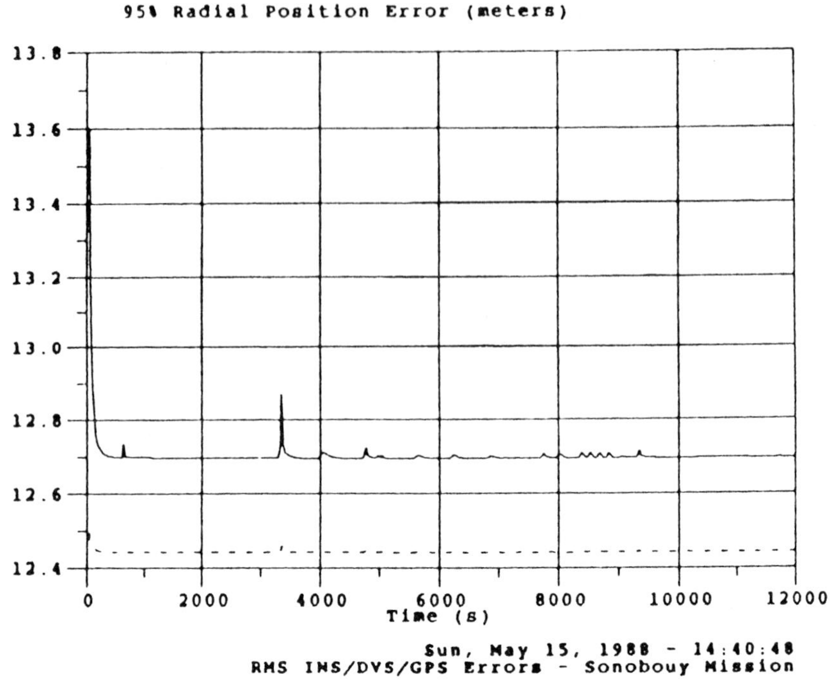

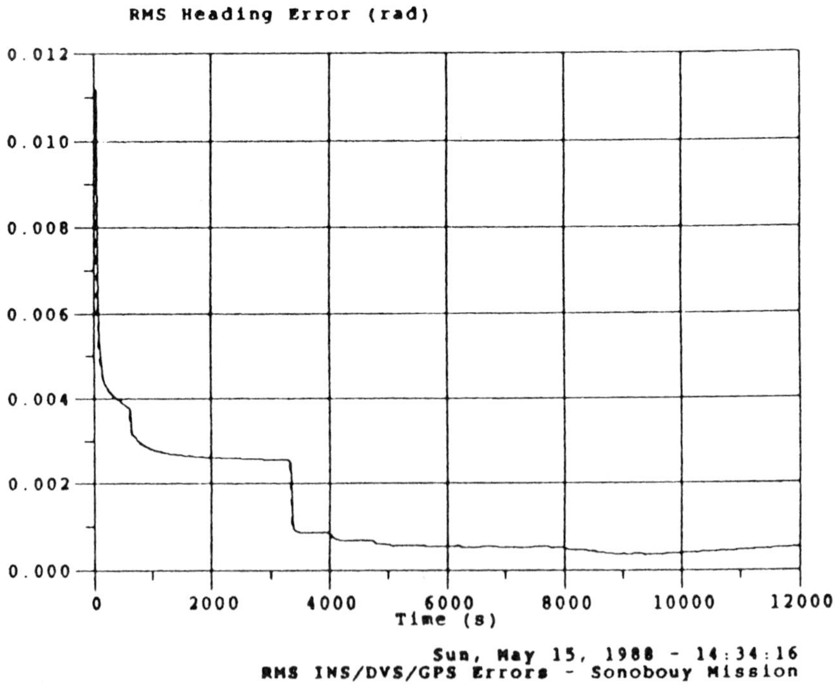

Figure 9: Performance of the full system INS/GPS/DVS/ADS

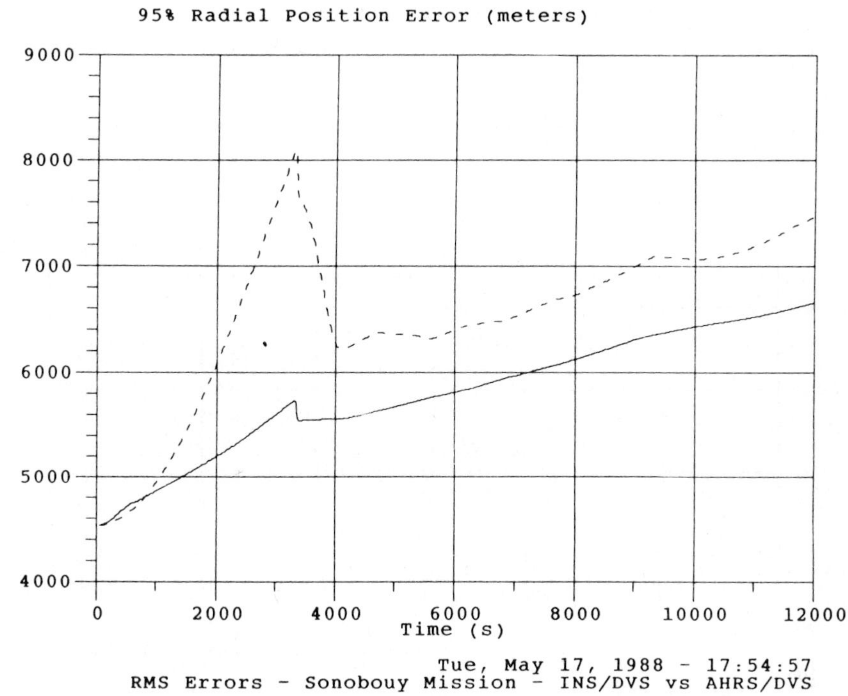

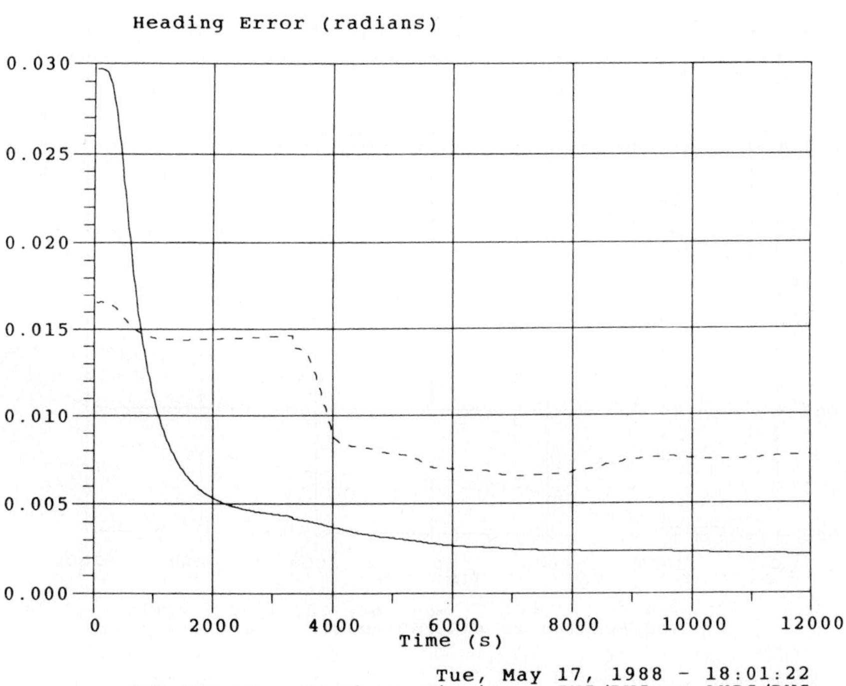

Figure 10: Performance Comparison:
INS/DVS/ADS/mag vs AHRS/DVS/ADS/mag

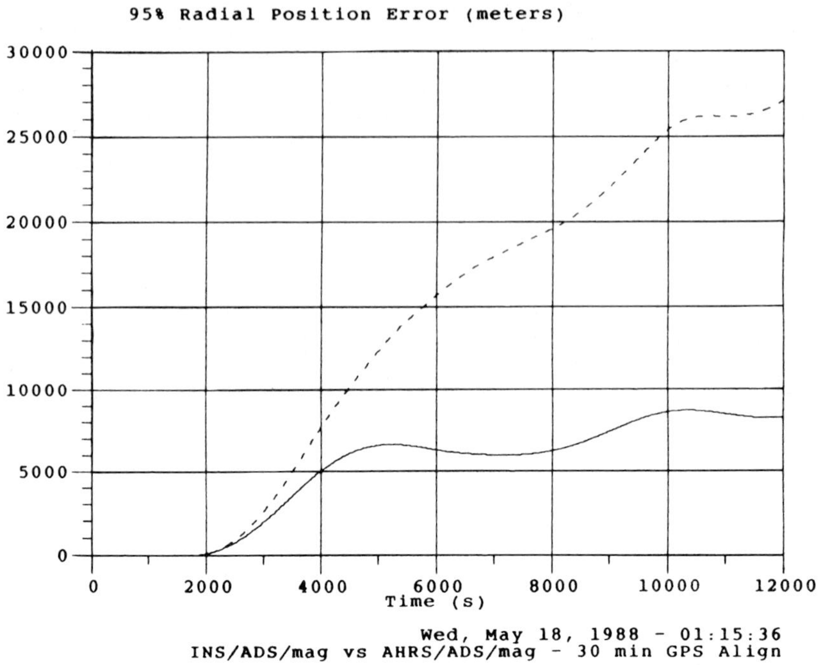

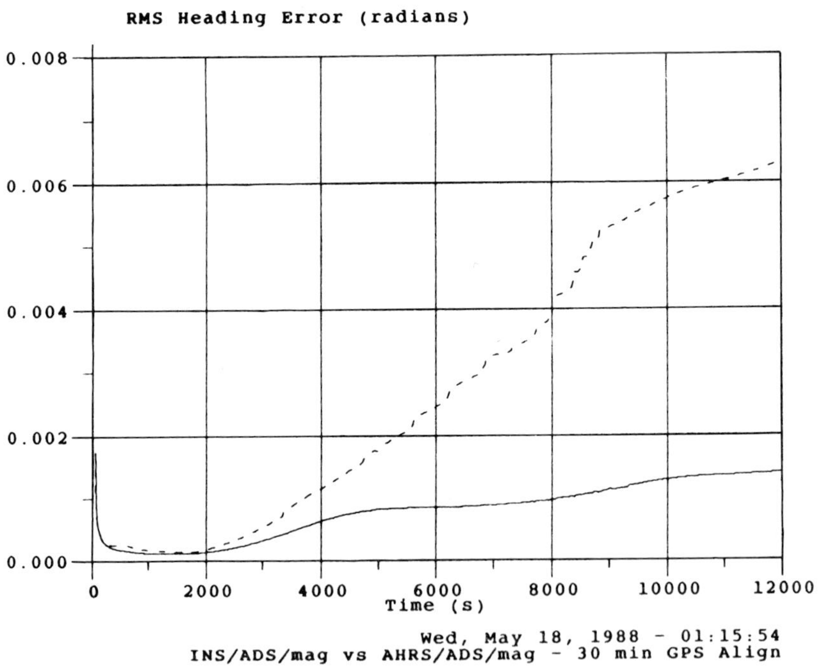

Figure 11: Performance Comparison:
INS/ADS/mag vs AHRS/ADS/mag after a 30 min In-Air GPS alignment

These results indicate that the ANS incorporating the H764 with Doppler damping and using the integration methods described above will satisfy the NH90 requirement of 1% of distance travelled (95%) over the majority of the sonobuoy mission if GPS fails. The performance compliance of the ANS configured with a DTG AHRS is marginal, even if the proposed integration methods are used. In fact, AHRS performance characteristics during the first hour of the mission, prior to the first significant heading change, are particularly poor due to the large uncompensated heading error (0.9 deg) which produces a position error rate of approximately 3.5% of distance travelled (95%).

5. Conclusion

Honeywell's approach to navigation systems integration, when coupled with proven RLG technology, will provide the level of performance, fault tolerance and reliability necessary to meet the demanding mission requirements of the maritime helicopter. Integrated system performance, weight, volume and reliability characteristics are expected to improve significantly in future years with continued improvements in the RLG and with the ever increasing power of microelectronics necessary for implementation of complex computational tasks.

Highly Stabilized Infrared Diode-Laser for Optical Pumping of Rb-Xe-NMR Gyroscope

Peter Scheufler, Klaus-Dieter Dünnebeil, Thomas Vetter, Michael Mehring
2. Physikalisches Institut, Universität Stuttgart

Summary

The properties and function of diode-lasers is discussed with special emphasis on mono-mode operation and selective optical pumping of Rb atoms. Several control loops are employed for temperature stabilization ($\Delta T \sim 10^{-3}$K) and current stabilization ($\Delta I/I = 10^{-5}$).

1. Introduction

The operation of NMR gyroscopes requires the use of light for optical spin polarization and for the detection of the precessing nuclear spins (see lecture by M.Mehring et.al. this volume)[1]. In the Stuttgart NMR gyro test setup the atomic gas mixture Rb/Xe is used. In order to optically spin-polarize the Rb atoms in the gyro sample cell, an intense light source is required, which selectively excites the D_1-line of the Rb atoms. The spectrum of Rb vapour is shown in Fig.1. A number of lines occur around 794.76nm. All lines are very narrow and are broadened by Doppler effect. The line width is typically 500MHz. Two isotopes, namely ^{85}Rb (72,8%) and ^{87}Rb (27,2%) contribute to the spectrum.

The operation of the NMR-gyro requires the preparation of a well defined spin-polarized state. Therefore the pumping process has to be performed very selectively. We choose e.g. the D_1-line of ^{87}Rb at 794.76nm. In order to avoid pumping of other transitions as well, a very narrow band light source is required, as is evident from Fig.1.

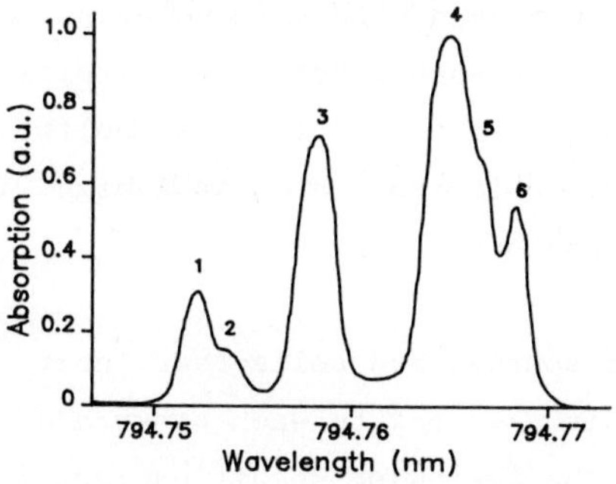

Fig.1 Structure of the D_1-absorption line of Rb
1,2,5,6: Transitions in ^{87}Rb
3,4: Transitions in ^{85}Rb

light source	opt. power	line width	advantages	disadvantages
current pumped spectral lamp	50μW	0.1nm	easy to use cheap	high noise level low power
RF-pumped spectral lamp	100μW	0.02nm	low noise	cell preparation low power
infrared-LED	20mW	30nm	easy to use cheap, robust	low spectral power
infrared diode-laser	3mW	<0.001nm	high power narrow line	complex stabilization very sensitive mode hopping
dye laser	100mW	<0.001nm	v. high power narrow line easy frequency tuning	very expensive pumplaser required maintenance complex stabilization

Table 1 Light sources for optical pumping of Rb atoms

Moreover, many photons are necessary to achieve a high spin polarization. The intensity of the light source must be correspondingly high. Typically several mW light power are required. Since part of the light is splitt off and used for optical detection of the spin-polarization, also a high amplitude- and frequency stability is mandatory.

In the Table a number of different light sources are collected. Their features, advantages and disadvantages with respect to each other are compared. From the foregoing discussion the following specifications for the light source are required for optical pumping of Rb atoms:

- Wavelength 794.76nm (tunable)
- high intensity (several mW)
- narrow line (<500MHz)
- large amplitude stability
- high frequency stability

A comparison of the different light sources according to the Table leads to the conclusion that (a) a highly stabalized infrared laser-diode or (b) a highly stabalized dye laser system fulfill the necessary requirements [2].
In the Stuttgart NMR gyro test setup we employ both sources. Here we discuss the details of the infrared diode-laser setup.

2. The Infrared Diode-Laser

Infrared laserdiodes are commonly made of GaAlAs structures and are currently available in large quantities due to their use in compact disc (CD) players and laser printers. They emit in the range 760-800nm. If a particular wavelength is required the appropriate diode must be selected from a larger number of diodes. Single longitudinal and single transverse modes can be excited if special precautions are taken. The light power ranges from 3 to 10mW for monomode continously (cw) emitting diodes. Diodes with larger output power are available, however, spectral purity

decreases drastically and mode interactions set in. The frequency noise vastly increases in diodes with higher power.
The commercially available diodes come with some usefull features like an integrated monitor diode for measuring and controlling the output power. The internal line width seems to be limited to about 10MHz due to the electron-hole recombination process [3].
The wavelength can be adjusted by varying the temperature and/or the injection current. The typical tuning parameters are:

$$\frac{d\lambda}{dT} \simeq 0.2\frac{nm}{K} \; ; \; \frac{d\lambda}{dI} \cong 0.002 \; \frac{nm}{K}$$

Fig.2 shows the dependence of the wavelength on temperature. It demonstrates quite clearly, that temperature tuning of the wavelength is feasable. However, there is a limit to temperature tuning because of mode hopping as is evident from Fig.2 [4].

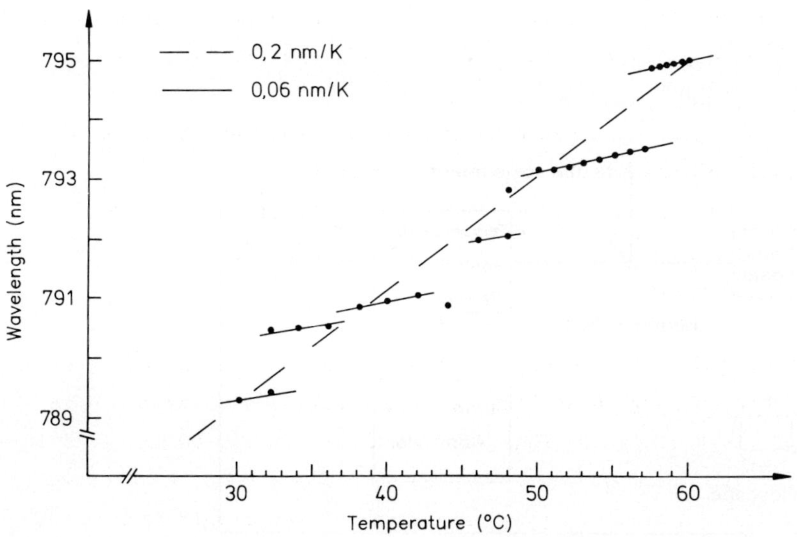

Fig.2 Temperature dependence of the emission wavelength of a GaAlAs laser diode type Hitachi HL 7801

Other features of infrared laserdiodes include (i) polarization: linear with ratio 1:50 up to 1:500, (ii) operating temperature -20 ℃ - +60 ℃, (iii) beam divergence: 10 °x30 ° (elliptical) which can, however, be collimated by an external lense system.

Laser diodes for optical pumping of Rb atoms are being selected according to the following procedure:

(a) select diode with $\lambda(25\,^{\circ}C) \simeq 793nm$.

(b) operate diode at appropriate temperature for $\lambda = 794.76nm$.

(c) stable single mode operation must be maintained up to necessary output power.

If the laserdiode fulfills these requirements it is acceptable for optical pumping of Rb atoms.

3. The Infrared Diode-Laser Setup

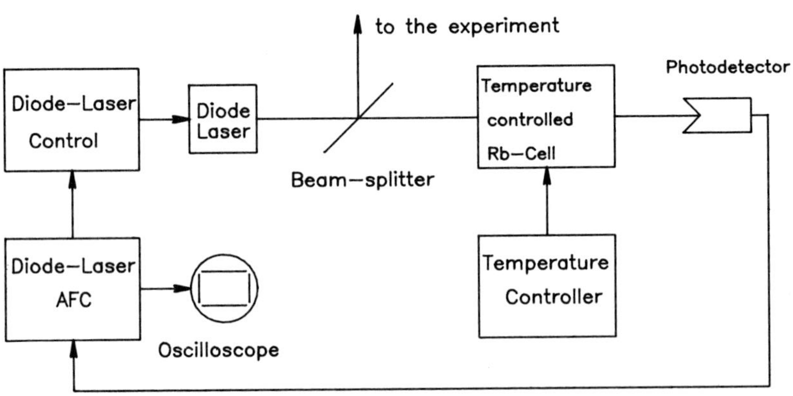

Fig.3 Diode-laser setup

The blockdiagramm of the infrared diode-laser setup is shown in Fig.3. The laser-diode control unit contains highly stabilized and temperature controlled electronic circuits, which stabilize the operating conditions of the laser diode to a high degree, namely to a temperature stability of ΔT=2mK and current stability of $\Delta I/I=10^{-5}$.
Part of the prestabalized laser light is split off by the beam splitter and traverses a temperature controlled Rb vapour cell before it is detected by a photodiode. The laserdiode AFC unit modulates the injection current with $I_{mod} \simeq 10\mu A$ at a frequency of 20KHz. This introduces a frequency and amplitude modulation of the laserlight. After traversing the Rb cell the light is phase synchronously detected which leads to the derivative of the absorption signal as shown in Fig.4.
This signal can be used as control voltage to lock the laserdiode to a specific wavelength in the Rb-spectrum.

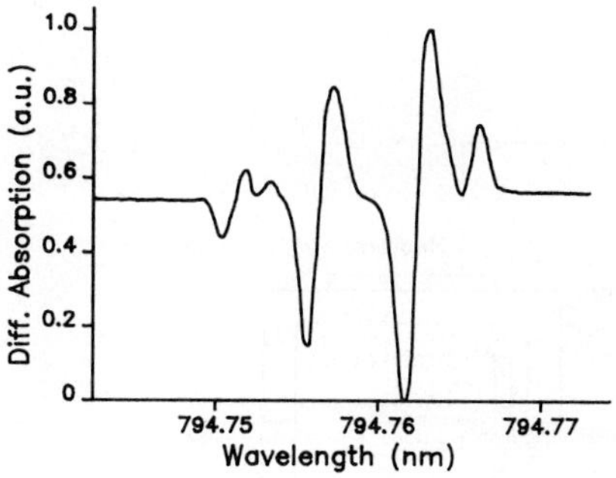

Fig.4 Derivative of the Rb-absorption signal

Let us now turn to the detailed features of the laserdiode control unit. The blockdiagramm is shown in Fig.5. The laserdiode itself is housed in a thermally stabilized metal compartment and is attached to a thermally stabilized copper block. Two separate control circuits regulate the temperature of these units. A constant temperature of the laserdiode with accuracy of 2mK can be obtained this way. The temperature tuning of the laserdiode is achieved by feeding a ΔT-signal into the inner control loop.

The current of the laserdiode is regulated to a relative stability of about 10^{-5} with a thermally stabalized control circuit. The modulation signal for the AFC is added to the control voltage. The output power is constantly monitored and kept constant by the current control loop. A safeguard circuitry takes care that the current is kept always below demage level.

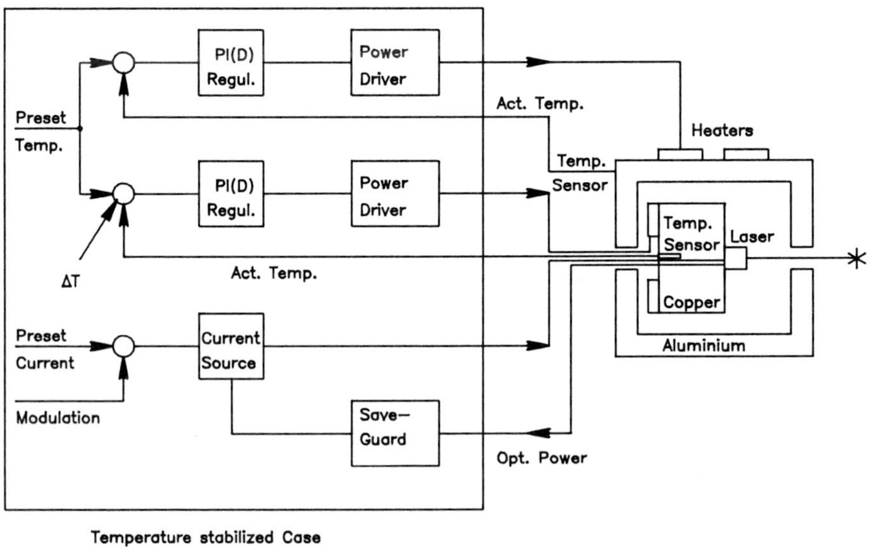

Fig.5 Diode-laser control unit

The blockdiagramm of the AFC-unit is shown in Fig.6. It is based on the frequency modulation of the laserlight with 10-100kHz, which after phase synchronous detection leads to a derivative signal as described before. If no offset is present this signal changes sign when passing through resonance. By feeding this signal with the appropriate phase back into the frequency control of the laserdiode a frequency locking of the laserdiode onto a particular Rb absorption line is achieved.

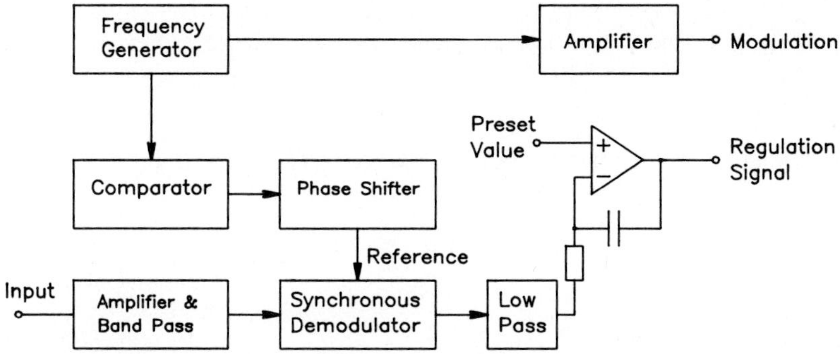

Fig.6 AFC-unit for frequency locking of the diode-laser

4. Operation of the Diode-Laser Unit

After turn on of the infrared diodelaser unit the appropriate temperature is chosen for the operation at a wavelength of 794.76nm within an accuray of $\Delta T<10mK$. The injection current is set to the appropriate value and regulated to $\Delta I/I<10^{-5}$. The operational stability is about 100MHz/h free running and less than 30MHz total when locked to the Rb D_1-line. The pumping power is sufficient for polarizing Rb to more than 50% in a $3cm^3$ cell. The unit has been used in a number of spin-polarizing experiments and has proven to be stable over long periods of time.

Acknowledgments: The Deutsche Forschungsgemeinschaft (SFB 228) has given a financial support.

References:

1. L.K.Lam, E.Phillips, E.Kanegsberg, G.Kamin, Proc.SPIE.Int.Soc.opt. eng. 412, 272 (1983).

2. J.C.Camparo, Contemp.Phys. 26, No.5, 443 (1985).

3. T.Takakura, K.Iga, T.Tako, Jpn.J.Appl.Phys. Vol 19, No.12, L725 (1980).

4. T.Okoshi, K.Kikuchi, Electr.Lett. Vol 16, No.5, 179 (1980).

Comparative Study of Inertial Measurement Systems for Geodetic Applications

Dietrich Schröder
Geodätisches Institut, Universität Stuttgart

Summary

Inertial measurement systems are being increasingly used for geodetic applications. For this purpose stabilized platform systems have been used almost exclusively up to now, but nowadays strapdown systems with the inertial sensors fixed to the vehicle are becoming more and more important. For both mechanizations mentioned above the basic mechanization equations are derived, which result in a system of coupled, non-linear and inhomogeneous differential equations. The equations are linearized, and with the help of the solved linear incremental system of differential equations the effects on point positioning of different sensor model parameters are discussed.

1. Introduction

The basic components of an inertial measurement unit (IMU) are the inertial sensors, at least three *accelerometer* and three *gyroscopes*. The output of the accelerometers contains after a double time-integration information about the position of the IMU in *geometry space* as well as in *gravity space*. Here we will confine ourselves to positioning in geometry space assuming the gravity field is known.

The two mechanizations discussed differ in the *determination of the orientation of the accelerometer input axes*. Considering the strapdown system, determination of orientation is performed by *measuring* the momentary angular velocity with the help of (optical or mechanical) gyros. Regarding the second mechanization, the gimbal suspended platform carrying the accelerometers and gyros, the gyros serve as orientation *stabilizing* elements, although the platform may be slaved by a feedback control system. Here we consider only the special case, where the commanded torques feeded to the gyros are chosen to maintain the accelerometer input axes parallel to the instantaneous horizontal frame.

2. Mechanization Equations

From the point of view of strapdown- as well as of platform system configuration the *centre of mass* P_d of the IMU is for the *translational motion* and the orthonormal triad $\underset{\approx}{d} = \{\underset{\sim}{d}_1, \underset{\sim}{d}_2, \underset{\sim}{d}_3\}$, located in P_d and orientated correspondingly to the ideal accelerometer and gyro input axes, for the *rotational motion* a natural describing system of the mechanic of rigid bodies. Besides the IMU fixed frame there are the three triads $\underset{\approx}{c} = \{\underset{\sim}{c}_1, \underset{\sim}{c}_2, \underset{\sim}{c}_3\}$, $\underset{\approx}{b} = \{\underset{\sim}{b}_1, \underset{\sim}{b}_2, \underset{\sim}{b}_3\}$, $\underset{\approx}{a} = \{\underset{\sim}{a}_1, \underset{\sim}{a}_2, \underset{\sim}{a}_3\}$ as orthonormal base vectors, which are located respectively in the *topocentre* P_c, in the *centre of mass* P_b of the earth and in the *centre of mass* P_a of the whole dynamical system. The frames $(P_d, \underset{\approx}{d})$, $(P_c, \underset{\approx}{c})$, $(P_b, \underset{\approx}{b})$ and $(P_a, \underset{\approx}{a})$ form the commutative diagram of figure 1.

inertial frame	$P_a, \underset{\approx}{a}$
	$\overrightarrow{P_aP_b} = \underset{\sim}{x}$ $R_{ba} = R_E(0,0,\vartheta) = \begin{bmatrix} \cos\vartheta & -\sin\vartheta & 0 \\ \sin\vartheta & \cos\vartheta & 0 \\ 0 & 0 & 1 \end{bmatrix}$
equatorial frame	$P_b, \underset{\approx}{b}$
	$\overrightarrow{P_bP_c} = \underset{\sim}{y}$ $R_{cb} = R_E(\lambda, \phi, 0) = \begin{bmatrix} \sin\phi\cos\lambda & -\sin\lambda & \cos\phi\cos\lambda \\ \sin\phi\sin\lambda & \cos\lambda & \cos\phi\sin\lambda \\ -\cos\phi & 0 & \sin\phi \end{bmatrix}$
local horizontal frame	$P_c, \underset{\approx}{c}$
	$\overrightarrow{P_cP_d} = 0$ $R_{dc} = R_q = 2\begin{bmatrix} q_0^2 + q_1^2 - 1/2 & q_1q_2 - q_0q_3 & q_1q_3 + q_0q_2 \\ q_1q_2 + q_0q_3 & q_0^2 + q_2^2 - 1/2 & q_2q_3 - q_0q_1 \\ q_1q_3 - q_0q_2 & q_2q_3 + q_0q_1 & q_0^2 + q_3^2 - 1/2 \end{bmatrix}$
IMU frame	$P_d, \underset{\approx}{d}$

Fig. 1: Commutative diagram of the reference frames, the basic or inertial frame $(P_a, \underset{\approx}{a})$ fixed in space and time, the equatorial frame $(P_b, \underset{\approx}{b})$, the south-east-vertical oriented horizontal frame $(P_c, \underset{\approx}{c})$ and the IMU frame $(P_d, \underset{\approx}{d})$

The orientation of the orthonormal triad $\underset{\approx}{c}$ is defined by introducing surface normal coordinates with respect to a biaxial ellipsoid (oblate rotational ellipsoid). We denote by λ the *longitude*, by ϕ the *latitude* and by h the *height* measured in the direction of the surface normal. Thus we arrive at the following representation of the coordinates of the position vector $\overrightarrow{P_bP_c}$

$$\underset{\sim}{y} = \left(\frac{a}{\sqrt{1-e^2\sin^2\phi}} + h\right)\cos\lambda\ \cos\phi\ \underset{\sim}{b}_1 + \left(\frac{a}{\sqrt{1-e^2\sin^2\phi}} + h\right)\sin\lambda\ \cos\phi\ \underset{\sim}{b}_2 + $$
$$+ \left(\frac{a(1-e^2)}{\sqrt{1-e^2\sin^2\phi}} + h\right)\sin\phi\ \underset{\sim}{b}_3 \tag{1-1}$$

where a denotes the semimajor axis, c the semiminor axis and $e = \sqrt{1-c^2/a^2}$ the eccentricity in the (1,3) plane. The base vectors of the frame $\underset{\approx}{c}$ are defined by

$$\begin{aligned} \underset{\sim}{c}_1 &:= -(\partial\underset{\sim}{y}/\partial\phi)/\|\partial\underset{\sim}{y}/\partial\phi\| \\ \underset{\sim}{c}_2 &:= (\partial\underset{\sim}{y}/\partial\lambda)/\|\partial\underset{\sim}{y}/\partial\lambda\| \\ \underset{\sim}{c}_3 &:= (\partial\underset{\sim}{y}/\partial h)/\|\partial\underset{\sim}{y}/\partial h\| \ . \end{aligned} \tag{1-2}$$

The orthonormal triad $\underset{\approx}{b}$ is defined by the base vector $\underset{\sim}{b}_3$, that is the normalized velocity $\underset{\sim}{\omega}_{ba}$ of the earth, the base vector $\underset{\sim}{b}_1$ "of Greenwich" and finally the orthonormal supplement $\underset{\sim}{b}_2$ of the base vectors $\underset{\sim}{b}_1$ and $\underset{\sim}{b}_2$. Neglecting the translational motion of the earth, we locate P_d in the earth's mass centre. The angular velocity of the earth $\underset{\sim}{\omega}_{ba} = \underset{\approx}{b}\ \underline{w}^b_{ba}$ is approximated by

$$\underline{w}^b_{ba} = [0,0,d\vartheta/dt]^T \quad \text{with } d^2\vartheta/dt^2 = 0 \ . \tag{1-3}$$

We want now to derive the proper equation of linear motion of an IMU. With the help of the *second Newtonial theorem* we obtain

$$\underset{\sim}{s} = d^2\underset{\sim}{y}/dt^2 - \underset{\sim}{\gamma} \ , \tag{1-5}$$

where $\underset{\sim}{s}$ is the vector of acceleration difference measured by the accelerometers but reduced to the mass center of the IMU, called vector of the *specific forces*, $d^2\underset{\sim}{y}/dt^2$ the vehicles linear acceleration with respect to the earth's center of mass, and $\underset{\sim}{\gamma}$ the vector of gravitation. Evaluating the time derivative of the position vector in (1-5) we arrive at

$$d^2\underset{\sim}{y}/dt^2 = \underset{\approx}{b}[d^2\underline{y}_b/dt^2 - 2\underline{\Omega}^b_{ba}d\underline{y}_b/dt - d\underline{\Omega}^b_{ba}/dt\underline{y}_b + (\underline{\Omega}^b_{ba})^2\underline{y}_b] \tag{1-6}$$

where $\underline{y}_b = [y_1,y_2,y_3]^T$ represents the Cartesian coordinates of the position vector $\underset{\sim}{y}$ with respect to the triad $\underset{\approx}{b}$. The antisymmetric matrix $\underline{\Omega}^b_{ba}$ is defined by

$$\underline{\Omega}_{ba}^{b} := -\underline{R}_{ba}\, d\underline{R}_{ab}/dt \ . \tag{1-7}$$

Dually assigned to the coordinates of the matrix $\underline{\Omega}_{ba}^{b}$ are the coordinates of the vector of angular velocity $\underset{\sim}{\omega}_{ba} = \underset{\approx}{b}\underline{w}_{ba}^{b}$, in particular $w_1 = \Omega_{23}$, $w_2 = \Omega_{31}$ and $w_3 = \Omega_{12}$.

To compare the strapdown and platform mechanization we have to represent their mechanization equations in the same reference frame, for example in the instantaneous local horizontal frame $(P_c, \underset{\approx}{c})$. We obtain in the frame $\underset{\approx}{c}$ the following representation of the relative acceleration vector

$$\begin{aligned} d^2\underset{\sim}{y}/dt^2 = \underset{\approx}{c}[&\partial(\underline{R}_{cb}d\underline{y}_b/dt)/\partial t - \underline{\Omega}_{cb}^{c}\underline{R}_{cb}d\underline{y}_b/dt - \\ &-2\underline{\Omega}_{ba}^{c}\underline{R}_{cb}d\underline{y}_b/dt - (\underline{R}_{cb}d\underline{\Omega}_{ba}^{b}/dt + \underline{R}_{cb}(\underline{\Omega}_{ba}^{b})^2)\underline{y}_b] \end{aligned} \tag{1-8}$$

The antisymmetric matrix $\underline{\Omega}_{bc}^{c}$ is defined correspondingly to (1-7), where the dual coordinates of the angular velocity $\underset{\sim}{\omega}_{bc}$ with respect to the triad $\underset{\approx}{c}$ are $w_1 = -(d\lambda/dt)\cos\phi$, $w_2 = -d\phi/dt$, $w_3 = (d\lambda/dt)\sin\phi$. We have used further the matrix transformation

$$\underline{\Omega}_{ba}^{c} = \underline{R}_{cb}\underline{\Omega}_{ba}^{b}\underline{R}_{bc} \ . \tag{1-9}$$

To describe completely the translational equation of motion (1-4) the vector of gravitation $\underset{\sim}{\gamma}$ as well as the vector of the reduced measured specific forces $\underset{\sim}{s}$ have to be described with respect to the local horizontal frame. The coordinates of the latter are given with respect to the IMU frame $\underset{\approx}{d}$, and thus have to be transformed into the frame $\underset{\approx}{c}$:

$$\underset{\sim}{s} = \underset{\approx}{d}\underline{s}_d = \underset{\approx}{c}\underline{R}_{cd}\underline{s}_d \tag{1-10}$$

The determination of the sensor frame's orientation is performed with the help of the gyroscopes, but the application of the gyro's differ for both mechanizations.

a) strapdown mechanization

The time dependent transformation matrix $\underline{R}_{cd}$ has to be known at each moment an observation of the specific forces is present. The matrix is obtained by numerical integration of the differential equation

$$d\underline{R}_{cd}/dt = -\underline{R}_{cd}\underline{\Omega}_{dc}^{d} \ . \tag{1-11}$$

The antisymmetric matrix $\underline{\Omega}^d_{dc}$ we obtain from

$$\underline{\Omega}^d_{dc} = \underline{\Omega}^d_{da} - \underline{\Omega}^d_{ca} = \underline{\Omega}^d_{da} - \underline{R}_{dc}\underline{\Omega}^c_{ca}\underline{R}_{cd} \tag{1-12}$$

where $\underline{\Omega}^d_{da}$ comprises the measured coordinates of angular velocity and $\underline{\Omega}^c_{ca}$ the earth rate as well as the transport rate. The latter we get by evaluating the translational equation (1-4) for the ellipsoidal geodetic coordinates. Substituting (1-12) into (1-11) we finally obtain

$$\underline{R}_{cd} = -\underline{R}_{cd}\underline{\Omega}^d_{da} + \underline{\Omega}^c_{ca}\underline{R}_{cd} \tag{1-13}$$

b) platform mechanization

The evaluation of the rotational equations is here performed by feeding additional torques to the gyros mounted on the platform. These torque cause the gyros to precess. The servo loop gyro-platform drives the platform to maintain the gyro pickoff angle at zero. The platform, i.e. the IMU frame $\underset{\approx}{d}$, has thus a wellknown angular velocity with respect to the inertial frame. If we choose the applied torques corresponding to the angular velocity

$$\underset{\sim}{\omega}_{da} = \underset{\sim}{\omega}_{ca} , \tag{1-14}$$

the platform is slaved with respect to the local horizontal frame. Proceeding as for the strapdown mechanization, we obtain the rotational equation

$$d\underline{R}_{cd}/dt = \underline{0} \tag{1-15}$$

and by the choice of appropriate initial values

$$\underline{R}_{cd} = \underline{I} . \tag{1-16}$$

3. Linearization

The *linear system of incremental differential equations* is derived from the non linear system by a Taylor series expansion. The mechanization equations are rearranged in a first order system of differential equations

$$d\underline{x}/dt = \underline{f}(\underline{x},\underline{y}) \tag{2-1}$$

where the column matrix $\underline{x}$ comprises the unknown rotational parameter as well as the geodetic coordinates $\{\lambda,\phi,h\}$ and as supplementary unknown parameters their time derivatives $\{d\lambda/dt, d\phi/dt, dh/dt\}$. The column matrix $\underline{y}$ comprises the coordinates of the translational observables, i.e. the specific forces, and the coordinates of the rotational observables, i.e. the angular velocity of frame $\underline{\underline{d}}$ with respect to the inertial frame $\underline{\underline{a}}$. Besides the model equation (2-1) we have the observation equation

$$\underline{y}^* + \underline{e}^* + \delta\bar{\underline{y}} = \underline{g}(\underline{x}), \tag{2-2}$$

with its first and second stochastic moments

$$E\{\underline{y}^*\} = \underline{g}(\underline{x}) \;\; ; \;\; C(\underline{y}^*(t),\underline{y}^*(\tau)) = C(\underline{e}^*(t),\underline{e}^*(\tau)) = \underline{\Sigma}_{ee}(t,\tau). \tag{2-3}$$

The column matrix $\delta\bar{\underline{y}}$ specifies the bias of the sensor output not yet modelled in the mechanization equations. Substituting the observation $\underline{y}^*$ into the mechanization equations, we obtain a stochastic system of differential equations

$$d\underline{x}^*/dt = \underline{f}(\underline{x}^*,\underline{y}^*) \; . \tag{2-4}$$

To avoid a stochastic Jacobi matrix in the linearization, we define the non random values

$$\bar{\underline{y}} = \underline{y}^* + \underline{0}^*$$

by formally adding "random zero column matrices" with the stochastic properties

$$E\{\underline{0}^*\} = \bar{\underline{y}} - E\{\underline{y}^*\} = \underline{0} \quad \text{and} \quad C(\underline{0}^*,\underline{0}^*) = \underline{\Sigma}_{ee}(t,\tau) \; . \tag{2-5}$$

Substituting the nonrandom values $\bar{\underline{y}}$ into the mechanization equations, we obtain a nonrandom system, from which we compute the nonrandom approximation values $\bar{\underline{x}}$. We proceed in expanding the stochastic equation (2-4) as well as the model equation into a Taylor series with the nonrandom approximation values $\bar{\underline{x}}$ and $\bar{\underline{y}}$. Combining these two equations and neglecting higher order terms we finally arrive at

$$d\Delta\underline{x}/dt = \underline{F}\Delta\underline{x} + \underline{G}\Delta\underline{y} \tag{2-6}$$

by setting

$$\underline{\Delta x} := \underline{x}^* - \underline{x} \ , \quad \Delta \underline{y} = \underline{y}^* - \underline{y} \ , \quad \underline{F} := \partial \underline{f} / \partial \underline{x}^T \Big|_{\underline{x} = \bar{\underline{x}}} \quad \text{and}$$

$$\underline{G} := \partial \underline{f} / \partial \underline{y}^T \Big|_{\underline{y} = \bar{\underline{y}}} \tag{2-7}$$

Applying this concept of linearization to the mechanization equations (1-4) and (1-13), we have in particular the state vector

$$\Delta \underline{x} = \{\varepsilon_1, \varepsilon_2, \varepsilon_3, \Delta v_1, \Delta v_2, \Delta v_3, \Delta\phi, \Delta\lambda, \Delta h\}$$

with the rotation vector $\underset{\sim}{\varepsilon} = \underline{\underline{c}}[\varepsilon_1, \varepsilon_2, \varepsilon_3]^T$ and the linear velocities

$$v_1 = -(a(1-e^2)(1-e^2\sin^2\phi)^{-3/2} + h) d\phi/dt$$

$$v_2 = (a(1-e^2\sin^2\phi)^{-1/2} + h)\cos\phi \ d\lambda/dt$$

$$v_3 = dh/dt \ ,$$

the disturbance vector

$$\Delta \underline{y} = \{\Delta y_1, \Delta y_2, \Delta y_3, \Delta\omega_{1da}, \Delta\omega_{2da}, \Delta\omega_{3da}\} \ ,$$

the dynamics matrix $\underline{F}$ given in fig. 2, and the disturbing vector coefficient matrix

$$\underline{G} = \begin{bmatrix} \underline{R}_{cd} & \underline{0} \\ \underline{0} & \underline{R}_{cd} \end{bmatrix}$$

with the 3×3 zero submatrices. In the case of the platform mechanization $\underline{G}$ reduces to

$$\underline{G} = \begin{bmatrix} \underline{I} & \underline{0} \\ \underline{0} & \underline{I} \end{bmatrix}$$

where $\underline{I}$ denotes the 3× 3 unity matrix.

We now have to specify the disturbance vector $\Delta\underline{y}$ by modelling the sensor output in detail. We choose as translational sensors three *rotational (pendulum) accelerometers*, and as rotational sensors three *Single-Degree-of-Freedom gyros*.

$$
\begin{pmatrix}
0 & (\dot\vartheta+\dot\lambda)\cdot\sin\phi & \dot\phi & 0 & \frac{1}{N+h} & 0 & -(\dot\vartheta+\frac{(N-M)\dot\lambda}{N+h})\sin\phi & 0 & -\frac{\dot\lambda\cos\phi}{N+h} \\
(\dot\vartheta+\dot\lambda)\cdot\sin\phi & 0 & -(\dot\vartheta+\dot\lambda)\cdot\cos\phi & -\frac{1}{M+h} & 0 & 0 & -\frac{3\dot\phi\tan\phi M(N-M)}{N(M+h)} & 0 & -\frac{\dot\phi}{M+h} \\
-\dot\phi & (\dot\vartheta+\dot\lambda)\cdot\cos\phi & 0 & 0 & -\frac{\tan\phi}{N+h} & 0 & -\dot\vartheta\cos\phi-\frac{(M+h)\dot\lambda-(N-M)\dot\lambda\cos^2\phi}{(N+h)\cos\phi} & 0 & \frac{\dot\lambda\sin\phi}{N+h} \\
0 & s_3 & -s_2 & -\frac{\dot h}{M+h} & 2(\dot\vartheta+\dot\lambda)\cdot\sin\phi & \dot\phi & (\dot\vartheta+\dot\lambda)^2(N+h)\cos^2\phi-(\dot\vartheta^2-\dot\lambda^2)(M+h)\sin^2\phi-\frac{3\dot h\dot\phi\tan\phi M(N-M)}{N(M+h)} & 0 & (\dot\vartheta^2-\dot\lambda^2)\sin\phi\cos\phi-\frac{\dot h\dot\phi}{M+h} \\
-s_3 & 0 & s_1 & -(2\dot\vartheta+\dot\lambda)\cdot\sin\phi & \frac{\dot\phi\tan\phi(M+h)-\dot h}{N+h} & -(2\dot\vartheta+\dot\lambda)\cdot\cos\phi & (\frac{\dot\lambda(N-M)}{N+h}+2\dot\vartheta)(\dot\phi\cos\phi(M+h)+\dot h\sin\phi)+\frac{(M+h)^2\dot\phi\dot\lambda}{(N+h)\cos\phi} & 0 & -\frac{\dot\lambda\dot\phi\sin\phi(M+h)+\dot\lambda\dot h\cos\phi}{N+h} \\
s_2 & -s_1 & 0 & -2\dot\phi & 2(\dot\vartheta+\dot\lambda)\cdot\cos\phi & 0 & -(\dot\vartheta+\dot\lambda)^2(N+h)\sin\phi\cos\phi-(\dot\vartheta^2-\dot\lambda^2)(M+h)\sin\phi\cos\phi-\frac{3\dot\phi^2\tan\phi M(N-M)}{N} & 0 & (\dot\vartheta^2-\dot\lambda^2)\cos^2\phi-\dot\phi^2 \\
0 & 0 & 0 & -\frac{1}{M+h} & 0 & 0 & \frac{-3\dot\phi\tan\phi(N-M)M}{N(M+h)} & 0 & -\frac{\dot\phi}{M+h} \\
0 & 0 & 0 & 0 & \frac{1}{(N+h)\cos\phi} & 0 & \frac{\dot\lambda\tan\phi(M+h)}{N+h} & 0 & -\frac{\dot\lambda}{N+h} \\
0 & 0 & 0 & 0 & 0 & 1 & 0 & 0 & 0
\end{pmatrix}
$$

Fig. 2: Dynamics matrix; $N=a(1-e^2\sin 2\ )^{-1/2}$, $M=a(1-e^2)(1-e^2\sin^2\ )^{3/2}$, $\dot{\ldots} := d\ldots/dt$

4.1 Accelerometer Model

A natural describing frame of the accelerometer sensitive element is the *body axes triad* $\underset{\approx}{g} = \{\underset{\sim}{g}_1, \underset{\sim}{g}_2, \underset{\sim}{g}_3\}$ located at the center of mass P_g and its base vectors coinciding with the principal inertia axes of the pendulum.

To the *mechanized rotation axis* of the pendulum we attach the two orthonormal frames $\underset{\approx}{f} = \{\underset{\sim}{f}_1, \underset{\sim}{f}_2, \underset{\sim}{f}_3\}$ and $\underset{\approx}{e} = \{\underset{\sim}{e}_1, \underset{\sim}{e}_2, \underset{\sim}{e}_3\}$ with their common origin at the intersection point of the rotation axis and the projection line of the mass center of the pendulum on the axis. The base vector $\underset{\sim}{f}_2 \equiv \underset{\sim}{e}_2$ we assume parallel to the rotation axis (output axis), the vector $\underset{\sim}{f}_3$ parallel to the projection line fixed to the pendulum and the vector $\underset{\sim}{e}_3$ coinciding with $\underset{\sim}{f}_3$ for a pickoff angle zero, but fixed to the case. The vectors $\underset{\sim}{f}_1$ and $\underset{\sim}{e}_1$ (input axis) are the orthonormal supplements of $\underset{\sim}{f}_2$, $\underset{\sim}{f}_3$ and $\underset{\sim}{e}_2$, $\underset{\sim}{e}_3$, respectively.

The constant misalignment angles between the frames $\underset{\approx}{g}$, $\underset{\approx}{f}$ and $\underset{\approx}{e}$, $\underset{\approx}{d}$ are introduced as small rotation angles, and the pickoff angle θ is assumed kept small by a suited rebalance loop. The resulting transformation matrices of the accelerometer model are given in the commutative diagram of figure 3.

IMU frame $P_d, \underset{\approx}{d}$

$$\overrightarrow{P_dP_e} = \underset{\sim}{z}$$

$$R_{ed} = R_C(\alpha_1, \alpha_2, \alpha_3) = \begin{bmatrix} 1 & \alpha_3 & -\alpha_2 \\ -\alpha_3 & 1 & \alpha_1 \\ \alpha_2- & \alpha_1 & 1 \end{bmatrix}$$

rotation axis frame fixed to case $P_e, \underset{\approx}{e}$

$$\overrightarrow{P_eP_f} = \underset{\sim}{0}$$

$$R_{fe} = R_C(0, \theta, 0) = \begin{bmatrix} 1 & 0 & -\theta \\ 0 & 1 & 0 \\ \theta & 0 & 1 \end{bmatrix}$$

rotation axis frame fixed to pendulum $P_f, \underset{\approx}{f}$

$$\overrightarrow{P_fP_g} = \underset{\sim}{u}$$

$$R_{gf} = R_C(\gamma_1, \gamma_2, 0) = \begin{bmatrix} 1 & 0 & -\gamma_2 \\ 0 & 1 & \gamma_1 \\ \gamma_2 & -\gamma_1 & 1 \end{bmatrix}$$

body axes frame $P_g, \underset{\approx}{g}$

Fig. 3: Commutative diagram of accelerometer model reference frames for accelerometer (1) (input axis almost parallel $\underset{\sim}{d}_1$)

As the commanded torque of the rebalance loop, and thus the output signal, is related to the pendulum rotation axis, we have to represent all angular moments and external torques in the frame $(P_e, \underset{\approx}{e})$. Starting with the rotational form of the second Newtonial theorem

$$d\underset{\sim}{D}_a/dt = \underset{\sim}{M} \tag{4-1}$$

where $\underset{\sim}{D}_a$ is the vector of angular momentum with respect to the inertial point P_a and $\underset{\sim}{M}$ the vector of external torques, we finally arrive at

$$\begin{aligned} &\underset{\approx}{e}\{-\underline{\Omega}^e_{gd}\underline{R}_{eg}\underline{T}\underline{w}^g_{ga}+\underline{R}_{eg}\underline{T}d\underline{w}^g_{ga}/dt-\underline{\Omega}^e_{da}\underline{R}_{eg}\underline{T}\underline{w}^g_{ga} + \\ &+ \underline{U}_e\underline{R}_{ea}(d^2\underline{x}_a/dt^2+d^2\underline{z}_a/dt^2\} = \underset{\approx}{e}\,\underline{M}_e \end{aligned} \tag{4-2}$$

with

$$\underline{w}^g_{ga} = \underline{R}_{gf}w^f_{fe} + \underline{R}_{gd}\underline{w}^d_{da} \tag{4-3i}$$

$$\underline{w}^e_{gd} = \underline{R}_{ef}\,\underline{w}^f_{fe} \; . \tag{4-3ii}$$

The antisymmetric matrix $\underline{U}_e$ comprises the coordinates of the position vector P_eP_g, and the matrix $\underline{T}$ denotes the coordinates of the pendulum's inertia tensor with respect to the triad $\underset{\approx}{g}$ and the reference point P_e.
In particular we have

$$\underline{w}^f_{fe} = [0,d\theta/dt,0]^T \quad ; \quad w^d_{da} = [w_1,w_2,w_3]^T$$

$$\underline{u}_f = [0,0,\ell]^T \quad \text{and} \quad \underline{z}_d = [z_1,z_2,z_3]^T \; .$$

As external torques we have the commanded torque of the torque generator $\underset{\sim}{M}_s$, a torque $\underset{\sim}{M}_g$ caused by the gravitational forces acting on the pendulum, which we lump together with the linear acceleration of P_d to the specific forces assuming a homogeneous gravitation field inside the IMU, an elastic torque $\underset{\sim}{M}_k$ proportional to the pickoff angle θ, and a damping torque $\underset{\sim}{M}_D$ proportional to $d\theta/dt$. As the output signal is a voltage signal proportional to the rebalance torque $\underset{\sim}{M}_s$, we have to introduce a scale factor and his error. The scale factor itself we put for simplicity to one, retaining the scale factor error κ. Neglecting products of the small misalignment angles, we finally obtain the equation of the output axis torque given in fig. 4.

$M_s =$	**output axis torque**
$+s_1\, lm + \kappa\, s_1\, lm - M_0 +$	**input linear acc., scale factor error and bias torque**
$+(T_{11} - T_{33})\, w_1 w_3 +$	**anisoinertia**
$+T_{22}\, dw_2/dt +$	**angular acceleration**
$+(d^2\theta/dt^2)T_{22} - (d\theta/dt)D +$	**angular dynamic**
$+z_1(w_2^2 + w_3^2)lm +$ $+z_2(w_1 w_2 + dw_3/dt)lm +$ $+z_3(w_1 w_3 + dw_2/dt)lm +$	**mounting**
$+\alpha_1((T_{33} - T_{11})w_1 w_2 + T_{22} dw_3/dt) +$ $+(\alpha_2 + \theta)(z_1(w_1 w_3 - dw_2/dt)lm +$ $+z_2(w_2 w_3 - dw_1/dt)lm + z_3(w_1^2 + w_2^2)lm +$ $+(T_{11} - T_{33})(w_1^2 - w_3^2)) - \alpha_2 s_3 lm - \theta k +$ $+\alpha_3(z_1(w_1 w_2 - dw_3/dt)lm - z_2(w_1^2 + w_3^2)lm +$ $+z_3(w_2 w_3 - dw_1/dt)lm + w_2 w_3(T_{11} - T_{33}) -$ $-T_{22} dw_1/dt + s_2 lm) +$	**misalignment and pickoff angle**
$+\gamma_1(w_1 w_2 + dw_3/dt)(T_{22} - T_{33}) +$ $+\gamma_3(w_2 w_3 - dw_1/dt)(T_{22} - T_{11})$	**dynamic unbalance**

Fig. 4: Pendulum accelerometer model

4.2 SDF-Gyro Model

The gyro model is referred to a rate integrating gyro for the sake of comparability of both simulated mechanizations. A rate integrating gyro can be used as a stabilization device in an inertial platform as well as a strapdown rate sensor. In the former case the gyro pickoff angle is maintained at zero by turning the platform around the gyro input axis. In the latter case the pickoff angle is zeroed by the gyro torquer, and the integral of the torquer's electrical current is a measure of the integral of the gyro input rate. Although the two servo loops are different, we chose the same error coefficients for both simulations to compare the effect of the different low frequency dynamical environment.

As a natural describing orthonormal triad of a SDF-gyro we introduce the triad $\underset{\approx}{g} = \{\underset{\sim}{g}_1, \underset{\sim}{g}_2, \underset{\sim}{g}_3\}$ located at the *rotor's center of mass* P_g and the base vectors coinciding with the rotor's principal axes of inertia. To the *mechanized spin axis* of the rotor we attach the two frames $\underset{\approx}{\bar{f}} = \{\underset{\sim}{\bar{f}}_1, \underset{\sim}{\bar{f}}_2, \underset{\sim}{\bar{f}}_3\}$ and $\underset{\approx}{f} = \{\underset{\sim}{f}_1, \underset{\sim}{f}_2, \underset{\sim}{f}_3\}$, both located at the *center of support* P_f, i.e. the resultant vector of the supporting forces. We introduce the base vectors $\underset{\sim}{f}_3 = \underset{\sim}{\bar{f}}_3$ coinciding with the spin axes, $\underset{\sim}{\bar{f}}_1$ fixed to the rotor and orthogonal to $\underset{\sim}{g}_2$ and $\underset{\sim}{\bar{f}}_3$, and $\underset{\sim}{f}_1$ fixed to the gimbal coinciding with $\underset{\sim}{\bar{f}}_1$ for t = 0. The vectors $\underset{\sim}{\bar{f}}_2$ and $\underset{\sim}{f}_2$ are the orthonormal supplements of $\underset{\sim}{\bar{f}}_1$, $\underset{\sim}{\bar{f}}_2$ and $\underset{\sim}{f}_1$, $\underset{\sim}{f}_2$, respectively. To the gimbal we attach the triads $\underset{\approx}{\bar{e}} = \{\underset{\sim}{\bar{e}}_1, \underset{\sim}{\bar{e}}_2, \underset{\sim}{\bar{e}}_3\}$ and $\underset{\approx}{e} = \{\underset{\sim}{e}_1, \underset{\sim}{e}_2, \underset{\sim}{e}_3\}$, located at the center of support. The base vectors $\underset{\sim}{e}_2 = \underset{\sim}{\bar{e}}_2$ coincides with the gimbal rotation axis, $\underset{\sim}{\bar{e}}_1$ is fixed to the gimbal and orthogonal to $\underset{\sim}{e}_2$ and $\underset{\sim}{f}_3$, whereas $\underset{\sim}{e}_1$ is fixed to the case with $\underset{\sim}{e}_1 = \underset{\sim}{\bar{e}}_1$ for a pickoff angle θ equal zero. The reference frames form the commutative diagram of figure 5.

IMU frame: $P_d, \underset{\approx}{d}$

$$\overrightarrow{P_dP_e} = \underset{\sim}{z}$$

$$R_{ed} = R_2(\theta)R_C(\alpha_1, \alpha_2, \alpha_3) = \begin{bmatrix} 1 & \alpha_3 & -\alpha_2 - \theta \\ -\alpha_3 & 1 & \alpha_1 \\ \alpha_2 + \theta & -\alpha_1 & 1 \end{bmatrix}$$

gimbal frame: $P_e, \underset{\approx}{e}$

$$\overrightarrow{P_eP_f} = \underset{\sim}{0}$$

$$R_{fe} = R_C(\beta_1, \beta_2, 0) = \begin{bmatrix} 1 & 0 & -\beta_2 \\ 0 & 1 & \beta_1 \\ \beta_2 & -\beta_1 & 1 \end{bmatrix}$$

spin axis frame fixed to gimbal: $P_f, \underset{\approx}{f}$

$$\overrightarrow{P_fP_{\bar{f}}} = \underset{\sim}{0}$$

$$R_{\bar{f}f} = R_3(\omega t) = \begin{bmatrix} \cos\omega t & \sin\omega t & 0 \\ -\sin\omega t & \cos\omega t & 0 \\ 0 & 0 & 1 \end{bmatrix}$$

spin axis frame fixed to rotor: $P_{\bar{f}}, \underset{\approx}{\bar{f}}$

$$\overrightarrow{P_{\bar{f}}P_g} = \underset{\sim}{u}$$

$$R_{g\bar{f}} = R_C(\gamma_1, \gamma_2, 0) = \begin{bmatrix} 1 & 0 & -\gamma_2 \\ 0 & 1 & \gamma_1 \\ \gamma_2 & -\gamma_1 & 1 \end{bmatrix}$$

body axes frame: $P_g, \underset{\approx}{g}$

Fig. 5: Commutative diagram of gyro model reference frames for gyro (1) (input axis almost parallel $\underset{\sim}{d}_1$)

Proceeding as for the accelerometer, we now relate the angular momentum and the external torques to the rotor's center of mass P_g. We have further to distinguish between the masses of the gimbal, introduced by the coordinate matrix $\underline{T}_G$ of its inertia tensor with respect to the triad $\underset{\approx}{e}$ and of the rotor,

$M_s =$	**output axis torque**
$-T_{33R}\omega\, w_1 - \kappa\, T_{33R}\omega\, w_1 - M_0 +$	**input gyro term, scale factor error and bias torque**
$+(\overline{T} + T_{11G} - T_{33R} - T_{33G})w_1 w_3 +$	**anisoinertia**
$+(\overline{T} + T_{22G})dw_2/dt +$	**angular acceleration**
$+(d^2\theta/dt^2)\,(\overline{T} + T_{22G}) - (d\theta/dt)\,D +$	**angular dynamic**
$+T_{12G}(dw_1/dt + w_2 w_3)+$ $+T_{13G}(w_3^2 - w_1^2)+$ $+T_{32G}(dw_3/dt - w_1 w_2)+$	**dynamic unbalance gimbal**
$+\alpha_1\ ((T_{33R} + T_{33G} - \overline{T} - T_{11G})w_1 w_2 + (\overline{T} + T_{22G})dw_3/dt)+$ $+(\alpha_2 + \theta)\ (\overline{T} + T_{11G} - T_{33R} - T_{33G})(w_1^2 - w_3^2)+$ $+T_{33R}\omega\, w_3)) - \theta k +$ $+\alpha_3\ (\overline{T} + T_{11G} - T_{33R} + T_{33G})w_2 w_3 - (\overline{T} + T_{22G})dw_1/dt -$ $-T_{33R}\omega\, w_2)+$	**misalignment case to gimbal and pickoff angle**
$+\beta_1((T_{33R} - \overline{T})(w_1 w_2 - dw_3/dt) - T_{33R}d\omega/dt)+$ $+\beta_2((T_{33R} - \overline{T})(w_3^2 - w_1^2) + T_{33R}\omega w_3)+$	**misalignment gimbal to rotor axis**
$+\gamma_1(T_{33R} - T_{22R})(\sin\omega t((\omega + w_3)^2 - w_1^2)-$ $-\cos\omega t(d\omega/dt + dw_3/dt + w_1 w_2)+$ $+\gamma_2(T_{33R} - T_{11R})(\cos\omega t((\omega + w_3)^2 - w_1^2)-$ $-\cos\omega t(d\omega/dt + dw_3/dt + w_1 w2)+$	**dynamic unbalance rotor**
$+\overline{T}\lambda\cos 2\omega t(2\omega w_1 + w_1 w_3 + dw_2/dt)+$ $+\overline{T}\lambda\sin 2\omega t(2\omega w_2 + w_2 w_3 + dw_1/dt)+$	**asymmetry of rotor**
$+u_1 s_3 m - u_3 s_1 m$	**static unbalance**

Fig. 6: Single-Degree-of-Freedom gyro model

introduced by the coordinate matrix $\underline{T}_R$ of its inertia tensor with respect to the triad $\underline{g}$. Denoting by

$$\bar{T} := (T_{11R} + T_{22R})/2$$

the "averaged" moments of inertia of the rotor, and by

$$\lambda := (T_{11R} - T_{22R})/(2\bar{T})$$

its asymmetry, we arrive at the output axis torque equation of figure 6. We neglected products of the gimbal deviation moments, products of $d\theta/dt$ and of the asymmetry parameter λ with the misalignment angles.

5. Simulation Results

Let us now compare the timelike propagation of the incrementals $\Delta\lambda$, $\Delta\phi$ and Δh of a strapdown mechanization with that of a platform mechanization. For this purpose, the incremental differential equation (2-6) was solved numerically for an assumed trajectory. The sketch of this trajectory is shown in figure 7 and 8. Proceeding from the trajectory the approximation values were computed.

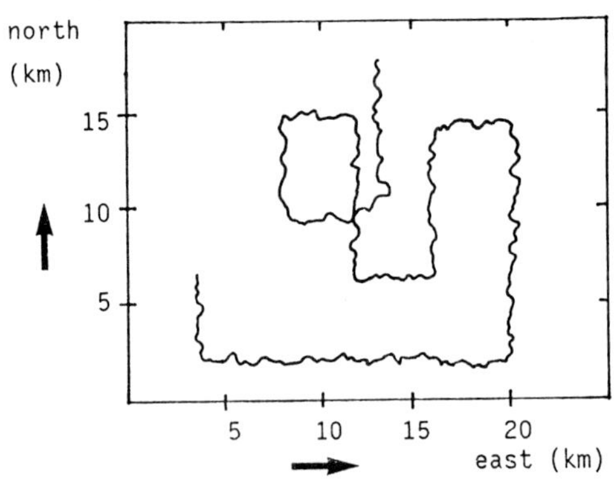

Fig. 7: Sketch of the horizontal covered course (start at $\phi = 45°$, $\lambda = 2°$)

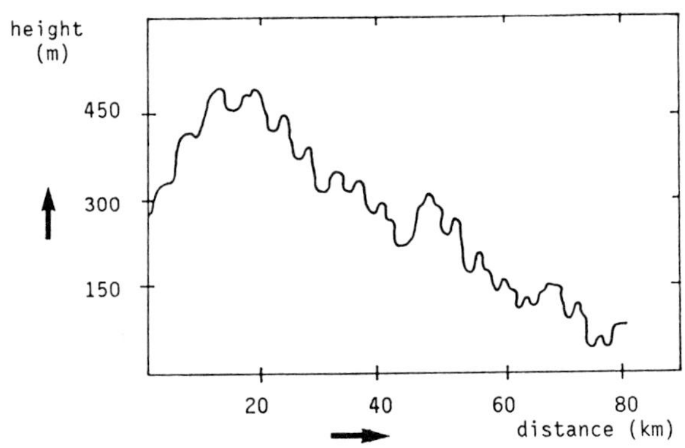

Fig. 8: Sketch of the height profile of the covered course

For a small time intervall the coefficients of the dynamics matrix were treated as constant. Thus the differential equation was solved by computing the matrix exponential of the dynamics matrix. The solution we got was used as initial values for the next time intervall.

Typical values of the sensor parameters used in the simulation are listed in table 1 and 2.

Parameter	symbol	Magnitude	Unit
mass	m	$1.5 \cdot 10^{-3}$	kg
distance center of mass			
to rotation axis	l	$7.5 \cdot 10^{-3}$	m
bias torque	$M_0/(lm)$	10^{-4}	m/sec^2
scale factor error	κ	10	ppm
mounting	z_1	$2 \cdot 10^{-3}$	m
	z_2	$2 \cdot 10^{-3}$	m
	z_3	$-9.9 \cdot 10^{-3}$	m
moment of inertia	T_{11}	$1.29 \cdot 10^{-7}$	kg m^2
	T_{22}	$1.12 \cdot 10^{-7}$	kg m^2
	T_{33}	$0.19 \cdot 10^{-7}$	kg m^2
misalignment	$\alpha_1, \alpha_2, \alpha_3$	5	''
dynamic unbalance	γ_1, γ_2	5	''

Table 1: Typical pendulum accelerometer parameter

Parameter	symbol	Magnitude	Unit
angular momentum			
of the rotor	$T_{33R}\ \omega$	$2 \cdot 10^{-2}$	kgm^2/sec
bias torque	M_0	$1 \cdot 10^{-10}$	kgm^2/sec^2
scale factor error	κ	50	ppm
moments of inertia	$\overline{T}$	$5 \cdot 10^{-6}$	kg m^2
of the rotor	T_{33R}	$8 \cdot 10^{-6}$	kg m^2
moments of inertia	T_{11G}	$1.96 \cdot 10^{-5}$	kg m^2
of the gimbal	T_{22G}	$2 \cdot 10^{-5}$	kg m^2
	T_{33G}	$1.7 \cdot 10^{-5}$	kg m^2
deviation moments	T_{12G}	$1.5 \cdot 10^{-7}$	kg m^2
of the gimbal	T_{13G}	$3 \cdot 10^{-8}$	kg m^2
	T_{32G}	$1.5 \cdot 10^{-7}$	kg m^2
misalignment			
case to gimbal	$\alpha_1, \alpha_2, \alpha_3$	5	''
misalignment			
gimbal to spin axis	$\beta_1, \beta_2,$	5	''
static unbalance	u_1, u_2	$1 \cdot 10^{-7}$	m
mass	m	$1 \cdot 10^{-1}$	kg

Table 2: Typical SDF gyro parameter

As far as only the propagation of the initial increments of the nine system parameter Δx is concerned, there is no difference for both mechanizations. The position increments show the typical oscillation of the Schuler frequency with a period of about 84 min, superposed by the long periodic oscillation of the earth's rate of about 24 h and of the Foucault oscillation with a period of 24 h sinϕ. Looking at the excitation functions of the sensor models, we have to distinguish between the specific forces and on the other hand the angular velocities and accelerations. Due to the comparatively small linear accelerations with respect to the earth fixed frame $\underset{\approx}{b}$ for a simulated geodetic application of an IMU in a land vehicle, the main part of the specific force vector is the share of the earth's gravitation vector. Thus for the platform mechanization almost only the vertical input channel is affected, whereas for the strapdown mechanization all three input channels are affected, depending on the IMU's orientation with respect to the local horizontal frame.

The angular velocities and the angular accelerations differ for both mechanizations about the factor 1000. Thus the effect of those coefficients depending only on the angular velocities and accelerations in the sensor models is much smaller for the platform mechanization. To qualify the different effects, the maximum value of the total position increment within the time intervall < 84 min is listed in table 3.

		Strapdown	Platform
accelerometer	bias torque	100 m	200 m
	mounting	80 m	0.005 m
	anisoinertia	7 m	0.003 m
	angular acceleration	5 m	0.004 m
	dynamic unbalance	4 m	0.003 m
	misalignment	130 m	180 m
	scale factor error	100 m	105 m
SDF gyro	bias torque	600 m	2400 m
	anisoinertia	80 m	0.3 m
	angular acceleration	6000 m	5 m
	dynamic unbalance	1800 m	0.1 m
	misalignment case to gimbal	2400 m	1000 m
	misalignment gimbal to spin axis	2400 m	1000 m
	scale factor error	1000 m	500 m
	static unbalance	500 m	1500 m

Table 3: Maximum value of total position increment for t < 84 min

A more critical examination has to take account not only of the low frequency dynamic of the vehicles covered path, but also of the higher frequency dynamic of the vehicles motor coupling. The latter leads especially for the strapdown mechanization to constant torques about the output axes.

As a conclusion of the effects listed in table 3 we remark that a careful determination of the sensor parameter and of the compensation of the output signals is more critical for the strapdown mechanization than for the platform mechanization.

Literature

Grafarend, E. (1981): From kinematical geodesy to inertial positioning. Bulletin Géodésique 55, 286-299, 1981.

Heitz, S. (1980/83): Mechanik fester Körper. Ferdinand Dümmler Verlag, 2 volumes, Bonn 1980/83.

Joos, D.K. (1977): Comparison of typical gyro errors for strapdown applications, Proceedings of the DGON Symposium 'Gyro Technology', 1977.

Rüeger, J.M. (1982/86): Inertial Sensors. University of Calgary, Surveying Engineering Report 30002 and 30008, Calgary, 1982/86.

Savage, P.G. (1978): Strapdown sensors. AGARD Lecture Series No. 95, 1978.

Schröder, D., Wiegner, S., Grafarend, E.W., Schaffrin, B. (1986): Simulation eines Geodätischen Inertialsystems. Schriftenreihe Studiengang Vermessungswesen, Universität der Bundeswehr München, Heft 22, 197-235, 1986.

Schröder, D., Chi Thong, N., Wiegner, S., Grafarend, E.W., Schaffrin, B. (1988): A comparative study of local level and strapdown inertial systems. manuscripta geodaetica 13/4, 224-248, 1988.

Stieler, B., Winter, H. (1982): Gyroscopic instruments and their application of flight testing. AGARDograph No. 160, Vol. 15, 1982.

Wiegner, S. (1985): Fehlersimulation eines Strapdown-Beschleunigungsmessers auf der Basis der Eulerschen Differentialgleichungen. Diplomarbeit, Universität Stuttgart, 1985.